Impingement of Man on the Oceans

IMPINGEMENT OF MAN ON THE OCEANS

Edited by

DONALD W. HOOD

Professor of Marine Science
Institute of Marine Science
University of Alaska

WILEY-INTERSCIENCE, a Division of John Wiley & Sons, Inc.
New York · London · Sydney · Toronto

Copyright © 1971, by John Wiley & Sons, Inc.

All rights reserved. Published simultaneously in Canada.

No part of this book may be reproduced by any means, nor transmitted, nor translated into a machine language without the written permission of the publisher.

Library of Congress Catalog Card Number: 74-151728

ISBN 0-471-40870-0

Printed in the United States of America.

10 9 8 7 6 5 4 3 2 1

Preface

The earth is being subjected to severe stresses by man and his ambitious enterprises. The upward spiral of human population with its attendant technology points to the shocking reality that the hour is fast approaching when the people of the earth will have exhausted nature's ability to adjust to the complexities of human attack. Society must cope with future problems in ways that are peculiar both to the past and to the present. Oceans are a very important element of this environment that must be protected from irreversible damage.

The world ocean is in many respects a giant flywheel that controls the climate; regulates the amount of oxygen, carbon dioxide, and other gases of the atmosphere; receives, metabolizes, and dilutes wastes; transports sediments as part of the major geological cycle; supplies moisture to the hydrologic cycle; acts as a major reservoir of nonrenewable resources; and provides a significant portion of the food for terrestrial animals from its bounty. An oceanless planet is untenable as a biological habitat; likewise, an ocean that is affected adversely in its important functions limits the earth's capacity for sustaining life.

The question "What is the effect of man's activities on the world-ocean's processes?" has become of great concern to many scientists, public officials, and others who are responsible for understanding, predicting, and controlling environmental impacts. World-ocean protection may be the most important issue for long-term survival of man on the earth. We need to know in great detail how the ocean works—what stresses it can tolerate with little or no damage and what stresses will erode and perhaps destroy processes vital to man. A thorough knowledge of the ocean's processes and the effect of contaminants on them is essential in order to assess where we stand.

It was the realization of the great importance of the oceans to the kind of world environment that we will have in the future that stimulated the preparation of this book. Acknowledgment is given to Dr. Edward D. Goldberg, Professor of Chemistry at Scripps Institution of Oceanography, and to Maurice Ancharoff, then Editor with Wiley-Interscience, who encouraged me to undertake the task of organizing and editing this book.

The book has six parts. Part I, "Transport Processes and Reservoirs," is concerned with the manner in which contaminants reach the oceans and then become distributed and retained. Part II, "Chemical Models of the Ocean," discusses some of the chemical systems that control the concentrations and equilibria status of chemicals added to the ocean. Part III, "Artifacts of Man," demonstrates that man's activities contribute chemicals to the ocean. In Part IV, the problem of "Man's Alteration on the Coastal Environment" is discussed. Part V, "Models for Studying Future Alterations of the Ocean by Input of Nonindigenous Substances," considers the well-known radioactive model in tracing ocean-water movements, geochemical cycling of man-made radioactive isotopes, and the use of added nutrients to better understand phytoplankton growth. The final part, "Implications of Man's Activities on Ocean Resource Development," describes legal implications, public policy, and philosophy of ocean uses. The last chapter reflects on the question of wise and appropriate use of ocean resources and summarizes other portions of the book.

The editorial policy followed was to select authors who are outstanding authorities on the subjects covered and to allow them maximum latitude in presenting the material. Editorial changes were made only to avoid repetition and to provide conformity of style, indexing, and organization.

This book is not intended to be a comprehensive document on the subject of man's effect on the oceans; it is directed, rather, at arousing serious thought on the subject. *Impingement of Man on the Oceans* is perhaps a first attempt at placing the vital subject of world-ocean pollution in survey perspective through single-volume presentation. It is my hope that this view is provocative and will lead to a better understanding of man's interaction with the ocean environment.

DONALD W. HOOD

College, Alaska
January 1971

Contents

1 Introduction: Man and the Ocean Environment   1
    Donald W. Hood

PART I   TRANSPORT PROCESSES AND RESERVOIRS

2 Rivers, Tributaries, and Estuaries   9
    Karl K. Turekian

3 Atmospheric Transport   75
    E. D. Goldberg

4 Horizontal and Vertical Mixing in the Sea   89
    Akira Okubo

5 Sediments   169
    G. D. Sharma

PART II   CHEMICAL MODELS OF THE OCEAN

6 Oceanic Residence Time and Geobiochemical Interactions   191
    Dayton E. Carritt

7	Anoxic versus Oxic Environments	201
	Francis A. Richards	
8	Thermodynamics	219
	Paul W. Schindler	

PART III ARTIFACTS OF MAN

9	Lead	245
	Claire Patterson	
10	Chlorinated Hydrocarbons	259
	R. W. Risebrough	
11	Carbon Dioxide—Man's Unseen Artifact	287
	Wallace S. Broecker, Yuan-Hui Li, and Tsung-Hone Peng	
12	Radioactivity—Chemical and Biological Aspects	325
	Theodore R. Rice and Douglas Wolfe	
13	Petroleum—the Problem	381
	James E. Moss	
14	Petroleum—Biological Effects in the Marine Environment	421
	D. K. Button	
15	Petroleum—Stable Isotope Ratio Variations	431
	Patrick L. Parker	
16	Municipal Wastes	445
	Ernst Føyn	

	17	Heavy—Metal Contamination	461
		MARGARET MERLINI	
PART IV		MAN'S ALTERATION OF COASTAL ENVIRONMENT	
	18	Coastal Engineering Practices	489
		CHARLES L. BRETSCHNEIDER	
	19	Forest Management and Agricultural Practices	503
		DALE W. COLE	
	20	Resource Exploitation—Living	529
		TIMOTHY JOYNER	
	21	Marine Mining and the Environment	553
		JOHN W. PADAN	
PART V		MODELS FOR STUDYING FUTURE ALTERATIONS OF THE OCEAN BY INPUT OF NON-INDIGENOUS SUBSTANCES	
	22	Radioactive Models	565
		YASUO MIYAKE	
	23	A Model of Nutrient-Limited Phytoplankton Growth	589
		RICHARD C. DUGDALE AND JOHN J. GOERING	
PART VI		IMPLICATIONS OF MAN'S ACTIVITIES ON OCEAN RESEARCH DEVELOPMENT	
	24	International and National Regulation of Pollution from Offshore Oil Production	603
		ROBERT S. KRUEGER	

25 Toward a Public Policy on the Ocean 635

E. W. SEABROOK HULL

26 Uses of the Ocean 667

D. W. HOOD AND C. PETER MCROY

Author Index 699

Subject Index 713

Impingement of Man on the Oceans

1

Introduction: Man and the Ocean Environment

DONALD W. HOOD, *Professor of Marine Science, Institute of Marine Science, University of Alaska, College, Alaska*

> There is a tide in the affairs of man,
> Which, taken at the flood, leads on to fortune;
> Omitted, all the voyages of their life
> Is bound in shallows and miseries
> On such a full sea are we now afloat;
> And we must take the current when it serves,
> Or lose our ventures.
>
> Shakespeare in *Julius Caesar*—Act IV, Scene III

Statement of the Problem

Since the oceans had their beginning, contamination, first from natural processes and later by man, has continuously occurred. Contamination* from man may have begun when he first floated timber, or hewed a tree to provide a means to negotiate the oceans surface. With small ships and limited equipment, man was for centuries compatible with his ocean environment much as today the Eskimos still live off and

---

* Contamination and pollution have separate meanings in this book. Contamination is the addition of a foreign substance to the ocean medium; pollution is the addition of a contaminant that is found harmful to an ocean use.

with the environment of the Northern Polar regions. The sailors' prayer, "Oh Lord, the sea is so big and my ship is so small," expressed man's awe of the infinite extent of the water and his own insignificant influence upon it. *Historically man has been relatively incapable of changing the oceans.*

In recent decades the awesome growth in technology initiated a new epic in the level of interaction between man and his ocean environment. A modern steel hull vessel filled with hydrocarbons does not appear so small, and its effects can be most significant. Insecticides, heavy metals, and debris of all kinds now reach the ocean, carried on the wind and water currents of the world. Industrial effluents, municipal wastes, and farm and forest runoff reach the sea through streams and rivers. Solid and liquid waste are being barged in enormous quantities offshore for deep sea disposal. Recently (Manheim et al., 1970), even cellulose fiber derived from toilet tissue has been found in the coastal waters of the eastern North American coast.

Since the beginning of human history, the world's oceans have generally been considered inexhaustible in all their resources, including the extent to which they favorably receive the wastes of man. Now we find this huge aqueous reservoir is under severe attack by human impingement with evidence accumulating daily that challenges this long held concept. Recent evidence has aroused oceanographers of the world to the realization that not only are the estuaries and near-shore regions subjected to severe stress, but even the ocean deeps show evidence of man's irrational behavior.

It is now quite clear that the capacity of the oceans to assimilate some of the more exotic artifacts of man is limited. Pesticides, halogenated aromatic hydrocarbons, and man-made radioactivity have been found in every sea organism examined and are thought to be present in all living biological beings on earth. Concentrations of lead, mercury, and perhaps even carbon dioxide are reaching alarming levels.

The effects of these and many other contaminants, natural or man-made, have not yet been clearly determined, but several have the capacity to significally change the ocean environment as a medium for biological life and ultimately to diminish its usefulness to the rapidly increasing human population. The tide is still at flood, and the choice is ours whether to preserve the ocean for mankind or fail in taking the current as it serves and thus "lose our ventures." *No longer is man an insignificant influence on the sea.*

Enlightenment is the first logical approach to the solution of impending problems. Information on the status of the ocean pollution problem has not been well documented. Many of the processes are not understood; the inventory of contamination is very incomplete; and conflicting attitudes and actions are prevalent. The time has come for knowledgeable individuals to set down the current status of the present level of damage, evaluate the ultimate effects of our present trends, develop guidelines for an international ethic concerning the ocean, and stimulate action to correct hazardous activities as needed to protect the ocean for multiple and proper use to mankind. This is the purpose of this book. Each of the authors is an authority in his subject and is deeply concerned with the ultimate effects of *Man's Impingement on the Ocean*.

The oceans have a volume of about 1.4 billion km^3. With the present world's population of about 3.5 billion people each human being has 0.4 km^3 (or 1 km^2 400 m deep) in the ocean for food, water, and disposal of wastes. Were all this readily available, the prospect for a continued high quality of life for people of the earth would be very promising indeed. However, the oceans have an average depth of about 3800 m, so that the surface area for each individual would only be 0.12 km^2. Also, only a few kilometers removed from the shore the oceans are, as a rule, veritable deserts, and mixing rates between the surface and deep water are on the order of hundreds of years. Even horizontal transport, while more rapid, tends to follow the great currents of the oceans, and considerable time is needed to reach homogeneity between water masses.

There are many ways that the dispersal of ocean waters, and thus the included wastes, can be enhanced. Ketchum and Ford (1952) studied the dispersal of an iron sulfuric acid waste in the New York Bite. This operation is still carried on with no apparent damage to the environment. In fact, the sport fishing has apparently improved in this area for reasons not fully clarified. Likewise, Hood et al. (1958) studied the dispersal of chlorinated aliphatic hydrocarbons, paper mill wastes, and other industrial materials behind a barge 110 miles into the Gulf of Mexico and found that even what would normally be considered very toxic materials or materials unsuitable for discharge into streams, lakes, or estuaries, could be safely disposed at sea if both a knowledge of the toxic levels to indigenous organisms and full advantage of the huge volume of the ocean were utilized. Sufficient mechanical mixing must be provided to reach acceptable concentration levels in a relatively short time period. Many examples of municipal and industrial discharges into the ocean are given in this book (Chapters 12, 13, 16, and 24) that apparently affect the environment in a very limited way, if at all, providing such outfalls are properly designed and operated.

The destruction of the microflora and fauna over small areas of the deep ocean is not a serious consequence providing an adequate seed population exists nearby to restock the area through turbulent diffusion processes. The basis for this statement rests in the fact that surface waters over the deep ocean are largely nutrient-limited and the production (growth) of organisms rises to a level in population in equilibrium with the cycling of the available nutrients. Removal of nutrients in the form of animal or plant body materials as is done by fishing or direct harvesting of plants may have more effect on the productivity of the ocean than does that of destruction of biota over limited regions of the sea, providing no *residual* materials remain that might be concentrated by organisms or cause *long range subtle effects*.

It is the problem of *nondegradable materials placed in the ocean that is of most concern*. The aromatic hydrocarbons, heavy metals, radioactive isotopes, and other man-made chemical compounds fit into this class. There is little concern about excessive fertilization of the deep ocean by components of ordinary human sewage (parasites, phages, and viruses accompanying sewage may be a problem). If thoroughly mixed, it is inconceivable that man could significantly raise the already existing levels on a world wide basis (see Chapters 23 and 26). *Locally, the problem is vastly different and must receive special attention.*

The fact that municipal wastes carry with them some nondegradable components limits the usefulness of the oceans for disposal of wastes as they now typically appear. Were these successfully separated, perhaps by a process which allows reuse of water, the problem of sewage disposal for the world's increasing population could be conveniently handled through sea disposal but only by radical changes in techniques currently in use. It is technically feasible to distribute wastes over large areas of the ocean, either by mechanical means or taking advantage of favorable current systems and thus dilute the materials to a level utilized by the metabolic capacity of the ocean biota through incorporation into body materials or to be degraded to carbon dioxide and water. Inadvertently, this has happened to most of the petroleum hydrocarbons and other materials which find their way into the sea by normal procedures of commerce (see Chapters 13 and 14).

Were it practical to carry wastes down the rivers through the estuaries across the continental shelf and disperse them in the deep ocean, the resource of the ocean for waste disposal could be advantageously utilized. Unfortunately, man is a land creature and produces his wastes in the regions of the rivers and seas least able to metabolize them. Perhaps one day we will learn how to economically utilize the ocean's potential for waste disposal, but for the foreseeable future we must concentrate our attention on the recovery of our rivers and estuaries from their presently damaged state and protect the waters of shore and deep ocean environment from devastation by man's activities.

The possibility of pollution of the world's oceans with nondegradable substances is a most serious problem. When man by his careless discharge of wastes into stream, lake, river, or even estuary effectively kills the body of water, for useful purposes a relatively small area is affected. The consequences may be severe locally, but in most cases recovery, though slow, is possible if the problem is promptly dealt with. Should the oceans become so affected, little hope for their recovery exists. Because of their size and the extremely long time cycles involved, *a poisoned ocean is untenable for man's existence on the earth.*

The record clearly shows man's inability to live with his environment. He has failed to recognize that the earth is itself an organism that has certain driving forces based on a specific, well-defined energy system. Man is a part of this system and he must desist from total dominance. Otherwise there will be no outcome but his demise from a plundered earth. To live within the capacity of the earth's resources demands a reversal of man's past performance. To avoid the ultimate he must direct his attention to intelligent utilization, conservation, and regrowth of renewable resources, and rationing of energy and nonrenewable resources to that of true need. Our attitude has been to exploit for temporary gain, increase population with little regard for quality of life, consume more per capita per year to bolster economy, and utilize our water, air, mineral, and energy resources at the lowest immediate cost possible with a flagrant disregard for the consequences.

We have rushed headlong toward personal, local, national, and international goals that have often been ill defined. We have opposed all forces, plundering as we rush, and we have neglected to consider the uses of these forces for compatible coexistence. Our debt to the organism earth is colossal; we must now begin to pay up.

The future need not be totally black. Man's great ingenuity, which has brought the human race to such fantastic levels of achievement, can be used to escape imminent disaster. Perhaps the most optimistic sign of recent times is that people of all levels of society are becoming concerned about the quality of their environment. The hour is late, but not too late, providing we are wise enough to correct our wrongs, take a new look at the environment that has sustained us so well for so long, and become a functioning partner with it in the years ahead. Now we are far off the road to compatible coexistence. So far the road has been largely downhill. We have discarded our wastes with great abandon. This requires little or no energy at very little or no cost. *For this earth, free water, free air, and free waste disposal have gone forever.* We must now pay the cost of the use of these commodities.

Thus far we have stated the problem of ocean contamination and pollution in as factual a manner as present data and interpretation will permit. Hopefully it will stimulate and direct programs that, contrary to our past and present patterns of behavior, will allow man to live compatibly with the natural processes of the oceans and the planet earth.

References

Dillingham Corporation. 1969. Marine disposal of solid wastes. Interim summary to Bureau of Solid Wastes Management, Department of Health, Education, and Welfare.

Hood, Donald W., Bernadette Stevenson, and Lela M. Jeffrey. 1958. Deep-sea disposal of industrial wastes. *Ind. Eng. Chem.*, **50**:885–888.

Ketchum, Bostwich H., and William H. Ford. 1952. Rate of dispersion in the wake of a barge at sea. *Trans. Amer. Geophys. Union*, **33**:680–683.

Manheim, F. T., R. H. Meade, and G. E. Bond. 1970. Suspended matter in surface of the Atlantic continental margin from Cape Cod to the Florida Keys. *Science*, **167**:371–376.

Part I

Transport Processes and Reservoirs

2

Rivers, Tributaries, and Estuaries

KARL K. TUREKIAN, *Professor of Geology and Geophysics, Department of Geology and Geophysics, Yale University, New Haven, Connecticut*

The writer of the Book of Ecclesiastes observed: "The rivers run into the sea and the sea is not full." This early statement of the material balance between rivers and the ocean has an analogous modern observation: "Chemical species are brought to the sea by rivers and yet the oceans seem to have retained the same saltiness, composition, and pH over long periods of time." In this chapter the nature of riverine supply of chemical species is explored and an attempt is made to guess at what happens to them when the river encounters the ocean.

We must, of course, consider the detrital, adsorbed, and dissolved loads of rivers. But in addition we require some idea of the sources of all of these materials and the relative controls on their distribution by atmospheric precipitation, weathering, and human modification as well as processes acting in the rivers and at their mouths.

Atmospheric Precipitation and Weathering

Rain and snow are the forms by which water is transferred from the oceans to the land. Atmospheric precipitation is nucleated in the air by the presence of aerosols—particles of sufficiently small size to remain aloft for long periods of time. The transport of inorganic and organic material from the oceans to the continents and across the continents is effected through the aerosols, whether as precipitation, dry fallout, or impacting. This mass transfer of chemicals of various kinds will influence the composition of streams.

Aerosols

There are two primary sources of very fine particles in the atmosphere: sea salts and continental dust. The former may be thought of as composed of highly soluble chloride and sulfate salts whereas the latter may include in addition to some soluble salts dominant silicate and refractory oxide components as discussed in Chapter 3.

The main portion of the tropospheric aerosol mass is represented by particles ranging in size between 0.1 and 20 μ radius. Particles smaller than 0.1 μ are detected indirectly by a "cloud chamber" type detector. These are called Aitken particles after the designer of the detection system. There are no chemical data on the Aitken particles, but because of their small mass contribution they may be ignored in the chemical discussion of aerosols and precipitation.

Marine aerosols, as has been demonstrated in the laboratory, are formed by the breaking of bubbles in the sea surface (see Junge, 1963 for a review and references). They are borne upward and form a uniform concentration distribution up to about 1 km over the sea after which the concentration drops markedly. Over continents marine aerosols are apparently transported to greater heights. The total mass of atmospheric sea salts per unit area of the earth may vary by no more than a factor of two over continents and oceans (5 to 11 mg salt/m^2 estimated by Eriksson, 1959; and Junge, 1963).

Junge (1963) believes that the gradients of decreasing chloride concentration in precipitation as one goes inward from the coast to the interior of a continent (Fig. 2.1) indicates not so much a particularly efficient scavenging near the coast line or a short residence time for the aerosols as it does a distribution of the marine aerosols to greater heights over the continents than over the oceans.

The Composition of Precipitation

The aerosols act as the nuclei for rain and snow falling on the land and over the oceans. The precipitation reflects the composition of the aerosol modified by chemical reactions to which the aerosol or the water drop may be subjected. Hence although marine aerosols and marine rains may, to a first approximation, replicate the proportions of ions

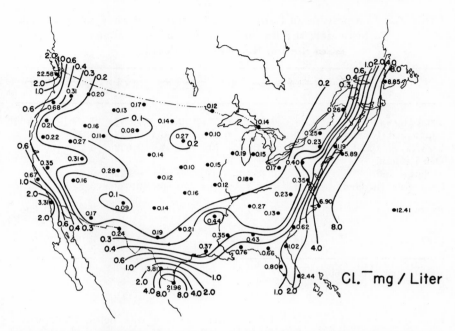

Figure 2.1 Chloride concentration in atmospheric precipitation over the United States. From Junge and Werby, 1958.

found in seawater, on closer examination a great deal of variation is found. The most important difference is the occurrence of nitrate and ammonium ions in high concentrations in the precipitation relative to sea water. Lesser differences are found in the ratios of the major cations to each other and to the anions. This is shown in Table 2.1 wherein data for coastal stations are compared with sea salt.

The concentrations of the ions in continental precipitation are considerably different from that expected of marine precipitation. Attempts have been made to assess the composition of continentally derived aerosols by subtracting out all the chloride and the proportionate amounts of the other ions on the basis of sea salt composition. This results in a component that can, to a first approximation, be identified as the continentally derived aerosol component.

The total flux of atmospherically transported material has been shown by Eriksson (1955, 1960) to be larger than that associated with net precipitation. His study of Scandinavian precipitation and river runoff

Table 2.1. Composition of Rains from Maritime Stations Compared with Seawater; the Ratios are of Gram Equivalents of the Ion Relative to Chloride[a]

	SO_4/Cl	NO_3/Cl	NH_4/Cl	Na/Cl	K/Cl	Mg/Cl	Ca/Cl
Lista, Norway	0.28	0.071	0.074	0.90	0.032	0.28	0.15
Rjupnahed, Iceland	0.10	0.013	0.015	0.86	0.028	0.23	0.11
Lerwick, Shetland	0.17	0.009	0.014	0.94	0.022	0.25	0.05
Den Heldern, Netherlands	0.21	0.052	0.030	0.95	0.022	0.30	0.09
Camborne, England	0.14	0.023	0.064	0.86	0.033	0.23	0.07
Average	0.18	0.034	0.039	0.90	0.027	0.26	0.094
Seawater	0.14	0.000035		0.56	0.020	0.067	0.021

[a] From Eriksson (1960).

indicated that about three times more chloride was being transported from areas by streams than was indicated by traps for wet precipitation. He suggested that impact of aerosols with trees and other vegetation during dry times with subsequent washout during times of rain can account for the additional chloride. The measured influx of salts by wet precipitation must be multiplied by three to obtain the total flux. This has been demonstrated in a direct fashion by Koyama et al. (1965) at a Japanese site.

Modification by Pollution

Aside from soot, modern industry releases, directly or indirectly, a number of chemicals that modify the composition of atmospheric precipitation.

Sulfur compounds and carbon monoxide resulting from fuel burning and hydrogen chloride gas may be mentioned. There is a strong increase locally of the chloride content of rains in areas of intensive industrial activity. The nitrate-to-chloride ratio also has been shown to be higher than ambient concentrations in aerosols from urban industrial areas near the oceans. A series of reactions performed in the laboratory (Robbins et al., 1959) indicates that this may be the result of the reaction of industrially produced NO_2 with sodium chloride in maritime aerosols:

Step 1: $3NO_2 + H_2O$ (liquid) $= 2HNO_3 + NO$

Step 2: $\quad\quad\quad HNO_3 + NaCl = HCl + NaNO_3$

Although some of the sulfur in the atmosphere may be due to direct pollution as the result of fossil fuel burning it appears that at least in some parts of the world the sulfur pollution is in part due to less direct effects of industrial and domestic pollution. Koyama et al. (1965) made a material balance for sulfur in precipitation and dry accumulation in Japan comparing the 1946 level with the 1959 level. The average sulfate concentration in 1946 rain was 1.34 mg/l and in 1959, 4.5 mg/l. The lower 1946 level is compatible with the low level of industrial activity in Japan following World War II and the 1959 level is used as an index of highly industrial activity.

The total wash-out of sulfur (as sulfate) for all of Japan including dry fallout and impact particles was estimated at 2.6×10^{12} and 8.7×10^{12} g/yr in 1946 and 1959, respectively. The estimates for contribution from various sources including fuel burning are given in Table 2.2. Their data show that about 75 per cent of the total sulfate deposited in 1959 was due to some sources other than those listed and that this proportion to the total fallout was about the same for 1946 and 1959. They infer from these results, as well as from isotopic arguments, that this excess sulfur is sulfate formed in the atmosphere from hydrogen sulfide produced by sulfate-reducing bacteria in near-shore polluted sediments.

Pollution of rain by trace metals certainly exists as is indicated by the wide distribution of lead in surface waters and snow fields but except possibly for this element (see Chapter 9) very little of a quantitative nature is available on this subject. Recent work by Brar et al. (1970) and Rancitelli and Perkins (1970) are representative of some of the newer attempts at filling this void.

Influence on Composition of Runoff

It is evident from what has been said about the composition of aerosols that a considerable amount of the ions carried by streams is from this source. The relative contribution of dissolved species from weathering when compared with the aerosol supply will vary from place to place.

Table 2.2. Atmospheric Sulfur Originated from Various Sources in Japan for 1959 and its Isotopic Composition

	SO_4^{2-} (ton/yr)	Contribution (per cent)	$\delta^{34}S$
Sea spray	0.3×10^6	3.4	$+20.2$
Volcano	0.1×10^6	1.1	$+3.8$
Fuel	1.8×10^6	20.7	$+9.1$
Other sources	6.5×10^6	74.8	$+9.1$[a]
Total	8.7×10^6	100	$+10.0$[a]

After Koyama et al. (1965).

[a] The value of $\delta^{34}S$ of $+10.0$ for the total is based on measurement of rainwater. The $\delta^{34}S$ of $+9.1$ for "other sources" is then calculated to complete the material balance.

It may be difficult to assess the relative importance of each unless a good deal is known about the important controlling factors.

A common procedure for assessing the role of weathering relative to marine aerosol supply is to normalize the composition of seawater ions relative to the chloride ion and to subtract the proportionate amount of each ion from the composition of the stream. This same technique has been used in determining the continental component of aerosols. On this basis it remains indeterminate as to what portion of a particular stream's ionic burden is of aerosol origin and what portion is due to weathering in its drainage basin. Certainly the continental aerosol is due to chemical weathering and to some degree mobility, but the direct relation to a particular drainage basin is obscure. Material balances in weathering studies have validity only when a simple maritime component can be subtracted from the total aerosol contribution.

Complicating the picture is the possibility that cations, and to some degree anions, originally part of the atmospheric precipitation component may, after hitting the ground, be tied up in particulate matter in the drainage basin thus depleting these ions in streams when chloride is used for normalization. Sugawara (1968) has summarized the work done by him and his co-workers on the composition of atmospheric precipitation and river water in Japan. Table 2.3, taken from his summary, shows that some of the trace elements in the surface runoff are actually depleted relative to rainwater falling on the drainage basin, indicating adsorption on particles or biological removal. Of the trace elements

Table 2.3. Average Chemical Composition of Precipitation and River Water in Japan

	Precipitation (ppm)	River Water (ppm)	River Water Relative to Precipitation Normalized to Cl Ratio = 1
Na	1.1	5.1	1
K	0.26	1.0	0.8
Mg	0.36	2.4	0.7
Ca	0.94	6.3	1.4
Sr	0.011	0.057	1.1
Cl	1.1	5.2	1
I	0.0018	0.0022	0.3
F	0.08	0.15	0.4
S	1.5	3.5	0.5
Si	0.83	8.1	0.5
Fe	0.23	0.48	0.4
Al	0.11	0.36	0.7
P	0.014		
Mo	0.00006	0.0006	2.1
V	0.0014	0.0010	0.2
Cu	0.0008	0.0014	0.4
Zn	0.0042	0.0050	0.2
As	0.0016	0.0017	0.2

From Sugawara (1968).

studied only molybdenum shows a factor of two increase over the rainwater supply.

Although it is evident that for the Japanese Islands as well as the maritime provinces of the continents the trace element as well as the major element concentration is controlled by the supply by atmospheric precipitation, it seems likely that continental weathering should have a major part in determining the trace element composition of large rivers.

Weathering

During the degradation of rocks by the action of organisms—a process called weathering—metals are released and eventually make their way to the ocean by way of streams.

It has long been known that not all elements are efficiently mobilized in solution by the weathering process. Aside from minerals that are strongly resistant to chemical attack, metals that quickly precipitate as insoluble oxides or other compounds after being released from a mineral also form residual deposits. This is the origin of lateritic (iron-oxide rich) and bauxitic (aluminum-oxide rich) soils. Various attempts have been made to determine the flux of metals carried away from the region of weathering by streams and ground water. The most successful of these calculations has been that of Garrels and Mackenzie (1967) using data for the Sierra Nevada's granitic terraine from Feth et al. (1964). There the aerosol contribution is small and clearly of maritime origin so subtraction of the maritime aerosol component from the dissolved burden is possible. The bicarbonate flux measured in springs is then used as a measure of the rate of removal in solution of the cations and silica derived from minerals; hence also the rate of weathering.

The method of Johnson et al. (1968) of attempting to subtract the river flux from the atmospheric flux for the various cations in a study of weathering in New Hampshire is less satisfactory because of the uncertainty in the composition and true total flux of the aerosol component. The very low bicarbonate concentration in the runoff causes a large uncertainty in estimating the amount of weathering since it is the one reliable independent indicator of the presence of weathering.

Methods based on measuring differences between original rock and weathered residue (Short, 1961; Harriss and Adams, 1966) may yield an index of the solubilized metals compared with the more refractory, but without a measure of the rate of weathering such studies cannot provide any indication of the flux of metals.

It appears then that we are still basically in a quandary as to the rate of supply of metals to streams from the weathering of rocks compared with aerosol contributions. We have a hunch that the weathering contribution cannot be trivial but as yet, except for a few major elements, we have no quantitative estimates.

Streams

Transport of material from the continents to the oceans is affected by four major agents: streams, wind, coastal erosion by waves, and glaciers. At the present time the role of streams seems paramount but the roles of the other forms of transport may not be trivial in any one locale. This precaution applies to the so-called dissolved load as well as the detrital load. For example, Schutz and Turekian (1965a) calculated that potentially an equal amount of soluble silica could be supplied to the oceans from Antarctica as from all the world streams because of the enhanced solubility of glacial debris as shown by Keller et al. (1963). For our present purpose, however, only stream supply will be considered.

We must assess the total runoff of the world and the mean concentrations of dissolved and particulate species in order to understand the chemical burden brought to the sea and speculate on its future path.

Discharge, Dissolved Load, and Sediment Load

Table 2.4 lists two recent estimates of surface runoff into the oceans and the dissolved load of streams, one by Livingstone (1963) and the other by Alekin and Brazhnikova (1961). For purposes of comparison

Table 2.4. **Estimates of the Total Runoff of the World and the Flux of Dissolved Solids**

	Livingstone (1963)			Alekin and Brazhnikova (1961)		
		Dissolved Solids			Dissolved Solids	
	Water Flux (10^{15} l/yr)	Concentration (mg/l)	Flux (10^{15} g/yr)	Water Flux (10^{15} l/yr)	Concentration (mg/l, including organic)	Flux (10^{15} g/yr)
North America	4.55	142	0.646	6.43	89	0.572
Europe	2.50	182	0.455	3.00	101	0.303
Asia	11.05	142	1.570	12.25	111	1.360
Africa	5.90	121	0.715	6.05	96	0.581
Australia	0.32	59	0.019	0.61	176	0.107
South America	8.01	69[a]	0.552	8.10	71[a]	0.575
Total	32.33	120	3.957	36.45	88	3.498

[a] Gibbs' (1967) data on the Amazon reduce this figure to 36 mg/l reducing estimates of the average concentration of dissolved solids and the flux by about 5 per cent.

the results of Alekin and Brazhnikova (1961) on dissolved load based on CO_3^{2-} have been converted to a HCO_3^- basis using their factor of 1.364 based on the average composition of river water. Nevertheless it can be seen that except for South America, where both compilations undoubtedly used the same source of information, there are some disparities in both the discharge and dissolved load concentration. Recent data on the Amazon from Gibbs (1967) are included to show the effect of his revised estimate.

Alekin and Brazhnikova used observed dissolved load data from the USSR as a function of different climatic and geomorphic types and extended their results for each type (except tropical) to the other parts of the world using the geographic distribution of vegetation types as an index of dissolved load.

In light of these uncertainties it may be useful as a first approximation for material balance calculations to note the following relationships:

Area of the oceans	3.6×10^{18} cm^2
Total discharge	$\sim 3.6 \times 10^6$ l/yr
Total dissolved load supply rate	$\sim 3.6 \times 10^{15}$ g/yr
Discharge per unit area of ocean	~ 10 l/(cm^2)(1000 yr)
Supply of dissolved load per unit area of ocean	~ 1 g/(cm^2)(1000 yr)

The sediment load of streams is even more difficult to estimate than the dissolved load. The sediment load of a particular stream varies with the discharge of that stream in a nonlinear fashion. It is thus difficult to estimate the total supply of solid material to the oceans by streams, unless the major streams are monitored over long periods of time.

A recent attempt at determining the sediment transport by streams and the regional denudation rates in the United States has been made by Judson and Ritter (1964). They used continuous data from gauging stations on the rivers of major regions of the United States as delineated in Fig. 2.2. Their summary is presented in Table 2.5.

If we take the ratio of dissolved load to suspended load typical of the United States and assume that it is approximated by the Earth as a whole, the average world-wide suspended load is 330 mg/l. This relation appears to be valid for the Amazon, a totally different drainage type from the Mississippi (Gibbs, 1967). Judson and Ritter (1964) believe that the correspondence of the accumulation rate of sediment in man-made Lake Mead, behind Hoover Dam on the Colorado River, and the suspended sediment supply by the Colorado River indicates that the traction or bed load is not very important in most rivers. They

Figure 2.2 Drainage region used in the construction of Table 2.2. From Judson and Ritter, 1964.

suggest 10 per cent of the suspended load as the maximum for the bedload. This raises the total supply of detrital load by streams to 365 mg/l. We will round this off to 400 mg/l for the world average. Holeman (1968) suggests 560 mg/l for a world average. It is doubtful that these figures, despite their tennous nature, will be found to be too high due to man's activities as suggested by Douglas (1967) and Judson (1968) and implied by Meade (1969). The really large sediment transporting streams appear to have a sediment carrying capacity independent of human activity.

Average Composition of River Water

The composition of the major components of the dissolved material in river waters is fairly well known from the compilation of Livingstone (1963). The trace element composition of streams is more difficult to ascertain. Not only is it probable that there are real regional variations depending on aerosol composition and both the rock type being weathered and the degree of weathering but also the likelihood of human contamination is no longer trivial. Nevertheless a summary of selected data is presented in Table 2.6.

Regional Variations

There are strong regional variations in the total dissolved solids of streams as indicated by Tables 2.4 and 2.5. Since the results in these

Table 2.5. Sediment Load of Streams in Various Regions of the United States[a] Compared with the Amazon[b]

Drainage Region	Drainage Area ($\times 10^3$ km^2)	Runoff ($\times 10^{15}$ l/yr)	Average Concentration Suspended Load (mg/l)	Average Concentration Dissolved Load (mg/l)	Average Annual Suspended Load (10^{15} g/yr)	Average Annual Dissolved Load (10^{15} g/yr)
Colorado	637	0.021	13,930	760		
Pacific Slopes, California	303	0.072	970	167		
Western Gulf	829	0.049	1880	770		
Mississippi	3240	0.555	604	248		
South Atlantic and eastern Gulf	735	0.291	131	171		
North Atlantic	383	0.188	156	128		
Columbia	679	0.309	106	138		
United States totals and weighted means	6807	1.485	602	214	0.894	0.318
Amazon	6300	5.5	100	36	0.499	0.232

[a] From Judson and Ritter (1964).
[b] From Gibbs (1967).

Table 2.6. The Composition of Streams[a]

	(µg/1)	Approximate Estimate (µg/1)	Region	Reference
Lithium	3.3	3	North America	Durum and Haffty (1960)
Beryllium				
Boron	13	10	U.S.S.R.	Konovalov (1959)
Carbon		11,496		Livingstone (1963)
Nitrogen		225.9		Livingstone (1963)
Fluorine	88	100	U.S.S.R.	Konovalov (1959)
	150		Japan	Sugawara (1968)
Sodium		6,300		Livingstone (1963)
Magnesium		4,100		Livingstone (1963)
Aluminum	360	400	Japan	Sugawara (1968)
Silicon		6,120.2		Livingstone (1963)
Phosphorus	19	20	Columbia River	Silker (1964)
Sulfur		3,739		Livingstone (1963)
Chlorine		7,800		Livingstone (1963)
Potassium		2,300		Livingstone (1963)
Calcium		15,000		Livingstone (1963)
Scandium	0.004	0.004	Columbia River	Silker (1964)
Titanium	2.7	3	Maine (U.S.A.) lakes and streams	Turekian and Kleinkopf (1956)
Vanadium	0.9	0.9	Japan	Sugawara et al. (1956)
Chromium	1.4	1	United States streams, Rhone, Amazon	Kharkar et al. (1968)
	0.3		Maine (U.S.A.) lakes and streams	Turekian and Kleinkopf (1956)
Manganese	12	7	U.S.S.R.	Konovalov (1959)
	4.0		Maine (U.S.A.) lakes and streams	Turekian and Kleinkopf (1956)
	4.8		Columbia River	Silker (1964)
Cobalt	0.19	0.2	United States streams, Rhone, Amazon	Kharkar et al. (1968)
Nickel	0.3	0.3	Maine (U.S.A.) lakes and streams	Turekian and Kleinkopf (1956)
Copper	0.9	7	Japan	Sugawara (1968)
	10		U.S.S.R.	Konovalov (1956)
	12		Maine (U.S.A.) lakes and streams	Turekian and Kleinkopf (1956)
	4.4		Columbia River	Silker (1964)
Zinc	5.0	20	Japan	Sugawara (1968)
	45.		U.S.S.R.	Konovalov (1956)
	16.		Columbia River	Silker (1964)

Table 2.6. (*Continued*)

	(μg/1)	Approximate Estimate (μg/1)	Region	Reference
Gallium	0.089	0.09	Saale and Elbe (Germany)	Heide and Ködderitzsch (1964)
Germanium				
Arsenic	1.7	2	Japan	Sugawara (1968)
	1.6		Columbia River	Silker (1964)
Selenium	0.20	0.2	United States streams, Rhone, Amazon	Kharkar et al. (1968)
Bromine	19	20	U.S.S.R.	Konovalov (1959)
Rubidium	1.1	1	United States streams, Rhone, Amazon	Kharkar et al. (1968)
Strontium	46	60	Eastern U.S.	Turekian et al. (1967) and this paper
Yttrium		0.7		Estimate[a]
Zirconium				
Niobium				
Molybdenum	0.6	1	Japan	Sugawara (1968)
	1.2		World-wide	Bertine (1970)
Ruthenium				
Rhodium				
Palladium				
Silver	0.39	0.3	United States streams, Rhone, Amazon	Kharkar et al. (1968)
Cadmium				
Indium				
Tin				
Antimony	1.1	1	United States streams, Rhone, Amazon	Kharkar et al. (1968)
Tellurium				
Iodine	7.1	7	U.S.S.R.	Konovalov (1959)
Cesium	0.020	0.02	United States streams, Rhone, Amazon	Kharkar et al. (1968)
Barium	11	10	Eastern U.S.	Turekian et al. (1967) and this paper
Lanthanum	0.2	0.2	Sweden	Landström and Wenner (1965)
		0.19	Columbia River	Silker (1964)
Cerium		0.06		Estimate[a]
Praseodymium		0.03		Estimate[a]
Neodymium		0.2		Estimate[a]
Samarium		0.03		Estimate[a]

Table 2.6. (*Continued*)

	(μg/l)	Approximate Estimate (μg/l)	Region	Reference
Europium		0.007		Estimate[a]
Gadolinium		0.04		Estimate[a]
Terbium		0.008		Estimate[a]
Dysprosium		0.05		Estimate[a]
Holmium		0.01		Estimate[a]
Erbium		0.05		Estimate[a]
Thulium		0.009		Estimate[a]
Ytterbium		0.05		Estimate[a]
Lutetium		0.008		Estimate[a]
Hafnium				
Tantalum				
Tungsten	0.03	0.03	Sweden	Landström and Wenner (1965)
Rhenium				
Osmium				
Iridium				
Platinum				
Gold	0.002	0.002	Sweden	Landström and Wenner (1965)
Mercury	0.074	0.07	Saale and Elbe (Germany)	Heide et al. (1957)
Thallium				
Lead	3.9	3	Saale and Elbe (Germany)	Heide et al. (1957)
	2.3		Maine (U.S.A.) lakes and streams	Turekian and Kleinkopf (1956)
Bismuth	?			
Thorium	0.096	0.1	Amazon	Moore (1967)
Uranium	0.06	0.3	Sweden	Landström and Wenner (1965)
	0.043		Amazon	Moore (1967)
	0.026		North America	Rona and Urry (1952)
	0.31		World-wide	Bertine et al. (1970), Turekian and Chan (1971)

After Turekian (1969).
[a] Estimate based on prorating rare-earth values in streams using La concentration in streams and the relative proportions of rare earths found in the oceans.

Table 2.7. Range of Trace Element Concentration Variations for Some United States and Foreign Streams.

	Number of Observations	Range (μg/l)	Average (μg/l)
Silver	13	0.10 –0.55	0.30
Antimony	9	0.27 –4.90	1.1
Chromium	9	0.10 –2.46	1.4
Cobalt	13	0.037–0.44	0.19
Rubidium	9	0.56 –1.61	1.1
Cesium	9	0.011–0.043	0.020
Selenium	9	0.11 –0.35	0.20
Molybdenum	9	0.44 –5.67	1.8

After Kharkar (1968).

tables are regional averages, the actual range from river to river is considerably greater.

The trace elements behave differently from the major components making up most of the dissolved fraction. Kharkar et al. (1968) showed that the concentration of silver, antimony, chromium, cobalt, rubidium, cesium, selenium, and molybdenum show variations (or lack of them) independent of the total dissolved solid concentration of streams. Their observed ranges are shown in Table 2.7. The most variable are antimony, chromium, and molybdenum and the least silver, rubidium, cesium, and selenium. It should be noted that these results do not include the strong downstream effects typical of regions with industrial pollution discussed below.

Durum and Haffty (1960, 1963) have also shown regional variations for the United States. Their results also show large variations with time for some of the elements at the individual gauging stations sampled, for which complete data were available (chromium and copper). The reasons for these variations are unknown.

Downstream Variations

Although it is useful to know the average composition of streams, it is also useful to have some idea of the degree of variation down the length of a stream. Not only will this latter information give an estimate of the confidence that a small number of analyses can be assigned, but it will also show the effects of reactions in the streams and the effect of multiple sources of element addition along the length of the stream.

We have studied a group of Connecticut rivers (Turekian, 1966) and the Neuse River of North Carolina (Turekian et al., 1967). The information from these two sets of studies can be used to make some generalizations about stream transport of soluble material.

Connecticut Streams

There are two major streams draining Connecticut: the Connecticut River and the Housatonic River. The Connecticut River's drainage is through the Triassic valley, for the most part receiving tributaries from the crystalline highlands on either side. Most of the tributaries pass over glacial debris as well as crystalline rock. From Middletown, Connecticut, to its mouth it cuts through crystalline rocks. The Housatonic

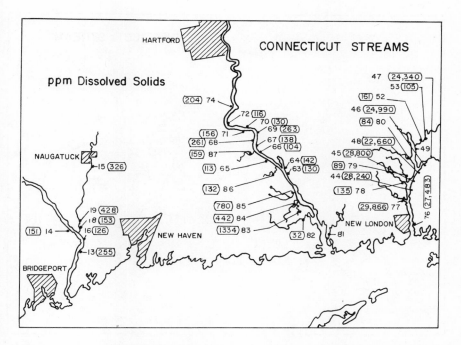

Figure 2.3 Concentrations of dissolved solids in some Connecticut streams. Concentration values are shown in the boxes.

River has its origin in the Berkshires and makes most of its traverse through a glaciated crystalline rock terrain. It represents a smaller supply of water than does the Connecticut.

The Naugatuck River is a small tributary of the Housatonic which is closely tied to the industrial development of the towns along its valley. The Thames is actually an inlet of Long Island Sound for most of its part with streams feeding into it to lower the salinity significantly. The industrial complex of the New London area is associated with the Thames. The Quinnipiac River empties into New Haven Harbor and drains industrial parts of Meriden, Wallingford, North Haven, and New Haven.

As can be seen from the description of the rivers that have been sampled in this study these are hardly virgin streams; hence the main aim of the study is to follow the path of natural and artificial injections of trace elements into streams. The data are presented in Fig. 2.3 to 2.10.

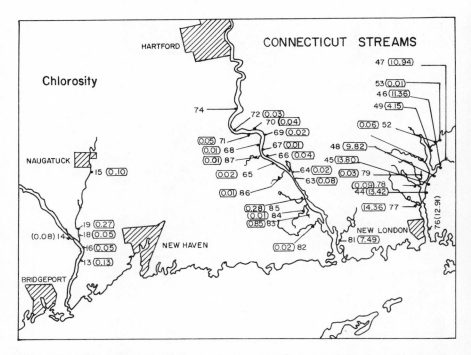

Figure 2.4 Chlorosity of some Connecticut streams. Concentration values are shown in the boxes.

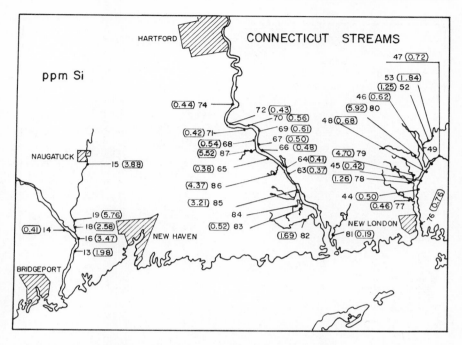

Figure 2.5 Concentrations of silica in some Connecticut streams. Concentration values are shown in the boxes.

Housatonic River and Naugatuck River. The Housatonic River is fairly free of strong industrial pollution until it reaches the towns of Derby and Ansonia. The Naugatuck River, on the other hand, is a gigantic sewer hopelessly polluted by the industry along its valley. It is a shallow stream generally black in color and putrid along its length at least south of Naugatuck until it joins the Housatonic.

The most startling fact that emerges from the study of the Naugatuck is that despite the injection of "soluble" silver and cobalt into the river, most of this is removed from the dissolved state in about 5 miles (7 km) of flow. In that distance silver decreases from 1.44 to 0.54 μg/l and cobalt decreases from 6.14 to 0.64 μg/l. The Housatonic River below the juncture does not sense the contaminated Naugatuck as far as soluble silver and cobalt are concerned. Hence the supply of "soluble" metals by the Housatonic to the Long Island Sound is about 0.46 μg Ag/l and 0.11 μg Co/l.

In contrast to these trace elements, the supply of silicon is increased in the Housatonic River after the addition from the Naugatuck. There

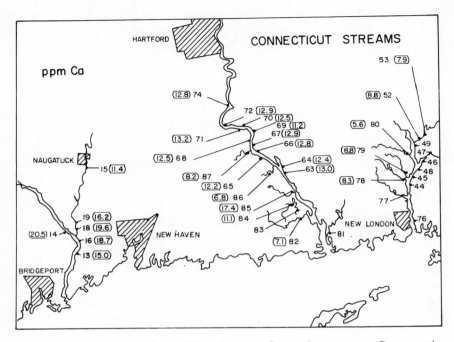

Figure 2.6 Concentrations of calcium in some Connecticut streams. Concentration values are shown in the boxes.

must be actual release of silicon into solution from a bound state to give the high value in the Housatonic south of the Naugatuck juncture since the discharge of the Housatonic is much greater than that of the Naugatuck.

Strontium also seems to behave like silica while the calcium concentrations are fairly constant for the whole system. The low barium concentration (2 μg/l) below the juncture is difficult to explain since both the Naugatuck and the Housatonic have concentrations of about 20 μg Ba/l. The sample of water just south of the town of Naugatuck is marked by extremely high concentrations of all the trace elements.

Connecticut River and Its Tributaries. The Connecticut River has its sources in the northern New England states, in particular New Hampshire. By the time it reaches Connecticut the contributions of the small streams emptying into it are not of great importance. Our sampling was done from south of Hartford to the mouth. Figure 2.11 is a plot of the data down the river.

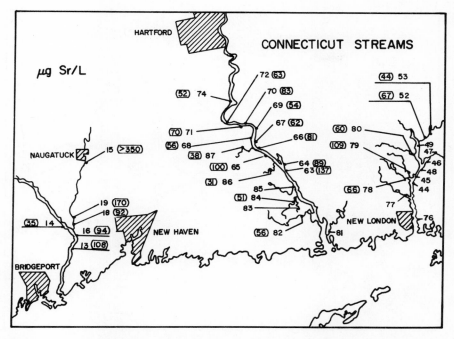

Figure 2.7 Concentrations of strontium in some Connecticut streams. Concentration values are shown in the boxes.

Calcium and barium show the least variation. The most interesting features are found in the other elements. Station 63 represents the southern-most sample of Connecticut River water without massive mixing with seawater. It has a chloride concentration of 80 ppm which is the highest for the river. Since the error of chloride determination is approximately 10 to 20 ppm by the technique used for the Connecticut rivers, variations further upstreams are not clearly defined. The general feature is that, proceeding downstream from station 69 to station 63, cobalt, silver, and silicon concentrations decrease while strontium increases.

This is not due to dilution by tributary streams changing the complexion of the Connecticut River as can be seen in Figs. 2.3 to 2.10. Hence it must be due to some process involving the regime of the Connecticut River itself uninfluenced by its tributaries.

The injection of soluble metal industrial pollutants between stations 70 and 69, as in the case of the Naugatuck, may be one explanation.

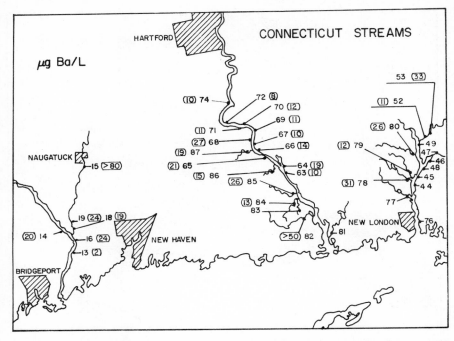

Figure 2.8 Concentrations of barium in some Connecticut streams. Concentration values are shown in the boxes.

This could explain the high cobalt and silver concentrations at the point of injection but strontium starts out low and increases downstream unlike the Naugatuck River.

The hypothesis that appears to hold the greatest potential for further exploration is that the Connecticut River upstream to Middletown from station 63 is mixed with seawater or marine aerosols, although very slightly. At station 63, a chloride concentration of 80 ppm, when related to the average chlorosity of the Long Island Sound of 16 parts per thousand, indicates that the water at this point is equivalent to 99.5 per cent fresh Connecticut River water and 0.5 per cent Long Island Sound water.

Table 2.8 gives the expected composition of pure Connecticut River water and the 99.5 per cent Connecticut River water. The average of four sulfate determinations of Connecticut River water reported by Pauszek (1958) is 16 ppm SO_4^{2-} and this is the value used. The other metals are as determined. The concentration of sulfate and strontium ions is most strongly affected by ,the mixing process. However station

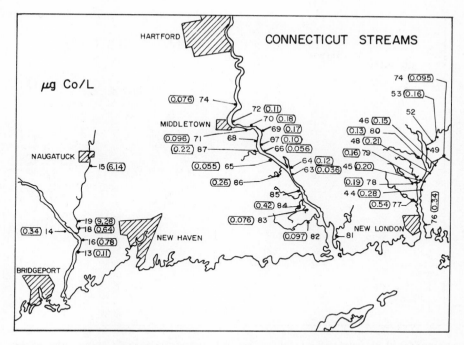

Figure 2.9 Concentrations of cobalt in some Connecticut streams. Concentration values are shown in the boxes.

Table 2.8. Compositions of Connecticut River, Long Island Sound, and Mixed Water

	100% Connecticut River (Station 69, ppm)	100% Connecticut River (0.08‰ chlorosity, Station 63, ppm)	100% Long Island Sound (16‰ chlorosity, ppm)	99.5% Connecticut River 0.5% Long Island Sound (ppm)
Si	0.61	0.37	0.2	0.61
Ca	11	13	345	12.7
SO_4^{2-}	16[a]	...	2230	27
Sr	0.052	0.130	6.81	0.086
Ba	10×10^{-3}	10×10^{-3}	14×10^{-3}	10×10^{-3}
Co	0.17×10^{-3}	0.037×10^{-3}	0.05×10^{-3}	0.17×10^{-3}
Ag	1.0×10^{-3}	0.40×10^{-3}	10×10^{-3}	1.0×10^{-3}

[a] From Pauszek (1958).

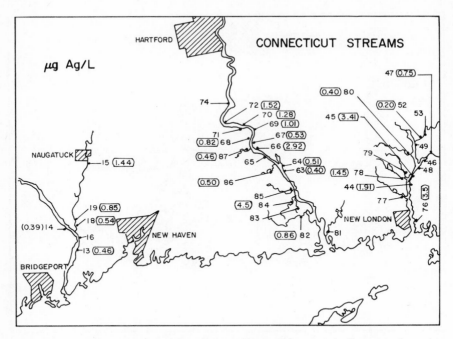

Figure 2.10 Concentrations of silver in some Connecticut streams. Concentration values are shown in the boxes.

63 indicates much lower concentrations of silver, cobalt, and silicon than is predicted by simple mixing or the addition of a marine aerosol of Long Island Sound composition. If the marine aerosol is derived from the open ocean rather than from the Sound, a lower concentration of silver and cobalt relative to chlorine might be expected but not sufficient to lower the value to the low observed values. It appears then that near the mouth the river, while still fresh water, is scavenged for silver and cobalt by some process as yet unknown. Perhaps the behavior of silicon can be taken as an analogy. Silica diminishes in concentration probably mainly because of biological activity. The chlorosity is too low and the initial silica content too low for inorganic reactions of the type implied by Bien et al. (1958) to be invoked. It is further evident that much of the stream-borne silica is fixed before it reaches the major encounter with the sea.

Thames Estuary and the Streams Feeding It. The analyses of the Thames water proper indicate that it is essentially Long Island Sound

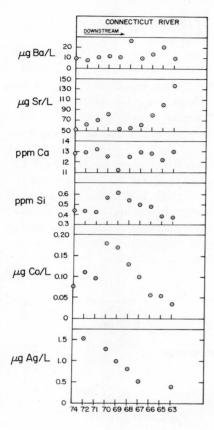

Figure 2.11 Downstream variation in composition in the Connecticut River.

water with some dilution by streams. The streams feeding it are lower in silver and cobalt than the estuarine waters. The Thames water itself is about 10 times higher in cobalt concentration than the average for Long Island Sound water indicating the possible effect of industrial pollution.

The Neuse River, North Carolina

The study of the Neuse River in North Carolina was undertaken to make a detailed study from source to mouth of a river which traverses several different geologic terrains and which has a minimum of pollution by heavy metal industries. In this sense the study was different from the study of the Connecticut rivers. It was hoped that such a program would give insight into the best method of sampling streams on a worldwide basis with assurance of a minimum of bias.

The Neuse River originates in the deeply weathered crystalline rocks of the Piedmont, runs across the Cretaceous sandstone, shales, and limestone and finally over Tertiary deposits, mainly limestones and marls, before entering Pamlico Sound. It skirts the major cities along its route, and major industry is not common, if present at all, along its shores. The only source of massive contamination could be due to agriculture and the associated pesticide usage.

The results of the analyses of the Neuse River are seen in Figs. 2.12 to 2.19. At the time of sampling, the Neuse River (November 1965) precipitation and runoff were probably at their lowest for the year.

Systematic decreases in sodium, chlorine, silicon, and calcium downstream are difficult to explain. Possible explanations are the modification by groundwater of continuously changing composition downstream or the removal of these elements by biologic or inorganic processes.

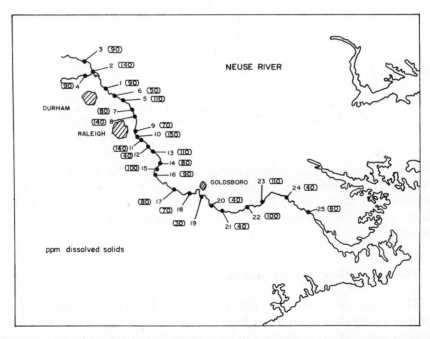

Figure 2.12 Concentrations of dissolved solids along the Neuse River. Concentration values are shown in the boxes.

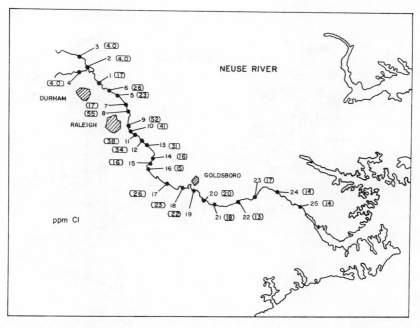

Figure 2.13 Concentrations of chloride along the Neuse River. Concentration values are shown in the boxes.

The concentrations of the trace elements strontium and barium have dispersions of about 24 and 33 per cent standard deviations, while cobalt and silver have dispersions of 56 and 49 per cent, indicating that for streams the size of the Neuse a single sample of water represents the trace element composition of the stream at that time within 25 to 50 per cent.

Trace Elements in the Suspended Load

The suspended load of streams is composed of mineral and organic material with the former commonly the dominant component. In the next section the transport of metals by adsorption on particles is discussed in detail on the basis of laboratory experiments. In this section we discuss the properties of the observed sediment loads of streams.

A detailed survey of the mineralogy, organic content, and ion exchange capacity of stream sediments of the United States has been made by

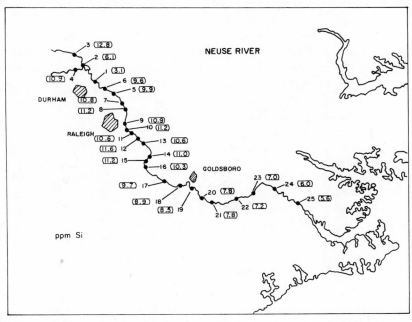

Figure 2.14 Concentrations of silica along the Neuse River. Concentration values are shown in the boxes.

Kennedy (1965). The results of his findings are summarized in Figs. 2.20 and 2.21.

The Mississippi River and those west of it draining into the Gulf of Mexico differ in several respects from the rivers of the eastern United States (Kennedy, 1965). The Mississippi River and central United States rivers have higher suspended loads and sediments rich in montmorillonite; the eastern rivers are characterized by lower suspended loads, larger proportions of kaolinite, illite, and "aluminum-interlayered clay" than the more western rivers, and sediments containing significantly more organic carbon than the average for United States streams.

The cation exchange capacity of the clay fraction from eastern United States streams is 14 to 28 meq/100 g, while that of central and west central United States stream clays is 25 to 65 meq/100 g (Fig. 2.21). Although there may be higher silt and sand contents of the western streams, the cation-exchange capacity of the total suspended load is still higher for the Mississippi and rivers west of it than for the eastern ones (Kennedy, 1965). The cation exchange capacity of the sediment of each stream does not appear to vary with rate of discharge or concentrations of suspended sediment.

The trace-element concentrations of the suspended sediments of some streams from the United States and two foreign streams are shown in Table 2.9. The eastern United States streams are higher in the concentration of most elements than the western United States streams even though the cation-exchange capacities of the eastern United States stream sediments are the lowest of the rivers sampled. Hence the major trace-element transport mode cannot be simple cation exchange. Qualitative spectrographic analyses of Brazos River bottom sediment showed no significant differences in the trace-element contents of the 2- to 20-, 0.2- to 2-, and less than 0.2-μ size fractions, indicating that the fine fraction of this stream sediment does not contain a large amount of trace elements relative to the coarser fraction despite a generally higher cation-exchange capacity of finer material.

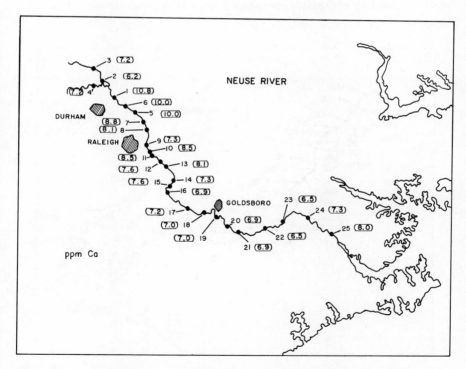

Figure 2.15 Concentrations of calcium along the Neuse River. Concentration values are shown in the boxes.

When the trace element concentration averages for the suspended load of the rivers west of the Mississippi in the United States and the Rhone River in France are compared with the average shale, except possibly for silver, the concentrations are similar, and seem to indicate that these rivers are essentially transporting material equivalent to shales that have been eroded without much chemical modification. The rivers of the eastern United States and possibly the Rio Maipo, draining the Andes, have much higher trace-element concentrations indicating other sources of trace elements.

The silver and cobalt concentrations of the stream waters are not reflected in the composition of the detrital load (Kharkar et al., 1968). Therefore, the trace elements carried by the stream sediments to a large degree are probably in inert positions.

The strong correlation of the trace elements with manganese may mean either co-precipitation of trace elements with iron and manganese oxides in weathering profiles, association with organic material from soils, or possibly industrial contamination. For normal stream concentra-

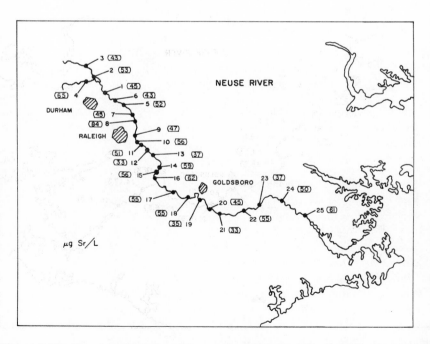

Figure 2.16 Concentrations of strontium along the Neuse River. Concentration values are shown in the boxes.

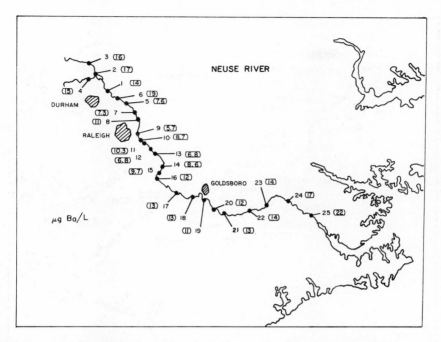

Figure 2.17 Concentrations of barium along the Neuse River. Concentration values are shown in the boxes.

tions of trace elements, aged iron and manganese oxides are not good cation absorbers, but fresh precipitates of iron or manganese oxide (hydroxide)—either in soil profiles or as the result of acid industrial wastes being neutralized—act as excellent scavengers.

The Estuarine Region

In this section an attempt will be made to understand what may be happening in estuarine environments as far as the dissolved species supplied by streams are concerned. Primarily data from our laboratory will be used to analyze the problem.

When a stream with its dissolved and detrital load reaches the ocean it creates a strong chemical gradient. In the context of this gradient certain reactions may occur. The salinity gradient results in ion exchange reactions especially on the fine-grained particles. Organic material from land will enhance the utilization of oxygen by marine organisms in the

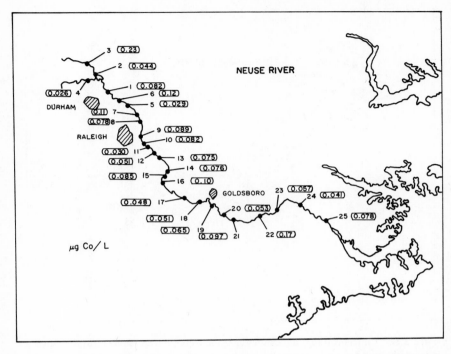

Figure 2.18 Concentrations of cobalt along the Neuse River. Concentration values are shown in the boxes.

deposited mud. In the resulting anaerobic muds the sulfate from seawater itself trapped in the sediment will be utilized by sulfate-reducing bacteria in the further exploitation of the organic material for food. The resulting hydrogen sulfide not only supplies sulfide ions to interstitial waters but controls the oxidation state of the mud and in some restricted environments eventually of the water itself. This environment is very different from the open ocean and the behavior of the metals in this environment will be unique.

The resulting material balance for metals in an estuary can then be made, in principle, if the fluxes and modes of removal are known.

Adsorption and Desorption Studies

Table 2.10 gives the results obtained in our laboratory (Kharkar et al., 1968) on the adsorption of cobalt, silver, molybdenum, and selenium

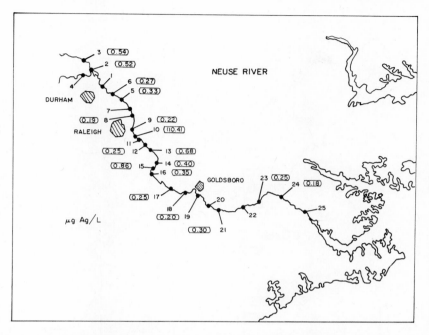

Figure 2.19 Concentrations of silver along the Neuse River. Concentration values are shown in the boxes.

on the clay minerals: montmorillonite, illite and kaolinite and the reagents, ferric oxide, manganese oxide, hydrated ferric oxide, and peat.

It is seen from Table 2.10 that montmorillonite and illite under conditions resembling those in streams adsorb about 90 per cent of the cobalt, 20–30 per cent of the silver and 30–50 per cent of the selenium present in solution. Reagent grade ferric oxide shows no detectable adsorption of cobalt and silver while adsorbing about 90 per cent of the selenium from solution. Manganese dioxide adsorbs 80–85 per cent of silver and selenium and 20 per cent of the cobalt. Kaolinite does not show any appreciable adsorption of silver and selenium. In the 1000 ppm suspended kaolinite concentration (most closely resembling the average value of suspended material in streams) the cobalt adsorption was not measurable; hence the high adsorption of the 10,000 ppm suspension is used as an upper limit. It was found that Cr^{6+} and Mo^{6+} were not adsorbed to any significant degree by any of the solids used.

Transfer of minerals to seawater results in release of about 40–70 per cent of the adsorbed cobalt, 20–30 per cent of the adsorbed silver,

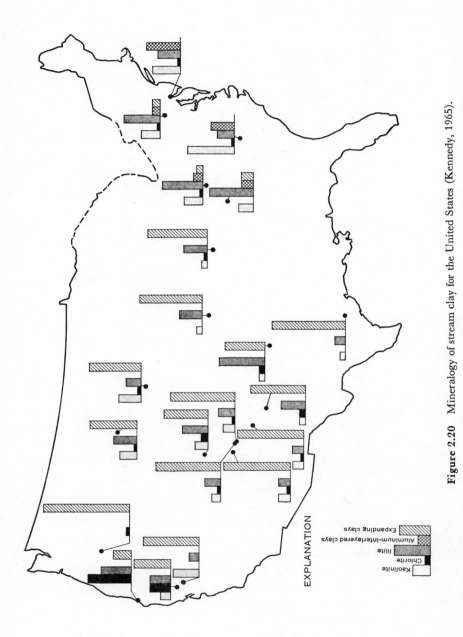

Figure 2.20 Mineralogy of stream clay for the United States (Kennedy, 1965).

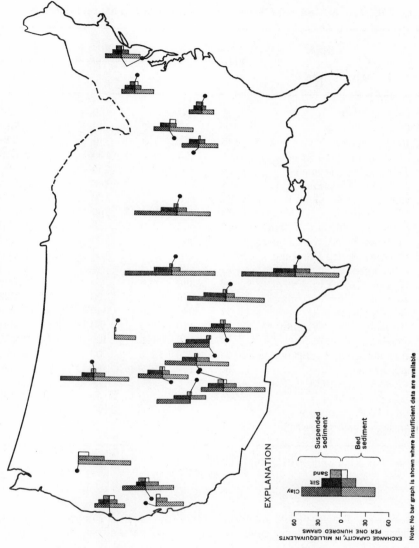

Figure 2.21 Cation-exchange capacity of stream sand, silt, and clay in United States streams (Kennedy, 1965).

Table 2.9. Trace-Element Composition of Suspended Material in Rivers.[a]

River and State	Suspended Load (mg/l)	Cr	Ag	Ni	Co	Mn
			(ppm)			
1. Brazos, Texas	954	100	0.4	30	20	690
2. Colorado, Texas	150	82	0.6	40	17	780
3. Red, Louisiana	436	37	0.3	6	7	320
4. Mississippi, Arkansas	185	150	0.7	100	33	2,300
5. Tombigbee, Alabama	25	220	1.0	200	31	5,900
6. Alabama, Alabama	54	150	4.0	100	34	3,700
7. Chattahoochie, Georgia	71	190	7.0	100	35	2,400
8. Flint, Georgia	12	210	1.0	100	39	5,100
9. Savannah, South Carolina	30	460	2.0	250	36	4,400
10. Wateree, South Carolina	37	200	1.5	100	34	7,000
11. Pee Dee, South Carolina	188	150	0.4	100	23	1,300
12. Cape Fear, North Carolina	61	130	0.7	70	21	1,700
13. Neuse, North Carolina	36	380	4.0	70	30	3,000
14. Roanoke, North Carolina	33	240	4.9	100	45	7,900
15. James, Virginia	41	290	7.0	300	60	15,000
16. Rappahannock, Virginia	28	140	1.0	80	46	2,200
17. Potomac, Virginia	34	170	1.5	400	94	7,700
18. Susquehanna, Pennsylvania	54	290	15.0	>1000	>500	12,000
Rhone, France Avignon, June 1966	296	150	0.7	60	29	820
Rio Maipo, Chile Puente Alto, South of Santiago, September 1966	41	68	1.0	40	76	2,400

[a] From Turekian and Scott (1967).

and 30–50 per cent of the adsorbed selenium. Freshly precipitated ferric hydroxide does not release much of adsorbed cobalt, silver, and selenium. Powdered MnO_2 releases almost the entire amount of the silver adsorbed in distilled water and about 60 per cent of the selenium but none of the cobalt.

A few experiments done with adsorption from seawater show that, except for freshly precipitated iron hydroxide, the adsorption is lower in seawater than in distilled water and is compatible with the observed release of metals to seawater when adsorbed in streams. These results

Table 2.10. Adsorption and Desorption Experiments of Trace Elements on Particles.[a]

Concentration (ppm)	^{60}Co Adsorbed from Distilled Water (per cent)	^{60}Co Desorbed in Seawater (per cent)	^{60}Co Adsorbed from Seawater (per cent)
Montmorillonite No. 22A, Mississippi			
200			
250			
1,000	86.3 ± 0.7	67.7 ± 2.4	24.2 ± 2.5
10,000	89.4 ± 0.8	99.5 ± 2.0	
100,000	96.6 ± 0.03	65.9 ± 2.1	
Illite, No. 35 Fithian, Illinois			
200			
250			
1,000	90.2 ± 0.6	48.1 ± 2.8	
10,000	93.9 ± 0.05	27.0 ± 5.7	
Kaolinite, No. 7 South Carolina			
200			
250			
1,000	no detectable adsorption		
10,000	41.5 ± 1.3	56.5 ± 3.7	
Ferric-oxide (Baker)			
200			
1,000	No detectable adsorption		
Manganese dioxide (Baker)			
200			
250			
1,000	25.0 ± 2.7	no detectable desorption	
Ferric hydroxide freshly precipitated at pH ~ 7.0			
200			
250			
1,000	(1) 86.1 ± 0.7	15.0 ± 7.8	94.6 ± 0.04
	(2) 99.2 ± 0.15	12.6 ± 5.0	
Peat, oyster pond, Massachusetts			
200			
250			
1,000			

Table 2.10. (*Continued*)

Concentration (ppm)	^{110}Ag Adsorbed from Distilled Water (per cent)	^{110}Ag Desorbed in Seawater (per cent)	^{110}Ag Adsorbed from Seawater (per cent)
Montmorillonite No. 22A, Mississippi			
200			
250			7.5 ± 2.0
1,000	33 ± 2.1	26.5 ± 3.0	
10,000	(1) 60 ± 1.5	(1) 30.7 ± 3.0	
	(2) 59.5 ± 1.5	(2) 35.0 ± 5.0	
100,000			
Illite, No. 35 Fithian, Illinois			
200			
250			
1,000	(1) 13.9 ± 2.6		
	(2) 28.0 ± 2.0		
10,000	(1) 74.6 ± 1.1	25.9 ± 4.5	
	(2) 83.0 ± 1.7		
Kaolinite, No. 7 South Carolina			
200			
250			
1,000	13.0 ± 1.4		
10,000	13.0 ± 1.4		
Ferric-oxide (Baker)			
200			
1,000	5.0 ± 0.2		
Manganese dioxide (Baker)			
200			
250			5.1 ± 2.0
1,000	(1) 81.4 ± 2.8	96.1 ± 3.0	
	(2) 89.8 ± 2.6		
Ferric hydroxide freshly precipitated at pH ~ 7.0			
200			
250			10.1 ± 2.0
1,000	59.0 ± 1.5	17.0 ± 4.9	
Peat, oyster pond, Massachusetts			
200			
250			6.6 ± 2.0
1,000			

Table 2.10. (*Continued*)

Concentration (ppm)	^{75}Se Adsorbed from Distilled Water (per cent)	^{75}Se Desorbed in Seawater (per cent)	^{75}Se Adsorbed from Seawater (per cent)
Montmorillonite No. 22A, Mississippi			
200	53.7 ± 0.7	27.6 ± 1.2	4.8 ± 0.5
250			
1,000			
10,000			
100,000			
Illite, No. 35 Fithian, Illinois			
200	32.0 ± 0.4	55.5 ± 1.2	12.3 ± 0.6
250			
1,000			
10,000			
Kaolinite, No. 7 South Carolina			
200	13.4 ± 0.6	70.5 ± 1.76	5.0 ± 0.56
250			
1,000			
10,000			
Ferric-oxide (Baker)			
200	86.5 ± 1.17	28.0 ± 0.95	40.0 ± 0.65
1,000			
Manganese dioxide (Baker)			
200	85.0 ± 1.12	60.0 ± 0.73	11.5 ± 0.6
250			
1,000			
Ferric hydroxide freshly precipitated at pH $\sim$ 7.0			
200	90.0 ± 1.2	18.3 ± 1.1	68.0 ± 0.81
250			
1,000			
Peat, oyster pond, Massachusetts			
200	14.6 ± 0.6	60.0 ± 1.76	no adsorption
250			
1,000			

[a] Kharkar et al. (1968) with corrected error estimates.

imply that where a trace element is adsorbed from solution it is always released to a greater or lesser extent on contact with seawater presumably because of displacement of the ions by magnesium and sodium ions present in seawater in such high concentrations. Hence stream-borne particles, rather than acting as sites for the removal of trace elements in seawater, actually contribute additional soluble amounts of some trace elements to the oceans.

Using these results we can estimate the relative importance of dissolved and desorbable ions in supplying various metals to the sea in a soluble form (Table 2.11). The desorbable cobalt load carried by streams ranges from a value equal to the dissolved load of streams to five times the dissolved load. Since the main minerals of deep-sea sediments are illite and montmorillonite, with lesser kaolinite and chlorite, a rough estimate for the worldwide contribution of desorbable cobalt would be about two times the dissolved load. Silver and selenium on the other hand are supplied mainly as dissolved species with the desorbable component making up an additional 10 per cent. One would not expect any desorbable chromium and molybdenum since so little chromium and molybdenum (as chromate and molybdate) is adsorbed in streams.

Other Tracer Studies

In the studies described above the time scales were shorter than would be expected for stream transport of sediments. It is possible that some ions are gradually fixed into ion exchange positions which are not easily affected by the major ions of seawater although replaceable by certain specific ions. Gross and Nelson (1966) report data showing that more than 80 per cent of ^{65}Zn and more than 60 per cent of the ^{60}Co from the Hanford reactor on the Columbia River are incorporated into the stream sediment. They assumed that both of these nuclides would not be released on encountering the ocean and on this basis determined the rate of long-shore transport of the ^{65}Zn and ^{60}Co labeled sediment off the mouth of the Columbia River.

Johnson et al. (1967) studied the adsorption state of ^{65}Zn and ^{54}Mn on Columbia River sediments and showed by simulating the encounter in the laboratory, that indeed ^{65}Zn is not released from river-borne sediments on encounter with seawater. Whereas almost no ^{65}Zn was released, 20 to 70 per cent of the adsorbed ^{54}Mn was released in the experiments. Up to about 50 per cent of the ^{65}Zn could be released, however, by a 0.05 M $CuSO_4$ solution. Hence although ^{65}Zn is not held in an irreversible

Table 2.11. Dissolved, Desorbable, and Total Soluble Supply of Cobalt, Silver, and Selenium to the Oceans.[a]

	Cobalt			Silver			Selenium		
	Dissolved in Stream (μg/l)	Adsorbed in Stream and Desorbed in Contact with Seawater (μg/l)	Total Soluble Load (μg/l)	Dissolved in Stream (μg/l)	Adsorbed in Stream and Desorbed in Contact with Seawater (μg/l)	Total Soluble Load (μg/l)	Dissolved in Stream (μg/l)	Adsorbed in Stream and Desorbed in Contact with Seawater (μg/l)	Total Soluble Load (μg/l)
Mississipp	0.11	0.50	0.61	0.24	0.02	0.26	0.113	0.03	0.14
Susquehann	0.35	0.71	1.06	0.365	0.02	0.39	0.325	0.06	0.39
Mad	0.14	0.59	0.73	0.26	0.02	0.28	...	...	...
Neuse	0.078	0.080	0.16	0.36	0.02	0.38	...	...	...
Rhone	0.10	0.37	0.47	0.38	0.03	0.41	0.154	0.03	0.18
Amazon	0.115	0.13	0.25	0.225	0.01	0.235	0.21	0.03	0.24

From Kharkar et al. (1968).
[a] The following average adsorption and desorption values were used:

	Cobalt		Silver		Selenium	
	Adsorbed (per cent)	Desorbed (per cent)	Adsorbed (per cent)	Desorbed (per cent)	Adsorbed (per cent)	Desorbed (per cent)
Montmorillonite	90.0	70.0	30.0	30.0	50.0	30.0
Illite	90.0	40.0	20.0	20.0	30.0	50.0
Kaolinite (upper limit)	40.0	60.0	10.0	...	10.0	70.0

adsorption site, the ions that can replace it clearly are not the common species of sea water.

Long Island Sound

Estuaries are characterized by flow of surface, less saline water from the coast seaward with transport from the sea landward at depth to maintain salt balance. Since the surface waters are in general more productive than deeper waters we expect transfer by sinking organisms and organic debris of a variety of substances from the surface to depth. Some of this may redissolve and some accumulate in the sediments. Therefore surface waters should be lower in phosphorus and silicon concentrations, for instance, and these elements should ultimately be trapped in the sediments if the regime is maintained over long periods of time.

It seems, then, that a good analogy of metal removal from the estuarine waters by organic matter and subsequent transfer to depth and to sediments is given by the behavior of silicon. It can be measured easily; there are already a number of studies on its behavior in streams and estuaries. The regime of silicon in Long Island Sound is first described as a very general analogy for the behavior of trace metals as the result of organic transport. Then alkalinity is discussed as an index to the degree of inorganic estuarine reactions presumably involving clay minerals. On the basis of insights gained from these two studies we proceed to a discussion of the results to date on the disposition of trace elements in Long Island Sound.

The Marine Cycle of Silicon in Long Island Sound

It has long been established that seawater is not saturated in silicon relative to the polymorph deposited by siliceous marine organisms. Most parts of the oceans are not even near saturation relative to quartz. Hence efficient removal of silicon by either biogenic or inorganic processes must provide the main regulating device for maintaining the observed low concentrations. Discussion of the inorganic removal of silicon has centered on the role of the continental margins. Bien et al. (1958) indicate that at the interaction of the Mississippi River with the waters of the Gulf of Mexico along the delta, silicon is removed by inorganic processes. Presumably the dissolved silicon, brought in by the freshwater, is adsorbed by suspended clay minerals on encountering the seawater. Stefansson and Richards (1963), on the other hand, observed only conservative mixing for silicon in the Columbia River estuary with no obvious removal with increasing salinity.

The role of biological processes has been identified clearly in fjords, the Antarctic Ocean, and arms of the sea such as the Gulf of California (Calvert, 1966). In these areas there are now forming large accumulations of siliceous diatom deposits. That the biological effects of silicon removal are important in near shore areas can be deduced from the observation on productivity in these marine environments. The importance may have been minimized because of the masking effect of rapidly accumulating detrital material in estuarine and coastal areas.

When we consider Long Island Sound as a large estuarine body with smaller estuaries associated with the streams draining into it we have a number of possible sources and sites of removal of silicon. Analyses for silicon down the lengths of the estuaries of the major rivers of Connecticut indicate that the variations observed can all be explained as the result of local effects of stream supply or biological removal. To a first approximation the pattern can be said to resemble most closely conservative mixing of the stream waters with Long Island Sound water. A conservative estimate of the concentration of silicon in these streams without any Sound water mixing is 0.5 ppm silicon for the Connecticut and 2.0 ppm for the Housatonic. Riley (1956) estimates a fresh water supply rate to Long Island Sound of 300×10^3 l/sec. If we assume that 10 parts of this is due to the Connecticut River and 1 part to the Housatonic, we get an average silicon concentration for the fresh water supply to the Sound of 0.6 ppm silicon. This corresponds to a silicon supply rate of 180 g Si/sec from streams. In light of the above discussion, we will assume that all this silicon reaches the Sound and none is removed in the estuaries of the rivers.

The distribution of silicon in an east-west section across Long Island and into the Atlantic Ocean is shown in Fig. 2.22. In simplest terms, Atlantic water enters the Sound at depth and returns at the surface, the rate of exchange at the Sound-ocean boundary being about $15,000 \times 10^3$ l/sec (Riley, 1956).

From Fig. 2.22 we see that the deep water (>40 ft) enters the Sound with an average silicon concentration of 0.16 ppm silicon and the surface water (0–40 ft) leaves with a concentration of about 0.08 ppm silicon, indicating a difference of 0.08 ppm silicon. This corresponds to a net transfer of 1200 g Si/sec from the open ocean to the Sound. This may be a high value because there is undoubtedly some transfer of silicon to depth at the Sound-Ocean boundary just as there is in the main part of Long Island Sound.

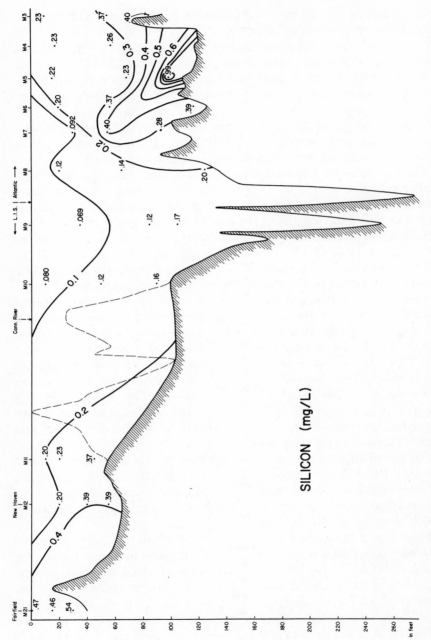

Figure 2.22 Profile of the length of Long Island Sound from M 21 to M 3 (see index map in Fig. 2.25) showing silica distribution.

Table 2.12. The Silicon Budget in Long Island Sound

Maximum total supply from streams	180 g Si/sec
Maximum total supply from ocean	1200 g Si/sec
Maximum total removed into sediments of Long Island Sound	1380 g si/sec
Area of Long Island Sound	2400×10^{10} cm^2
Maximum average deposition rate	1.8×10^{-3} g Si/(cm^2)(yr)
Estimated detrital accumulation rate in western Long Island Sound[a]	0.9 g/(cm^2)(yr)
Expected concentration of amorphous silicon if this rate is the same throughout Long Island Sound	0.2 per cent Si
Measured concentration of amorphous silicon off New Haven, Connecticut (10 per cent Si at surface of sediment, 1 per cent Si at about 50 cm depth in core)[b]	1 to 10 per cent Si

[a] Personal communication, Dr. S. Schaffel, City University of New York.
[b] Personal communication, Dr. D. P. Kharkar, Yale University.

The calculated rate of deposition of silicon in Long Island Sound is shown in Table 2.12. The average deposition rate is an upper limit and may actually be considerably lower but it is not inconsistent with all the other data for the Sound. We thus see that silicon is being "pumped" from the open ocean into the sediments of Long Island Sound presumably by biological removal.

Alkalinity in Long Island Sound

The specific alkalinity of seawater is the ratio of the alkalinity, however determined, to the salinity or some measure of it such as the chlorinity or chlorosity. In the open sea a change in specific alkalinity is due to the addition or subtraction of calcium carbonate, but in nearshore waters the specific alkalinity can be altered by the addition of bicarbonate associated with other strong bases, in particular sodium. A simple dilution of seawater by pure rainwater would not greatly alter the specific alkalinity, but the addition of high alkalinity stream waters would increase the specific alkalinity because of the excess of dissolved bicarbonate salts relative to chlorides.

A plot of the chlorosity versus the specific alkalinity for Long Island

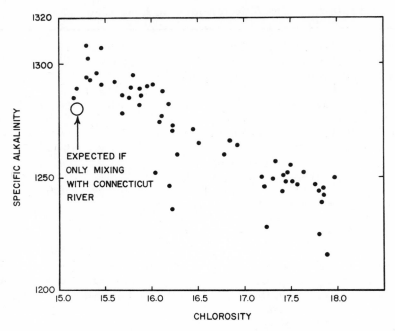

Figure 2.23 Plot of specific titration alkalinity against chlorosity for Long Island Sound water.

Sound waters is shown in Fig. 2.23. It is evident that with decreasing chlorosity there is increasing specific alkalinity, showing the influence of dilution by streams. Figure 2.24 shows a specific alkalinity profile across the length of Long Island Sound.

The aim of the study, however, was not merely to confirm a stream dilution model for an estuarine body of water like Long Island Sound but to find some clue to the locus of removal of the excess alkalinity from seawater supplied by streams. It is clear that this must be done or else the oceans would resemble the alkali lakes of the Great Basin as pointed out by several observers (see Garrels and Mackenzie, 1967, for example).

A decrease in alkalinity due to the back reaction of sodium bicarbonate with clay minerals will result in lower specific alkalinities for a given chlorosity than that predicted from simple conservative mixing. The average bicarbonate concentration of the Connecticut River is 28 ppm (Pauszek, 1958) which corresponds to a mixing specific alkalinity for 15.2 parts per thousand chlorosity of about 0.128. The Housatonic has

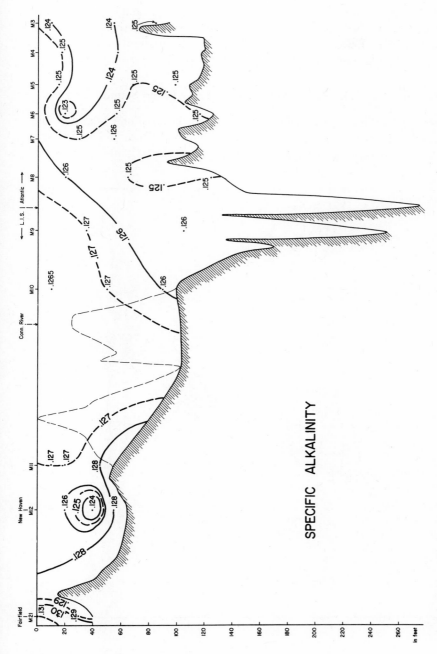

Figure 2.24 Profile of the length of Long Island Sound from M 21 to M 3 (see index map in Fig. 2.25) showing specific titration alkalinity distribution.

about 100 ppm bicarbonate, and its contribution, although smaller than the Connecticut, will tend to raise this value. Hence if there is removal of alkalinity in Long Island Sound by reactions involving clay minerals it is not detectable on the basis of specific alkalinity measurements.

The Fate of Cobalt, Nickel, and Silver in Long Island Sound

A study of the supply of several trace elements to Long Island Sound and their disposition was made to see if more definite judgments can be made about the marine fate of trace elements. The Connecticut rivers,

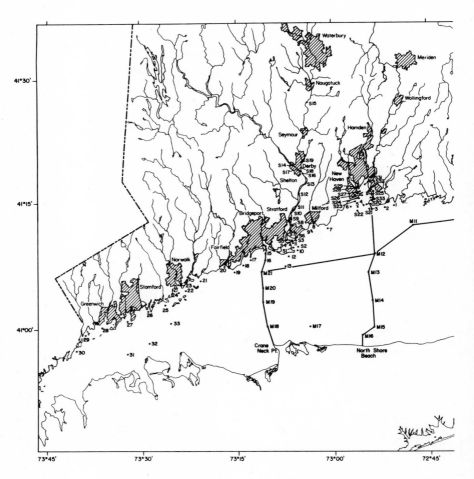

Figure 2.25 Index map of Long

especially the Housatonic (due primarily to its one tributary, the Naugatuck), carry, in addition to their natural trace element load, a considerable load of industrially injected metals that can be used as tracers.

What we learn about the behavior of certain trace elements in Long Island Sound can be used as the basis for predicting expected patterns in other estuarine near-shore areas.

The stream samples discussed in an earlier section (S-series) and the off-shore seawater samples were collected in the summer of 1965. In the summer of 1964 a series of profiles were made in Long Island Sound aboard the *Manning,* a research vessel operated by the Hudson Labora-

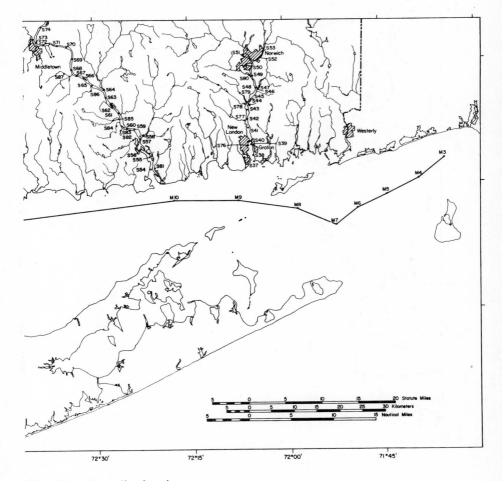

Island Sound sampling locations.

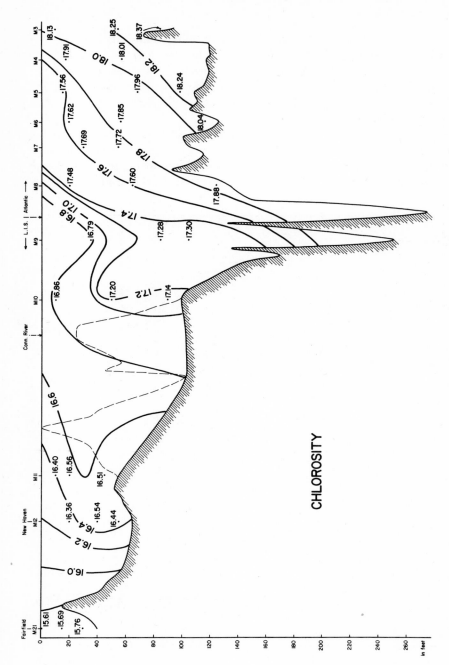

Figure 2.26 Profile of the length of Long Island Sound from M 21 to M 3 showing chlorosity.

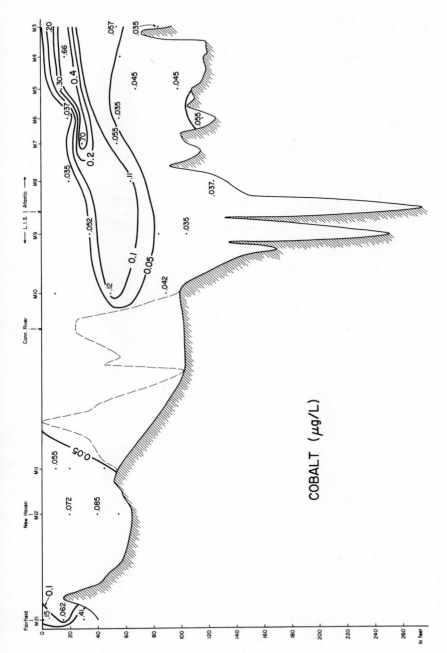

Figure 2.27 Profile of the length of Long Island Sound from M 21 to M 3 showing cobalt.

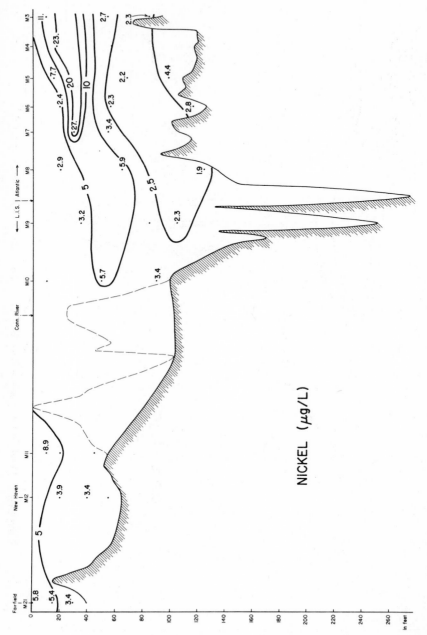

Figure 2.28 Profile of the length of Long Island Sound from M 21 to M 3 showing nickel.

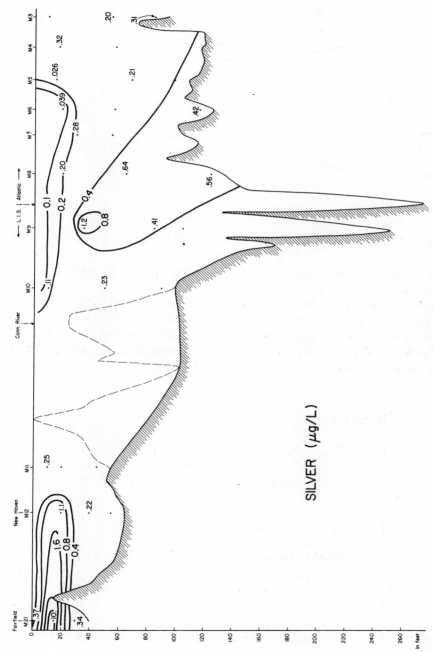

Figure 2.29 Profile of the length of Long Island Sound from M 21 to M 3 showing silver.

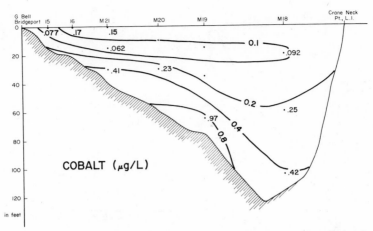

Figure 2.30 Profile from G Bell in Bridgeport to Crane Neck Pt., Long Island cobalt.

tories of Columbia University (the M-series). These are marked on the map in Fig. 2.25.

All samples were stored in polyethylene bottles and filtered through 0.45-μ Millipore filters prior to freeze-dry procedures for neutron activation analysis. The procedures for seawater are described in Schutz and Turekian (1965a,b) and the procedures for stream waters are described in Turekian (1966) and Kharkar et al. (1968).

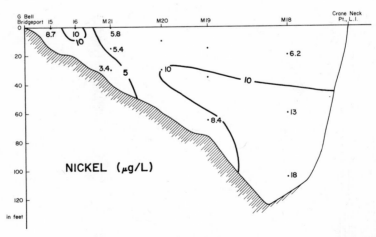

Figure 2.31 Profile from G Bell in Bridgeport to Crane Neck Pt., Long Island showing nickel.

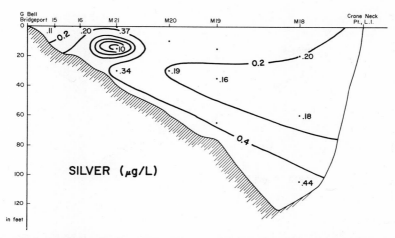

Figure 2.32 Profile from G Bell in Bridgeport to Crane Neck Pt., Long Island showing silver.

The seawater samples were analyzed for cobalt, nickel, and silver but the stream samples could be analyzed only for cobalt and silver because the latter samples were irradiated with a neutron flux with a smaller fast neutron component by which the ^{58}Ni (n,p) ^{58}Co reaction occurs. All determinations were made only once and have previously

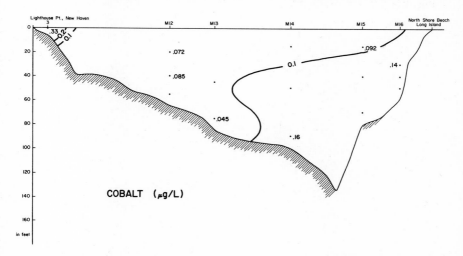

Figure 2.33 Profile from Lighthouse Point, New Haven to North Shore Beach, Long Island showing cobalt.

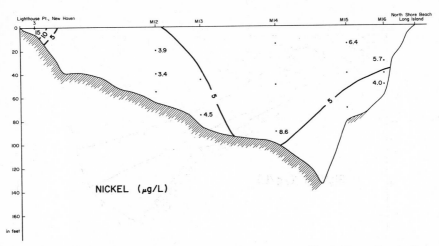

Figure 2.34 Profile from Lighthouse Point, New Haven to North Shore Beach, Long Island showing nickel.

determined errors of about 20 per cent coefficient of variation for the concentration ranges encountered.

The sampling locations are shown in Fig. 25 which also shows the lines for the cross section of Figs. 2.26 to 2.35. Figures 2.26 to 2.29 show the distribution of chlorosity and the trace elements in a cross section along the length of Long Island Sound close to the Connecticut side.

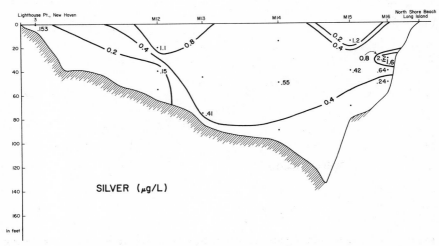

Figure 2.35 Profile from Lighthouse Point, New Haven to North Shore Beach, Long Island showing silver.

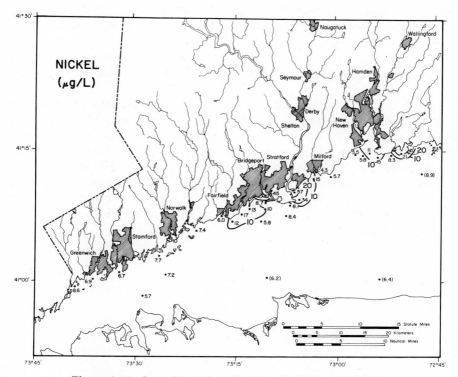

Figure 2.36 Long Island Sound surface distribution of nickel.

All the seawater values are normalized to a salinity of 35 parts per thousand. Within the experimental errors it is evident that correction for salinity is not significant. Figures 2.30 to 2.35 are similar sections across the width of the Sound.

The relationship between the concentrations along the lower Housatonic River (and its industrially polluted tributary, the Naugatuck River) and Long Island Sound are presented in Figs. 2.36 to 2.38. The contours drawn for cobalt and silver are 0.05, 0.1, 0.2, 0.4 (μg/l), and for nickel 2.5, 5, 10, 20, 40 (μg/l), so as to include only variations outside the expected error.

The distribution patterns of the trace elements in the streams and land are undoubtedly ephemeral and depend upon injection rates of the metals into these water bodies as well as oceanographic parameters. Nevertheless the distributions do give a representative picture of the processes operating in the Long Island Sound system. The patterns for cobalt, nickel, and silver shown in the cross sections of Figs. 2.30 to

2.35 are similar, in their major features, to a nearby profile presented in Schutz and Turekian (1965a) based on samples collected in the summer of 1963.

The sequence of events which appears to be required to explain the trace-element distribution patterns shown in the figures is the following:

1. Trace elements injected into the stream in soluble form are adsorbed rapidly but on suspended particles. This is best seen in the case of the Naugatuck River where the high cobalt and silver concentration as the result of industrial injection north of Naugatuck diminishes downstream so that where the Naugatuck River joins the Housatonic River it is as low in trace-element concentration as the Housatonic.

2. Some trace elements seem to be injected by industry either directly into the Sound or through small streams before adsorption on suspended material takes place.

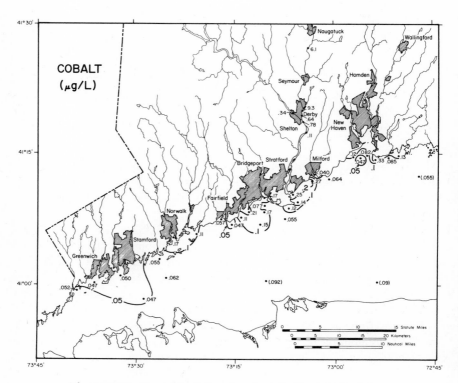

Figure 2.37 Long Island Sound surface distribution of cobalt.

3. At the mouth of the Housatonic, as the salinity of the Sound is approached, the trace-element content increases dramatically. This may be interpreted to mean either that there is release of the trace elements adsorbed by the suspended material on contact with seawater as shown by the experiments of Kharkar et al. (1968) discussed above or that there is peculiarly high release of these metals into the mouth of the Housatonic by local industry. Of the two explanations, the first seems more probable because the effect is seen by all the elements and industry is not uniquely associated with the mouth of this river.

4. A short distance away from the sources of injection of trace elements into the Sound, by whatever means, they are apparently removed from the dissolved state by planktonic organisms.

5. Some of the plankton undoubtedly disintegrates on the bottom or on the way to the bottom, and local pockets of high trace-element concentration in the water are encountered at depth on the western side of the Sound.

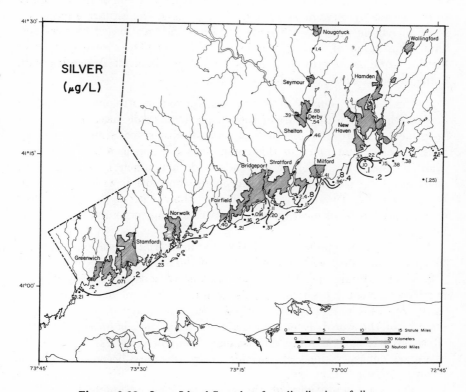

Figure 2.38 Long Island Sound surface distribution of silver.

6. Most of the trace elements, however, are removed and retained in the reducing sediments of the Sound since the solubility of the sulfides of the metals would be exceeded in the sediments of the Sound with the production of H_2S. The sediments of Long Island Sound are very high in silt, yet the silver content is about 1 ppm which implies some concentration above average shales (0.1 ppm).

7. The water leaving Long Island Sound (that above 20 ft for a large part of the Sound, Riley, 1956) is less than or equal to the concentration of trace elements in the deeper water entering the Sound from the ocean, indicating that little or none of the trace-element load supplied by streams (reinforced by the strong contribution from industry) leaves the Sound but is trapped in the sediments depositing there.

Summary

We must now draw together the results of these various studies in order to make some judgment about the general patterns of metal supply to the oceans by streams and the modifications at the estuarine interface. Most of our conclusions are tentative since they are based on very few data. The value of such projections lies as much in showing areas of further work as in providing bases for actions in matters dealing with human modifications of analogous systems.

1. The natural sources of elements dissolved in streams are aerosols and weathering. In the heart of continents the aerosols have a significant "continental" aerosol component compared with a maritime component. It is not clear how much of this component is indigenous to a particular drainage basin on which it descends and how much is a generalized continental aerosol with no correlation in composition with the terraine over which the rivers flow. We do not know with any accuracy the exact contribution of most metals from weathering compared with aerosol fluxes.

2. Metals are supplied in solution by industrial operations and can be clearly identified within a mile or so of the input.

3. Most cationic species of trace metals are adsorbed effectively on particles carried by streams. This results in a tendency to maintain fairly constant values of the cationic trace elements among the major streams of the world. Even industrial contamination is obscured by this effect so that several miles downstream from the artificial injection the dissolved concentrations of cationic trace elements become as low as in an unpolluted tributary.

4. The wide variations of the concentrations of anionic species in

streams such as chromate, molybdate, and antimonate are due to the lack of efficient adsorption of these anions by mineral particles carried by streams.

5. In addition to the adsorbed load carried by the detrital load of streams, the trace-element concentrations of the stream-borne debris reflect the source of soil. The suspended load of streams of the eastern United States, for example, have higher organic contents, higher $^{230}Th/^{232}Th$ ratios (Scott, 1968), and higher trace-element concentrations than streams debouching into the Gulf of Mexico even though the cation exchange capacity of the latter's sediments are higher than the eastern United States sediments.

6. Near the mouth of a river, just before the zone of strong mixing gradient with seawater, there are changes in concentrations of the trace elements and some of the major elements. To a first approximation the chemical concentration gradients resemble those expected from the mixing of river water with a small amount of seawater, or more probably of maritime aerosol. But imposed on this mixing effect there is the effect of removal of trace metals like silver and cobalt from the river water, thereby attenuating the dissolved trace-element flux of streams.

7. In the more saline parts of an estuary where there is an approach to seawater concentration of salts an effect, which we have interpreted to resemble desorption, takes place and the highest concentration of some metals for the river-estuarine system are encountered (our study included silver, cobalt, and nickel—all showing the effect to some degree).

8. By removal of the metals from solution or by conservative mixing with incoming seawater the high concentrations do not extend in the surface waters for more than a mile or so from the mouth of the river. In profile, however, a large estuarine embayment like Long Island Sound shows a patchwork of high values at depth near the source of injection.

9. Despite the injection of trace metals by streams (whether natural or artificial) very little of the metal in a dissolved form leaves the estuarine environment but presumably is trapped in the sediments in the estuary. The trapping ultimately is due to the action of organisms. One reasonable model following the model for silicon is that trace metals are incorporated into organic material which falls to the bottom. During decay in the sediment under anaerobic conditions the metals are trapped as sulfides or as reduced oxides.

10. This analysis suggests that most cationic trace metals have very little chance of leaving an estuary in solution, most of the stream input ultimately finding its way into the estuarine sediments.

Anionic species are not subject to the same efficient scavenging, although clearly if chromium, uranium, or molybdenum are buried in estuarine sediments or interstitial water or associated with organic matter, they may be reduced to insoluble forms.

Manganese and iron are solublized under reducing conditions and may actually be supplied to the deep ocean from estuaries by release from the sediments in an estuary and transport proceeding farther than oxidation and flocculation.

References

Alekin, O. A., and L. V. Brazhnikova. 1961. The discharge of soluble matter from dry land of the Earth. *Gidrokhim. Materialy,* **32**:12–24 (translated into English by M. Fleischer, U.S.G.S.).

Bertine, K. K. 1970. The marine geochemical cycle of chromium and molybdenum. Ph.D. Dissertation, Yale Univ.

Bertine, K. K., L. H. Chan, and K. K. Turekian. 1970. Uranium determinations in deep-sea sediments and natural waters using fission tracks, *Geochim. Cosmochim. Acta,* **34**:641–648.

Bien, G. S., D. E. Contois, and W. H. Thomas. 1958. Removal of soluble silica from fresh water entering the sea. *Geochim. Cosmochim. Acta,* **14**:35–54.

Brar, S. S., D. M. Nelson, J. R. Kline, P. F. Gustafson, E. L. Kanabrocki, C. E. Moore, and D. M. Hattori. 1970. Instrumental analysis for trace elements present in Chicago area surface air. *J. Geophys. Res.* **75**:2939–2945.

Calvert, S. E. 1966. Accumulation of diatomaceous silica in the sediments of the Gulf of California. *Geol. Soc. Amer. Bull.,* **77**:569–596.

Douglas, I. 1967. Man, vegetation and the sediment yields of rivers. *Nature,* **215**:925–928.

Durum, W. H., and J. Haffty. 1960. Occurrence of minor elements in water. *U.S. Geol. Surv. Circ.,* **445**.

Durum, W. H., and J. Haffty. 1963. Implications of the minor element content of some major streams of the world. *Geochim. Cosmochim. Acta,* **27**:1–11.

Eriksson, Erik. 1955. Air borne salts and the chemical composition of river waters. *Tellus,* **7**:243–250.

Eriksson, Erik. 1959. The yearly circulation of chloride and sulfur in nature; meteorological, geochemical, and petrological implications. Part I. *Tellus,* **11**:375–403.

Eriksson, Erik. 1960. The yearly circulation of chloride and sulfur in nature; meteorological, geochemical, and petrological implications. Part II. *Tellus,* **12**:63–109.

Feth, J. H., C. E. Roberson, and W. L. Polzer. 1964. Sources of mineral constituents in water from granitic rocks: Sierra Nevada California and Nevada. *U.S. Geol. Surv. Water Supply Papers,* **1535.**

Garrels, R. M., and F. T. Mackenzie. 1967. Origin of the chemical compositions of some springs and lakes. *In* Equilibrium concepts in natural water systems. Amer. Chemical Soc. *Advan. Chem. Ser.,* **67:**222–242.

Gibbs, R. J. 1967. The geochemistry of the Amazon River system: Part I. the factors that control the salinity and the composition and concentration of the suspended solids. *Geol. Soc. Amer. Bull.,* **78:**1203–1232.

Gross, M. G., and J. L. Nelson. 1966. Sediment movement on the continental shelf near Washington and Oregon. *Science,* **154:**879–885.

Harriss, R. C., and J. A. S. Adams. 1966. Geochemical and mineralogical studies on the weathering of granitic rocks. *Amer. J. Sci.,* **264:**146–173.

Heide, F., and H. Kodderitzsch. 1964. Der galliumgehalt des Saale-und Elbewassers. *Naturwissenschaften,* **51:**104.

Heide, F., H. Lerz, and G. Bohm. (1957). Gehalt des Saalewassers an blei und quecksilber. *Naturwissenschaften,* **44:**441–442.

Holeman, J. N. 1968. The sediment yield of major rivers of the world. *Water Resources Res.,* **4:**737–747.

Johnson, N. M., G. E. Likens, and F. H. Bormann. 1968. Rate of chemical weathering of silicate minerals in New Hampshire. *Geochim. Cosmochim. Acta,* **32:**531–545.

Johnson, V., N. Cutshall, and C. Osterberg. 1967. Retention of ^{65}Zn by Columbia River sediment. *Water Resources Res.,* **3:**99–102.

Judson, S. 1968. Erosion of the land. *Amer. Sci.,* **56:**356–374.

Judson, S., and D. F. Ritter. 1964. Rates of regional denudation in the United States. *J. Geophys. Res.,* **64:**3395–3401.

Junge, C. E. 1963. *Air chemistry and radioactivity,* Academic Press, New York, 382 pp.

Junge, C. E., and R. T. Werby. 1958. The concentration of chloride, sodium, potassium, calcium, and sulfate in rain water over the United States. *J. Meteorology,* **15:**417–425.

Keller, W. D., W. D. Balgord, and A. L. Reesman. 1963. Dissolved products of artificially pulverized silicate minerals and rocks, Part I. *J. Sed. Petrol.,* **33:**191–204.

Kennedy, V. C. 1965. Mineralogy and cation-exchange capacity of sediments from selected streams. *Geol. Surv. Professional Paper,* **433-D.**

Kharkar, D. P., K. K. Turekian, and K. K. Bertine. 1968. Stream supply of dissolved silver, molybdenum, antimony, selenium, chromium, cobalt, rubidium, and cesium to the oceans. *Geochim. Cosmochim. Acta,* **32:**285–298.

Konovalov, G. S. 1959. The transport of microelements by the most important rivers of the U.S.S.R. *Dokl. Akad. Nauk USSR,* **129:**912–915 (translated into English by M. Fleischer, U.S.G.S.).

Koyama, T., N. Nakai, and E. Kamata. 1965. Possible discharge rate of hydrogen sulfide from polluted coastal belts in Japan. *J. Earth Sci.*, **13**:1–11.

Landström, O., and C. G. Wenner. 1965. Neutron-activation analysis of natural water applied to hydrogeology. *Aktiebolaget Atomenerg* (Sweden), AE-204.

Livingstone, D. 1963. Chemical composition of rivers and lakes. *Geol. Sur. Professional Paper* No. **440-G**.

Meade, R. H. 1969. Errors in using modern stream-load data to estimate natural rates of denudation. *Bull. Geol. Soc. Amer.*, **80**:1265–1274.

Moore, W. S. 1967. Amazon and Mississippi River concentrations of uranium, thorium, and radium isotopes. *Earth and Planetary Sci. Lett.*, **2**:231–234.

Pauszek, F. H. 1958. Chemical and physical quality of water resources in Connecticut, 1955–1958. *Conn. Water Res. Bull.* **1**.

Rancitelli, L. A., and R. W. Perkins. 1970. Trace element concentrations in the troposphere and lower stratosphere. *J. Geophys. Res.*, **75**:3055–3064.

Riley, G. A. 1956. Oceanography of Long Island Sound, 1952–1954. *Bull. Bingham Oceanog. Collection*, **15**:15–46.

Robbins, R. C., R. D. Cadle, and D. L. Eckhardt. 1959. The conversion of sodium chloride to hydrogen chloride in the atmosphere. *J. Meteorol.* **16**:53–56.

Rona, E., and W. D. Urry. 1952. Radioactivity of ocean sediments, VIII: Radium and uranium content of ocean and river waters. *Amer. J. Sci.*, **250**:241–262.

Schutz, D. F., and K. K. Turekian. 1965a. The investigation of the geographical and vertical distribution of several trace elements in seawater using neutron activation analysis. *Geochim. Cosmochin. Acta.* **29**:259–313.

Schutz, D. F., and K. K. Turekian. 1965b. The distribution of cobalt, nickel, and silver in ocean water profiles around Pacific Antarctica. *J. Geophys. Res.*, **70**:5519–5528.

Scott, M. R. 1968. Thorium and uranium concentrations and isotope ratios in river sediments. *Earth and Planetary Sci. Lett.*, **4**:245–252.

Short, N. M. 1961. Geochemical variations in four residual soils. *J. Geol.*, **69**:534–571.

Silker, W. B. 1964. Variations in elemental concentrations in the Columbia River. *Limnol. Oceanog.*, **9**:540–545.

Stefánsson, U., and F. A. Richards. 1963. Processes contributing to the nutrient distributions off the Columbia River and Strait of Juan de Fuca. *Limnol. Oceanog.*, **8**:394–410.

Sugawara, K., H. Naito, and S. Yamada. 1956. Geochemistry of vanadium in natural waters. *J. Earth Sci. Nagoya Univ.*, **4**:44–61.

Sugawara, K. 1968. Secular change of the chemical composition of sea water. In *Origin and distribution of the elements*, L. H. Ahrens (ed.), Pergamon, Oxford, 1017–1021.

Turekian, K. K. 1966. Trace elements in sea water and other natural waters. Annual Report A.E.C., Contract No. AT(30-1)2912, Yale Publ. 2912-12.

Turekian, K. K. 1969. The ocean, streams, and atmosphere. In *Handbook of geochemistry*, K. H. Wedepohl et al. (eds.), Springer-Verlag, New York, Chap. 10.

Turekian, K. K., R. C. Harriss, and D. G. Johnson. 1967. The variations of Si, Cl, Na, Ca, Sr, Ba, Co, and Ag in the Neuse River, North Carolina. *Limnol. Oceanog.*, **12**:702–706.

Turekian, K. K., and M. D. Kleinkopf. 1956. Estimates of the average abundance of Cu, Mn, Pb, Ti, Ni, and Cr in surface waters of Maine. *Bull. Geol. Soc. Amer.*, **67**:1129–1132.

Turekian, K. K., and M. R. Scott. 1967. Concentrations of Cr, Ag, Co, and Mn in suspended material in streams. *Environ. Sci. Tech.* **1**:940–942.

Turekian, K. K., and L. H. Chan. 1971. The marine geochemistries of the uranium isotopes, ^{230}Th and ^{231}Pa. In *NATO Advanced Study Group in Neutron Activation Report of Meeting, September 6–13, 1970*, Univ. Oslo Press, Blindern.

3

Atmospheric Transport

EDWARD D. GOLDBERG, *Professor of Chemistry, Scripps Institution of Oceanography, La Jolla, California*

The transport of materials from the continents to the marine environment primarily takes place through wind, river, or glacial systems. The activities of man have added two other paths—introduction, both by intent and by accident, from ships, and discharges from domestic and industrial waste outfalls.

Aerial transport can result in both the rapid and widespread dispersal of solids, liquids, and gases. Some open ocean marine deposits, as those in the North Pacific, most probably receive the dominant portion of their solid phases as a result of the eolian transport of detrital solids from the continents (Griffin et al., 1968; Windom, 1969).

The atmospheric entry to the marine environment of substances as a result of the industrial, social, and agricultural activities of our civilization can be of the same order of magnitude as that of natural processes. The input of lead to the oceans, following its release to the atmosphere from internal combustion engines where it is used as an antiknock agent, appears to be of a similar level as that introduced by the rivers of the world (see Chapter 9 by Patterson). *The carbon-containing gases, resulting from the burning of fuels are introduced into the atmosphere at a rate of 10^{16} g/yr, which is within an order of magnitude of the amount of carbon fixed by photosynthesis in the oceans.*

The substances vented directly into the atmosphere, such as stack gases or agricultural sprays, can be translocated over such vast distances that their site of entry into the oceans can be far removed from their source of origin on the continents. Radioactive debris from the May 14, 1965 explosion of a Chinese nuclear device was detected in its travels around the world at sampling sites in Tokyo (140°E 36°N) and Fayetteville, Arkansas (94°W 36°N). The site of the nuclear detonation was at Lop Nor (90°E 40°N). ^{89}Sr and ^{90}Sr were sampled from rains and were detected in June and July, indicating that a part of the materials had circled the world twice. The average velocity of the wind transport was about 16 m/sec. The movement of material took place in the tropospheric jet streams (Cooper and Kuroda, 1966). Part of the DDT sprayed upon agricultural crops in Africa precipitates in the Barbados in the Caribbean (Risebrough et al., 1967). It is of interest to note that Stickel (1968) suggests that less than 50 per cent of the aerially applied pesticides actually reach their targets; the remainder is carried from the site of application by wind systems.

A description of the characteristics of the major wind systems will provide an entry to a general consideration of atmospheric transport. This section is followed by an examination of the processes and of the times involved in the transfer of materials from the atmosphere to the earth's surface. Finally, background data to consider the fates of the atmospherically introduced contaminants to the marine environment will be sought from their source functions and chemistries as well as from the observed geographical distributions of continental dusts which, following wind transport, have become incorporated in marine sediments.

Wind Systems

The global dissemination of atmospherically introduced substances can be initially considered through the main air mass movements: the equatorial easterlies (the trades), the temperate westerlies (the jets), and the polar easterlies. In the northern hemisphere the trade winds issue from the northeast and in the southern hemisphere from the southeast. In both hemispheres the strong westerly jets blow from west to east and increase in intensity with altitude in the troposphere, unlike the trades whose strengths decrease with increasing height. Around 60°N and 60°S latitudes are centered the zones of the polar easterlies—winds which also diminish in speed with increasing height. Superimposed upon such patterns (Fig. 3.1) are the smaller scale winds, often arising from the differing effects of continental and oceanic masses upon air movements, such as the sea breezes and the monsoons.

An example of the transport of a solid phase over long distances by a major wind system is reflected in the quartz concentrations in deep-sea sediments. Rex and Goldberg (1958) pointed out the rather remarkable latitudinal dependence of this mineral in Pacific deposits where maximum contents were found at mid-latitudes, between 30° and 40°N and

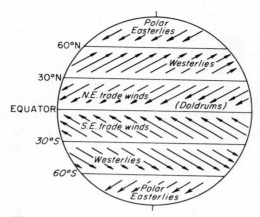

Figure 3.1 The planetary winds of the earth.

somewhat less distinctly around 35°S (Fig. 3.2). Such covariances could not be explained by a transport by oceanic currents. The highest quartz contents in the deposits were often furthest from land. Several lines of evidence lead to the hypothesis of the atmospheric transport of quartz. There is the correlation of quartz distributions in the sediments to the areal exposures of arid land areas (Fig. 3.2). The high wind velocities of the main jet streams in the upper troposphere make them the logical

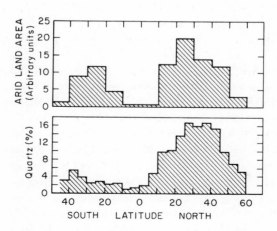

Figure 3.2 Quartz concentrations in Pacific deep-sea sediments (values on a calcium carbonate free basis) and arid land areas as a function of latitude. From Rex and Goldberg, 1958.

transporting agency. Also, near-surface filter-feeding plankton, collected from a number of Pacific Ocean areas below the jet streams, contained quartz in their guts, indicating recent dust falls.

Similarly, Delaney et al. (1967) related dust collected on the island of Barbados to a source in the European-African continents with a transport by the Northeast Trades on the basis of its size distribution and its mineralogy. Such dusts had markedly similar properties to the solid phases of the marine sediments underlying the wind system. The occurrences of freshwater diatoms in the wind-transported solids confirmed their continental origin and the presence of one species, *Denticula elegans*, which only inhabits cold running waters, indicated a contribution from mountainous regions.

Transfer Processes from the Atmosphere to the Oceans

Particulate Matter

The two principle modes of transfer of particles from the atmosphere to the oceans are gravitational settling (dry fallout) and removal by rain, snow, or other forms of precipitation. The former is especially effective for particles whose sizes are greater than tens of μ where the fallout speed is significant in comparison with atmospheric eddy velocities. There appears to be a certain uniformity in the size distributions of tropospheric aerosols in areas remote from continental sources where the number, $n(r)$ with a given radius, r, is given by

$$n(r) = 0.26 \left(\frac{r_0}{r}\right)^{3.5}$$

where r_0 from field studies has a value of 10^{-4} cm (Junge, 1968). This formulation is in accord with recent field observations on size distributions of atmospheric dusts which indicate the importance of the larger particles. Game (1964) found that over 80 per cent of the atmospheric dust collected 700 km southwest of the Canary Islands had particles in the size range between 5 and 30 μ. The Spanish Sahara was proposed as a source for this dust. About 75 per cent of the atmospheric dust collected in the monsoon winds in the Bay of Bengal in May, 1968, was in the greater than 2-μ-size class, whereas the sediments in the open Bay, primarily river-borne, had about 75 per cent of their material in the less than 2-μ-size class. The sediments underlying the Northeast Trades between Africa and the Caribbean are richer in the smaller size components (less than 2 μ) than the atmospheric dusts above, from which they receive a most substantial portion of their detrital solids, Delaney et al. (1967). Such results strongly suggest that atmospheric dusts are, in general, coarser than the nonbiogenous phases of pelagic marine sediments. The marine environment receives a large input of clay size materials (under 2 μ) by way of the rivers and is transported over wide distances in the oceans before being accommodated in the sedimentary deposits.

The widespread distribution of the talc minerals in atmospheric dusts is in accord with an atmospheric dissemination following their use as carriers for pesticides in spraying operations on the basis of their particle size range of 0.1 to 25 μ (Scotton, 1965). Windom et al. (1967) found talc in river, glacier, and atmospheric samples; talc occurred primarily in the less than 15-μ-size fractions.

Gases

The removal processes for gases from the atmosphere are probably controlled by their solubilities in water, the more soluble species being removed by rain or snow and the less soluble species by molecular exchange at the air-sea interface. The uptake of carbon dioxide and perhaps other gases by vegetation and the sorption or interaction of gases as HF with continental solids may also be of importance.

Such insoluble vapors, as the noble gases, would be removed from the atmosphere by molecular-exchange and probably enjoy a relatively long residence time there. For carbon dioxide, a relatively soluble gas, it appears to be of the order of 7 yr (Lal and Peters, 1962).

The very soluble substances that may be released to the atmosphere as volatiles would spend short passage times in the atmosphere and would accumulate in surface seawaters. The mean time they spend in the atmosphere is probably the period between rains, 10 days or so. An illustration of such substances may be found in the work of Corwin (1970) who found acetone, butyraldehyde, and 2-butanone (methyl-ethyl ketone) in surface waters from the Florida Straits, eastern Mediterranean (no butyraldehyde found), and the Amazon estuary. These aldehydes and ketones have several characteristics in common: low molecular weights, relatively low boiling points, high solubilities in water, and high utilization and production in the chemical industry. Corwin points out that formaldehyde and ethyl alcohol, substances also expected in surface ocean waters on such a basis, were not detectable at the lowest concentration levels of 400–410 μg/l that he used as standards. The absence of acetaldehyde was not explained.

Some species can undergo chemistries that limit their residences in the atmosphere. For example, sulphur dioxide in the air is oxidized to sulphur trioxide which can interact with water to give sulphuric acid or with ammonium or calcium to give the corresponding sulphates. These salts can be absorbed by precipitation and rapidly removed from the air. Residence times of 43 days or less are cited by Haagen-Smit and Wayne (1968).

Time Parameters of Atmospheric Transport

The fluxes of materials transported through the atmosphere to the marine environment can be considered for time periods extending from days to a million years or so. Direct flux measurements, obtained from monitoring rain washout or dry fallout of materials, are useful for studies involving short time intervals, periods of the order of a day or days. Oceanic sediments, which accumulate at rates of millimeters to centimeters per thousand years, can be used for flux measurements integrated over longer time periods. The input rates of clay minerals, transported from the continents to the deep sea floor through the atmosphere and the ocean, have been determined for the past half million years (Griffin et al., 1968).

The introduction of the time parameter into pollution problems through marine sedimentation rates has been made by Risebrough et al. (1968), who were investigating the transport of chlorinated hydrocarbons, such as DDT, across the Atlantic from a source in Africa. The pesticides, accompanying the dust in the winds, were at average concentrations of 41 ppb (a range of 1 to 164 ppb) parts of solids or 7.8×10^{-14} g/m^3 of air.

The input of the pesticides to the equatorial Atlantic by atmospheric transport was calculated in the following way. The area involved in dissemination by the trade winds was taken as that between the equator and 30°N latitude, 1.94×10^{17} cm^2. The reported rate of sedimentation for this area is 0.6 mm/1000 yr of solids (on a calcium carbonate free basis), equivalent to an annual input of dust solids of 9.70×10^{12} g (density of dust, 2.5 g/cm^3, water content of sediments, 50 per cent). Assuming the atmospheric dusts completely make up the nonbiogenous portions of the sediments (see subsequent discussions), such solids would bring with them 600 kg/yr of pesticides to the Atlantic.

For the entry of the products of man's activities the determination of annual fluxes appears to be of great significance, and for this purpose glacial snow fields have proven a most attractive recorder (Windom,

1969). They accumulate at rates of centimeters to tens of centimeters per year and have allowed integrated annual fluxes to be obtained in a variety of cases. The time of accumulation of various levels of the fields can be obtained by the lead-210 dating technique or by ice/firn stratigraphy. The glacial log has been developed for 100-yr periods on a year-by-year basis. Finally, the glaciers exist over a wide range of latitudes and can be used to study all of the principle wind systems.

To utilize the glacial record certain assumptions must be invoked. First, the system must be closed to gains or losses of the materials under study at all levels. A sequential accumulation must have taken place. Finally, the materials must be chemically stable over the time period, or their rates of degradation known as a function of time and temperature.

Windom et al. (1967) noted the appearance of talc in glacier samples from the state of Washington in 1940 and related the introduction of this mineral to its involvement in agriculture where it is utilized as a carrier and diluent for pesticides. The glacial ice laid down in 1964 on Mount Olympus, Washington contained 0.3 ppb of DDT, but none was detected in ice levels before 1944 (Metcalf, 1964). Increases in the rate of dust accumulated in Cacausus glacier has been related to the mechanization and industrialization of eastern Europe (Fig. 3.3, Davitaia, 1968). The dust-accumulation rate clearly shows marked in-

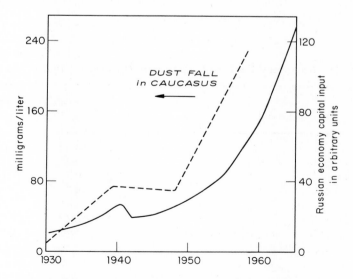

Figure 3.3 Dust fall in high Caucausus and the capital input into the Russian economy.

creases beginning in 1950 which parallels the growth of the Russian economy.

Source Functions

The input of materials to the atmosphere as a result of man's activities is incompletely documented. Two general areas of concern appear worthy of consideration: the products of fuel combustion and volatile industrial materials. Table 3.1 gives the estimated injections of substances resulting from the combustion of fuels such as coal, oil, and natural gas, for both the United States and for the world. As a first approximation, it appears that the United States is responsible for the production of between one-third and one-half of such emissions. With respect to that group of these substances considered potentially harmful (all but carbon dioxide), the automobile is responsible for about 60 per cent of the input, while power plants and industry for about 30 per cent. The remainder can be accounted for by space heating and refuse disposal.

Recent reports of losses of volatile organic compounds to the environment have suggested a more general consideration of the evaporation of substances used in industry. About 2.5 per cent of the total production of gasoline in the United States is lost by evaporation during transfer processes, from production site to vehicles and to storage tanks, and through vaporization from the automobile gas tank and carburetor. This amounts to 1.75×10^{12} g/yr (Duprey, 1968).

Table 3.1. Estimated Rates of Injection of Materials into the Atmosphere[a]

Material	United States Rate (g/yr)	World Rate (g/yr)
CO	7×10^{13}	2×10^{14}
Sulphur oxides	2×10^{13}	8×10^{13}
Hydrocarbons	2×10^{13}	8×10^{13}
Nitrogen oxides	8×10^{12}	5×10^{13}
CO_2		9×10^{15}
Smoke particles	1×10^{13}	2×10^{13}
Tetra-ethyl lead combustion products	3×10^{11}	7×10^{11}

[a] Collated from the literature.

Table 3.2. United States Production of Synthetic Organic Chemicals and Their Properties

Substance	Annual Production ($\times 10^{12}$ g)	Boiling Point (C)	Water Solubility[a]
Boiling Points Between <0 and 30 C			
Formaldehyde	1.7	−21	s
Ethylene oxide	1.0	13–14	s
Vinyl chloride	1.1	−14	ss
Acetaldehyde	0.59	21	vs
Ethyl chloride	0.31	13	ss
Dichloro-difluoro methane (Freon-12)	0.13	−29	s
Methyl chloride	0.12	−24	s
Glycerol tri-ether	0.077	0.95	
Methyl amine	0.050	−6	vs
Boiling Points Between 30 and 60 C			
Acetone	0.59	56	vs
Carbon disulphide	0.34	45	s
Propylene oxide	0.32	35	vs
Methylene chloride	0.12	40–41	ss
Diethyl ether	0.050	35	s
Boiling Points Between 60 and 100 C			
1, 2 dichloro ethane	1.6	84	ss
Methanol	1.5	65	vs
Cyclohexane	0.86	81	i
Ethanol	0.86	78.5	vs
Isopropyl alcohol	0.77	82	vs
Acrylonitrile	0.33	90–92	s
Carbon tetrachloride	0.30	77	i
Vinyl acetate	0.28	71–72	i
Trichloroethylene	0.22	87	ss
Ethyl methyl ketone	0.18	80	vs
Methyl chloroform	0.11	74	i
Chloroform	0.068	61	s
Ethyl acrylate	0.059	100	
Ethyl acetate	0.055	77	s

From Anonymous (1968).
[a] vs, very soluble; s, soluble; ss, slightly soluble; i, insoluble.

The evaporation of dry cleaning solvents attains levels of about 1800 g/d per capita in the United States (Duprey, 1968) or about 3.6×10^{11} g/yr. This number is of the same order as the production rate of perchloroethylene, the most widely used dry cleaning agent, 2.1×10^{11} g/yr. The production of such materials is compensated by an equal loss during their utilization.

Table 3.2 gives the annual production of synthetic organic chemicals on the basis of rank and boiling points (Anonymous, 1968). With production rates of the order of 10^{10} to 10^{12} g/yr, losses in the per cent range to the atmosphere could be of significance to a pollution problem locally and in the marine environment.

Metals, such as mercury and arsenic that have volatile forms, are entering the atmosphere, and subsequently the oceans, as a result of mining, extractive metallurgical, industrial, and agricultural operations. High concentrations of atmospheric mercury accompanies the smog in the San Francisco Bay region (Williston, 1968). High arsenic contents of Japanese rain waters have been attributed primarily to the smelting of sulfide ores and fuel combustion, with minor contribution from clastic solids suspended in the air (Kanamori and Sugawara, 1965).

Quantitative information on injection rates of inorganic compounds is sparse.

Wind Transport Processes: Diagnostic Criteria

In oceanic areas where the influence of river-borne solids is weak, the reflection of a wind system upon the sediments through the fallout of its dust burden can be quite distinctive. Such criteria as comparisons of the mineral biological assemblages in the sediments and in the atmosphere, and their size distributions, forms, habits, isotopic, or chemical compositions or ages can be applied to establish transport paths and the sources of the solid phases in conjunction with the prevailing meteorology.

Delaney et al. (1967) related the dusts collected from the Northeast Trades on the Island of Barbados to a source in the European-African continents. The dust traveled at least 6000 km in the winds across the Atlantic before being captured in the collecting nets. The clay mineralogies of the dusts and of the sediments underlying the Northeast Trades are quite similar and the results suggest a major portion of the sediments were wind-transported. The higher montmorillonite concentrations in the sediments west of the mid-Atlantic Ridge result from

	Mont-morillonite	Illite	Kaolinite	Chlorite
Samples	(per cent in less than 2-μ size class)			
Barbados dusts (10)	16	41	32	10
Sediments, east of mid-Atlantic Ridge (6)	18	44	26	12
Sediments, on Ridge and flanks (14)	18	39	30	13
Sediments, west of mid-Atlantic Ridge (3)	31	32	26	10

the alteration of volcanic debris introduced from the tectonically active Antilles arc area.

The quartz concentrations in the sediments (on a calcium-carbonate-free basis) and in the dusts are both around 10 per cent and both samples show a monotonous increase in quartz percentages with increasing size class.

Size Class (μ)	Barbados Dusts		Sediments East of Mid-Atlantic Ridge	
<2	5.6	8.0	7.0	6.2
2–4	12.8	13.5	22.3	20.1
4–8	17.0	19.5	25.0	30.5
8–16	23.6	21.0	39.0	39.6
16–40	a	a	56.6	46.8

a Insufficient sample for analysis.

The Northeast Trades are also defined in the Atlantic Ocean deposits through the occurrences of dolomite. This carbonate was found in the sediments, occurring as perfect to corroded rhombohedra in all size fractions above 2 μ. There is a very broad geographic distribution under the Northeast Trades in the sediments with the highest values occurring near the North African coast and a clear decrease going from east to west. The source of the dolomite could have been the African deserts as samples from such arid regions all contained this mineral in the silt-sized fraction.

The Barbados collecting nets also contained a great variety of biological forms ranging from bacteria to fragments of vascular plants, such as phloem fibers, sclerids, and tracheary elements. Siliceous shells of both marine and freshwater diatoms were also present, the latter strikingly accounting for approximately one-third the total number. One of the most abundant of the freshwater species was *Melosira granulata* which was previously reported in deep-sea sediments of the tropical belt of the Atlantic (Kolbe, 1957). Occasionally the freshwater species *Denticula elegans* was present, which has a specific habitat in the running water of cold mountainous regions such as the Alps, indicating that the dusts arriving at Barbados could have come, in part and at various times, from such an area.

Three winds encountered in the Bay of Bengal (Goldberg and Griffin, 1969) could be characterized by their dust loads and mineralogies: (*a*) the southwest monsoon, prevailing during the month of May, whose direction was quite uniform at about 250°; (*b*) an east wind, prevailing during the winter monsoon in January; and (*c*) a south wind, which occurred during the same time period as the east wind in the winter monsoon.

The heaviest dust loads, with values exceeding 8 μg/m³ were found in the southwest monsoon, which also had the highest velocities; the winds from the east had the lightest burdens, ranging between 0.59 and 0.94 μg/m³, whereas the winds from the south were distinctly higher than their easterly counterparts with values between 0.82 and 2.70 μg/m³.

Both the silt and clay sizes in the solids collected from the atmosphere were unique for each of the three winds. The clay size fractions will serve satisfactorily to illustrate this point (Table 3.3).

Table 3.3. Clay Mineralogies of Atmospheric Dusts Collected in the Bay of Bengal;[a] Ranges are Given in Parentheses.

Samples	Montmorillonite	Illite	Kaolinite	Chlorite
Southwest winds	14(8–23)	48(39–56)	11(8–16)	27(21–32)
East winds	12(7–18)	48(42–54)	28(22–36)	13(8–15)
South winds	39(33–46)	24(19–30)	28(24–31)	9(5–13)

From Goldberg and Griffin (1969).
[a] In per cent of the less than 2-μ-size fractions.

Although at the present time we can not definitively establish the sources of these solids, it is evident that the source areas are different. However, the wind directions do place restrictions upon where the atmospheric loads were picked up.

Such examples as those cited in this section indicate that the natural inputs to winds may be used as tracers to the sites of injection. Inevitably, the studies on man introduced pollutants will develop our knowledges of wind transport to an even greater degree.

References

Anonymous, 1968. Synthetic organic chemicals, U.S. Production and Sales, 1966. *U.S. Tariff Comm. Publ.*, **248**.

Cooper, W. W., and P. K. Kuroda. 1966. Global circulation of nuclear debris from may 14, 1965 nuclear explosion. *J. Geophys. Res.*, **71**:5471–3.

Davitaia, F. F., 1968. Cited by R. A. Bryson at AAAS National Meeting, Dallas, Texas, 1968.

Corwin, J. F. 1970. Volatile organic materials in sea water. In *Organic matter in natural waters*, D. W. Hood (ed.), University of Alaska, Institute of Marine Science Occ. Publ. No. **1**, 625 pp.

Delany, A. C., A. C. Delany, D. W. Parkin, J. J. Griffin, E. D. Goldberg, and B. E. F. Reimann. 1967. Airborne dust collected at Barbados. *Geochim. Cosmochim. Acta*, **31**:885–909.

Duprey, R. L. 1968. Compilation of air pollutant emission factors. *U.S. Public Health Ser. Publ.*, No. 999-AP-42.

Game, P. M. 1964. Observations on a dust fall in the Eastern Atlantic, February, 1962. *J. Sediment. Petrol.*, **34**:355–9.

Goldberg, E. D., and J. J. Griffin. 1969. The sediments of the Bay of Bengal. Manuscript in preparation.

Griffin, J. J., H. Windom, and E. D. Goldberg. 1968. The distribution of clay minerals in the world ocean. *Deep-Sea Res.*, **15**:433–59.

Haagen-Smit, A. J., and L. G. Wayne. 1968. Atmospheric reactions and scavenging processes. In *Air pollution*, Vol. I, 2nd Ed., A. C. Stern (ed.) Academic Press, New York, pp. 149–186.

Junge, C. E. 1968. Airborne dust at Barbados and its relation to global tropospheric aerosols. *Geochim. Cosmochim. Acta*, **32**:1219–22.

Kanamori, S., and K. Sugawara. 1965. Geochemistry of arsenic in natural waters. I. arsenic in rain and snow. *J. Earth Sci. Nagoya Univ.* **13**:23–25.

Kolbe, R. W. 1957. Fresh-water diatoms from Atlantic deep-sea sediments. *Science*, **126**:1053–1056.

Lal, D., and B. Peters, 1962. Cosmic ray produced isotopes and their application to problems in geophysics. *Progr. Elem. Particle Cosmic Ray Phys.*, **6**:1–74.

Metcalf, R. L., 1964. Report on National Academy of Science Traveling Symposium on Pesticides, No. 15-21, 1964. Cited by W. E. Westlake and Francis A. Gunther in Occurrence and mode of introduction of pesticides in the environment. *Advan. Chem.*, Ser. **60**, American Chemical Society, Washington, D.C. 1966.

Rex, R. W., and E. D. Goldberg. 1958. Quartz contents of pelagic sediments of the Pacific Ocean. *Tellus*, **10**:153–9.

Risebrough, R. W., R. J. Huggett, J. J. Griffin, and E. D. Goldberg. 1967 Pesticides: Transatlantic movements in the Northeast Trades. *Science,* **159**:1233–1236.

Scotton, J. W. 1965. Atmospheric transport of pesticide residues. *U.S. Dept. of Health, Education and Welfare, Office of Pesticides.* Manuscript.

Stickel, L. F. 1968. Organochlorine pesticides in the environment. *U.S. Spec. Sci. Rept.—Wildlife,* No. 119.

Williston, S. H. 1968. Mercury in the atmosphere. *J. Geophys. Res.*, **73**:7051–5.

Windom, H., J. J. Griffin, and E. D. Goldberg. 1967. Talc in atmospheric dusts. *Environ. Sci. Tech.*, **1**:923–6.

Windom, H. 1969. Atmospheric dust records in permanent snowfields: Implications to marine sedimentation. *Bull. Geol. Soc. Amer.* **80**:761–82.

4

Horizontal and Vertical Mixing in the Sea

AKIRA OKUBO, *Chesapeake Bay Institute, The Johns Hopkins University, Baltimore, Maryland*

In this book the term "oceanic mixing" applies to a multitude of physical processes occurring internally in the sea which tend to produce uniformities in properties such as temperature. Such mixing is essentially irreversible in the sense that the mixture by itself can not return to its original state.

There are two processes which contribute substantially to oceanic mixing (Bowden, 1965a): "advection" (or "convection") and "diffusion." In advective processes, (large-scale) regular patterns of water movement, for example, ocean currents, carry a property with them, thus producing a local change of its concentration. In diffusive processes, on the other hand, (small-scale) irregular movements of water called "turbulence," together with molecular diffusion, give rise to a local exchange of the property without any net transport of water. Since the diffusive power of oceanic turbulence is very much greater than that of molecular diffusion, the diffusive processes are often called "turbulent diffusion."

The term "diffusion" is also used synonymously with "mixing." Strictly speaking, both "diffusion" and "turbulent diffusion" are equivalent to "mixing" only when there is no mean velocity in the fluid. In a loose sense, however, we use "diffusion," or "turbulent diffusion" as the *result* of mixing of property; when we wish to stress the *process* of mixing, however, we use the term "diffusive process," instead of simply "diffusion." For the spread or intermingling of immiscible particles for which the effects of molecular diffusion are entirely negligible, we use the term "dispersion."

The computation of advection requires a detailed knowledge of the field of mean velocities. Such information is incomplete for most regions of the ocean. The characteristics of the field of random velocities are even less well known than those of mean velocities. Worst of all, the division between the part of the water motion assigned to the advective processes and the part which is characterized as the diffusive processes is not always self-evident. The oceanic motion consists of a more or less continuous spectrum containing scales from the molecular free path up to the ocean-wide general circulation.

In principle, of course, one could assign all the complex motions in oceans to "advection" and leave the molecular diffusion for "diffusive process," so that the ambiguity mentioned previously would never occur. Ultimate understanding of the mixing process must be accomplished along this line of thought. However, this concept is neither convenient nor useful in dealing with practical problems of oceanic mixing. For these we must still rely heavily upon an approach in which certain averaging procedures are inevitable. Thus the division between the two processes may be somewhat arbitrary, depending primarily on time-and-spatial scales of mixing in connection with our observation. In other words, the division is also determined by precisely how one wishes to describe the process of oceanic mixing in time and space.

As an introduction to the subsequent sections, we qualitatively consider a case where a contaminant is discharged as a local source over a short period of time into the sea. For simplicity, assume that the contaminant is dynamically passive, that is, the presence of the contaminant does not affect the dynamics of the ambient fluid. The local current will then transport a patch of contaminant away from the discharge area, while turbulent motion will spread the contaminant about its moving center. A detailed picture would be far more complex.

It is customary to analyze turbulent motion into Fourier components and then to identify the wave-numbers of the components with length scales of "eddies" in the fluid motion. As a result, the term "Fourier components" is often understood to mean the water motion rather than the artifact of the analysis. We use it in this sense when no ambiguity results. Fourier components with length scales much larger than the cloud of substance advect it as a whole, and those with scales much smaller than the patch produce the relative spread of contaminant in the patch. Eddies with length scales of the order of the size of the cloud significantly influence the deformation of the cloud as well as making the greatest contribution to the relative spread. It is these eddies in the midrange that produce the shears or spatial variations in the mean velocity field which are effective in diffusing a patch of contaminant. They deform the patch and a deformed patch will have a more convoluted boundary than a smoother, more compact patch. It will thus be subject to more efficient diffusion by turbulent eddies whose length scales are smaller than the overall patch size. At microscale the molecular diffusion, which is always present, contributes to the same end. As a result, the cloud tends to grow in size and the substance in the cloud to mix more and more uniformly into the surrounding water.

As the patch increases in size, the boundary between the eddies contributing to advection and those responsible for diffusion shifts toward the larger scale. Spectral measurements in the ocean show that the larger scale eddies characteristically contain more kinetic energy than the smaller scale eddies. As a result the apparent power of the mixing which acts on a diffusing patch increases with patch size or, to put it another way, with the time elapsed since the patch first began to mix. It should be apparent, therefore, that a model on the Fickian diffusion using a constant diffusivity is inappropriate for oceanic mixing from an instantaneous source. This suggests that the oceanic mixing processes may not be properly described by Fickian diffusion with a constant diffusivity.

The above discussion can be applied chiefly to the mixing of the patch in horizontal directions. The contaminant can also spread vertically. Vertical motions occur in the ocean. The discussion may be applied equally well to mixing in the vertical direction if one remembers that the structure of turbulence of the vertical velocity differs, in certain aspects, from that of the horizontal velocity. Vertical mixing in general is a weak process compared with horizontal mixing since the scale of oceanic motions is much larger horizontally than vertically. Nevertheless, the vertical mixing can by no means be ignored if a proper understanding of the mechanism of oceanic mixing is to be gained. An attempt in this direction by Bowles et al. (1958) created the concept of the "shear-effect." This so-called shear-effect is in fact a measure of the combined action of two processes, one the vertical shear in the *horizontal* currents and the other the *vertical* mixing. Together they produce an effect which appears as an augmented *horizontal mixing*.

An analysis of the mixing of a dynamically active property is much more complicated than the corresponding analysis for a dynamically passive one. A property is said to be "dynamically active" if it either changes the density field or acts on the velocity field. In analyzing dynamically active properties one must deal with the dynamical coupling between the velocity field and the field of the property. Nevertheless, useful solutions to practical problems involving dynamically active properties may, within limits, be secured by treating the active property as though it were passive, that is, by simply ignoring the dynamical coupling with the velocity field. Thus the fate of a weak vortex embedded in a strong current can be discussed by approximating the vortex as a passive quantity. On the contrary, the fate and mixing of a blob of water of uniform temperature and salinity introduced into a stably stratified sea of different temperature and salinity otherwise at rest could not be properly studied without the consideration of the velocity field created by the presence of the blob of water.

Our discussion hereafter is restricted primarily to the mixing of a passive or approximately passive property.

Mechanism of Mixing

Before plunging ourselves into the ocean, we consider the mechanism of mixing in a more common place, a cup of coffee. When cream is added to coffee, three more or less distinct stages can be recognized (Eckart, 1948): (a) In the initial stage, the blob of cream is distinctly visible; the interface between cream and coffee is sharp. The gradients are low except across the interface (Fig. 4.1a). If the coffee is undisturbed this stage may last a little while. (b) The intermediate stage is commonly brought on by stirring the liquid. The blob of cream is deformed into a complex pattern, accompanied by a rapid increase in the interfacial region which is characterized by high gradients of cream concentration (Fig. 4.1b). (c) The final stage appears to be a homogeneous mixture of cream and coffee. The increase of the interfacial regions is subjected very effectively to molecular diffusion (Fig. 4.1c).

As Eckart explains, the potentially long duration of the first stage is to be ascribed to slow molecular diffusion operating across the relatively limited interfacial area. In the second stage, the action of stirring, or advection, gives rise to the relative motion of different parts of the liquid and results in an increase in the interfacial area. During the third stage, homogeneity results from molecular diffusion, whose effect has been amplified because advection has increased the opportunity for it to act.

The entire problem can be discussed more precisely with the aid of the differential equation governing the instantaneous variation of a conservative property.

$$\frac{\partial \theta}{\partial t} + u_i \frac{\partial \theta}{\partial x_i} = D \frac{\partial^2 \theta}{\partial x_l^2} \qquad (1)$$

MECHANISM OF MIXING

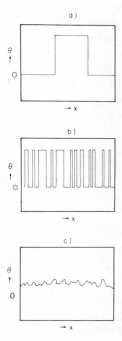

Figure 4.1 Schematic sketch of mixing of a property θ on a cut: a) initial stage; b) intermediate stage; and c) final stage.

where θ is the concentration of the property, t the time, x_i ($i = 1, 2, 3$) the coordinate system, u_i the velocity of the fluid, and D the molecular diffusivity. Tensor notation together with the summation convention for repeated indices is used. Equation 1 is an Eulerian representation.

From (1) we can derive an equation for the mean square of the θ-gradient averaged over a volume, V, whose boundary, B_n, moves with the fluid. To this end, differentiate (1) with respect to x_j, multiply by $\partial \theta / \partial x_j$, and average over the volume. The result is

$$\frac{d\langle G^2 \rangle}{dt} = -2 \left\langle \frac{\partial u_i}{\partial x_j} \frac{\partial \theta}{\partial x_i} \frac{\partial \theta}{\partial x_j} \right\rangle + 2D \int_{B_n} n_j \frac{\partial \theta}{\partial x_j} \frac{\partial^2 \theta}{\partial x_l^2} d\sigma - 2D \left\langle \left(\frac{\partial^2 \theta}{\partial x_i^2} \right)^2 \right\rangle \quad (2)$$

where $G^2 \equiv (\partial \theta / \partial x_i)^2$, **n** represents the unit outward vector normal to the boundary, and $\langle \ \rangle$ designates the mean value over the volume.

Assume, for the moment, that the surface integral vanishes. This is the case whenever the boundary passes only through points at which no flux of θ takes place. Certainly the wall of the coffee cup is such a boundary, provided no cracks are present. We shall first examine the last term on the right side of (2). The term $-2D\langle(\partial^2\theta/\partial x_i^2)^2\rangle$ represents the effect of molecular diffusion on the variation of the mean square gradient. Since this term is nonpositive, the molecular diffusion always tends to dissipate the mean gradients of θ inside the volume. The first term on the right-hand side of (2), $-2\langle(\partial u_i/\partial x_j)(\partial\theta/\partial x_i)(\partial\theta/\partial x_j)\rangle$, indicates the rate at which advection or fluid motion changes the mean square gradient. The change may be either an increase or a decrease. The velocity gradient tensor, $\partial u_i/\partial x_j$, can be resolved into symmetric and asymmetric parts.

$$\frac{\partial u_i}{\partial x_j} = \underbrace{\frac{1}{2}\left\{\frac{\partial u_i}{\partial x_j} + \frac{\partial u_j}{\partial x_i}\right\}}_{\text{Rate of deformation tensor}} + \underbrace{\frac{1}{2}\left\{\frac{\partial u_i}{\partial x_j} - \frac{\partial u_j}{\partial x_i}\right\}}_{\text{Vorticity}} \tag{3}$$

By substituting (3) into the first term, one can see that the vorticity of the fluid motion does not contribute at all to the change of the mean square gradient. Therefore, the advective term can be written as

$$-\left\langle\left\{\frac{\partial u_i}{\partial x_j} + \frac{\partial u_j}{\partial x_i}\right\}\frac{\partial\theta}{\partial x_i}\frac{\partial\theta}{\partial x_j}\right\rangle \tag{4}$$

That the rate of change of the mean square gradient is completely unaffected by the vorticity seems a little surprising. On the other hand, it is intuitively appealing that the rate of deformation (stretching and shearing) should contribute most effectively to the variation of the mean square gradient by drawing out and twisting a blob of substance.

The effect of the velocity gradients or more exactly the rate of deformation tensor is usually to *increase* the mean gradient. Therefore, the advective term (4) might be expected to be always positive. This can not be proved, however. Certain fluid motions may possibly bring back a well-deformed blob of substance into a more regular shape so that the velocity gradients themselves can decrease the mean gradient of θ. Despite this, a least expectation is that the direct effect of advection ultimately increases the magnitude of the gradient so that (4) is positive. At the same time, the increase of the mean θ-gradient by advection accelerates the dissipation rate of the θ-gradient due to molecular diffusion. For large time values, the dissipation becomes predominant over the production of the gradient by advection, and hence the mean θ-gradient in an isolated system will ultimately decrease to result in a homogeneous mixture. Increase of $\langle G^2\rangle$ will cease when the characteristic length of the spatial variation of the θ-gradient is reduced to the order $D^{1/2}(\overline{\omega^2})^{-1/4}$, where $(\overline{\omega^2})^{1/2}$ is the root-mean square vorticity.

In the ocean, when an arbitrary unconfined volume of water is considered, the physical conditions for the surface integral in (2) to vanish may not be realized. The exchange of θ through the boundary may play an important role in the variation of θ and its gradients within the volume of interest. Thus the mean θ-gradient over a body of water in the ocean may increase or decrease with time (or may even attain a steady-state value), depending upon the rate of supply of the θ-gradient through the boundary.

Our discussion has been somewhat general. A simple example of mixing by a uniform rate of strain field will provide an aid in further clarifying the mixing mechanism. For simplicity, consider a two-dimensional pure straining motion, $\mathbf{u} = (\alpha x_1, -\alpha x_2)$. Then the exact solution of (1) for an instantaneous point source of unit intensity at $\mathbf{x} = 0$ is given by

$$\theta(t, \mathbf{x}) = \frac{1}{2\pi D\alpha^{-1}(e^{2\alpha t} - 1)^{1/2}(1 - e^{-2\alpha t})^{1/2}} \exp - \left[\frac{x_1^2}{2D\alpha^{-1}(e^{2\alpha t} - 1)} + \frac{x_2^2}{2D\alpha^{-1}(1 - e^{-2\alpha t})} \right] \quad (5)$$

From (2) the rate of change of $\langle G^2 \rangle$ can be calculated by choosing V sufficiently large so that the θ-flux through the boundary may be ignored for the time interval of interest. A long-time situation can properly be discussed. For $t \gg \alpha^{-1}$, that is, for times long compared with a characteristic time of the straining motion, one obtains

$$\frac{d}{dt} \langle G^2 \rangle = \alpha \langle G^2 \rangle - (2\sqrt{2} + \tfrac{1}{2})\alpha \langle G^2 \rangle \quad (6)$$

Equation 6 tells us that the convective effect represented by the first term on the right-hand side tends to increase $\langle G^2 \rangle$ exponentially with time, whereas the molecular diffusion effect, the second term, has been so much enhanced by the stretching of a blob of θ that it reduces the value of $\langle G^2 \rangle$ exponentially and reduces it faster than the advection increases it. As a result, the θ-blob eventually fills up the entire space more and more uniformly.

Thus far we have not asked whether the velocity field is laminar or turbulent. The preceding discussion is applicable both to laminar and to turbulent flows, though a turbulent flow is usually more powerful than a laminar flow in mixing.* No attempt is made here to give a precise definition of turbulence or of the features peculiar to oceanic turbulence. These will be found in the next section. It will be sufficient, for the moment, to

---

*For photographic demonstration of the contrast between turbulent and laminar mixing, see Corrsin, 1961a.

note that turbulence refers to an irregular condition of flow in which the fluid properties vary with time and space in such a complex manner that only statistically averaged values are useful. This implies that any discussions of mixing in a turbulent flow should inevitably involve some kind of averaging procedure. Mathematically speaking, the flow is so complicated that any hope of solving (1) analytically for the instantaneous (nonaveraged) value of θ can hardly be expected. We therefore use a statistical method to approach the problem.

Suppose that we have a precise record of the one-dimensional spatial distribution of the concentration, θ, of a property observed at an instant t', that is, $\theta = \theta(x; t')$. Let $L_0 = AA'$ be the observational scale of θ (Fig. 4.2a). The pattern of θ is too complicated to handle. We shall smooth it. A common way of smoothing the record is to take a running mean centered on the point x with respect to an interval $2l$ (the "measurement scale"), that is, from $x - l$ to $x + l$. Mathematically, the smoothing or the averaging procedure can be expressed as

$$\bar{\theta}(x; l, t') \equiv \frac{1}{2l} \int_{x-l}^{x+l} \theta(\xi; t') \, d\xi; \qquad (A + l \leq x \leq A' - l) \qquad (7)$$

Note that the smoothed or averaged value depends not only on the position x but also on the measurement scale l. We may point out in passing that it also depends on t' since the time t' serves to select the particular space record being analyzed. As schematically shown in Figs. 2.4b–d, the larger the value of the measurement scale, the smoother the resulting record becomes.

Once the averaging procedure is defined, one can consider the difference between the unsmoothed value and the averaged value

$$\theta'(x; l, t') \equiv \theta(x; t') - \bar{\theta}(x; l, t') \qquad (8)$$

Note that the residue or deviation from the mean, θ', has the same dependence as $\bar{\theta}$. Explicitly, the measurement scale is involved. We shall call $\bar{\theta}$ the spatially averaged value or mean value of θ, and θ' the fluctuation or turbulent-component* of θ. Similarly, a time record made at a fixed point, $\theta(t; x')$, can be time-averaged in an analogous way.

The magnitude of the measurement scale should be determined primarily by how precisely one can describe the pattern of mixing or, alternatively, how precisely one wants to. The measurement scale can be chosen

---

* The reader should be on his guard with the term "turbulent component." In common usage "turbulence" means the *high-frequency* fluctuations. Whether θ' is restricted to high frequencies or not depends on the strength of the smoothing, that is, on the size of l. In the extreme case where $2l = L_0$, θ' is θ except for the reference level.

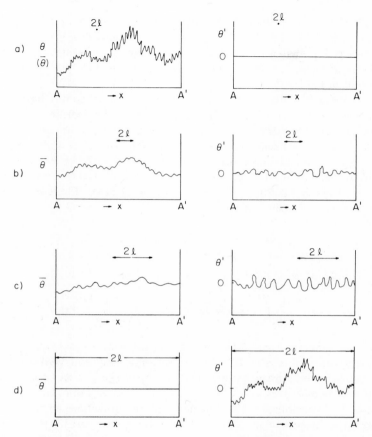

Figure 4.2 Schematic presentation of smoothing a record. θ: unsmoothed record, $\bar{\theta}$: smoothed record, θ': residue, and $2l$: smoothing interval (measurement scale).

as large as L_0, that is, the entire interval of observation. If one does so $\bar{\theta}$ becomes constant and there is no hope of discussing the spatial variations of the averaged pattern of mixing. On the other hand, if one chooses the measurement scale as small as the microstructure of the θ-variation, there will be no smoothing of θ.

If the original pattern consists of a relatively rapid variation superimposed on a slowly varying one, we can deal with the latter component by choosing a measurement scale that is sufficiently large compared with the scale of the rapid variation but sufficiently small in comparison with the scale of the slow variation. But if the scale of the two variations is close together there may be some difficulty. And if the original pattern contains a great variety of variations, a certain arbitrariness exists in the choice of the averaged quantity that we wish to consider. Fortunately, in practical problems of oceanic mixing such a choice can usually be made without too much difficulty.

The averaged patterns thus created will be, one would hope, simple enough for further treatment. Some details must be sacrificed in order to obtain practical solutions capable of describing the averaged situations.

Let us consider briefly the mathematics of the averaging procedure. Let $\phi_\infty(x)$ be a physical quantity defined for all x, $-\infty < x < \infty$ at some instant of time, and let $\phi(x)$ be $\phi_\infty(x)$ observed on the finite interval $0 \leq x \leq L_0$. Outside $0 \leq x \leq L_0$ define $\phi(x) \equiv 0$. Then $\phi(x)$ can be expressed by a Fourier integral.

$$\phi(x) = \frac{1}{\sqrt{2\pi}} \int_{-\infty}^{\infty} \Psi(k) e^{ikx} \, dk$$

The average of $\phi(x)$ over the measurement scale $2l$ can then be written

$$\bar{\phi}(x; l) \equiv \frac{1}{2l} \int_{x-l}^{x+l} \phi(\xi) \, d\xi = \frac{1}{\sqrt{2\pi}} \int_{-\infty}^{\infty} \frac{\sin kl}{kl} \Psi(k) e^{ikx} \, dk$$

This integral is identical with the Fourier integral for $\phi(x)$ except for the weighting function, $\sin kl/kl$, which determines the relative importance of each Fourier mode in the averaging procedure. For $|k|l \gg 1$ the weighting function becomes very small. For $|k|l \ll 1$ it is nearly one. In other words, small-scale variations, that is, Fourier components whose wave number, k, is large in comparison with l^{-1} are strongly attenuated by the averaging procedure, and only slowly varying fluctuations whose wave number is small compared with l^{-1} retain their relative importance. Hence the averaged value $\bar{\phi}$ is smoother with respect to x than is ϕ itself.

When we construct infinitely extended mathematical models of oceanic mixing, the Fourier transform of an oceanic property considered as a random function of position and time may not exist in the Riemann sense. This purely mathematical difficulty may be met by using either the Fourier-Stieltjes representation which is unfamiliar to many of us, or the Generalized Fourier representation. Since any information about the real ocean will always be limited by the observational scale, the transforms always exist in the ordinary sense so that we need not trouble ourselves. A brief summary of the generalized Fourier transform of a random function is provided by Kinsman (1965), especially for oceanographers' convenience.

In the following an averaged value is designated in general by an overscore without otherwise specifying how it has been taken. For the mathematical analysis of mixing problems, we must assume that any averaging procedures chosen satisfy "Reynolds rules": thus, if A and B are random functions,

(i) $\quad\overline{A + B} = \bar{A} + \bar{B}$

(ii) $\quad\overline{AB} = \bar{A} \cdot \bar{B}$

(iii) $\quad\overline{cA} = c\bar{A} \qquad (c \text{ constant}),$

(iv) $\quad\dfrac{\overline{\partial A}}{\partial \eta} = \dfrac{\partial}{\partial \eta} \bar{A} \qquad (\eta \text{ a variable on which } A \text{ depends})$

Note that rules i and iii are common properties of integrals and consequently hold for $\bar{\theta}$ as defined by (7). However, (ii) will not hold for arbitrary A and B while (iv) may be suspect if A is not sufficiently smooth. In the section on vertical and horizontal mixing from specified sources we talk about ensemble averages, with which Reynolds rules are satisfied for a much larger class of functions.

We now discuss the mixing in a turbulent field more analytically. The instantaneous value of θ at $\mathbf{x}$ and t can be represented by its average value and a superposed fluctuation.

$$\theta = \bar{\theta} + \theta' \tag{9}$$

Similarly, the instantaneous velocity can be decomposed into its average and superposed fluctuation, or turbulence.

$$u_i = \bar{u}_i + u'_i \tag{10}$$

In general, $\bar{\theta}$ and $\bar{u}_i$ vary with $\mathbf{x}$ and t, so also do θ' and u'_i. Assume the fluid is incompressible so that

$$\frac{\partial u_i}{\partial x_i} = 0 \tag{11}$$

Substituting (9) and (10) into (1), taking averages and making use of (11) and Reynolds rules, we obtain the equation for $\bar{\theta}$.

$$\frac{\partial \bar{\theta}}{\partial t} + \bar{u}_i \frac{\partial \bar{\theta}}{\partial x_i} = -\frac{\overline{\partial u'_i \theta'}}{\partial x_i} + D \frac{\partial^2 \bar{\theta}}{\partial x^2_i} \tag{12}$$

Note that (12) is formally identical with the original equation 1 for θ with the exception of extra terms on the right-hand side of (12).

The term $\overline{u'_i\theta'}$ represents a flux of θ at a point **x** and a time t due to turbulence, that is, arising from the averaging procedure. If one defines a constant (one hopes nonnegative) coefficient of turbulent diffusion by

$$K \equiv \frac{\overline{u'_i\theta'}}{-(\partial\bar{\theta}/\partial x_i)} \qquad i = 1, 2, 3$$

then equation 12 can be expressed as

$$\frac{\partial\bar{\theta}}{\partial t} + \bar{u}_i\frac{\partial\bar{\theta}}{\partial x_i} = (K + D)\frac{\partial^2\bar{\theta}}{\partial x^2_i} \qquad (13)$$

By the use of this device the incomplete formal identity of (12) and (1) is reduced to an exact formal identity between (13) and (1). Usually the value of K exceeds that of the molecular diffusivity, D, by several orders of magnitude.

It should be emphasized, however, that there is an essential difference between the mechanisms of molecular and turbulent diffusion. The transfer of a property by molecular diffusion takes place always from the higher concentration to the lower concentration in harmony with the second law of thermodynamics. The turbulent transfer, however, need not always be in the direction of the mean gradient nor can it always be described as diffusion. The analogy between (13) and (1) is superficial. Above all the ultimate mixing could not be completed without molecular diffusion.

The role of molecular diffusion in mixing can be examined more clearly in a Lagrangian framework, since the mixing process involves the displacements of fluid particles essentially. The differential equation for θ expressed in Lagrangian form (Corrsin, 1962) is

$$\frac{\partial\theta}{\partial t} = D\{[(\theta, \xi_2, \xi_3), \xi_2, \xi_3] + [\xi_1(\xi_1, \theta, \xi_3), \xi_3] + [\xi_1, \xi_2, (\xi_1, \xi_2, \theta)]\}$$
$$\equiv D\nabla_L^2\theta \qquad (14)$$

where $\boldsymbol{\xi}$ denotes the coordinates at a time t of a fluid particle which is identified by its coordinates **a** at $t = 0$, and $\theta = \theta(\mathbf{a}, t)$. Triples of the form $[A_1, A_2, A_3]$ stand for $\partial(A_1, A_2, A_3)/\partial(a_1, a_2, a_3)$, following the notation of Monin (1962). The convective terms disappear in the Lagrangian equation since fluid particles are followed; instead, the irregular displacement field of fluid particles is blended into the molecular diffusion terms. Thus the direct interaction between advection and molecular diffusion is subsumed into a single quantity in this equation.

By expressing the gradients of displacement, $\partial\xi_i/\partial a_j$, in terms of the strain components, $\frac{1}{2}[(\partial\xi_i/\partial a_j) + (\partial\xi_j/\partial a_i)]$ and the rotation, $\frac{1}{2}[(\partial\xi_i/\partial a_j) -$

$(\partial \xi_j/\partial a_i)$], one can see that, for small straining motions, the stretching and shearing actions have the primary effect on the molecular diffusion, whereas the effect of the rotation is of secondary importance. This is equivalent to what we have found in the Eulerian representation.

Multiplication of ξ_i^m by (14) and integration over a result in

$$\frac{d}{dt}\int \xi_i^m \theta \, d\mathbf{a} = \frac{d}{dt}\int \xi_i^m \theta_0 \, d\mathbf{a} + Dm(m-1)\int \xi_i^{m-2}\theta \, d\mathbf{a} + \int \frac{\partial \xi_i^m}{\partial t}\gamma \, d\mathbf{a} \quad (15)$$

where θ_0 is the initial distribution of substance and $\gamma \equiv \theta - \theta_0$ (Okubo, 1967a). The physical meaning of (15) is self-evident. The left-hand side is simply the rate of change of the mth moment associated with a blob of substance taken relative to the origin ($\boldsymbol{\xi} = \mathbf{a} = 0$). The three terms on the right-hand side of (15) represent, respectively, the rate of change of the mth moment associated with a patch of fluid particles that is initially coincident with the blob of substance, the direct effect of molecular diffusion, and the rate of change of the mth moment due to the indirect effect arising from the interaction of advection and molecular diffusion. The direct effect of molecular diffusion becomes zero for $m = 0$ and 1 and $2D$ for $m = 2$.

Without molecular diffusion, $\theta = \theta_0$ ($\mathbf{a}, t = 0$), a constant at all times. The fluid particles initially coincide with a patch of substance conserve the property as they wander in any irregular manner. In the presence of molecular diffusion, the initial fluid particles begin to exchange the property with the ambient fluid particles. The complex motion or turbulence would enhance the rate of exchange by increasing the mean spatial gradient of the property. Thus the concept of mixing by turbulence is easier to understand from a Lagrangian description (Fig. 4.3).

Equation 14 is an instantaneous ("nonaveraged") equation, in which the displacements are governed by the equation of continuity and the equations of motion in their Lagrangian forms. For a problem of small-scale mixing, like that in the coffee cup, they are given by

$$[\xi_1, \xi_2, \xi_3] = 1$$

$$\frac{\partial^2 \xi_1}{\partial t^2} = -\left[\frac{P}{\rho}, \xi_2, \xi_3\right] + \nu \nabla_L^2 \left(\frac{\partial \xi_1}{\partial t}\right) \quad \text{and so on}$$

where ν is the kinematic viscosity and P is the modified pressure (Corrsin, 1962; Batchelor, 1967).

Although the complete Lagrangian equations are more severely nonlinear than the Eulerian equations, a perturbation analysis still

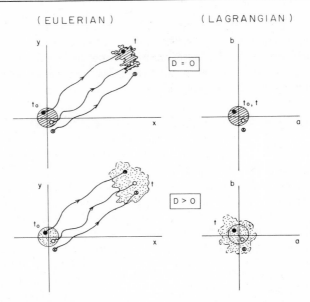

Figure 4.3 Eulerian and Lagrangian descriptions of turbulent mixing. D: molecular diffusivity.

yields a set of linear equations and an approximate solution of these linear equations may be a good start toward a correct solution of the problem (Pierson, 1962). The combined effect of turbulence and molecular diffusion on the mixing of a substance can be studied by means of the zero- and first-order equations in the perturbation scheme (Okubo, 1967a). As a result, it is shown that the effect of interaction between the molecular diffusion and the advection initially reduces the rate of spread of substance relative to the origin, while initially increasing the mixing of substance relative to the centroid of the patch of substance. Though this result is derived for small times, the basic physical concepts seems to be valid for all times. Asymptotic behavior of the interaction for large times is characterized by the Prandtl number, ν/D, in the case when the diffusing property is heat and by the Schmidt number, ν/D, in all other cases. When $\nu/D \gg 1$, which is always the case for the ocean, the interaction makes a very small contribution to mixing. This may justify the semi-empirical approach used in the Eulerian diffusion equation that assumes that the effect of molecular diffusion can be ignored in comparison with the effect of turbulent exchange.

An exceptional case where the molecular processes undoubtedly remain potentially important is seen in oceanic mixing associated with the "salt-finger" phenomenon (Stommel et al., 1956; Turner, 1967). Salt-fingers arise from the difference in molecular diffusivity between heat and salt. In seawater, the molecular diffusivity for heat exceeds that of salt by more than a hundred times. Typical values are 1.49×10^{-3} cm^2/sec for heat and 1.29×10^{-5} cm^2/sec for salt (Montgomery, 1957).

Imagine a layer of warm and saline water resting on top of cold and less saline water. Let the stratification be statically stable initially. If a long tube of heat-conducting material were lowered from the top layer to the depth of the cold water, and the deep water were very slowly pumped to the upper layer through the tube, what would happen? The slow motion would allow the water inside the tube to reach the same temperature as the surrounding water, whereas the salinity of the water would remain unchanged. Hence the density of the rising water would become less than that of its surroundings outside the tube at the same level, and the water would continue to flow upwards, by itself, even if the pump were disconnected. If the direction of the pump were reversed, the water initially flowing downward would continue to sink through the tube on account of its excess in density over that of the surrounding water at the same level. This is the idea of the perpetual salt fountain proposed by Stommel et al. (1956).

Actually the tube and pump are unnecessary. Nature can take care of the mechanism by herself, provided the conditions are favorable. The pattern of motion created consists of long narrow convective cells called salt-fingers moving alternately up and down. A similar potential instability may occur when a layer of cold fresh water is resting on top of warm salty water under statically stable conditions. In this case, however, the instability results in a series of convecting layers separated by sharp interfaces (Turner and Stommel, 1964.)

Environmental Features of the Sea for Advective and Diffusive Processes

As the combined effect of advection and diffusion gives rise to the mixing of cream and coffee in a cup, so does it to mixing in the ocean, except, of course, with a tremendous difference in scale. An attempt is made, in this section, to summarize the essential environmental features of the sea in relation to advective and diffusive processes for contaminants.

Ocean Currents*

In many parts of the oceans, it is possible to conveniently distinguish three layers: an upper layer and a deep layer characterized by nearly uniform temperatures separated by a comparatively thin layer of rapidly

* General references: Bowden, 1965a; Neumann, 1968; Pritchard et al., 1969a.

changing temperature called the thermocline. The upper layer extends from the sea surface, or naviface (Montgomery, 1969), down to a depth ranging from some tens of meters to a few hundred meters, where the relatively stable thermocline separates it from the deep layer beneath. The upper or mixed layer is relatively homogeneous, although its properties, temperature, for example, show considerable seasonal variation. Within it are found the most intense currents and turbulent motions of the ocean. The thermocline represents a sort of discontinuity layer in the vertical gradient of density. The deep layer occupied by cold water extends beneath the thermocline to the bottom of the ocean.

The general features of the mean navifacial current pattern of the world ocean are relatively well recognized. Figure 4.4 shows a map of the mean surface currents in the world ocean.

The general features of the circulation are somewhat similar in all the oceans. One striking common feature is the east-west asymmetry of the currents. Thus most strong currents are found along the western boundaries of the oceans; the Kuroshio in the North Pacific and the Gulf Stream in the North Atlantic are examples. These strong currents are narrow in width (less than a few hundred kilometers) and semi-permanent in character. A characteristic maximum velocity would be about 150 cm/sec (3 knots) at the naviface. The currents extend vertically to depths of 1000 m or more.

The equatorial region is characterized by a complex of zonal flows. The North and South Equatorial Currents flow westwards with a speed of 1 knot in the surface layer. Between these currents is found the Equatorial Counter Current flowing eastwards with a navifacial speed of 1–3 knots. A remarkable feature in the equatorial current system is the existence of a jet-like undercurrent flowing eastwards, opposite to the overlying South Equatorial Current, with speeds up to 150 cm/sec in its core and with a thickness of 400 to 500 m.

All three oceans open out at their southern ends into the Antarctic, where the Antarctic Circumpolar Current encircles the Antarctic Continent. This current with the navifacial speed of less than 1 knot reaches depths greater than 1000 m.

These predominant currents are classified as more or less permanent in their appearance. There are, however, some seasonal variations and other irregular fluctuations in certain regions. For instance, the Gulf Stream, after leaving the coastal area off Cape Hatteras, often takes a meandering course, and the amplitude of the meanders may amount to as much as a few hundred kilometers. In particular, when a conspicuous meandering occurs, the neck-like portion of the meander ends by breaking off to form

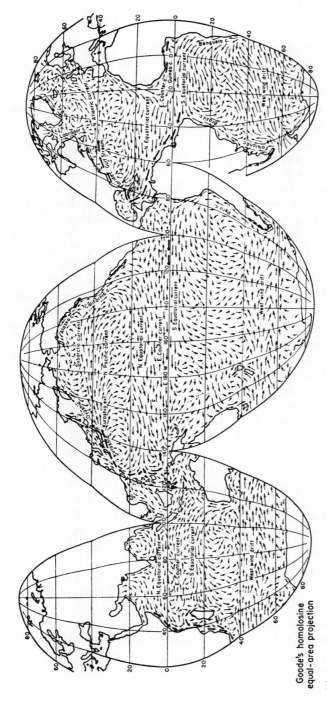

Figure 4.4 Surface currents of the ocean. From Somerville and Woodhouse, 1950. By permission of The University of Chicago, Goode Base Map Series, Department of Geography. Copyright by the University of Chicago.

a clearly defined eddy. Depending upon its initial size, the eddy thus produced may exist for more than 6 months (Fuglister, 1967). Similar eddies have been found in the Kuroshio.

Elsewhere in the major interior area of the ocean, away from these strong currents, one typically finds speeds of the order of 10 cm/sec. The flows in the interior areas are much more erratic and show little of the quasi-permanent appearance in the major currents.

There are a number of different physical sources which may possibly drive ocean currents. One, the direct effect of wind over the naviface, is limited, on the average, to shallow depths, characteristically to the upper 200 m. The wind does not directly affect the deep currents. They are closely related to the distribution of density. Throughout the bulk of the deep ocean the currents are geostrophic representing an equilibrium between the distribution of density and the Coriolis effect.

The geostrophic currents in the Atlantic Ocean at 800 and 2000 m depth are shown in Figs. 4.5a and 4.5b. Current directions at 800 m depth are essentially similar to those of the surface flow with some minor discrepancy. The speeds are generally less than 30 cm/sec. The pattern of flow at 2000 m depth, on the other hand, shows a striking departure from that at the surface and intermediate depths. Thus a southward flow obtains beneath the northward flowing Gulf Stream. The current speeds at 2000 m are generally less than 20 cm/sec. Before any measurements were made, Stommel (1957) predicted the possible existence of the southward flow from theoretical considerations. Swallow and Worthington (1957) made the first direct observations of the southward flow in the deep layer along the western boundary of the North Atlantic.

While the movement of the surface layer is generated by wind stresses at the naviface, the deep water movements are maintained primarily by thermohaline processes. In the North Atlantic, for example, a main source of deep water is located in the surface region near Greenland, where sinking water of low temperature provides the North Atlantic Deep Water. Other major sources of deep water are found in the Antarctic region.

In comparison with the current systems in the open sea, current patterns in coastal areas are far more complex. There the permanent currents are often weak. Tidal currents, local wind-drift, density currents due to the supply of fresh water by rivers and by runoff from land, and even transport due to waves play an important role in the mixing of properties.

A very few direct measurements of vertical velocity have been carried out in the ocean.* At present, almost all the information on the vertical

---

* An instrument has been designed by Webb (see Webb and Worthington, 1968) which is capable of recording vertical water movement of less than 1 m/hr (0.028 cm/

current velocity is obtained from indirect methods. Nevertheless, there are unmistakable indications that the water moves vertically in some regions of the ocean.

In the Irminger Sea between Iceland and Greenland, and in the Labrador Sea, water of a relatively high salinity when mixed with cold water from the Polar Sea attains a relatively high density by winter cooling so that it is able to sink into the deep layer. Close to the Antarctic Continent the freezing process gives rise to the formation of a homogeneous water that has a higher salinity, and hence a higher density, than the surrounding water, and accordingly flows down the continental slope to the bottom of the ocean.

In some adjacent seas like the Mediterranean Sea and the Red Sea, water of high salinity results from excess evaporation and needs only slight cooling to increase its density enough to form bottom water. When such a sea is in communication with the open ocean, the bottom water flows over the sill and spreads out while descending to an intermediate depth. Sinking of surface water also occurs when converging currents are present.

The sinking of water in the ocean must be compensated, on the average, by upwelling. While the sinking of water masses takes place in relatively restricted areas, the slow, gradual ascending motion is considered to occur generally in the deep layers at subtropical latitudes. Thus a depth of 1000 m is regarded as the depth where the upward motion reaches a maximum value estimated at 10^{-5} cm/sec in mid-latitudes and 10^{-4} cm/sec near the equator.

Ascending motion occurs also in regions of diverging currents. In particular, a conspicuous upwelling takes place along the western coasts of the continents, where prevailing winds tend to carry the surface water away from the coasts. The ascending motion, in this case, is limited in general to depths less than a few hundred meters.

Oceanic Turbulence*

The currents mentioned previously imply a more or less regular pattern of flow in the ocean. There exists, however, a very wide range of motions in the sea which are so irregular that one can not handle them without statistical considerations. Such irregular complex motion is called "turbu-

---

sec). The instrument was used in the deep water of the Cayman Basin of the Caribbean Sea (Webb and Worthington, 1968) and in the New England Slope Water (Voorhis, 1968) with some encouraging results.

* General references: Bowden, 1964, 1970; Phillips, 1966; Ozmidov, 1968a.

108　　　　HORIZONTAL AND VERTICAL MIXING

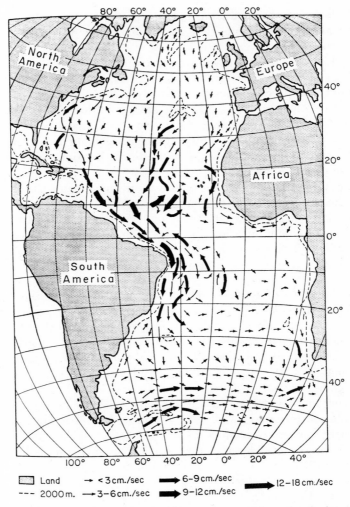

Figure 4.5a Current field at a depth of 800 m, computed from the absolute topography of the 800-dbar surface. Reprinted with permission from A. Defant, *Physical Oceanography*, Volume 1, 1961, Pergamon Press, New York.

lence." Hinze (1959) defines turbulence as "an irregular condition of flow in which the various quantities show a random variation with time and space coordinates, so that statistically distinct average values can be discerned."

Hinze's definition can be clarified by the following: turbulence, in general, (a) possesses a random distribution of vorticity in which there is

ENVIRONMENTAL FEATURES OF THE SEA 109

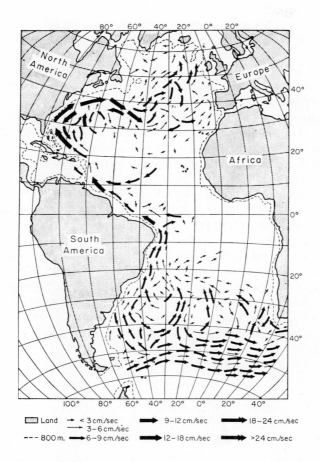

Figure 4.5b Current field at a depth of 2000 m, computed from the absolute topography of the 2000-dbar surface. Reprinted with permission from A. Defant, *Physical Oceanography*, *Volume 1*, 1961, Pergamon Press, New York.

no unique relation between the frequency and wave number of the Fourier modes or "eddies" (Phillips, 1966)*; and (b) is diffusive and capable of transporting or mixing properties far more effectively than molec-

---

* From this it follows that motion resulting from waves, which has a unique frequency-wave number relation, is not turbulent.

ular motions. (c) Oceanic turbulence is anisotropic except for the very small-scale motions which may be isotropic. Anisotropy is imposed inevitably by the disparity between the horizontal and vertical scales of the ocean basins. A single sheet of onion-skin typing paper very closely represents the proportions of the scales in an ocean. The anisotropy is reinforced by the stable stratification of the water, and (d) has intermediate scales which may derive their energy directly from local storms and tidal currents and still smaller scales which may be fed directly by waves (Stommel, 1949; Ozmidov, 1965).

In the upper ocean above the thermocline, where the water column is either neutral or weak stable, one can almost always expect turbulent motion. Turbulence can also be expected in shallow waters with strong velocity-shears such as coastal regimes and estuaries. For neutrally stratified water a Reynolds number, which may be interpreted as the ratio of the inertial forces to the viscous forces, determines whether or not a given flow will be turbulent. Thus in an upper layer of the thickness of say 100 m, a mean flow of 10 cm/sec gives a Reynolds number of about 10^7, which suggests that the flow must be fully turbulent.

For a stably stratified layer, on the other hand, the Reynolds number is rarely the important decisive parameter. It is stability that determines whether or not a given volume of water is turbulent (Stewart, 1959a). When a column of water is stable, the buoyancy forces tend to remove turbulent energy which is produced by the velocity shear. The stability effect is usually considered in terms of the gradient Richardson number, R_i, defined by

$$R_i = - \frac{g}{\bar{\rho}} \frac{\partial \bar{\rho}}{\partial z} \bigg/ \left(\frac{\partial \bar{U}}{\partial z}\right)^2 \equiv N^2 \bigg/ \left(\frac{\partial \bar{U}}{\partial z}\right)^2$$

where g is the acceleration of gravity, $\bar{\rho}$ the density of water, $\bar{U}$ the mean horizontal velocity and z the vertical coordinate taken positive upwards, and N is the Brunt-Väisälä frequency.

Stability analysis indicates that laminar motion is stable for R_i in excess of $+\frac{1}{4}$ (Taylor, 1931a). This does not mean that existing turbulence vanishes at that value of R_i; an instability due to velocity shear or to other sources must first generate large-scale turbulence and the Richardson number then determines whether the turbulent motion can be self-supporting (Lumley and Panofsky, 1964). Under certain conditions turbulence may exist even at large Richardson numbers (Ellison, 1962; Stewart, 1959b).

So far there has been no really convincing evidence that the water below the upper mixed layer is in a state of fully developed turbulence. In fact, there have been some observations that turbulence in the thermocline is highly intermittent or sporadic. Grant et al. (1968a) measured turbulence and temperature microstructure in the open sea. At a depth of 10 to 30 m, which is well above the thermocline, it is almost continuously turbulent. In and below the main thermocline, on the other hand, turbulence occurs in patches; temperature microstructure exists only when turbulence is present. When the turbulence disappears, molecular conduction effects seem to erase the microscale temperature structure very quickly.

Exciting direct observations of the fine structure in a thermocline have been reported by Woods (1968a, b). Investigations by skin divers, who film the flow patterns of small quantities of fluorescein dye, have revealed that the thermocline is divided into a series of relatively uniform "layers" a few meters thick, separated by thin interfacial regions or "sheets" a few centimeters thick which contain laminar-flow regimes at their cores. Generally the sheets are stable, but gravity waves have been observed to cause local breakers, which rapidly degenerate into patches of turbulence (Fig. 4.6).

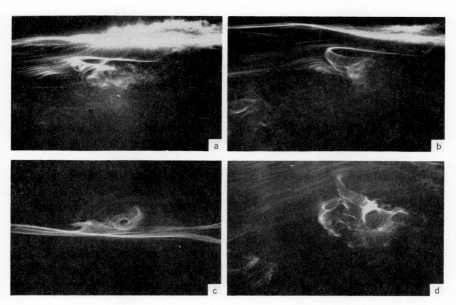

Figure 4.6 Stages in the growth of a breaker on a dyed thermocline "sheet." From Woods, 1968b.

In internal waves of the lowest modes, the maximum shear occurs at the point in the thermocline where the Brunt-Väisälä frequency is a maximum ($= N_m$). When the thickness of "sheets" is small compared with the wavelength of the internal waves, the local Richardson number is given by

$$R_i = \left(\frac{f}{N_m k a}\right)^2$$

where f, k, and a are the frequency, wave number, and amplitude of the internal waves, respectively (Phillips, 1966). Hence the limiting slope of the internal waves is

$$(ka)_c = \frac{2f}{N_m} \ll 1$$

when the critical Richardson number is $\frac{1}{4}$.

Beyond this limiting slope, a local instability may develop into a breakdown of the internal waves. As a result, a patch of weak turbulence is produced. After being thrown out into ambient water of weak stability, that is, into a "layer," each turbulent patch would tend to spread horizontally, forming a lens of homogeneous water with weak turbulence. A precise analysis of these lenses would be quite important in understanding the fine structure of the thermocline.

The "layer" between the "sheets" is characterized by weak turbulent motions. Direct (in situ) investigations of these motions using dye as a tracer have shown that turbulence in the layer has typical length and time scales of about 25 cm and 300 sec, respectively. From these values, Woods (1968a) estimates an effective vertical diffusivity of about 2 cm^2/sec. The temporary and local apertures on the thermocline sheets afforded by turbulent patches of breakers permit the vertical transfer of properties through the sheets with a vertical diffusivity of the order of 10 cm^2/sec. Elsewhere on the sheets the downward flux should be carried by molecular processes. The mean fractional coverage that would lead to a balance between the heat fluxes through sheets and layers is estimated to be about 2 per cent.

Although the findings of Woods concerning the thermocline structure are based only on observations of the summer thermocline around Malta, similar processes may be expected to occur commonly in the thermocline elsewhere. Stommel and Fedorov (1967) have found, from continuous soundings of salinity and temperature against depth, that the oceanic thermocline is actually made up of hundreds of superposed "laminae" or "layers" from 2 to 30 m thick, each fairly homogeneous in temperatures and salinity and separated from one another by sharp interfacial regions of large gradients in temperature and salinity. These laminae probably extend horizontally from 2 to 20 km. Stommel and Fedorov could not identify microvariations of the velocity structure that might coincide with similar small-scale variations in water properties, but there is some evidence that an extreme shear of the order of 10^{-2} sec or even higher does exist in a small interval of depth. More recently, Copper and Stommel (1968) have discovered that the main thermocline of the ocean off Bermuda consists of a remarkably regular series of steps of homogeneous water 3 to 5 m thick alternating with transition layers 10 to 15 m thick in which the gradients of temperature and salinity are sharp. These structures are believed to be ubiquitous in the main thermocline of the Sargasso Sea, and probably of all subtropical gyres, and to be relevant to the vertical mixing process. The actual mixing process, however, remains unexplained.

The "layering" structure seems to be common even in deep water well below the main thermocline. Hamon (1967) found a medium-scale temperature and salinity (T-S) structure in otherwise stably stratified water as deep as 1200 m in the Indian Ocean. Layers of almost neutral stability extended vertically from 10 to 100 m, and appreciable structures persisted horizontally for 5 to 10 km, but never to 40 km. Hamon has also noted on T-S diagrams that the medium-scale laminae are most marked in the σ_t range 27.0 to 27.4. This σ_t range corresponds well to water of Red Sea origin, so that this medium-scale structure is probably associated with the spreading and mixing of Red Sea water into the Indian Ocean.

In the Northeast Atlantic Ocean, Tait and Howe (1968) discovered similar laminae at a depth between 1280 and 1500 m, that is, immediately below the Mediterranean water intrusion. The thickness of laminae was between 15 and 30 m. It is suggested that this layer formation might be due partly to "salt-fingering" occurring in the sea. If this is really so, we must acknowledge that molecular diffusion processes play a more significant role in oceanic phenomena than is normally considered to be the case.

The microscale structure of temperature and salinity similar to that of Stommel and Fedorov (1967) has been discovered by Pingree (1969) also at some depths in the deep water of the Bay of Biscay. The microstructure observed shows phase relationship between temperature and salinity variations that differ above and below the Mediterranean salinity maximum. Pingree suggests that the microstructure might be generated by local mixing, perhaps by a breaking internal wave (Woods, 1968a, b), and horizontal mixing followed.

Very little is yet known about turbulence in deep waters. Since the main body of the ocean is approximately in a geostrophic balance, it may be instructive to ask whether or not the geostrophic current alone is capable of producing turbulence (Stewart, 1968). From the equation of motion for a geostrophic flow in combination with the hydrostatic equation, one can obtain

$$\frac{\partial \bar{V}_g}{\partial z} \sim -\frac{g}{\lambda \bar{\rho}}\frac{\partial \bar{\rho}}{\partial h} = -\frac{1}{\lambda}\frac{g}{\bar{\rho}}\frac{\partial \bar{\rho}}{\partial z.\partial h} \bigg/ \frac{\partial \bar{\rho}}{\partial z} = \frac{1}{\lambda}N^2 i_\rho$$

where $\bar{V}_g$ is the geostrophic velocity, λ is the Coriolis parameter, h is a horizontal coordinate, and i_ρ denotes the isopicnal slope. Hence the Richardson number associated with the geostrophic flow becomes

$$(R_i)_g = -\frac{g}{\bar{\rho}}\frac{\partial \bar{\rho}}{\partial z} \bigg/ \left(\frac{\partial \bar{V}_g}{\partial z}\right)^2 = \left(\frac{\lambda}{N i_\rho}\right)^2$$

Taking representative values: $\lambda/N = 10^{-1}$, $i_\rho = 10^{-3}$, one obtains

$$(R_i)_g = 10^{+4}$$

which indicates that the water column is too stable in a geostrophic balance to maintain a fully developed turbulence.

However, this does not rule out the possibility of turbulence in deep waters. Although the velocity shear associated with the general circulation (geostrophic flow) is too small for instability to occur, the density stratification in deep waters is not so strong as in the thermocline. This suggests that even a weak shear from some other source could be sufficient to give rise to an instability. For example, shears associated with internal tide waves might produce instability (Munk, 1966). A more promising source, however, might be internal planetary waveshear.

The characteristic scale of the processes responsible for the generation of turbulence determines the initial scale of turbulence. Thus the worldwide prevailing wind systems feed their energy to the large-scale ocean circulation. Large eddies of, say 100 km in diameter, are developed in the main currents. Those eddies tend to transfer their energy, presumably as a result of shears in the velocity field, into motions with smaller scale. This sequence of events gives rise to a cascade of eddies of ever-decreasing size. The kinetic energy of eddies or turbulent motions is eventually dissipated into heat through the viscosity of the water.

In the ocean, a characteristic length-scale of the energy-dissipating eddies is of the order of 1 cm. Between the large-scale ocean circulation and the energy-dissipating eddies, a wide spectrum of eddies is thus produced as a result of the cascade processes. Without direct supplies of energy to intermediate scales, a part of such eddies, if they were isotropic, would be expected to lie in the "inertial subrange" as developed in the theory of local isotropic turbulence (Kolmogorov, 1941).

It should, however, be noted that oceanic turbulence is not strictly isotropic for the most of these intermediate eddies. Small-scale eddies, say those less than 1 m, may be isotropic. Larger eddies may be horizontally isotropic with the horizontal scales much larger than the vertical scales. At the largest scales, those of the ocean basins, they can not be isotropic in any sense.

Energy is fed to the largest scale of oceanic turbulence by the general atmospheric circulation but it may also be fed directly into eddies of intermediate scale. Ozmidov (1965) speculated that, in spite of this situation, one might expect that the inertial subrange exists locally for the eddies between two length scales where significant energy inputs occur.

The use of a hot-wire instrument or a hot-film flow-meter enables us to study the small-scale high-frequency components of turbulence in the sea. For example, Grant et al. (1962), using a hot-film technique, obtained energy spectra for eddies ranging from 1 mm to ~10 m approximately. The spectra follow closely a $-\frac{5}{3}$ law for the wave number in a range of approximately 1 to 10^{-2}/cm, as predicted by the theory of local isotropy. For the higher wave numbers, that is for sizes smaller than 1 cm, viscous effects become significant. Grant et al. succeeded in measuring only one horizontal component of the velocity. A three-component turbulent velocity meter has been developed at the Chesapeake Bay Institute, The Johns Hopkins University. The meter works on a Doppler-shift principle. The instrument has proved to be useful in studying the three-dimensional structure of turbulence in nearshore waters for scales down to a few centimeters (Pritchard et al., 1969b; Wiseman, 1969).

In order to investigate the intermediate- or large-scale turbulent fluctuations having periods up to several days or even weeks, long-term current observations must be carried out. The development of moored buoys (Richardson et al., 1963) that can be instrumented and left unattended to measure local variations over long periods of time has introduced a new dimension into the study of long-term oceanic motions and other properties (Fofonoff, 1967; Webster, 1967). Using the technique of spectral analysis on long-term current records made at depths of 50 and 100 m south of Bermuda, Day and Webster (1965) obtained kinetic energy density spectra for horizontal currents. The spectra exhibit dominant spectral peaks corresponding to the local inertial period (25.47 hr at 28°07′N). Both spectra also show a peak toward zero frequency or very large-scale eddies, and a broad low peak centered at the semi-diurnal tidal period. Inertial-period motion has been observed over an extensive range of latitudes and depths in the ocean. Published reports of such motions have been summarized by Webster (1968). According to him, inertial motions have a transient nature, with generation and decay times of a few days. Munk and Phillips (1968) have recently attempted to explain some structures of the inertial motion by the theory of wave motion in a weakly stratified rotating ocean of constant depth. See also Webster (1968).

Quantitative information on oceanic turbulence is still far from complete, especially in the deeper waters. Among the important parameters to be measured are (a) the rate of energy dissipation per unit mass of water, ϵ, (b) the intensities of turbulence, that is, the root-mean-square values of the turbulent velocity components, and (c) the integral scales of turbulence. If Kolmogorov's similarity theory of turbulence can be applied to oceanic turbulence, the most important parameter is the rate of energy dissipation.

Even the knowledge of the single parameter ϵ is limited in the ocean. There is, however, an indication that the value of ϵ decreases with depth. The local rate of dissipation near the naviface, if waves are breaking, is generally much higher than in the subsurface water (Stewart and Grant, 1962). Near the naviface the expected increase of ϵ with wave height is noticed. Between 1 and 2 m in depth a typical value of ϵ amounts to 3×10^{-2} cm²/sec³. In the similarity region below the direct effect of surface waves when the water is neutrally stratified, the local dissipation-rate becomes of the order of $|\overline{uw}|^3/z$, where $\overline{uw}$ denotes the stress at the naviface (Phillips, 1966). Thus the value of ϵ at 100 m depth is typically of the order of 10^{-4} cm²/sec³.

Woods (1968b) estimates a value of ϵ in the summer thermocline of the Mediterranean Sea on the basis of the thermocline structure. In the "layer" characteristic of weak turbulence, he gives $\epsilon = 3 \times 10^{-4}$ cm²/sec³ as a typical value.

Estimates of ϵ for deep waters are far more scarce and indirect. From the dispersion of neutrally buoyant floats, Takenouti et al. (1962) obtained $\epsilon = 3.3 \times 10^{-5}$ cm²/sec³ at 500 m. Munk (1966) computed the rate of change of potential energy of a column of water due to the effect of vertical diffusion, from which he obtained 1.8×10^{-6} erg/(g) (sec) as the energy required for diffusion per unit mass of water averaged over the depth from 0 to 3.3 km. If this rate of energy should be balanced by dissipation associated with turbulent kinetic energy, ϵ would be 1.8×10^{-6} cm²/sec³ for the mean value between 0 and 3.3 km in depth. In general the value of ϵ in abyssal depths, between 1 and 4 km, would be of the order of 10^{-5} to 10^{-6} cm²/sec³. According to Knauss (1956), an estimate of the average rate of total kinetic energy added to the ocean was 30 to 100 erg/(cm)² (sec). If we take 4 km to be the mean depth of the ocean, $\bar{\epsilon} = 7.5 \times$

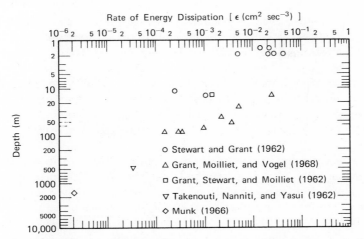

Figure 4.7 Depth variation of the rate of turbulent energy dissipation per unit mass of seawater.

10^{-5} to 2.5×10^{-4} cm²/sec³ for the ocean. Figure 4.7 summarizes some estimates of ϵ in the sea.

Basic Equations for Oceanic Mixing

In this section we discuss some basic equations for oceanic diffusion problems. Diffusion or mixing in general is concerned with the displacement field of fluid particles rather than with their velocity field. This suggests a Lagrangian point of view to be a natural one in describing the mixing process.

From a mathematical standpoint, however, a Lagrangian diffusion equation is far more complex in form (see the section on Mechanism of mixing) and more difficult to solve than the corresponding equation in an Eulerian representation. Above all we are interested, in many cases, in certain statistical properties of mixing such as the mean concentration at a fixed point. In practice, Eulerian data are more accessible than Lagrangian. Therefore, most mathematical approaches in oceanic mixing use an Eulerian or quasi-Eulerian formulation for the basic equation.

Equations for Mean Concentration

For the mean concentration of a passive contaminant, one has

$$\frac{\partial \bar{s}}{\partial t} + \bar{u}_i \frac{\partial \bar{s}}{\partial x_i} = - \frac{\partial \overline{u_i' s'}}{\partial x_i} + F(\bar{s}, \mathbf{x}, t) \tag{16}$$

where $\bar{s}$ denotes a mean concentration of contaminant at a point **x** and a time t, s' is the instantaneous fluctuation of contaminant concentration from the mean, and F represents the nonconservative processes such as radioactive decay. The origin of the coordinate system is taken either at a fixed point or at a point moving with a known velocity; in the latter case, $\bar{\mathbf{u}}$ represents the mean velocity relative to the moving origin. No term representing the molecular diffusion appears in (16) since in most cases of oceanic mixing for the mean concentration the molecular diffusion can be ignored.

The basic equation describes a kinematic event for $\bar{s}$ in the sense that the mean velocity $\bar{\mathbf{u}}$ is prescribed and the statistical properties of the turbulent field $\mathbf{u}'$ are known in principle. Initial and boundary conditions appropriate for the problem then specify the solution of (16).

Unfortunately (16) is not of closed form. The three covariance functions, $\overline{u_i's'}$, stand as unknowns. The addition of an equation for $\overline{u_i's'}$ derived by the same methods used for (16) would not remove the difficulty since it would involve unknown triple correlation functions, for example, $\overline{u_i'u_j's'}$. Equations for the triple correlations would create quadruple correlations, and so on. The creation of new unknown functions never ends, the method never breaks the circle, and the system of equations is never closed. One possible way to break out treats the covariance functions a priori with assumptions which relate $\overline{u_i's'}$ to the mean concentration $\bar{s}$. Semiempirical theories of turbulent diffusion relate $\overline{u_i's'}$ to the gradients of the mean concentration by

$$\overline{u_i's'} = -K_{ij}\frac{\partial \bar{s}}{\partial x_j} \qquad (17)$$

where the second-order tensor K_{ij} is called the "eddy diffusivity tensor" or "eddy diffusion coefficient." It has the dimensions of length squared per time. The components of the tensor may in general be a function of x and t. To be strict, the eddy diffusivities take different values for different oceanic variables. For example, the vertical eddy diffusivity for salt may differ in value from that for heat.* However, while functional relations like (17) based on empirical grounds are useful working tools, they can not be granted the same weight and status as relations derived from first principles. Their inclusion in any theory makes that theory provisional.

As a mathematical expression (17) tells us nothing about the possible values of K_{ij}. From the physical standpoint of diffusion, however, the

---

* The common name for the "eddy diffusivity of heat" is the "eddy conductivity."

diagonal components of K_{ij} must be nonnegative. If K_{ij} ($i = j$) had negative values the flux of substance due to diffusive processes could be directed from a region of lower concentration to one of higher concentration which contradicts to the concept of diffusion. There is no physical restriction on the off-diagonal components. Since definition 17 is essentially a formal one, we must not be surprised if the values of K_{ij} ($i = j$) estimated from measurements are sometimes negative. Because of anisotropy in the oceanic turbulent field, the principal axes of the eddy diffusivity tensor do not necessarily coincide with horizontal and vertical axes. Coincidence provides the simplest form of the eddy diffusivity tensor capable of describing the gross features of turbulent diffusion in the sea. When this is the case

$$K_{ij} = \begin{pmatrix} K_x & 0 & 0 \\ 0 & K_y & 0 \\ 0 & 0 & K_z \end{pmatrix} \qquad (18)$$

where K_x and K_y are horizontal eddy diffusivities and K_z is the vertical eddy diffusivity. The tensor form equivalent to (18) is

$$K_{ij} \doteq K_x \delta_{ij} + (K_y - K_x) m_i m_j + (K_z - K_x) n_i n_j \qquad (18')$$

where δ_{ij} is the unit diagonal tensor, which takes the values $\delta_{ij} = 1$ when $i = j$, and $\delta_{ij} = 0$ when $i \neq j$, and m and n denote the unit vectors in the y and z directions, respectively. Under these assumptions, (16) can be written as

$$\frac{\partial \bar{s}}{\partial t} + \bar{U} \frac{\partial \bar{s}}{\partial x} + \bar{V} \frac{\partial \bar{s}}{\partial y} + \bar{W} \frac{\partial \bar{s}}{\partial z} = \frac{\partial}{\partial x}\left(K_x \frac{\partial \bar{s}}{\partial x}\right)$$
$$+ \frac{\partial}{\partial y}\left(K_y \frac{\partial \bar{s}}{\partial y}\right) + \frac{\partial}{\partial z}\left(K_z \frac{\partial \bar{s}}{\partial z}\right) + F \qquad (19)$$

where $\bar{U}$, $\bar{V}$, and $\bar{W}$ are the mean velocity components in the x, y, and z directions, respectively.

Although the very concept of eddy diffusivity may not be valid in a turbulent field (Corrsin, 1957), (19) is a convenient point of departure for an exploration of the variations of mean concentration. Fundamental discussions about the use of eddy diffusivity tensors have been offered by various investigators (see, for instance, Calder, 1965; Saffman, 1963; Kamenkovich, 1967). They all conclude that in the present defective state of our understanding of the detailed mechanisms responsible for turbulent diffusion it is wise to prefer the simplest productive assumptions to those which are more complex.

As was pointed out in the previous section, a very wide spectrum of horizontal motions exists in the sea. As a result, the value of the horizontal eddy diffusivity usually increases with the scale of mixing considered. Any theoretical model of horizontal mixing must take into account this dependence of the eddy diffusivity on the scale of phenomenon. Details are given in the section on mixing from specified sources.

In contrast to the horizontal diffusion in the sea, the process of vertical diffusion is controlled primarily by small-scale motions characteristic of stably stratified water. The stability in a column of water plays a dominant role in the structure of turbulence in the vertical direction. This, in turn, suggests that the value of the vertical eddy diffusivity, K_z, should depend on the density stratification as well as, to a lesser degree, on the scale of diffusion. A comprehensive theory for the dependence of K_z on the stratification or, rather, on a Richardson number, is still lacking.

Several semi-empirical relations between K_z and R_i have been proposed. Most of them are presented in the form

$$K_z = A_0(1 + \beta R_i)^{-m}$$

where A_0, being dependent in general on **x**, denotes the vertical eddy diffusivity at neutral stability and β and m are empirical constants. For example, Munk and Anderson (1948) took $\beta = 3.33$ and $m = \frac{3}{2}$. The influence of stability on K_z has also been considered by Mamayev (1958) who proposed the form

$$K_z = A_0 \exp(-0.8\, R_i)$$

A mixing-length theory can be applied to express A_0 in terms of the velocity shear and the geometry

$$A_0 \sim l_0^2 \left| \frac{d\bar{U}}{dz} \right|$$

where l_0 is the mixing length in neutrally stratified water. Kent and Pritchard (1959) tested the mixing-length theory on the salinity distribution in the James River estuary. In the formulation of l_0, Kent and

Pritchard included the effects of surface wind waves as well as of the geometry of the waterway. The latter effect had been considered by Montgomery (1943). Pritchard (1960) proposed empirical expressions for the vertical eddy diffusivity in estuarine diffusion.

For a summary of values of K_z in the sea the reader should consult Sverdrup et al. (1942) and Defant (1961). By and large, the estimated values of K_z in the ocean lie between 1 and 100 cm²/sec. Values larger than 100 cm²/sec may be required for the upper part of the mixed layer. In the deep layers below the main thermocline a typical value of K_z is 1 cm²/sec. In the thermocline or in any layer of strong stability, exceptionally small values of K_z obtain. Folsom (1967) reported a result derived from radioactivity distributions following an underwater nuclear test in the Northeastern Pacific. The distributions were in a form of intense contaminations within very thin layers (1 to 10 m) at depths between 100 and 400 m. Resulting values of the vertical diffusivity were 6×10^{-3} and 3.5×10^{-2} cm²/sec for two patches after 3 days of diffusion. Kullenberg (1970) made a series of dye diffusion experiments in the thermocline region of the Baltic Sea. On averaging results from 10 sets of experiments, he estimated the mean vertical diffusivity as 8×10^{-2} cm²/sec, ranging from 4×10^{-2} to 1.5×10^{-1} cm²/sec, after a few hours of diffusion.

An apparent paradox arises when one compares the estimates of K_z in the thermocline which are based on the distributions of oceanic variables such as temperature with the estimates based on diffusing tracers—1 cm²/sec as against 10^{-2} cm²/sec. A difference of two orders of magnitude between two methods of estimating the same physical process is enough to shake our confidence. The resolution of the paradox may lie in the large time- and space-scales involved in estimating K_z from oceanic distributions. It is possible that the situation is similar to that encountered in the evaluation of horizontal eddy diffusivity. In a long-term average, part or even all of the horizontal transport processes may be parametrically disguised as vertical eddy diffusion. A short-term tracer experiment, on the other hand, more nearly represents a pure small-scale diffusive process in the vertical direction.

Finally, under certain conditions, the salt-fingering process may have something to do with the mechanism of vertical mixing which gives rise to the effective vertical diffusion. Turner (1967) computes a salt flux of 10^{-7} g/(cm²) (sec), arising from the salt-fingering process for the case of the Mediterranean outflow into the Atlantic. This flux would correspond to an effective vertical diffusivity of 5 cm²/sec. Oceanographic implications of the salt-fingers, especially for the mechanism of mixing, have recently been discussed by Stern (1967, 1968). This problem needs further investigation.

Up to the present time, no argument about the influence of the density stratification on the horizontal eddy diffusivity has gained general acceptance. Taylor (1931b) suggested, from results of meteorological observations, that a stable stratification of density reduced the lateral diffusion as well as the vertical diffusion. Later, Parr (1936) remarked that the data used by Taylor were referred to extremely small-scale processes, and hypothesized that large-scale lateral mixing might increase with vertical stability. However, Sverdrup et al. (1942) considered the effect to be negligibly small. Kirwan (1965) has demonstrated that vertical eddy diffusivity shows a minimum and lateral diffusivity a maximum in layers of high stability. Further studies are necessary. In particular, thought should be given to the possibility that a combination of vertical shear in the horizontal current and vertical diffusion may produce an effective horizontal mixing which could have been interpreted as lateral diffusion (see the section on mixing from specified sources).

The Turbulent Field for a Contaminant

The mean concentration field, $\bar{S}$, describes only one gross feature of the mixing. One could easily imagine a particular situation in which the mean concentration of the contaminant was uniform everywhere in the region of interest, yet in which fluctuations in contaminant concentration about the mean existed.

In order to study more precisely the structure of the contaminant field, one must have some measure of the fluctuations in the scalar field. From analogy with a turbulent-velocity field, one can take the correlations between varying scalar quantities at two points of the field as a fundamental measure describing geometrical relations in the structure. Define the correlation function by

$$Q(\mathbf{r}, \mathbf{x}; t) = S'(\mathbf{x} - \tfrac{1}{2}\mathbf{r}; t)\, S'(\mathbf{x} + \tfrac{1}{2}\mathbf{r}; t)$$

where S' is the fluctuation of the contaminant concentration about its mean value.

To obtain a differential equation for $Q(\mathbf{r}, \mathbf{x}; t)$ we start with the equation for S'.

$$\frac{\partial S'}{\partial t} + \bar{u}_i \frac{\partial S'}{\partial x_i} + u_i' \frac{\partial \bar{S}}{\partial x_i} + u_i' \frac{\partial S'}{\partial x_i} - \frac{\overline{\partial u_i' S'}}{\partial x_i} = D \frac{\partial^2 S'}{\partial x_i^2} \qquad (20)$$

S being conservative.* Take (20) referred to point $A \equiv \mathbf{x} - \tfrac{1}{2}\mathbf{r}$ and multiply it by S' at point $B \equiv \mathbf{x} + \tfrac{1}{2}\mathbf{r}$. Similarly take (20) referred to B and multiply it by S' at A. Addition of the two resulting equations yields

$$\frac{\partial Q}{\partial t} + \frac{1}{2}[(\bar{u}_i)_A + (\bar{u}_i)_B]\frac{\partial Q}{\partial x_i} + [(\bar{u}_i)_B - (\bar{u}_i)_A]\frac{\partial Q}{\partial r_i}$$

$$+ \left(\frac{1}{2}\frac{\partial}{\partial x_i} + \frac{\partial}{\partial r_i}\right)\{\overline{(u_i')_B S_A'} \cdot (\bar{S})_B + \overline{(u_i')_B S_A' S_B'}\}$$

$$+ \left(\frac{1}{2}\frac{\partial}{\partial x_i} - \frac{\partial}{\partial r_i}\right)\{\overline{(u_i')_A S_B'} \cdot (\bar{S})_A + \overline{(u_i')_A S_A' S_B'}\}$$

$$= \frac{1}{2}D\frac{\partial^2 Q}{\partial x_l^2} + 2D\frac{\partial^2 Q}{\partial r_l^2} \tag{21}$$

Again, (21) is not in closed form. The covariances and triple correlations are still unknown.

Let us consider some special cases.

1. Let the two points A and B coincide ($\mathbf{r} = 0$).

Q becomes $\overline{S'^2}$, that is, the mean square fluctuation. Equation 21 reduces to

$$\frac{\partial \overline{S'^2}}{\partial t} + \frac{\partial}{\partial x_i}\left[\bar{u}_i\,\overline{S'^2} + \overline{u_i' S'^2} - 2D\,\overline{S'\frac{\partial S'}{\partial x_i}}\right] + \overline{u_i' S'}\frac{\partial \bar{S}}{\partial x_i} = -D\,\overline{\left(\frac{\partial S'}{\partial x_i}\right)^2} \tag{22}$$

If now we integrate (22) over a large volume V, neglecting a surface integral in the same ways as we did in the section on mechanism of mixing, we obtain

$$\frac{d\langle \overline{S'^2}\rangle}{dt} + \left\langle \overline{u_i' S'}\frac{\partial \bar{S}}{\partial x_i}\right\rangle = -D\left\langle \overline{\left(\frac{\partial S'}{\partial x_i}\right)^2}\right\rangle \tag{23}$$

If we use (17) with (18), (23) becomes

$$\frac{d\langle \overline{S'^2}\rangle}{dt} = \sum_{j=1}^{3}\left\langle K_j\overline{\left(\frac{\partial \bar{S}}{\partial x_j}\right)^2}\right\rangle - D\left\langle \overline{\left(\frac{\partial S'}{\partial x_i}\right)^2}\right\rangle$$

In other words, the turbulent process increases the amount of inhomogeneity, while molecular diffusion tends to smear out the inhomogeneity.

* The reader should be acutely aware that while the molecular diffusion has been neglected in the mean equation, (19), it is very much present in the fluctuation equation, (20), being incorporated in the term $D(\partial^2 S'/\partial x_l^2)$.

Note that we have assumed that the turbulent fluxes can be expressed in terms of the gradients of the mean concentration and that the diffusivities are positive.

2. Let the turbulence be homogeneous and isotropic. Under these conditions all the derivatives of the correlations with respect to x_i vanish, and $\bar{u}_i$ and $\bar{S}$ are uniform. The correlation functions depend only on $r \equiv |\mathbf{r}|$ and t. Equation 21 then becomes

$$\frac{\partial Q(r, t)}{\partial t} - 2R(r, t) = 2D \frac{1}{r^2} \frac{\partial}{\partial r} \left(r^2 \frac{\partial}{\partial r} Q \right) \tag{24}$$

where $R(r, t)$ may be interpreted as an exchange function whose source is the interaction of the velocity and concentration fluctuations at the two points. Equation 24 can be transformed into the equation for the spectrum function by taking its Fourier transform (Hinze, 1959, Chapter 3)

$$\frac{\partial E_s(k, t)}{\partial t} - F_s(k, t) = -2Dk^2 E_s \tag{25}$$

where E_s is the spectrum function for the fluctuation of the contaminant concentration, and F_s represents the contribution to E_s due to transfer mechanisms. Note that

$$\int_0^\infty E_s \, dk = \overline{S'^2}$$

and

$$\int_0^\infty F_s \, dk = 0$$

To make (25) determinate, an additional relation between F_s and E_s is necessary.

Integrating (25) over the entire range of wave numbers, k, one obtains

$$\frac{d}{dt} \int_0^\infty E_s \, dk = -2D \int_0^\infty k^2 E_s \, dk \equiv -\chi$$

The quantity $\overline{S'^2}$, that is, $\int_0^\infty E_s \, dk$, must ultimately be dissipated by molecular diffusion. This dissipation takes place essentially at high wave numbers.

Let us consider the small-scale components of passive physical properties like temperature. The "small-scale" is taken here to mean that the components concerned have characteristic length-scales small compared with the length-scale, L, of the eddies containing the bulk of the kinetic energy of the turbulent motion. For this small-scale structure, that is, the large wave-number region, the scalar spectrum E_s can be divided into spectral ranges each characterized by a physical process.

Because of the nonlinear interaction of the fluctuating components of velocity and the scalar property represented by the transfer function F_s, a transfer of energy, or more exactly $\overline{S'^2}$-stuff, from component to component is possible (Batchelor, 1959). For any part of the spectrum to attain statistical equilibrium, the interactions transferring $\overline{S'^2}$-stuff from wave number to wave number must have a time scale very much less than the overall decay time of the scalar field of the property. This condition rules out an equilibrium range in the vicinity of the wave number L^{-1} characteristic of the energy-containing eddies, where the spectrum E_s attains a maximum value. In other words, one can expect an equilibrium range in the spectrum for wave numbers much larger than L^{-1}. Toward the other end of the spectrum, that is, for very large wavenumbers, all the dissipation of $\overline{S'^2}$-stuff takes place as a result of the action of molecular diffusion, the total rate of the dissipation being given by χ as was shown previously.

For the spectrum of the scalar contaminant, E_s, there are only two physical properties of S' which are relevant in the equilibrium range in the absence of local k-space variations in S'. They are χ, the total rate of dissipation of S', and D, the molecular diffusivity of S'. However, changes in E_s may also be brought about by local k-space changes in the turbulent velocity field since the turbulent motion determines the structure of the scalar distribution. For the turbulent energy (velocity squared) spectrum the relevant physical parameters are ϵ, the rate of viscous dissipation of the turbulent kinetic energy per unit mass, and ν, the kinematic viscosity of the fluid.

If the Reynolds number of the turbulence is so large that viscous and diffusion effects are unimportant, at least at small wave numbers in the equilibrium range, dimensional arguments lead to

$$E_s \propto \chi \epsilon^{-1/3} k^{-5/3} \tag{26}$$

This part of the equilibrium range has been called the "inertial-convective subrange" (Corrsin, 1961b). When $\nu \gg D$, as in the sea, Batchelor (1959) showed that (26) is valid for $L^{-1} \ll k \ll (\epsilon/\nu^3)^{1/4}$. Thus the scalar spectrum E_s here has the same dependence on k as the velocity spectrum E in the inertial subrange specified by $L^{-1} \ll k \ll (\epsilon/\nu^3)^{1/4}$.

$$E \propto \epsilon^{2/3} k^{-5/3} \quad \text{(Batchelor, 1956)}$$

When $\nu \gg D$, molecular diffusion effects occur at wave numbers much higher than $(\epsilon/\nu^3)^{1/4}$, that is, the wave number at which the turbulent energy spectrum starts to be cut off by viscosity. The form of the spectrum E_s at wave numbers much larger than $(\epsilon/\nu^3)^{1/4}$ is determined by the combination of semi-permanent, uniform straining motion arising from the small-scale turbulent components and of the smoothing action of molecular diffusion (Batchelor, 1959). The basic mechanism can be represented by a differential equation analogous to the one from which solution 5 is derived. As a result, Batchelor gives

$$E_s \doteq 2\chi\,\nu^{1/2}\,\epsilon^{-1/2}\,k^{-1}\,\exp\,(-2D\,\nu^{1/2}\,\epsilon^{-1/2}\,k^2)$$

for the wave numbers $k \gg (\epsilon/\nu^3)^{1/4}$.

In particular there exists a subrange called the "viscous-convective subrange" in the range of wave numbers specified by $(\epsilon/\nu^3)^{1/4} \ll k \ll (\epsilon/\nu D^2)^{1/4}$ (Corrsin, 1961b). The spectrum in the viscous-convective subrange obtains

$$E_s \doteq 2\chi\,\nu^{1/2}\,\epsilon^{-1/2}\,k^{-1}$$

For wave numbers larger than $(\epsilon/\nu D^2)^{1/4}$, the spectrum of the scalar contaminant, E_s, starts to be cut off by molecular diffusion. Corrsin (1961b) calls the range for which $k \gg (\epsilon/\nu D^2)^{1/4}$ the viscous-diffusive subrange. Figure 4.8 shows the scalar contaminant spectrum and the velocity spectrum for the various subranges in the case where $\nu \gg D$.

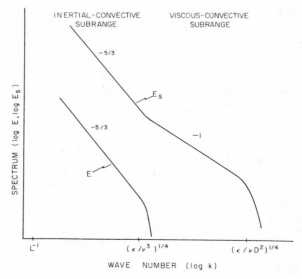

Figure 4.8 Spectra of velocity and passive-scaler fluctuations in the equilibrium ranges ($\nu \gg D$).

Corrsin (1961b) generalized the spectrum of a passive physical property by allowing a first-order chemical reaction. For very large Schmidt numbers, Corrsin deduced the form of the spectrum in three ranges: the inertial-convective, viscous-convective, and viscous-diffusive. Some modifications to the form of Batchelor's spectrum were recognized, which depended upon the intensity of the chemical reaction.

Voorhis and Perkins (1966) made an experimental estimate of the spatial (horizontal) spectrum of the temperature fluctuations in the near-surface summer thermocline off Bermuda. The temperature was measured at a depth of about 100 m with a ship-towed temperature detector. The resulting spatial spectrum for horizontal variations of temperature decreases rapidly in level in a wave length band from 7000 to 200 m. In this wave number range, the spectrum can be resonably well described by the $k^{-5/3}$ power law.

In the upper layer of the sea characteristic wave numbers are $k = (\epsilon/\nu^3)^{1/4}$ in the range from 1 to 10/cm and $k = (\epsilon/\nu D^2)^{1/4}$ in the range from 10 to 100/cm. To examine Batchelor's spectrum for small-scale temperature fluctuation, Grant et al. (1968b) used a platinum film resistance thermometer by means of which they were able to measure temperature fluctuations with frequencies up to 1000 Hz. The measurements were made in offshore waters as well as in the Discovery Passage area. The resulting one-dimensional temperature spectra agree fairly well with Batchelor's theory. Two subranges with power law k-dependencies of $-\frac{5}{3}$ and -1 were detected. Velocity records made at the same time have turbulent energy spectra which show a $-\frac{5}{3}$-power dependence on k in the inertial subrange.

Richardson's Equation for Diffusion

Richardson (1926) recognized that for diffusion in the atmosphere the Fickian diffusion equation could not properly describe the relative spread of a cluster of particles with respect to its center of mass. Since the turbulent regime contains a wide range of eddying motions, the effective power of dispersing a pair of particles increases with the distance between the pair. Thus with increasing separation a greater range of eddy sizes contributes to the relative diffusion. Richardson considered the important independent variable to be not the position of a particle but its separation from its neighbors.

The diffusion problem formulated in terms of separation instead of position has for its important parameters the projective separation, l, of a marked pair of particles and, q, the number of neighbors per unit of l, rather than position and point concentration. Richardson defined

$$q(l, t) \equiv \int_{-\infty}^{\infty} S(x, t)S(x + l, t)dx$$

where $S(x, t)$ is the concentration of particles at x and t when projected on the direction x. Thus obtained, q satisfies Richardson's equation for diffusion.

$$\frac{\partial q}{\partial t} = \frac{\partial}{\partial l}\left[F(l)\frac{\partial q}{\partial l}\right] = \alpha \frac{\partial}{\partial l}\left(l^{4/3}\frac{\partial q}{\partial l}\right)$$

where the last expression for $F(l)$ was found empirically by Richardson from sets of observations of relative diffusion in the atmosphere.

According to Batchelor (1952), it is more reasonable to suppose that F is a function of statistical quantities like $\overline{l^2}$, for example, $F = F\{\overline{l^2}(t)\}$. Richardson's neighbor concentration q can be regarded in one way as the fraction of the total number of particles in an instantaneous distribution of a single trail for which there is another particle whose relative position lies within given limits.

A number of tests made in the ocean verified Richardson's $\frac{4}{3}$ law. Ichiye and Olson (1960) showed that Richardson's law fits well to the data covering the range $l = 10$ to 10^8 cm.

Although there is no doubt that Richardson's ideas have contributed to our understanding of oceanic diffusion (Stommel, 1949) the older distribution of concentration rather than Richardson's neighbor concentration remains much more easily accessible to measurement at sea. There have appeared a number of theoretical solutions for the older distribution of concentration (e.g., Joseph and Sendner, 1958).

Horizontal and Vertical Mixing from Specified Sources

Mixing of a dynamically passive contaminant from a given source may be separated into two classes by the nature of the release. If the entire volume of contaminant is released in a very brief time we speak of mixing from an "instantaneous" source. If the contaminant is released over a long time we call it mixing from a "continuous" source. The intensity of a continuous source may vary with time, and its position need not be fixed. The sizes of both kinds of source may be treated as either infinitesimal or finite depending on the scales of the mixing.

For the theory of diffusion in laminar flows, solutions for instantaneous point-sources serve as the fundamental building blocks. By integrating the fundamental solutions with respect to time for appropriate source intensities one obtains the solution for the continuous point-source. Solutions of a large number of problems can be formulated immediately from the fundamental solution. Unfortunately, such a fundamental solution for diffusion problems does not exist, in a rigorous sense, for a turbulent field such as the sea. Thus strictly speaking, mere knowledge of the concentration distribution for an instantaneous point-source is not sufficient to obtain the distribution of substance from a continuous point-source in the sea. In diffusion from a continuous point-source, one must not only take into consideration diffusion relative to the center of mass of individual elementary patches, but must also consider variabilities in location of the center of mass of the individual patches.

In practical problems, however, we often approximate the concentration from a continuous source by superposing an infinite number of diffusing patches each from an instantaneous point-source and all moving with the mean velocity. It is sometimes possible to improve the concentration approximation by including another approximation of about the same quality to describe the fluctuations in position of the superposed elementary point-source patches.

Instantaneous Releases

The horizontal scale of motions in the sea is usually so much greater than the vertical scale that, for many purposes, their effect on mixing may be considered separately. Revelle et al. (1955) give an example of the mixing of radioactive material released below the thermocline. The radioactive water spread out over an area of about 100 km^2 while maintaining a thickness of the order of a few meters. In another instance, Folsom and Vine (1957) describe the spread of a radioactive tracer over a horizontal area of 40,000 km^2 in 40 days. During this time it mixed vertically through only 60 m. This mixing corresponds to virtual eddy diffusivities of the order of 10^7 cm^2/sec in the horizontal direction and of 1 cm^2/sec in the vertical direction.

These experimental facts suggest that the horizontal components of oceanic turbulence may alone play the essential role in the horizontal spread of material. This suggestion has been explored by a number of investigators. In this approach we first assume that the substance is subject to horizontal mixing within a sufficiently thin homogeneous layer so that all vertical variations in both concentration and velocity may be ignored. We further assume that the actual distribution of contaminant may be replaced by a circular distribution with center at the center of contaminant mass. This model, the "radially symmetric solution," has achieved some success in predicting the diffusion of introduced substances (Joseph and Sendner, 1958, 1962; Ozmidov, 1958; Okubo, 1962a).

The importance of vertical diffusion to adequate prediction, however, cannot be neglected even though the diffusing contaminant is, for practical purposes, confined to a very thin layer. Apparent vertical homogeneity in concentration may well exist in conjunction with vertical inhomogeneity in the velocity. The combined action of vertical shear in the mean flow and small vertical mixing may produce a considerable effective streamwise diffusion. As a matter of fact, this "advection-diffusion solution" based on a sheared velocity gives results which are about the same as the radially symmetric solution based on vertically uniform concentration and uniform velocity (Bowles et al. 1958; Bowden, 1965b; Carter and Okubo, 1965; Okubo, 1968a). However, it is more attractive for many applications and is to be preferred as a step forward to a better physical insight into the mixing process. It should be noted that the two models are not in essential conflict and may be used to supplement each other.

Radially Symmetric Solutions for Horizontal Diffusion. We shall consider the two-dimensional diffusion of a passive conservative contaminant which is introduced instantaneously at a point in the sea. The motion of a patch of diffusing substance consists of an overall wandering of its center of mass and an irregular spreading about its center. Our main interest will be in the horizontal distribution of the contaminant about its moving center.

At a time t after release, the shape of the concentration isolines of a patch will be very irregular, and if the release is repeated, the details will show little resemblance (Fig. 4.9). The patch of substance is usually elongated in one direction or another. We shall now carry out (conceptually!) an infinite number of releases under identical oceanographic conditions using in each case the same amount of contaminant. Let the centers of mass of the patches be superposed while their observed orientations are preserved and let the average concentration over all the patches be formed at some common time after release, t. If the mean flow pattern is two-dimensionally isotropic, it is reasonable to expect that the average so formed would be radially symmetric with a center of mass coincident with centers of mass of the patches. This construction can be viewed in the reverse sense. With strict horizontal isotropy in the diffusion, the contaminant distribution would be circular and increasing with time, $\bar{S}(r, t)$. Individual records from the ensemble of experiments can be regarded as particular realizations. That individual realizations depart

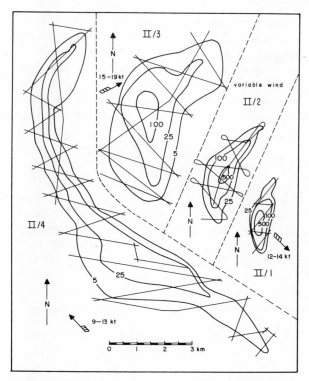

Figure 4.9 Patterns of dye patch on a horizontal plane. The wind speed and direction are also shown. From Joseph et al., 1964.

so widely from circularity indicates that the process is isotropic only in the ensemble mean. One never has infinite ensembles of the sort considered conceptually. One seldom has as many as five records made under sensibly identical oceanographic conditions. In short, one usually has a unique record and ensemble averages are impossible. Still, a single record may provide statistics for comparison with the circular distribution. Such statistics are created by contouring the concentrations at some time t, measuring the area enclosed by each isoline, and replacing the contour with a circle of equal area centered on the observed point of mass concentration.

The basic equation of horizontal mixing for a radially symmetric distribution of the contaminant as obtained from (19) is

$$\frac{\partial \bar{S}_h}{\partial t} = \frac{1}{r}\frac{\partial}{\partial r}\left(r\, K_h(r,t)\,\frac{\partial \bar{S}_h}{\partial r}\right) \qquad r = 0, \quad t = 0 \qquad (27)$$

where $K_h(r, t)$ denotes a horizontal eddy diffusivity. The mean velocity $\bar{U}$ relative to the center of mass is taken as zero, since a radially symmetric solution has no concern with the mean velocity. The subscript implies that the radially symmetric distribution is two-dimensional on a horizontal plane.

For an instantaneous point release of contaminant amount M within a homogeneous layer of thickness H, the initial condition is expressed by

$$\bar{S}_h(r, 0) = \frac{MH^{-1}}{\pi r} \delta(r) \tag{28}$$

where $\delta(r)$ is a Dirac function. A knowledge of $K_h(r, t)$ is required to solve (27) subject to (28). Different forms of K_h lead to different solutions of (27) and before describing them it will be interesting to discuss a similarity form for the radially symmetric distribution.

Assume that the distribution maintains its similarity in the course of time, that is, the pattern of concentration relative to the maximum concentration remains similar as the characteristic length of mixing grows with time. One may express this pattern formally as follows.

$$\bar{S}_h(r, t) = \bar{S}_0(t) G\left\{\frac{r}{b(t)}\right\}$$

where $\bar{S}_0(t)$ represents the peak concentration at the center of patch, $b(t)$ is the characteristic length of mixing, and G is a decreasing function taking the value of unity at $r = 0$.

The condition that the total amount of substance introduced must be conserved determines $\bar{S}_0(t)$. Further, $b(t)$ can be expressed in terms of the standard deviation for a radially symmetric distribution σ_{rc}, defined by

$$\sigma_{rc}(t) \equiv \left\{\frac{1}{MH^{-1}} \int_0^\infty r^2 \bar{S}_h(r, t)\, 2\pi r\, dr\right\}^{1/2} \tag{29}$$

The similarity solution can then be written as

$$\bar{S}_h(r, t) = \frac{MH^{-1}}{2\pi\phi\alpha^2\sigma_{rc}^2(t)} G\left\{\frac{r}{\alpha\sigma_{rc}}\right\} \tag{30}$$

where ϕ and α are numerical constants, their value being dependent upon the form of G.

The similarity solution may or may not satisfy the diffusion equation 27 which has so far not entered the argument. It can be shown that the similarity form (30) satisfied (27) only if the horizontal diffusivity takes the form

$$K_h(r, t) = A\left(\frac{r}{\alpha \sigma_{rc}}\right) B(\sigma_{rc}, t) \qquad (31)$$

where B is subject to the condition that $d\sigma_{rc}^2/dt = B$.

The radially symmetric solutions proposed thus far are in the class of similarity solutions. All the proposed solutions for the radially symmetric case can be derived systematically from (27) by choosing the class of possible forms for $K_h(r, t)$ given by (31) (Schönfeld and Groen, 1961; Okubo, 1962a).

A novel approach to radially symmetric diffusion has been devised by Schönfeld (1962). The essence of his attack is a Fourier representation of (16) and the creation of the concept of "integral diffusivity" to close the equation set. Schönfeld's solutions are worthy of the reader's attention but are not discussed here.

The radially symmetric solution may be separated into two classes according to whether they are expressed in terms of a "diffusion velocity," $P([LT^{-1}])$, or in terms of the "rate of turbulent energy transfer-parameter," $\gamma([L^{2/3}T^{-1}])$. The Joseph and Sendner (1958) solution is a representative of the former group, and the Ozmidov (1958) solution represents the latter group.

$$\bar{S}_h = \frac{MH^{-1}}{2\pi P^2 t^2} \exp\left\{-\frac{r}{Pt}\right\} \qquad \text{(Joseph-Sendner)} \qquad (32)$$

$$\bar{S}_h = \frac{MH^{-1}}{6\pi \gamma^3 t^3} \exp\left\{-\frac{r^{2/3}}{\gamma t}\right\} \qquad \text{(Ozmidov)} \qquad (33)$$

Solution 32 was first obtained by MacEwan (1950). Later Joseph and Sendner derived it, independently, on a somewhat more sophisticated basis. P-class solutions are characterized by maximum concentrations that decrease with time as t^{-2} and the horizontal variances, σ_{rc}^2, that increase as t^2, while γ-class solutions find that they are t^{-3} and t^3, respectively. The solutions in each group differ only in their spatial distributions. Constancy of each characteristic parameter would be desirable for the purpose of practical application of the proposed solutions to the horizontal mixing of contaminants. Actually the values of the parameters vary with the scale of mixing in a more or less regular fashion as well as with the environmental conditions such as the intensity of turbulence. A table of the values is contained in Okubo (1962a).

The basic concept involved in γ-class solutions is that the eddies responsible for the horizontal spread of substance lie in the inertial subrange. The statistical properties of these eddies thus depend only on the rate of energy transfer which must be equal to the rate of energy dissipation if the kinetic energy of the eddies is to remain constant. Under natural conditions, however, there is often the possibility of energy being fed locally to certain ranges of eddies commonly included in the inertial subrange, as mentioned in the section on environmental features of the sea. These excited eddies destroy the concept of the inertial subrange, so that results derived from that concept would not necessarily be applicable. Ozmidov (1965), nevertheless, postulates the existence of *local* inertial subranges between eddies at which influxes of external energy take place. This, in turn, suggests that γ-class solutions may be used locally with different values of γ for each local range. As the scale of mixing increases one should expect γ to decrease since local supplies to energy are usually transferred from the larger to the smaller eddies.

P-class solutions, on the other hand, are based on the intuitive concept that the rate of change of the local variance should depend only on a characteristic velocity, that is, on the diffusion velocity. Since oceanic mixing processes result from oceanic turbulence, a close relationship should exist between the diffusion velocity and the root-mean-square value of the fluctuations in the velocity of water. The estimated value of the diffusion velocity is of the order of 1 cm/sec. Joseph and Sendner (1958) obtained the value of 1 ± 0.5 cm/sec for P, and later (1962) they presented a diffusion-velocity spectrum which indicated a weak dependence of the diffusion velocity on the scale of mixing.

Fortunately, we have a considerable body of data accumulated since 1960 on which we can draw for tests of our solutions. The data represent the fruits of a field technique developed by Pritchard and Carpenter (1960) which depends on fluorescence. The data cover time scales of mixing from 1 hr to 1 month and length scales from 30 m to nearly 100 km. One of the first uses to which the data were put was to guide the selection of a reasonable functional form for $\sigma_{rc}^2(t)$. A diffusion diagram plotting the horizontal variance, σ_{rc}^2, against the diffusion time, t, is shown in Fig. 4.10. A diagram based on data accumulated to 1962 by a variety of techniques shows nearly the same behavior for $\sigma_{rc}^2(t)$ (Okubo, 1962b). Irrespective of the detailed oceanographic conditions, σ_{rc}^2 exhibits a general trend. Specifically, it increases with time at a power between 2 and 3, that is, at a power which lies between those predicted by P- and γ-class solutions. However, this finding does not exclude the possibility that P-class solutions with a t^2 dependence or γ-class solutions with a t^3 depen-

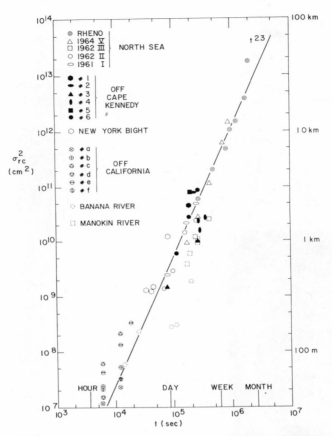

Figure 4.10 Variance, σ_{rc}^2, versus diffusion time, t, from dye-diffusion experiments (For details, see Okubo, 1968b).

dence may be locally valid for some time or length scales. All that is required is that P and γ be allowed to vary with time or length scale (Ozmidov, 1968b; Okubo, 1968b; Okubo and Ozmidov, 1970).

Figure 4.10 is constructed from data taken exclusively from diffusion experiments in the upper mixed layer. Because of operational difficulties only a very few tracer experiments have been made either in deep water or in the thermocline.

It is certainly surprising that any have succeeded. However, experiments in a thermocline (Kullenberg, 1970) and in deep water (Schuert, 1970) have revealed that the rate of horizontal mixing is smaller by an order of magnitude than that of the upper mixed layer. A typical value of the diffusion velocity P in both layers is of the order of 0.1 cm/sec compared with 1 cm/sec in the upper layer.

It is quite important to study the variation of diffusion characteristics in connection with the environmental factors. Although some efforts have been made (Foxworthy et al., 1966; Kullenberg, 1970), further studies are necessary before anything reliable can be said.

Advection-Diffusion Solutions. Many aerial photographs of dye patterns from an instantaneous source reveal that dye patches become more or less elongated. Figure 4.11 shows an example of an elongated dye patch; the dye, having been introduced as a vertical line source, began to elongate during the first 2 hr or so roughly in the direction of the local wind.

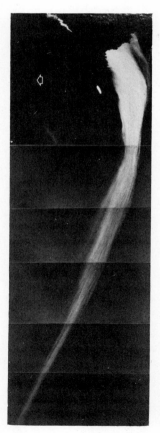

Figure 4.11 Aerial photography of a dye patch (1 hr 27 min after release from an instantaneous small source off Cape Kennedy). This photo was taken by Mr. R. W. Linfield, Chesapeake Bay Institute, The Johns Hopkins University, on August 15, 1962.

In the photo, the smoke bomb on the sea surface indicates the wind direction. North is shown by a white arrow. The leading portion ("head") of the dye patch contained a higher dye concentration than the trailing portion ("tail"). Observations of this dye patch reveal the head at or near the naviface and the tail at depths of several meters. From the leading edge toward the tail, the dye was found at successively deeper and deeper levels. The elongated appearance of the dye is an obvious result of the vertical shear in the horizontal current.

An equally prominent feature of the dye patch is the clockwise curvature of the tail when one looks from the head. In all Northern Hemisphere cases in which an elongated dye patch was curved, the curvature was clockwise. A few observations exist for the Southern Hemisphere but those do show anticlockwise curvature (Katz et al., 1965). This suggests that the curved tail may be due to the effect of the vertically differential advection by the wind-driven Ekman current which appears as a spiral on a horizontal projection.

The elongation of the patch due to the differential velocity in the mean flow gives rise to an effective mixing in the longitudinal direction when combined with transverse diffusion due to small-scale random motions. Bowles et al. (1958), noticing the importance of the process in horizontal mixing, called it the "shear effect" by which they meant the distortion of a vertical column of contaminant due to the variation of mean velocity with depth combined with the vertical diffusion. Any gradient of mean velocity, however, leads to an effective mixing. Thus the shear effect, in a broad sense, may be associated with tidal currents, inertial currents, density currents, wind-driven currents, and currents arising from wave motions as examples.

A very simple model provides a clue to an understanding of the shear effect. Consider a shallow basin of constant depth H (Fig. 4.12). The mean horizontal velocity is taken to vary linearly with depth only. At the surface its value is $2\bar{U}$ and at the bottom, zero. The model alternates the shearing effect and the diffusing effect in discrete steps. The first step supposes that shear alone distorts the contaminant patch during a time τ. The second step assumes that at the end of time τ the shear ceases for an instant and vertical mixing immediately spreads the contaminant vertically to produce a uniform concentration throughout the depth H. In reality, vertical homogeneity is not attained instantly. For this reason the τ allowed for the shearing effect must be related to a reasonable time for vertical mixing. Repetition of the two steps over intervals of time, τ, mixes the substance in the direction of $\bar{U}$ and at each repetition a new cell of length $2\bar{U}\tau$ is added to the patch.

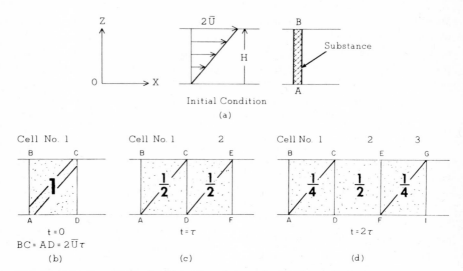

Figure 4.12 Longitudinal mixing of a substance by the shear effect.

It can be shown (Okubo and Carter, 1966) that, after a time $t = n\tau$, the distribution of substance over the cells is represented by a binomial distribution with variance $n(\bar{U}\tau)^2 = \bar{U}^2\tau t$. Since the time interval τ is a measure of the time required for a given bounded system to attain substantial vertical homogeneity of substance, τ must be of the order of $H^2/\pi^2 K_z$. Hence the downstream variance becomes of the order of $\bar{U}^2 H^2 t/\pi^2 K_z$; a more exact derivation gives $\bar{U}^2 H^2 t/15 K_z$ (Saffman, 1962). The longitudinal distribution of contaminant approaches a Gaussian distribution with an effective longitudinal diffusivity of $\bar{U}^2 H^2/30 K_z$ for large times according to the central limit theorem.

This simple model implies that horizontal mixing can occur even without horizontal components of random movements of water. And the value of the variance due to the shear diffusion may well account for the scale of mixing observed in shallow waters (Bowden, 1965b). For example, if we take $\bar{U} = 10$ cm/sec, $H = 3$ meters and $K_z = 10$ cm²/sec then an effective horizontal diffusivity due to shear diffusion amounts to 3.0 × 10⁴ cm²/sec and the size of the patch becomes several hundred meters by $t = 10^4$ sec. Fischer (1967) has discovered that lateral shears are far more important than vertical shears in the longitudinal spread of contaminants in natural streams like rivers and estuaries.

The shear effect, however, appears in a different manner when the region is effectively unbounded because the substance is now allowed to diffuse indefinitely in the direction of the shear, that is, perpendicular to the mean velocity. The effect of shear on dispersion will be far more marked in an unbounded region than in a bounded one. For a uniform-shear field the longitudinal variance increases as t^3 in an unbounded region (Corrsin, 1953), whereas it goes as t^1 in the bounded case. Because the oceanic area may, in many cases, be regarded as semi-infinite as far as mixing from an instantaneous source is concerned, we first consider a shear diffusion in an infinite sea and later take into account the possible effect of boundaries.

In the radially symmetrical model no allowance is made explicitly for factors such as shears in the mean flow. The shear-diffusion model, on the other hand, focuses attention on the actual pattern of flow which can contribute in combination with random motions of small-scale eddies to the resultant mixing. For simplicity, assume a mean velocity field having lateral and vertical shears. They are to be considered steady and homogenous, that is, the mean velocity field is given by

$$\bar{U} = U_0(t) + \Omega_y y + \Omega_z z$$

and

$$\bar{V} = \bar{W} = 0$$

where we take x, y, and z axes, respectively, in the longitudinal, lateral, and vertical directions; $U_0(t)$ denotes the time-dependent mean velocity on the line $y = z = 0$, and Ω_y and Ω_z are the lateral and vertical shears, respectively. We further assume that eddy diffusivities are constant in order to make the problem tractable.

The basic equation for shear diffusion is then formulated as

$$\frac{\partial \bar{S}}{\partial t} + (U_0 + \Omega_y y + \Omega_z z)\frac{\partial \bar{S}}{\partial x} = K_x \frac{\partial^2 \bar{S}}{\partial x^2} + K_y \frac{\partial^2 \bar{S}}{\partial y^2} + K_z \frac{\partial^2 \bar{S}}{\partial z^2} \quad (34)$$

and the solution for an instantaneous point source at $x = y = z = 0$ is given by

$$\bar{S}(t,x,y,z) = \frac{M}{8\pi^{3/2}(K_x K_y K_z)^{1/2} t^{3/2}(1 + \phi_3^2 t^2)^{1/2}}$$

$$\exp - \left[\frac{\{x - \int_0^t U_0(t')\,dt' - \frac{1}{2}(\Omega_y y + \Omega_z z)t\}^2}{4k_x t(1 + \phi_3^2 t^2)} + \frac{y^2}{4K_y t} + \frac{z^2}{4K_z t}\right] \quad (35)$$

where $\phi_3^2 \equiv \frac{1}{12}\left(\Omega_y^2 \frac{K_y}{K_x} + \Omega_z^2 \frac{K_z}{K_x}\right)$. One interprets ϕ_3^{-1} as the time at which the velocity shears begin to affect the mixing of substance significantly.

Since (35) has an exponent quadratic in x, y, and z, the contours of the concentration are a set of ellipsoids whose principal axes all have the same directions and whose origins coincide and move with the velocity $U_0(t)$. The orientation of the principal axes with respect to the coordinate axes varies with time. The elongation of the patch of contaminant is explained by (35). In the main period of shear diffusion, that is, $t \gg \phi_3^{-1}$, (35) shows the following:

1. The maximum concentration decreases as $t^{-2.5}$.
2. The horizontal variance in the major principal axis increases as t^3.
3. The mean horizontal variance, σ_{rc}^2, increases as t^2.
4. The virtual coefficient of horizontal mixing, defined by $\sigma_{rc}^2/2t$, is proportional to t^1. If one considers the coefficients of x^2, y^2, and z^2 in (35), it is evident that in the main period, $t \gg \phi_3^{-1}$, the patch will be greatly elongated in the x direction. In contrast, during the initial period, $t \ll \phi_3^{-1}$, the shear effect will not have had time to make itself felt and a contaminant introduced as a compact patch will retain its compactness.

Equation 35 describes a three-dimensional advection-diffusive process under shears in an infinite ocean. For a release in the surface layer or near the bottom of the sea we have to take into consideration the boundary. The method of images (Carslaw and Jaeger, 1959) can not be used in a trivial way. Difficulties arise because the longitudinal distribution of contaminant becomes asymmetrical near a boundary. It has been found that near the naviface the contaminant concentration exhibits a sharp front and decreases toward the trailing edge to small concentration gradients. Aside from this the essential features of shear-diffusion remain the same in the infinite and semi-infinite cases. Saffman (1962) showed that the longitudinal variance for the semi-bounded case differs only by a numerical factor from that for the infinite case.

In shallow waters the proximity of both the naviface and bottom must be considered after an appreciable amount of contaminant has diffused throughout the depth. Vertical shear then becomes relatively ineffective having little or no vertical gradient of concentration upon which to act and the horizontal effects of lateral shear become dominant. The solution for the depth-mean concentration under the effect of lateral shear can be expressed as

$$\bar{S}_h = \frac{MH^{-1}}{4\pi(A_xK_y)^{1/2}t(1 + \phi_2^2t^2)^{1/2}}$$

$$\times \exp - \left[\frac{\left\{x - \int_0^t \bar{U}_0(t')\,dt' - \tfrac{1}{2}\Omega_v yt\right\}^2}{4A_xt(1 + \phi_2^2t^2)} + \frac{y^2}{4K_yt}\right] \quad (36)$$

where $\phi_2{}^2 \equiv \frac{1}{12}(K_y/A_x)\Omega_y{}^2$, and A_x is the sum of K_x and the effective longitudinal coefficient of mixing due to the vertical shear-effect. One interprets ϕ_2^{-1} as a time at which the lateral shear begins to affect the mixing of the contaminant significantly.

Essential features of the shear diffusion are the same for the two-dimensional case as for the three-dimensional case. The only exception is that the maximum concentration decreases as t^{-2} in the two-dimensional regime; the more rapid decrease with time of the peak concentration in three-dimensional mixing is a simple consequence of the fact that the contaminant is free to diffuse in one more dimension.

Carter and Okubo (1965) and Pritchard et al. (1966) have analyzed dye release experiments made in the Cape Kennedy area solely on the basis of shear-diffusion models with considerable success. The offshore release (three-dimensional experiments) showed decreases with time in the maximum concentration proportional to $t^{-1.5}$ and $t^{-2.5}$ as required by the model (Fig. 4.13). The nearshore and inshore releases (two-dimensional experiments) also conformed with time regimes of $t^{-1.0}$ and $t^{-2.0}$. For both the deep- and shallow-water cases, the horizontal variance increases as t^2 for the main period of diffusion. Based on these data, representative values for the shears and eddy diffusivities in the offshore region of the Cape Kennedy area are

	Winter	Summer
$K_x \simeq K_y$	4×10^3 cm²/sec	4×10^3 cm²/sec
K_z	19 cm²/sec	1.3 cm²/sec
Ω_z	1.8×10^{-3}/sec	6.6×10^{-3}/sec
Ω_y	$<10^{-4}$/sec	$<10^{-4}$/sec
ϕ_3^{-1}	4 hr	8 hr

As may be seen, there is an interesting resemblance between the shear-diffusion solutions and the P-class solutions for the radially symmetric distribution. Both models give the same rate for the horizontal variance; $\sigma_{rc}{}^2 \propto t^2$. As a matter of fact, a combined parameter, say $\Omega_z^{1/2}(K_y K_z)^{1/4}$, which appears in the expression of $\sigma_{rc}{}^2$ in the three-dimensional shear-diffusion solution has the dimensions of velocity. Using the shears and eddy diffusivities estimated for Cape Kennedy, we obtain a value of the combined parameter of the order of 1 cm/sec which is comparable to the value of the diffusion velocity estimated for the radially symmetrical solutions. In other words, the diffusion velocity may be estimated from environmental factors such as the shears. However, the shear-diffusion

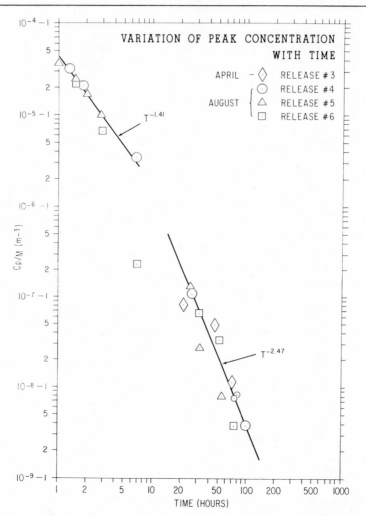

Figure 4.13 Variation of max concentration with time for instantaneous dye-release experiments in the sea off Cape Kennedy.

model describes the physical process of mixing more explicitly than does the radially symmetrical model.

The shear-diffusion model as developed here views the spectrum of oceanic turbulence as separable into two major parts: the large-scale eddies which appear as shears and the small-scale eddies responsible for eddy diffusion. The spectrum of oceanic turbulence, however, actually contains a wide variety of eddies and is not easily separable in this way. Thus as a patch ages and spreads, the line of demarcation between the part of the turbulent spectrum properly assignable to shears (advective processes) and the part assignable to eddy diffusion (diffusive processes) tends toward smaller wave numbers and frequencies. Even if the spectrum presents no natural separation, we may assume that the concept of shear diffusion is still valid locally in time and that the eddies responsible for shear diffusion lie in the inertial subrange characterized by ϵ. In these circumstances we can derive $\sigma_{rc}^2 \sim \epsilon t^3$ (Okubo, 1968a), which is identical with the result for γ-class solutions.

Both the concentration variance and the general shape of a patch can be affected by eddies from many parts of the spectrum according to our shear-diffusion mechanism. The unidirectional shear assumed in the simple shear-diffusion models will no longer serve us well. A patch of contaminant will not become so elongated as these solutions predict because the eddies in the spectrum assigned to the shear will have many variations in direction and magnitude with both space and time. Thus the shears they produce are similarly variable and the net effect on the patch less highly directional. One may anticipate a long-time patch length $\frac{1}{3}$ to $\frac{1}{10}$ of the length predicted by the unidirectional shear-diffusion models if there is a broad continuous spectrum of eddies. At small scales, say patches less than 5 km in size or diffusion times less than 2 days, the simple shear-diffusion solutions here presented generally provide fairly good information on the shape of patch as well as on its rate of diffusion.

A number of models which attempt a higher contact with the real world than the simple shear-diffusion models provide have been constructed. An advection-diffusion model including the rate of deformation and the vorticity was created by Okubo (1966). The solutions which result show different characteristics according as the rate of deformation or the vorticity is dominant. A model incorporating the effect on horizontal mixing of shear in an oscillatory current has been studied by Bowden (1965b) and Okubo (1967b). It was shown that the shear effect due to a periodic flow gives rise asymptotically to a constant effective diffusivity in the longitudinal direction even though the regime be unbounded. In a bounded region, a difference in the nature of the solutions results from the relative sizes of the time scales of diffusion with shear direction, τ, and of the oscillation, T. If $\tau/T \gg 1$, the shear effect of the oscillation may be neglected in comparison with the shear effect of the unidirectional residual flow. However, if $\tau/T \leq 1$, both effects must be retained. The effect of Ekman current shear on horizontal mixing has been studied by Bowden (1965b) and more extensively by Csanady (1969).

The concept of the advection diffusion can also be applied to the horizontal dispersion of floatable particles. In this case a collecting power of convergence in the velocity field suppresses the growth of a batch of floatables. A steady-state situation may eventually be established in which a group of floatable particles attains a certain limiting size. The theoretical aspect of this problem has been discussed by Okubo (1970).

Continuous Release from a Fixed Source

The solution for an instantaneous source is not necessarily fundamental to the construction of models for a continuous source. This has already been pointed out in the section on horizontal and vertical mixing from specified sources. We could deal directly with a continuous release by the use of the diffusion (19) under appropriate boundary conditions. Brooks (1960), using a simplified version of (19), successfully employed this approach. He took the lateral (cross-stream) eddy diffusivity to be a function only of the longitudinal (streamline) coordinate. His solutions have been applied to ocean outfall problems with encouraging results (Yudelson, 1967). However, the proper specification of $\bar{\mathbf{U}}$ and K_i for the general fixed source-continuous release problems usually leads to irrelevant and intractable mathematical difficulties so that it is of considerable practical interest to see what can be constructed from instantaneous solutions.

Let us consider "plume diffusion models" such as that of Frenkiel (1953). They have been widely used in practical applications. The ideal plume model is assembled by superposing an infinite number of overlapping instantaneous patches as elements, each patch being released sequentially from a fixed origin, and each being translated by the mean velocity $U_0(t)$. Such a model is illustrated in Fig. 4.14a, b. The concentration of contaminant from a continuous source may be formulated from elementary instantaneous releases as

$$\bar{C}(t, x, y, z) = \int_0^t q(t - t')\bar{S}_i\left(t', x - \int_0^{t'} U_0(t'')\,dt'', y, z\right) dt' \quad (37)$$

where $\bar{C}(t, x, y, z)$ denotes the concentration of contaminant at a point (x, y, z) at a time t elapsed since the start of release, $q(t)$ is the rate of release at the source, and $\bar{S}_i$ represents a concentration distribution for an instantaneous release of a unit amount of contaminant.

Any solution for an instantaneous release may be used for S_i but in only a few cases can one integrate (37) analytically. For example, if S_i is a two-dimensional radially symmetrical P-class solution, (37) can be integrated in closed form. The solution is a plume in which the concentra-

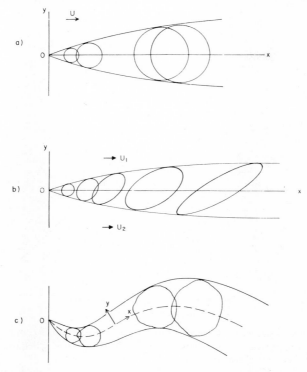

Figure 4.14 Schematic plan views of model plumes: a) ideal plume in a uniform flow; b) ideal plume in a shear flow; c) menadering single plume.

tion maximum tends asymptotically ($t \to \infty$) to an inverse variation with the distance from the source (Hecht, 1964). In contrast, if S_i is the shear-diffusion solution 35, (37) can not be solved analytically. For this latter case, Okubo and Karweit (1969) obtained a numerical solution which showed that (a) the plume exhibits a pronounced asymmetry with respect to the streamwise axis, x, (b) the cross-stream distribution is skewed to the side where the flow velocity is smaller, (c) as a result of (a) and (b) the location of the maximum concentration shifts to the side where the flow velocity is greater, (d) the maximum concentration decreases as the reciprocal of the distance both near the source and at great distances from it, (e) at intermediate distances from the source the maximum concentration shows a slight hump, and (f) the overall rate of decrease in the maximum concentration varies very nearly as the reciprocal of the distance from the source. Many dye-tracer, diffusion experi-

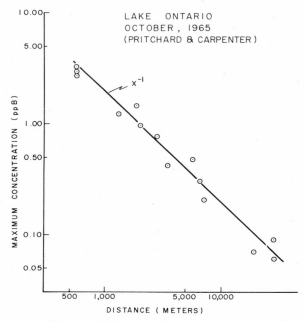

Figure 4.15 Maximum concentration versus distance from a continuous source. The data were provided by D. W. Pritchard and J. H. Carpenter, Chesapeake Bay Institute, The Johns Hopkins University.

ments in seas and lakes agree with the inverse-distance law, for example Fig. 4.15.

Figure 4.14 shows some characteristic plumes created by superposition. The individual elementary patches spread and deform and their centers of mass meander in reality. The meandering is the result of eddies large in comparison with the patch sizes. Complicated plume models can be built in this way but none are so complex as a real plume. Still, the model plume can be a useful approximation to a real plume and the application is facilitated if one uses patch-attached time-varying axes (Fig. 4.14c).

If instead of tracing the development of a single ideal plume the ensemble average taken over many realizations of an ideal plume is considered, the ensemble mean concentration at any fixed point will be found to be smaller. According to Gifford (1959), the ensemble mean concentration, $\bar{C}_m$, can be expressed as

$$\bar{C}_m(t, x, y, z) = \int_{-\infty}^{\infty} \bar{C}(t, x, y - y_0, z) g(x, y_0) \, dy_0 \tag{38}$$

where $g(x, y_0)$ denotes the frequency function associated with the meandering of the center line of the ensemble of plumes about the x axis, that is, the direction of the overall mean velocity, $\bar{U}_m$. At present, our knowledge of the g-function is limited. If it is assumed Gaussian $(0, \sigma_g(x))$, then for any physically appropriate solution, $\bar{C}$, integral (38) exists although it may not be possible to express it in a closed analytic form. Whether $g(x, y_0)$ is Guassian or not, Csanady (1963) demonstrated experimentally that its standard deviation is proportional to the distance from the source.

Evolving plumes which do not yet approximate their terminal states are of the most urgent practical interest. In a continuous discharge of contaminant the leading edge of the plume advances as a well-defined front at a velocity nearly that of the mean water velocity. While this front is well defined it is not inordinately sharp since longitudinal diffusion softens it. In no case will an asymptotic solution describe the region of the advancing contaminant front. At only a little distance ahead of the front no detectable amount of contaminant is to be found. Interpretation of data from plume experiments by applying asymptotic solutions can be extremely misleading if these facts are not remembered. (For further comment see Carter and Okubo, 1966.) Evolving plumes from continuous releases in estuaries and coastal waters show additional features. The oscillating tidal currents whip the growing plume back and forth about the source so that the front of the plume advances only to the limit of the tidal excursion after which the nontidal current modifies the contaminant motion.

The Effects of Horizontal and Vertical Mixing on the Distributions of Properties in the Ocean

The study of horizontal and vertical mixing in the sea has as its goal the explanation or interpretation of the distributions of oceanographic parameters, such as temperature and salinity, in terms of the velocity field together with initial and boundary conditions. As long as a complete knowledge of oceanic motions is lacking, the understanding of mixing processes in the sea must remain incomplete and, at best, approximate.

Although some progress has been made toward a proper understanding of the horizontal mixing process from an instantaneous source, our knowledge is still primitive. The principal difficulties of all problems of mixing in turbulent media are statistical. Even though the instantaneous equation, (16), for the conservation of a dynamically passive contaminant is essentially linear, the introduction of averaging inevitably produces nonlinearity. The closure problem which arises when the equations for turbulent *motion* are averaged arises in the same way in turbulent *diffusion*. Thus, the equation for the mean concentration of a substance contains the turbulent flux tensors which can not be strictly determined from a knowledge of the mean concentration field alone.

The semi-empirical concept of eddy diffusivity, although serviceable, will never lead to a full explanation of mixing processes in the sea. Any empirical theory can only hope for an explanation of the distributions of oceanic variables to the extent that the estimates of the eddy diffusivities are valid. Once eddy diffusivities are established for a few oceanic parameters they can be used to predict or describe the gross pattern of oceanic mixing of other parameters including man-made contaminants such as radioactivity. This extrapolation is useful only if the mean velocity field, the external influences, and the nonconservative processes occurring internally are known. The path to an understanding of oceanic distributions offered by this approach has not yet been as widely exploited as would seem desirable. In its complete form (19) is too complicated to be treated analytically. Perhaps a powerful computer will relieve the difficulty. Numerical models developed for atmospheric diffusion, for example, Davidson et al. (1966), and for oceanic circulation, for example, Bryan and Cox (1967), have proved encouraging. In this connection, the assault by Riley (1951) on the distributions of oxygen, phosphate, and nitrate in the Atlantic Ocean on the basis of a finite difference form for the complete diffusion equation without the aid of the high-speed digital computer is remarkable and instructive.

In many cases, a preliminary examination of a problem enables us to simplify the governing equation enough to make it analytically tractable while still retaining the main features of the oceanic mixing. A large number of approximate solutions have been found by this method. There are often, in fact, several different solutions to the same problem since there is no routine method for simplification of the basic equation. It is a matter of art rather than of mechanics. One method—the method of scaling on the equation—will often give clues which determine the terms in the equation which may be expected to represent the dominant processes. For example, in horizontal mixing from a continuous source placed in a substantial current the longitudinal diffusive process may usually be ignored in comparison with the lateral process. However, a similar judgment between the lateral and vertical diffusive processes can not, in general, be made a priori. In some circumstances they may be equally important. For success with this method experience, intuition, and luck are essential.

Yet another approach, the "reservoir" or "box" model, is a formulation of the diffusion equation as a difference equation. It is widely used, particularly among chemical oceanographers and geochemists, for studies of chemical tracers.

Steady-State Distributions of Oceanic Parameters Produced by the Vertical Mixing Process

The Pacific Ocean, if one excludes the regions substantially affected by its boundaries, is characterized by water so horizontally uniform in its properties that the advective and diffusive processes in the horizontal direction may be ignored in comparison with those in the vertical direction. This fact makes possible a simple model of vertical mixing which fits the observed distributions of oceanic variables rather better than one would expect (Munk, 1966).

For steady-state, (19) reduces to

$$\bar{W}\frac{d\bar{S}}{dz} = \frac{d}{dz}\left(K_z \frac{d\bar{S}}{dz}\right) + F \quad (39)$$

where z is positive upward. For any conservative property, $F \equiv 0$ and the symbol $\bar{S}$ may be interpreted physically as you please. For example, you may make $\bar{S}$ the mean salinity in one instance and $\bar{S}$ the mean temperature in another. Since the equation is formally the same in each case, the solution for the distribution of mean salinity with depth will be the same as the solution for the distribution of mean temperature with depth except for scale factors and additive constants. This holds for any reasonable functional forms for $\bar{W}$ and K_z. Thus the depth distributions of mean conservative properties must be linearly related, for example, (mean salinity)$_z = A + B$(mean temperature)$_z$, where A and B are constants.

For our exploration of (39) we do not want to restrict ourselves to conservative properties for which $F \equiv 0$. As a first step, we assume that $\bar{W}$ and K_z are constant with respect to the vertical coordinate. Equation 39 then becomes

$$\frac{d^2\bar{S}}{dz^2} - \frac{\bar{W}}{K_z}\frac{d\bar{S}}{dz} = -\frac{F}{K_z} \quad (40)$$

If the property $\bar{S}$ is conservative, the resulting equation from (40) can be solved analytically. The distribution of $\bar{S}$ is exponential. Munk (1966) found that distributions of potential temperature and salinity in the

Mindanao Deep and off California are consistent with (40) with $F \equiv 0$ when $\bar{W}/K_z = 1.3 \times 10^{-5}$/cm and 1.1×10^{-5}/cm, respectively.

In some cases information about nonconservative properties can be derived in explicit form from (40). Suppose that some function is specified for F, the production or destruction of the nonconservative property. As an example, for radioactivity the customary decay function $F = -\lambda \bar{S}$, where λ is the decay constant. Equation 40 then becomes

$$\frac{d^2\bar{S}}{dz^2} - \frac{\bar{W}}{K_z}\frac{d\bar{S}}{dz} = \frac{\lambda}{K_z}\bar{S} \qquad (41)$$

The observed vertical distributions of radiocarbon in the Pacific are consistent with the solutions of (41) if $\bar{W} = 1.4 \times 10^{-5}$ cm/sec (1.2 cm/day) and $K_z = 1.3$ cm²/sec (Munk, 1966). This numerical value for K_z suggests that some appreciable turbulence may exist deep within the interior of the Pacific although the value provides no insight into the processes responsible for vertical mixing there. Perhaps the shears associated with internal waves are a sufficient source for the turbulent energy required for such a turbulent diffusion.

Estimates of the vertical velocity with regard to the steady-state oceanic thermocline circulation have been made theoretically by Robinson and Stommel (1959), by Robinson and Welander (1963), and by others. These studies showed, in agreement with Munk (1966), that a vertical diffusivity of about 0.7 cm²/sec and an upward mean velocity of about 10^{-5} cm/sec led to a realistic temperature profile at mid-latitudes.

In the Antarctic, 2.1×10^{19} g of pack ice forms annually. Munk (1966) hypothesizes that this ice formation is associated with bottom water formation in the ratio of 43 to 1. An estimate of 4×10^{20} g of Pacific bottom water formed each year follows. If this were so, Antarctic ice formation alone would account for about two-thirds of the entire annual vertical mass flux in the Pacific excluding its marginal seas estimated on the basis of $\bar{W} = 1.2$ cm/day.

Wyrtki (1962), using the same model of vertical advection and diffusion, (40), together with an oxygen consumption function, $F(z)$, calculated curves of vertical distribution of oxygen in the oxygen minimum layer which compared well with observations.

Overstreet and Rattray (1969) use (39) with $F \equiv 0$ to obtain steady-state solutions of the vertical heat balance for various assumed vertical profiles of $\bar{W}$ and K_z. The solutions are applied to discuss the roles of $\bar{W}$ and K_z in maintaining a thermocline. Relations have been found for the depth and thickness of the thermocline in terms of parameters that characterize the vertical velocity and eddy conductivity. Two of their solutions agree with observations in regions having respectively divergent and convergent Ekman mass transport.

Steady-State Distributions of Properties of Seawater Produced by Horizontal Advection and Vertical Diffusion

Suppose that a mean current flows everywhere parallel to the x axis, and that the horizontal diffusion is negligible. For the steady-state distribution of a conservative property, (19) then simplifies to

$$\bar{U}\frac{\partial \bar{S}}{\partial x} = \frac{\partial}{\partial z}\left(K_z \frac{\partial \bar{S}}{\partial z}\right) \qquad (42)$$

where $\bar{U}$ and K_z depend in general on depth. If, further, we assume that K_z is constant with depth, (42) becomes

$$\bar{U}\frac{\partial \bar{S}}{\partial x} = K_z \frac{\partial^2 \bar{S}}{\partial z^2} \qquad (43)$$

Equations 42 and 43 are often used to evaluate K_z from the current profile and the spatial distribution of property $\bar{S}$ (see Chapter VII of Proudman, 1953).

Model 43 can be applied to the mixing of any layer of water sandwiched between upper and lower water layers of distinct characters. An interesting feature of solutions to mixing problems of this sort is the appearance of tongue-like distributions of properties in a vertical plane. Suppose that at $x = 0$ the intermediate layer extends from $z = -h$ to $z = +h$ and the flow is uniform everywhere so that $\bar{U} = U$ (constant). Further, suppose that the boundary values of $\bar{S}$ at $x = 0$ are $\bar{S}(0, z) = S_1$, $(z > h)$ in the upper layer, $\bar{S}(0, z) = S_0$, $(h \geq z \geq -h)$ in the intermediate layer, and $\bar{S}(0, z) = S_2$, $(-h > z)$ in the lower layer. Under these conditions one may ask what vertical distribution to expect for $\bar{S}$ as one moves downstream from $x = 0$. In order to get a tongue-like distribution in x, z-space we must impose the additional restriction, $0 < (S_2 - S_0)/(S_1 - S_0) < \infty$. Equation 43 then has the solution

$$\bar{S} = \tfrac{1}{2}\left[S_1 + S_2 + (S_1 - S_0)\,\mathrm{erf}\left(\frac{z - h}{2K_z^{1/2}U^{-1/2}x^{1/2}}\right) \right. \\ \left. + (S_0 - S_1)\,\mathrm{erf}\left(\frac{z + h}{2K_z^{1/2}U^{-1/2}x^{1/2}}\right)\right]$$

An extremum of $\bar{S}$ is to be found at z_m where

$$z_m = \frac{K_z x}{Uh}\ln\frac{S_2 - S_0}{S_1 - S_0} \qquad (44)$$

From 44 it is evident that the depth at which the extremum of $\bar{S}$ is found depends on the ratio $(S_2 - S_0)/(S_1 - S_0)$ and need not be at $z = 0$. The extremum may be displaced either upward or downward according to the relative magnitude of S_0, S_1, and S_2. It may even lie outside the intermediate layer and the axis of the tongue need not lie parallel to the flow direction of U. If the extremum is a maximum, $z_m(x)$ increases with x if $S_1 > S_2$ and decreases if $S_1 < S_2$ and conversely if it is a minimum. If and only if $S_1 = S_2$ is $z_m(x)$ a straight line parallel to the flow direction.

Long ago, Thorade (1931) derived a similar result by assuming S_1 and S_2 constant along the bounding x, z planes at $z = \pm h$. Thorade's model seems somewhat more artificial than ours since it is difficult to conceive physical processes capable of maintaining the required constancy of S_1 and S_2 on the horizontal boundaries within the sea.

Ivanov (1965) replaced the constant concentrations along the boundaries by prescribed flux functions for $-K_z(\partial \bar{S}/\partial z)\big|_{z=h}$ and $-K_z(\partial \bar{S}/\partial z)\big|_{z=-h}$. He also required that the vertical profiles of $\bar{U}$ and $\bar{S}$ be similar at $x = 0$. His solution showed, in addition to a curving $z_m(x)$, a second extremum of opposite sense developing downstream. Ivanov also noted some modification in the curving $z_m(x)$ due to the presence of shear in the mean velocity, $\bar{U}(z)$. The effect of a uniform shear on the curving of the axis of tongue was discussed in the section on continuous release from a fixed source.

In view of these results, the "core" method of Wüst (1935) must be used with great care when drawing conclusions about currents from observed tongue-like distributions of other properties. Similar results can be applied to the case of horizontal tongues by replacing the vertical diffusion term in (42) and (43) by the lateral diffusion term.

An actual example of the vertical shifts of the position of extrema can be seen in the intermediate depths of the sea northeast of Japan, where the Subarctic Water carried by the cold Oyashio meets the warm Kuroshio, and a number of tongues of low-salinity develop between the upper and lower saline waters (Masuzawa, 1956a, b; Ichiye, 1956). The Subarctic Water is characterized by low temperature, low salinity, and high oxygen, while the Kuroshio water has high temperature, high salinity, and low oxygen. It is noteworthy that the depths at which the extrema of salinity, temperature, and oxygen are found do not necessarily coincide. In the vicinity of the coast of Japan, they often do to the extent that our measurements permit us to judge. As the entire water column moves, on the average, eastward away from the coast of Japan, all three extrema slowly weaken and the depths at which they lie increasingly diverge. For North Pacific intermediate water the oxygen minimum is found slightly above the salinity minimum (Reid, 1965) and the temperature minimum, when it is measurable, lies below. The weakening of the temperature minimum as the water moves eastward is so marked that it can usually be detected only near the coast of Japan. These observations can be interpreted on the basis of an advection-vertical diffusion model with different sets of boundary conditions for salt, temperature, and oxygen. An interpretation of this sort has suggested that North Pacific Intermediate Water may be formed in high latitudes by vertical mixing through the pycnocline (Reid, 1965).

Horizontal Mixing Along Isopycnals: Isentropic Analysis

The horizontal advection-vertical diffusion model with its tongue-like vertical distributions of properties has been developed on the assumption that horizontal diffusion is negligible. In some situations horizontal diffusion may play a significant role and any model which neglects it will be inadequate. The tongue-like distributions observed in the deep water of the South Atlantic Ocean may be the result of dominant horizontal advection and vertical diffusion, but if the horizontal eddy diffusivity is some 10^8 times the vertical eddy diffusivity, they may equally well be the result of horizontal and vertical diffusion with no advection (Sverdrup et al., 1942).

Surfaces of constant potential-density in the ocean can be considered as nearly equivalent to the isentropic surfaces in a dry atmosphere and, for that reason, σ_t-surfaces can be called "quasi-isentropic" surfaces. This implies that interchange or mixing of water masses along σ_t-surfaces brings about only small changes in the potential energy and entropy of the body of water (Sverdrup et al., 1942). Note that this does not mean that mixing is necessarily confined to the directions of the σ_t-surfaces.

One may expect situations in which the effects of horizontal and vertical diffusion are of the same order. However, remember that a vertical shear in the mean flow combined with vertical diffusion can produce an *effective* horizontal mixing. Thus a large horizontal mixing is not necessarily indicative of a horizontal diffusion of major importance. In any particular case, a very critical scrutiny of the measured distributions is required before any judgement can be attempted. Montgomery (1940) gives examples of oceanic mixing in which a clear choice between horizontal diffusion (along isopycnals) and vertical diffusion (across isopycnals) could be made if only we could be more certain about the kinematics.

Few studies of the fundamental physics of isentropic processes in the sea exist and few of these few attempt turbulence. This could be surprising since isentropic analysis and the study of mixing along σ_t-surfaces were introduced into oceanography long enough ago to have become classic (Rossby, 1936; Montgomery, 1938, 1940; Sverdrup, 1939) and descriptive oceanographers have frequently taken advantage of the idea. However, one must remember the difficulties. A satisfactory and complete solution of the problem of isentropic processes at sea requires, first, a full elucidation of the mixing problem on the basis of the complete dynamic equations, and second, the application of nonequilibrium thermodynamics. The task seems formidable and the promise of a successful outcome for a reasonable investment of effort slight.

Box-Model Solutions to Oceanic Mixing Problems

Box models offer a simple way of estimating first-order approximations for contaminant residence times. The diffusion equation, (19), is expressed for continuous functions while oceanographic data are usually discrete point-measurements. Thus the data can be used directly only in a finite difference analog of (19). Such an analog is equivalent to a "reservoir" or "box" model.

A box model is constructed by conceptually dividing the water into boxes within which one or more of the data values are located. The properties of the water are then estimated for each box from the data which are included within it. A box model treats mixing "averaged" over each box and attempts to see changes only as between boxes. It affords some clues on the amount of time that water spends in each box. A box model is the model of choice only when the precise mechanism of turbulent diffusion is beyond our knowledge and some sort of answer is urgent. For example, Craig (1957), Arnold and Anderson (1957), and Revelle and Suess (1957) used box models to discuss carbon dioxide mixing while Broecker (1963), Eriksson (1962), Bolin and Stommel (1961), and Keeling and Bolin (1967, 1968) used them for ocean circulations as well as mixing.

A paper by Keeling and Bolin (1967) contains an extensive exposition of the theoretical foundation of box models. Suppose that (16) is integrated over some fixed volume V (the box) and that Gauss's theorem is applied. Then

$$\frac{\partial \langle \bar{S} \rangle}{\partial t} + \frac{1}{V} \int_A (\bar{S}\bar{U}_j + \overline{S'u_j'})n_j \, dA = \langle F \rangle$$

where A is the surface area of the box on which the integral is to be evaluated. Box models require that the integral be expressed in terms of the volume average. The quantities representing advective and diffusive fluxes across the surface of any box are simply approximated by linear combinations of the volume averages. The end result is a set of linear simultaneous equations for the quantities $\langle \bar{S} \rangle$, one for each box. The mixing processes at the interfaces of the boxes are parametrically disguised as exchange- or transfer-rate constants with the dimensions $[T^{-1}]$. The exchange rate constants in the box model have thus no clearly defined physical basis, but, to be honest about it, neither do the eddy diffusivities.

The exchange-rate constants have been interpreted variously (Craig, 1957; Schönfeld and Groen, 1961; Keeling and Bolin, 1967). The box model customarily assumes that properties are well-mixed within each box. However, Keeling and Bolin (1967) have shown the assumption unnecessary for a successful application of a box model. They have also related the transfer-rate constants of reservoir models to the advective velocity and eddy diffusivities appropriate for continuous models.

How one is to set about dividing an ocean and an estuary into boxes is not always clear. What one selects depends heavily on ones intuition and what one already knows. For an example, in the section on environmental features of the sea we made a separation of the ocean into three layers chosen on the basis of the thermal structure. It was our hope that the water within each layer was somewhat more uniform in its properties than water from different layers would be. In estuaries sometimes the boxes can be made to correspond to tidal excursions, or perhaps the salinity will help us. Division into boxes needs not be restricted to the water. The atmosphere may be taken as a box filled with gases. Here the rate of exchange of the gas between the atmosphere and ocean plays a controlling role in the mixing of atmospheric gases into the sea. For some substances, exchange between sediments and sea, deposition and redissolution, may be important and in that case the ocean bottom should also be treated as a box.

The basic equation set for box models may be expressed in the following way. Let V_i be the volume of box i and $\bar{q}_i$ be the mean concentration of a property within box i. Let $k_{i \to j}$ denote the exchange-rate constants for transfer of the property from box i to box j. Further, let $\bar{D}_i$ and $\bar{P}_i$ represent the mean rate of decay or consumption of the property and the mean rate of production of the property within box i, respectively. As an example, a radioactive substance has $\bar{D}_i = \lambda V_i \bar{q}_i$, where λ is the radioactive decay constant. The system of equations for box models is then given by

$$\frac{dV_i \bar{q}_i}{dt} = \sum_{j=1}^{N} (k_{j \to i} V_j \bar{q}_j - k_{i \to j} V_i \bar{q}_i) - \bar{D}_i + \bar{P}_i \qquad i = 1, 2, \ldots, N \qquad (45)$$

where N is the total number of boxes and $k_{i \to i}$ is assumed to be identically zero. One use for box models is to estimate the values of $k_{i \to j}$ from a steady-state distribution of $\bar{q}_i$ with known $\bar{D}_i$ and $\bar{P}_i$ functions. With the $k_{i \to j}$ now known we can use (45) to predict the nonsteady distribution of the property, $\bar{q}$, for which the steady state $k_{i \to j}$ were found. Using the $k_{i \to j}$ for a particular $\bar{q}$ we may even extrapolate to the distributions of other oceanic parameters. A detailed discussion of this approach is provided by Dorrestein (1960) for estuarine box models.

Often one needs distributions of two or more properties in order to estimate $k_{i \to j}$. In this case, however, one must watch out for a pitfall hidden in the numerical solutions of (45) for two or more properties. Some of the equations may be so nearly identical that the solutions for $k_{i \to j}$ become highly sensitive to small changes in the mean values, $\bar{q}_i$, $\bar{D}_i$, and $\bar{P}_i$—changes so small as to be entirely acceptable when we consider the accuracy to which we can reasonably estimate the means. In other words, the value of the determinant made of the coefficients of $k_{i \to j}$ is nearly zero. A check must be made as suggested by Bolin and Stommel (1961).

From among the many box models which have been constructed for oceanic mixing, we shall choose two as representative.

A "Cyclic Box Model." Craig (1958, 1963) divided the exchange reservoirs into an atmosphere box ($i = 1$), a mixed-layer box ($i = 2$), and a deep-sea box ($i = 3$). The division is a stab at a model for the globe in which box 2 represents the well-mixed surface layer which lies in middle or low latitudes but disappears at high latitudes where the atmosphere, box 1, and the "deep sea," box 3, are in direct contact across the navíface (Fig. 4.16). Craig took $k_{1 \to 2} = 3k_{1 \to 3}$ under the assumption that 75 per cent of the surface area of the sea is covered by the mixed-layer box and atmospheric exchange rates per unit navifacial area are equal everywhere.*

Using the amounts of stable and radioactive carbons known to exist in these three reservoirs, Craig calculated the residence times of CO_2 in the three reservoirs. The results showed that the residence time of CO_2 in the atmosphere is 15 ± 5 yr, in the mixed layer is 21 ± 9 yr, and in the deep sea is 2400 ± 1300 yr. Craig pointed out that the numbers one secures for residence times from a box model may change somewhat with the way in which the boxes are chosen but that the changes will not be inordinately large.

---

* Craig (1963) reports that Broecker privately criticized his equal exchange-rate assumption as unrealistic. The assumption is unattractive but, in the absence of information, it is hard to see how to do better.

THE EFFECTS OF HORIZONTAL AND VERTICAL MIXING 157

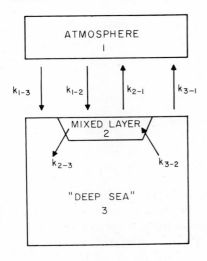

Figure 4.16 A cyclic box model.

A more sophisticated cyclic box-model of the CO_2 exchange (Broecker et al., 1960) found an atmospheric residence time of about 7 yr, a mixed layer residence time from 10 to 30 yr, and a "deep sea" residence time of the order of 1000 yr. Another cyclic model with one atmospheric and three oceanic boxes has recently been investigated by Keeling and Bolin (1968) for application to mixing in the North and South Pacific Oceans.

A Box Model for the Origin and Rate of Circulation of Deep Ocean Water. If one has a comparatively few carbon-14 measurements in addition to the distributions of temperature and salinity, a box model can yield information about the deep ocean circulation which is much more comprehensive than can be derived from a box model only for the distributions of any tracers, introduced or natural. For example, Bolin and Stommel (1961) using such a box model, discussed that origin of the Common Water, the formation of Antarctic Bottom Water, and the rate of flow of Intermediate Water towards the north in the Atlantic Ocean.

The most extensive water mass in the world is the deep water of the Indian and Pacific Oceans which Montgomery (1958) named "Common Water." Although Common Water maintains its identity there must be some mixing with immediately overlying waters and some geothermal heating from the ocean bottom. We compute the rate of flow of the Common Water using box models.

Bolin and Stommel (1961) divide the oceans into four boxes: box 1 ∼ North Atlantic Deep Water, box 2 ∼ Antarctic Bottom Water, box 3 ∼ Pacific and Indian Intermediate Water, and box 4 ∼ Common Water.

Let $k_{i \to j} V_i \equiv E_{i \to j}$ be the volume flux from box i to box j, and $E_{4 \to T}$ be the total volume flux out of box 4 into other boxes.

A set of steady-state equations for mass, heat, salt, and radiocarbon for box 4 can be written as

$$E_{1 \to 4} + E_{2 \to 4} + E_{3 \to 4} - E_{4 \to T} = 0$$

$$E_{1 \to 4}\theta_1 + E_{2 \to 4}\theta_2 + E_{3 \to 4}\theta_3 - E_{4 \to T}\theta_4 = -HV_4$$

$$E_{1 \to 4}\gamma_1 + E_{2 \to 4}\gamma_2 + E_{3 \to 4}\gamma_3 - E_{4 \to T}\gamma_4 = 0$$

$$E_{1 \to 4}C_1^* + E_{2 \to 4}C_2^* + E_{3 \to 4}C_3^* - E_{4 \to T}C_4^* = \lambda^* V_4 C_4^*$$

where θ_i, γ_i, and C_i^* denote the temperature, salinity, and the concentration of radiocarbon in box i, respectively. H represents the rate of gain of heat due to geothermal flux through the bottom of the oceans.

Using numerical estimates of θ_i, γ_i, HV_4, λ^*, and C_i^* they obtain $E_{1 \to 4} = 3.6$ Sv, $E_{2 \to 4} = 9.6$ Sv, $E_{3 \to 4} = 2.6$ Sv, and $E_{4 \to T} = 16.2$ Sv (Sv $\sim$ Sverdrup: 1 Sv $\equiv 10^6$ m^3/sec). The residence time in the Common Water is $V_4/E_{4 \to T} = 1200$ yr. Since $E_{1 \to 4}$ is the greatest of the $E_{. \to 4}$ quantities, it is apparent that the main source of the Common Water is the Antarctic Bottom Water.

Bolin and Strommel also computed the flow of Antarctic Intermediate Water using another box model: box 1 $\sim$ Antarctic Surface Water, box 2 $\sim$ Atlantic Surface Water, box 3 $\sim$ Atlantic Deep Water, and box 4 $\sim$ Atlantic Intermediate Water. They found $E_{1 \to 4} = 3.6$ Sv, $E_{2 \to 4} = 1.4$ Sv, $E_{3 \to 4} = 0.4$ Sv, and $E_{4 \to T} = 5.4$ Sv. The residence time for Intermediate Water is about 250 yr. By assuming that the water during this time moves northwards uniformly, they estimated the mean velocity at about 0.05 cm/sec.

Final Remarks

This chapter has attempted to describe our progress in the field of oceanic mixing and to indicate some of the areas in need of further investigation. The reader should be aware that space limitations have forced the author to omit many aspects of horizontal and vertical mixing. One important lacuna arises from our concentration on the mixing of dynamically passive contaminants. Such contaminants permit us to use the relatively simple kinematic point of view. However, this gain in facility of analysis has been paid for by the weakening of our contact with the real world where dynamically active contaminants introduced into the ocean urgently demand our deeper understanding. For example, sewage, which ordinarily has a density near that of freshwater, will hardly be dynamically passive when discharged into seawater and buoyancy will have to be included in our problem. When an effluent is discharged from a pipe lying on the sea bed it has been found that the initial dilution near the pipe is dominated by buoyant-jet mixing. Ocean engineers have recently been concentrating much of their attention on problems of this sort, for example, Wiegel (1964) and Fan (1967). As another example in which buoyancy is important, consider the discharge of heated water from power plants. Power plants, both conventional and nuclear, withdraw water from estuarine or coastal waters, use it as a coolant, and return it to the natural environment hotter but otherwise unaltered. A plume from a heated discharge has a tendency to spread over the environmental water and the excess temperature of the plume will be lost not only by mixing with the underlying water, but also through exchanges with the atmosphere across the water surface. Some progress has been made in the prediction of the distribution of excess temperature in such a plume (Pritchard and Carter, 1965; Harleman et al., 1968). The problem is urgent since society's need for power must be intelligently balanced against the damage inflicted on delicately balanced ecological systems. If we do not understand the mechanisms of heated discharges well enough to predict their behavior with reasonable certainty we will be in for some most unpleasant, and possibly irremediable, surprises.

Since oil is lighter than water and immiscible with it, the dispersion of oil spills and pollution from runaway wells is primarily a pseudo two-dimensional problem restricted to the naviface (for exception, see Forrester, 1970). Little can be said with certainty. The oil dispersion pattern appears to be more patchy than the patterns for soluble contaminants. Possibly wind and wave actions will prove to be as important as turbulence and currents. Surface tension can play an important role in the behavior of thin surface films of oil and pollutants. We do not know the detailed mechanism. However, it should be obvious that with this problem it is one minute to midnight.

I close this chapter by saying *diffusion is confusion*. No one but Maxwell's Demon really knows what's going on.

Acknowledgment

I wish to express my sincere appreciation to Dr. Blair Kinsman who spent many many hours reviewing most of the preliminary draft, improving the manuscript, and giving valuable comments and suggestions.

This work was supported partially by the Office of Naval Research under Contract No. Nonr 4010(11) and partially by the U.S. Atomic Energy Commission under Contract No. AT(30-1)-3109.

References

Arnold, J. R., and E. C. Anderson. 1957. The distribution of carbon-14 in nature. *Tellus*, **9**:28–32.

Batchelor, G. K. 1952. Diffusion in a field of homogeneous turbulence. II. The relative motion of particles. *Proc. Cambridge Phil. Soc.*, **48**:345–362.

Batchelor, G. K. 1956. *The theory of homogeneous turbulence*, Cambridge University Press, 197 pp.

Batchelor, G. K. 1959. Small-scale variation of convective quantities like temperature in turbulent field. Part I. General discussion and the case of small conductivity. *J. Fluid Mech.*, **5**:113–133.

Batchelor, G. K. 1967. *An introduction to fluid dynamics*. Cambridge University Press, 615 pp.

Bolin, B., and H. Stommel. 1961. On the abyssal circulation of the world ocean-IV. Origin and rate of circulation of deep water as determined with the aid of tracers. *Deep-Sea Res.*, **8**:95–110.

Bowden, K. F. 1964. Turbulence. *Oceanogr. Mar. Biol. Ann. Rev.*, **2**:11–30.

Bowden, K. F. 1965a. Currents and mixing in the sea. In *Chemical Oceanography*, Academic Press, London, Vol. 1, Chap. 2, pp. 43–72.

Bowden, K. F. 1965b. Horizontal mixing in the sea due to a shearing current. *J. Fluid Mech.*, **21**:83–95.

Bowden, K. F. 1970. Turbulence II. *Oceanogr. Mar. Biol. Ann. Rev.*, **8**:11–32.

Bowles, P., R. H. Burns, F. Hudswell, and R. T. P. Whipple. 1958. Sea disposal of low activity effluent. *Proc. Conf. Peaceful Uses At. Energy, 2nd, Geneva*, **18**:376–389.

Broecker, W. S., R. Gerard, W. M. Ewing, and B. C. Heezen. 1960. Radiocarbon in the Atlantic Ocean. *J. Geophys. Res.*, **65**:2903–2931.

Broecker, W. S. 1963. Radioisotopes and large-scale oceanic mixing. In *The sea*. Interscience Publishers, New York, Vol. 2, pp. 88–108.

Brooks, N. H. 1960. Diffusion of sewage effluent in an ocean-current. In *Waste disposal in the marine environment*, Pergamon Press, New York, pp. 246–267.

Bryan, K., and M. D. Cox. 1967. A numerical investigation of the oceanic general circulation. *Tellus*, **19**:54–80.

Calder, K. L. 1965. On the equation of atmospheric diffusion. *Quart. J. Roy. Met. Soc.*, **91**:514–517.

Carslaw, H. S., and J. C. Jaeger. 1959. *Conduction of heat in solids*, Oxford University Press, London, 510 pp.

Carter, H. H., and A. Okubo. 1965. A study of the physical processes of movement and dispersion in the Cape Kennedy area. Final Report under the U.S. Atomic Energy Commission, Chesapeake Bay Institute, The Johns Hopkins University. Rept. No. NYO-2973-1.

Carter, H. H., and A. Okubo. 1966. Comments on paper by G. T. Csanady. Accelerated diffusion in the skewed shear flow of lake currents. *J. Geophys. Res.*, **71**:5012–5014.

Cooper, J. W., and H. Stommel. 1968. Regularly spaced steps in the main thermocline near Bermuda. *J. Geophys. Res.*, **73**:5849–5854.

Corrsin, S. 1953. Remarks on turbulent heat transfer. In *Proc. First Iowa Symp. Thermodyn.*, Iowa State University.

Corrsin, S. 1957. Some current problems in turbulent shear flow. Naval hydrodynamics. *Natl. Acad. Sci.—Natl. Res. Council, Publ.* **515**:373–400.

Corrsin, S. 1961a. Turbulent flow. *Amer. Scientist*, **49**:300–325.

Corrsin, S. 1961b. The reactant concentration spectrum in turbulent mixing with a first-order reaction. *J. Fluid. Mech.*, **11**:409–416.

Corrsin, S. 1962. Theories of turbulent dispersion. In *Mécanique de la Túrbulence*, Centre National de la Recherche Scientifique, Paris, pp. 27–52.

Craig, H. 1957. The natural distribution of radiocarbon and the exchange time of carbon dioxide between atmosphere and sea. *Tellus*, **9**:1–17.

Craig, H. 1958. A critical evaluation of radiocarbon techniques for determining mixing rates in the oceans and the atmosphere. *Proc. Second U.N. Int. Conf. on Peaceful Uses of Atomic Energy*, **18**:358–363.

Craig, H. 1963. The natural distribution of radiocarbon: mixing rates in the sea and residence times of carbon and water. In *Earth Science and Meteorites*, North-Holland Publishing Co., Amsterdam, 103–114.

Csanady, G. T. 1963. Turbulent diffusion in Lake Huron. *J. Fluid Mech.*, **17**:360–384.

Csanady, G. T. 1969. Diffusion in an Ekman layer. *J. Atmospheric Sci.*, **26**:414–426.

Davidson, B., J. P. Friend, and H. Seitz. 1966. Numerical models of diffusion and rainout of stratospheric radioactive materials. *Tellus*, **18**:301–315.

Day, C. G., and F. Webster. 1965. Some current measurements in the Sargasso Sea. *Deep-Sea Res.*, **12**:805–814.

Defant, A. 1961. *Physical oceanography*, Pergamon Press, New York, Vol. 1, 729 pp.

Dorrestein, R. 1960. A method of computing the spreading of matter in the water of an estuary. *Disposal of radioactive wastes, Intern. Atomic Energy Agency*, **2**:163–166.

Eckart, C. 1948. An analysis of the stirring and mixing processes in incompressible fluids. *J. Mar. Res.*, **7**:265–275.

Ellison, T. H. 1962. Laboratory measurements of turbulent diffusion in stratified flow. *J. Geophys. Res.*, **67**:3029–3031.

Eriksson, E. 1962. Oceanic mixing. *Deep-Sea Res.*, **9**:1–9.

Fischer, H. B. 1967. The mechanics of dispersion in natural streams. *J. Hydraulics Div., Proc. Amer. Soc. Civil Eng.*, **93**:187–216.

Fofonoff, N. P. 1967. Measurement of ocean currents. Selected papers from the Governor's Conference on Oceanography at the Rockefeller Univ. The State of New York and the New York State Science and Technology Foundation, 95–107.

Folsom, T. R., and A. C. Vine. 1957. On the tagging of water masses for the study of physical processes in the oceans. In *Natl. Acad. Sci.-Natl. Res. Council, Publ.* **551**:121–132.

Folsom, T. R. 1967. Persistence of radioactive contaminations in thin layers in the ocean (unpublished).

Forrester, W. D. 1970. Distribution of suspended oil particles following wreck of tanker "Arrow." Preprint for "The Ocean World," Tokyo, September 13–25, 1970.

Foxworthy, J. E., R. B. Tibby, and G. M. Barsom. 1966. Dispersion of a surface waste field in the sea. J. Water Pollution Control Federation, Washington, D.C., 1170–1193. See also Allan Hancock Foundation. 1965. An investigation of the fate of organic and inorganic wastes discharged into the marine environment and their effects on biological productivity. State Water Quality Control Board, Sacramento, California, Publ. **29**.

Frenkiel, F. N. 1953. Turbulent diffusion. *Advan. Appl. Mech.*, **3**:61–107.

Fuglister, F. C. 1967. Cyclonic eddies formed from meanders of the Gulf Stream. Paper presented at the Amer. Geophys. Union, April, 1967.

Grant, H. L., R. W. Stewart, and A. Moilliet. 1962. Turbulence spectra from a tidal channel. *J. Fluid Mech.*, **12**:241–268.

Grant, H. L., A. Moilliet, and W. M. Vogel. 1968a. Some observations of the occurrence of turbulence in and above the thermocline. *J. Fluid Mech.*, **34**:443–448.

Grant, H. L., B. A. Hughes, W. M. Vogel, and A. Moilliet. 1968b. The spectrum of temperature fluctuations in turbulent flow. *J. Fluid Mech.*, **34**:423–442.

Hamon, B. V. 1967. Medium-scale temperature and salinity structure in the upper 1500 m in the Indian Ocean. *Deep-Sea Res.*, **14**:169–181.

Harleman, D. R. F., L. C. Hall, and T. G. Curtis. 1968. Thermal diffusion of condenser water in a river during steady and unsteady flows with application to the T. V. A. Browns Ferry Nuclear Power Plant. Hydrodynamics Lab. Rept. No. 111, Department of Civil Engineering, Massachusetts Institute of Technology.

Gifford, F. 1959. Statistical properties of a fluctuating plume dispersion model. *Advan. Geophys.*, **6**:117–137.

Hecht, A. 1964. On the turbulent diffusion of the water of the Nile floods in the Mediterranean Sea. Bulletin No. 36, Sea Fisheries Res. Station, Dept. of Fisheries, Ministry of Agriculture, State of Israel.

Hinze, J. O. 1959. *Turbulence*, McGraw-Hill Book Co., New York, 586 pp.

Ichiye, T. 1956. On the properties of the isolated cold and fresh water along the edge of the Kuroshio. *Oceanogr. Mag.*, **8**:53–64.

Ichiye, T., and F. C. W. Olson. 1960. Über die neighbour diffusivity im Ozean. *Dt. Hydrogr. Z.*, **13**:13–23.

Ivanov, Yu. A. 1965. The role of boundary conditions and advection in the formation and depth distribution of extreme values of oceanological characteristics. *Oceanology*, **5**:27–31 (translation).

Joseph, J., and H. Sendner. 1958. Über die horizontale Diffusion im Meere. *Dt. Hydrogr. Z.*, **11**:49–77.

Joseph, J. and H. Sendner. 1962. On the spectrum of the mean diffusion velocity in the ocean. *J. Geophys. Res.*, **67**:3201–3205.

Joseph, J., H. Sendner, and H. Weidemann. 1964. Untersuchungen über die horizontale Diffusion im der Nordsee. *Dt. Hydrogr. Z.*, **17**:57–75.

Kamenkovich, V. M. 1967. On the coefficients of eddy diffusion and eddy viscosity in large-scale oceanic and atmospheric motions. *Izv. Atmospheric and Oceanic Physics*, **3**:777–781 (English edition).

Katz, B., R. Gerard, and M. Costin. 1965. Response of dye tracers to sea surface conditions. *J. Geophys. Res.*, **70**:5505–5513.

Keeling, C. D., and B. Bolin. 1967. The simultaneous use of chemical tracers in oceanic studies. I. General theory of reservoir models. *Tellus*, **19**:566–581.

Keeling, C. D., and B. Bolin. 1968. The simultaneous use of chemical tracers in oceanic studies. II. A three-reservoir model of the North and South Pacific Oceans. *Tellus*, **20**:17–54.

Kent, R. E., and D. W. Pritchard. 1959. A test of mixing length theories in a coastal plain estuary. *J. Mar. Res.*, **18**:62–72.

Kinsman, B. 1965. *Wind waves*, Prentice-Hall, Inc., Englewood Cliffs, N.J., 676 pp.

Kirwan, A. D. 1965. On the use of the Rayleigh-Ritz method for calculating the eddy diffusivity. *Proc. Symp. Diffusion in Oceans and Fresh Water*, September, 1964, Lamont Geological Observatory, Columbia University, New York, 86–92.

Knauss, J. A. 1956. An estimate of the effect of turbulence in the ocean on the propagation of sound. *J. Acoust. Soc. Amer.*, **28**:443–446.

Kolmogorov, A. N. 1941. The local structure of turbulence in incompressible viscous fluid for very large Reynolds' numbers. *Comptes Rend. Acad. Sci., U.S.S.R.*, **30**:301–305.

Kullenberg, G. 1970. Measurements of horizontal and vertical diffusion in coastal waters. *Kungl. Vetenskaps-och Vitterhets-Samhället, Göteborg, series Geophysica* **2**.

Lumley, J. L., and H. A. Panofsky. 1964. *The structure of atmospheric turbulence.* Interscience Publishers, New York, 239 pp.

MacEwan, G. F. 1950. A statistical model of instantaneous point and disc sources with application to oceanographic observations. *Trans. Amer. Geophys. Union*, **31**:35–46.

Mamayev, O. I. 1958. Influence of stratification on vertical turbulent mixing in the sea. *Izv. Akad. Nauk SSSR, geofiz.*, No. 7, 870–875.

Masuzawa, J. 1956a. A note on the Kuroshio farther to the east of Japan. *Oceanogr. Mag.*, **7**:97–104.

Masuzawa, J. 1956b. On the cold belt along the northern edge of the Kuroshio. *Oceanogr. Mag.*, **8**:151–156.

Monin, A. S. 1962. On the Lagrangian equations of the hydrodynamics of an incompressible viscous fluid. *J. Appl. Math. and Mech.*, **26**:458–468 (English edition).

Montgomery, R. B. 1938. Circulation in upper layers of southern North Atlantic deduced with use of isentropic analysis, *Papers in Physical Oceanography and Meteorology*, Massachusetts Inst. of Technol. and Woods Hole Oceanogr. Institution, Vol. 6, No. 2.

Montgomery, R. B. 1940. The present evidence on the importance of lateral mixing processes in the ocean. *Bull. Amer. Meteorol. Soc.*, **21**:87–94.

Montgomery, R. B. 1943. Generalization for cylinders of Prandtl's linear assumption for mixing length. *Ann. N.Y. Acad. Sci.*, **44**:89–103.

Montgomery, R. B. 1957. 2k. Oceanographic Data, *American Institute of Physics Handbook.* McGraw-Hill Book Co., New York, pp. 123–132.

Montgomery, R. B. 1958. Water characteristics of Atlantic Ocean and of world ocean. *Deep-Sea Res.*, **5**:134–148.

Montgomery, R. B. 1969. The words naviface and oxyty. *J. Mar. Res.*, **27**:161–162.

Munk, W. H., and E. R. Anderson. 1948. Note on a theory of the thermocline. *J. Mar. Res.*, **7**:276–295.

Munk, W. H. 1966. Abyssal recipes. *Deep-Sea Res.*, **13**:707–730.

Munk, W. H., and N. Phillips. 1968. Coherence and band structure of inertial motion in the sea. *Rev. Geophys.*, **6**:447–472.

Neumann, G. 1968. *Ocean current*, Elsevier Publishing Co., Amsterdam. 352 pp.

Okubo, A. 1962a. A review of theoretical models of turbulent diffusion in the sea. *J. Oceanogr. Soc. Japan, 20th Anniv. Vol.*, pp. 286–320.

Okubo, A. 1962b. Horizontal diffusion from an instantaneous point-source due to oceanic turbulence. *Chesapeake Bay Inst. Tech. Report* **32**.

Okubo, A. 1966. A note on horizontal diffusion from an instantaneous source in a nonuniform flow. *J. Oceanogr. Soc., Japan*, **22**:35–40.

Okubo, A. 1970. Horizontal dispersion of floatable particles in the vicinity of velocity singularities such as convergences. *Deep-Sea Res.*, **17**:445–454.

Okubo, A., and H. H. Carter. 1966. An extremely simplified model of the "shear effect" on horizontal mixing in a bounded sea. *J. Geophys. Res.*, **71**:5267–5270.

Okubo, A. 1967a. Study of turbulent dispersion by use of Lagrangian diffusion equation. *Phys. Fluids*, Suppl. 1967, S72–75.

Okubo, A. 1967b. The effect of shear in an oscillatory current on horizontal diffusion from an instantaneous source. *Int. J. Oceanol. Limnol.*, **1**:194–204.

Okubo, A. 1968a. Some remarks on the importance of the "shear effect" on horizontal diffusion. *J. Oceanogr. Soc., Japan*, **24**:60–69.

Okubo, A. 1968b. A new set of oceanic diffusion diagrams. *Chesapeake Bay Inst. Tech. Rept.* **38**.

Okubo, A., and M. J. Karweit. 1969. Diffusion from a continuous source in a uniform shear flow. *Limnol. Oceanog.*, **14**, 514–520.

Okubo, A., and R. V. Ozmidov. 1970. An empirical relationship between the horizontal eddy diffusivity in the ocean and the scale of the phenomenon. *Izv. Atmospheric and Oceanic Physics*, **6**:534–536.

Overstreet, R., and M. Rattray, Jr. 1969. On the roles of vertical velocity and eddy conductivity in maintaining a thermocline. *J. Mar. Res.*, **27**:172–190.

Ozmidov, R. V. 1958. On the calculation of horizontal turbulent diffusion of the pollutant patches in the sea. *Doklady Akad. Nauk, SSSR*, **120**:761–763.

Ozmidov, R. V. 1965. Energy distribution between oceanic motions of different scales. *Izv. Atmospheric and Oceanic Physics Series*, **1**:257–261 (English edition).

Ozmidov, R. V. 1968a. Horizontal turbulence and turbulent exchange in the ocean. Institute of Oceanography, Academy of Sciences of the USSR, Moskow. 196pp (in Russian).

Ozmidov, R. V. 1968b. The dependence of the horizontal turbulent exchange coefficient in the ocean on the scale of the phenomenon. *Izv. Atmospheric and Oceanic Physics*, **4**:703–704.

Parr, A. E. 1936. On the probable relationship between vertical stability and lateral mixing processes. *Conseil Perm. Intern. p. l'Expl. de la Mer, J. Cons.*, **11**:308–313.

Phillips, O. M. 1966. *The dynamics of the upper ocean.* Cambridge University Press, New York, 26 pp.

Pierson, W. J. 1962. Perturbation analysis of the Navier-Stokes equations in Lagrangian form with selected linear solutions. *J. Geophys. Res.*, **67**:3151–3160.

Pingree, R. D. 1969. Small scale structure of temperature and salinity near Station Covall. *Deep-Sea Res.*, **16**:275–295.

Pritchard, D. W. 1960. The movement and mixing of contaminants in tidal estuaries. *Proc. first int. conf. on waste disposal in the marine environment*, July 1959, University of California, Berkeley, Pergamon Press, New York, pp. 512–525.

Pritchard, D. W., and J. H. Carpenter. 1960. Measurements of turbulent diffusion in estuarine and inshore waters. *Bull. Int. Assoc. Sci. Hydrol.*, 20, 37–50.

Pritchard, D. W., and H. H. Carter. 1965. On the prediction of the distribution of excess temperature from a heated discharge in an estuary. *Chesapeake Bay Inst. Tech. Rept.* **33,** 45 pp.

Pritchard, D. W., A. Okubo, and H. H. Carter. 1966. Observations and theory of eddy movement and diffusion of an introduced tracer material in the surface layers of the sea, In *Disposal of Radioactive Wastes into Seas, Oceans, and Surface Waters.*, Int. Atomic Energy Agency, Vienna, pp. 397–424.

Pritchard, D. W., R. O. Reid, A. Okubo, and H. H. Carter. 1969a. Physical processes of movement and mixing, In *Radioactivity in the Marine Environment*, Natl. Acad. Sci.—Natl. Res. Council. In press.

Pritchard, D. W., W. J. Wiseman, and R. M. Crosby. 1969b. A three-dimensional current meter for estuarine applications. *IEEE Geosci. Electron. Symp.*, April, 1969, Washington, D.C. (abstract only).

Proudman, J. 1953. *Dynamical oceanography*, Methuen & Co., Ltd., London.

Reid, J. L. 1965. *Intermediate Waters of the Pacific Ocean*. The Johns Hopkins Oceanographic Studies No. 2, The Johns Hopkins Press, Baltimore, 85 pp.

Revelle, R., T. R. Folsom, E. D. Goldberg, and J. D. Isaacs. 1955. In *Nuclear Science and oceanography*. United Nations Int. Conf. on the Peaceful Uses of Atomic Energy, Geneva, **8**:paper 277.

Revelle, R. and H. E. Suess. 1957. Carbon dioxide exchange between atmosphere and ocean and the question of increase of atmospheric CO_2 during the past decades. *Tellus*, **9**:18–27.

Richardson, L. F. 1926. Atmospheric diffusion shown on a distance-neighbour graph. *Proc. Roy. Soc. London*, **A110**:709–727.

Richardson, W. S., P. B. Stimson, and C. H. Wilkins. 1963. Current measurements from moored buoys. *Deep-Sea Res.*, **10**:369–388.

Riley, G. A. 1951. Oxygen, phosphate, and nitrate in the Atlantic Ocean. *Bull. Bingham Oceanogr. Coll.*, **12**:1–126.

Robinson, A. R., and H. Stommel. 1959. The oceanic thermocline and the associated thermohaline circulation. *Tellus*, **11**:295–308.

Robinson, A. R., and P. Welander. 1963. Thermal circulation on a rotating sphere; with application to the oceanic thermocline. *J. Mar. Res.*, **21**:25–38.

Rossby, C.-G. 1936. Dynamics of steady ocean currents in the light of experimental fluid mechanics. *Papers Phys. Oceanog. Meteor.*, **5**: No. 1, Massachusetts Institute of Technology and Woods Hole Oceanographic Institution, 43 pp.

Saffman, P. G. 1962. The effects of wind shear on horizontal spread from an instantaneous ground source. *Quart. J. Roy. Meteorol. Soc.*, **88**:382–393.

Saffman, P. G. 1963. Reply to discussion. *Quart. J. Roy. Meteorol. Soc.*, **89**:288–289, 293–295.

Schönfeld, J. C., and P. Groen. 1961. Mixing and exchange processes. In *Radioactive Waste Disposal into the Sea*, Int. Atomic Energy Agency, Vienna, Appendix VI.

Schönfeld, J. C. 1962. Integral diffusivity. *J. Geophys. Res.*, **67**:3187–3199.

Schuert, E. A. 1970. Turbulent diffusion in the intermediate waters of the North Pacific Ocean, *J. Geophy. Res.*, **75**:673–682.

Somerville, B. T., and A. F. B. Woodhouse. 1950. Ocean passages for the world. Dept. Admiralty, Hydrogr., London.

Stern, M. E. 1967. Lateral mixing of water masses. *Deep-Sea Res.*, **14**:747–753.

Stern, M. E. 1968. T-S gradients on the micro-scale. *Deep-Sea Res.*, **15**:245–250.

Stewart, R. W. 1959a. The natural occurrence of turbulence. *J. Geophys. Res.*, **64**:2112–2115.

Stewart, R. W. 1959b. The problem of diffusion in a stratified fluid. *Advan. Geophys.*, **6**:303–311.

Stewart, R. W., and H. L. Grant. 1962. Determination of the rate of dissipation of turbulent energy near the sea surface in the presence of waves. *J. Geophys. Res.*, **67**:3177–3180.

Stewart, R. W. 1968. Discussion during Symposium on Turbulence in the Ocean, Int. Assoc. Physical Sci. Ocean, held June 11–13, 1968, Vancouver, B.C.

Stommel, H. 1949. Horizontal diffusion due to oceanic turbulence. *J. Mar. Res.*, **8**:199–225.

Stommel, H. 1957. A survey of ocean current theory. *Deep-Sea Res.*, **4**:149–184.

Stommel, H., A. B. Arons, and D. Blanchard. 1956. An oceanographical curiosity: the perpetual salt fountain. *Deep-Sea Res.*, **3**:152–153.

Stommel, H., and K. N. Fedorov. 1967. Small scale structure in temperature and salinity near Timor and Mindanao, *Tellus*, **19**:306–325.

Sverdrup, H. U. 1939. Lateral mixing in the deep water of the South Atlantic Ocean. *J. Mar. Res.*, **2**:195–207.

Sverdrup, H. U., M. W. Johnson, and R. H. Fleming. 1942. *The oceans*, Prentice-Hall, Inc., Englewood Cliffs, N.J., 1087 pp.

Swallow, J. C., and L. V. Worthington. 1957. Measurements of deep currents in the western North Atlantic. *Nature*, **179**:1183–1184.

Tait, R. I., and M. R. Howe. 1968. Some observations of thermohaline stratification in the deep ocean. *Deep-Sea Res.*, **15**:275–280.

Takenouti, Y., T. Nan'niti, and M. Yasui. 1962. The deep-current in the sea east of Japan. *Oceanogr. Mag.*, **13**:89–101.

Taylor, G. I. 1931a. Effects of variation in density on the stability of superposed streams of fluid. *Proc. Roy. Soc., London*, **A 132**:499–523.

Taylor, G. I. 1931b. Internal waves and turbulence in a fluid of variable density. *Rapports et Proces-Verbaux du Conseil Perm. Int. pour Exp. de la Mer*, **LXXVI**:35–43.

Thorade, H. 1931. Strömung und zungenförmige Ausbreitung des Wassers. *Beiträge z. Geophysik*, **34**:57–76.

Turner, J. S. and H. Stommel. 1964. A new case of convection in the presence of combined vertical salinity and temperature gradients. *Proc. Nat. Acad. Sci.*, **52**:49–53.

Turner, J. S. 1967. Salt fingers across a density interface. *Deep-Sea Res.*, **14**:599–611.

Voorhis, A. D. 1968. Measurements of vertical motion and the partition of energy in the New England Slopewater. *Deep-Sea Res.*, **15**: 599–608.

Voorhis, A. D. and H. T. Perkins. 1966. The spatial spectrum of short-wave temperature fluctuations in the near-surface thermocline. *Deep-Sea Res.*, **13**:641–654.

Webb, D. C., and L. V. Worthington. 1968. Measurements of vertical water movement in the Cayman Basin. *Deep-Sea Res.* **15**:609–612.

Webster, F. 1967. A scheme for sampling deep-sea currents from moored buoys. Trans. 2nd Int. Buoy. Tech. Symp., Marine Technological Soc., Washington, D.C., September 18–20, 1967, 419–431.

Webster, F. 1968. Observations of inertial-period motions in the deep sea. *Rev. Geophys.*, **6**:473–490.

Wiegel, R. L. 1964. *Oceanographical engineering*, Prentice-Hall, Inc., Englewood Cliffs, N.J., Chapter 16. 527 pp.

Wiseman, W. J. 1969. On the structure of high-frequency turbulence in a tidal estuary. Chesapeake Bay Institute Technical Report **59**.

Woods, J. D. 1968a. An investigation of some physical processes associated with the vertical flow of heat through the upper ocean. *Meteorol. Mag.*, **97**:65–72.

Woods, J. D. 1968b. Wave-induced shear instability in the summer thermocline. *J. Fluid Mech.*, **32**:791–800.

Wüst, G. 1935. Die Stratosphäre. Deutsche Atlantische Exped. Meteor 1925–1927, *Wiss. Erg.*, Part 1, 2. Lief, Vol. **6**.

Wyrtki, K., 1962. The oxygen minima in relation to ocean circulation. *Deep-Sea Res.*, **9**:11–23.

Yudelson, J. M. 1967. A survey of ocean diffusion studies and data. W. K. Kech Laboratory of Hydraulics and Water Resources Tech. Memorandum No. 67-2, California Institute of Technology, Pasadena, Calif.

5

Sediments

G. D. SHARMA, *Associate Professor of Marine Science, University of Alaska, College, Alaska*

As the technical skill of man advances, his ambitions become bolder and his undertakings become greater with each step. When human ambitions are directed to control the environment, nature must alter its course to compensate for man's own design. It is ironic that man has always struggled to master his environment rather than comply with its restrictions; however, at his final hour of victory, he has realized that during this struggle he has caused enough environmental changes to pose new battles, this time not for victory but for survival.

Shortsighted goals and indiscriminate use of natural resources have caused imbalance in the natural order. The increasing human population and industrial complex need larger supplies of freshwater and cheap sources of electric power. The obvious solution to these needs was the damming of major rivers, which would not only eliminate devastating floods, but also supply water for irrigation and human consumption and, as a bonus, electricity. In reality, however, the consequences of such construction have often been more adverse than fruitful. The recent completion of the Aswan Dam on the Nile River is an example. It is realized now that the occasional floods of the Nile River deposited a nutrient-rich layer of sediments, thus replenishing soil fertility. It is also feared that damming of the river has cut off a major source of nutrients for the biota in the Mediterranean Sea.

The importance of sediment to ocean-water chemistry cannot be overemphasized. The chemical composition of the oceans is the result of complex interaction between sediments and water. For example, the larger quantity of silica in the Pacific Ocean compared to that of the Atlantic Ocean is explained on the basis of sediment-seawater interaction. Although the total amount of surface water draining into the Atlantic Ocean far exceeds that of the Pacific Ocean, the silica concentration and other trace elements such as nickel and silver is far greater in the Pacific Ocean than in the Atlantic Ocean. Therefore, the cycling and amount of element in the ocean are not controlled simply by the amounts added in solution but by other factors as well.

Since the amount of individual elements in the sea contributed by sediment over geological ages is large, the influence of man may not be apparent for some time. However, because of the complex interdependence of parameters it is difficult to observe subtle changes in chemistry of seawater resulting from man's activities. In localized areas these effects become apparent quickly, but large-scale changes in the oceans are slow. The course of some changes when set in motion is often irreversible.

Sediment Cycle in Oceans

The transformation of parent igneous rocks into sedimentary rocks in the oceans is basically a cycle of differentiation whereby minerals undergo solution and are reconstituted into more stable minerals in the marine environment. The continental environment mainly involves weathering which is accompanied by hydrolysis and acid attack by CO_2-charged waters. The elements are taken directly into solution or leached differentially from silicates leaving residues as soil. Thus weathering is characterized by the absorption of hydrogen ion by silicates and release of silica, alkalies, and bicarbonate ion to the oceans. During leaching of minerals the interlayer cations held by electrostatic attraction are progressively released, allowing the separation of corresponding layers through a series of stages. Most weathering products, together with certain amounts of unweathered continental rock material, are carried by rivers, wind, and glaciers into the oceans. During the stages of transport, the rocks and mineral particles undergo further mechanical and chemical disintegration. In the transition from the continental to the marine environment, physicochemical conditions change radically. This applies in particular to fine-grained sediments, which are strongly reactive because of their relatively large surface area. In addition, these fine particles remain in suspension in the marine environment for a long time and distance because of their low settling velocities. The term "halmyrolysis," introduced by Hummel (1922), encompasses all chemical and physicochemical processes which occur during transportation and the preburial stages of sediments in the marine environment.

In the marine environment the silicate mineral layers contract and tend to resume their original dimensions. Sedimentary products formed in oceans include reconstituted silicates, carbonate rocks, salts, and chert. In addition to removing elements by precipitation, the sediments remove elements and compounds from the seawater by absorption. Some elements and their compounds carried to the oceans are toxic; if allowed to accumulate, these would have made the marine environment uninhabitable for most of the species present today. Sediments continually remove toxic material through adsorption and formation of complexes.

The extent and nature of sediment-seawater interaction have been studied by many investigators in recent years. Evidence for probable alteration of montmorillonite and degraded illite to chlorite has been offered by Powers (1954) from studies of the surface sediments of the Chesapeake Bay area and the Gulf Coast, respectively. The alteration of clay minerals by ionic influences in the marine environment has also been postulated. For example, Grim et al. (1949) concluded that kaolinite is undergoing slow transformation, possibly to illite or chlorite in marine sedimentary deposits of the Gulf of California and adjacent waters of the Pacific Ocean. These changes may not necessarily alter the structural lattices of the clay size sediments, but they do profoundly change the chemical composition of these sediments.

As noted by Siever et al. (1965), it is reasonable to suppose, as a first approximation, that under ideal conditions the interstitial fluid composition may be a function of the mineralogy and the physical characteristics (grain size, packing, porosity, permeability, etc.) of the mineral phases with which it is in contact. In other words, the composition of the interstitial waters reflect the long-ranged chemical equilibrium with the surrounding sediments. Therefore the study of interstitial water is an integral part in an attempt to elucidate the role of sediments in the oceans. Studies of interstitial waters from recent marine sediments of the Black Sea (Shishkina and Poznaniyu, 1959) and in the Atlantic Ocean (Siever et al., 1961; Friedman et al., 1968) indicate that the buried waters are geochemically different from the overlying seawater. Friedman et al. (1968) reported higher chlorinity and $Ca:Cl$, $K:Cl$, and $Rb:Cl$ in interstitial waters from cores obtained from the inner and outer shelf than in the overlying waters. A slight increase in $K:Cl$ with depth in interstitial water (Siever et al., 1965) is probably due to dissolution of K-feldspars (Garrels and Howard, 1959). A decrease in $Mg:Cl$ with depth in cores has been reported by Siever et al. (1965), and McKinney and Friedman (1967). Powers (1954, 1957) showed by the study of interstitial water and absorbed cations that magnesium is absorbed by clay preferentially over potassium. Irregular variations in sodium ions with depth have been reported by Siever et al. (1961, 1965). There have been attempts to interpret paleosalinities from interstitial water analyses and thereby elucidate the environment of deposition of the enclosing sediments (Debyser, 1952; Kullenberg, 1952; Mikkelsen, 1956; Zhelezhova, 1958; and Siever et al, 1961).

The extent of sediment-seawater interaction, or hydrodiagenesis, in nature is influenced by many factors, such as (*a*) changes in the environment, (*b*) the rate of sedimentation, and (*c*) the length of exposure of detrital sediment in suspension. On the basis of laboratory experiments, Rae and Knowles (1965) concluded that mineral-metal exchange is indeed a dynamic process and that the time of exposure before deposition is a fundamental parameter in this process. Initial ion exchange is related to the tendency to reach equilibrium with a new environment: this type of change would be expected in areas where large amounts of detrital sediments are deposited close to the source.

Sediment-water interaction is often obscured in sediments transported to the ocean by agencies other than glaciers. For example, sediments transported by rivers are constantly exposed to highly variable chemical and physical environments, which renders the quantitative study of such interaction impractical. The interstitial water compositions reported by many investigators are from deep water, generally distant from the shelf area. Since terrigenous sediments interact with seawater during transport, the farther they are transported from the source, the closer they come to sediment-seawater equilibrium prior to deposition (Sharma and Burrell, 1970). The slow rate of deposition in deep water further prolongs sediment-seawater interaction near the depositional interface. Therefore, the influence of deep-sea terrigenous sediments on the composition of trapped water should be much less than that of sediments deposited near the shore. It is postulated that large amounts of glacially transported sediments, with little or no chemical weathering before being deposited in the marine environment will undergo the maximum possible sediment-seawater interaction as expressed in changes in the composition of interstitial water.

The coast of southeast Alaska, with active glaciers to provide large amounts of sediments which are locally deposited, offers a unique opportunity to study the interaction between sediments and seawater during and after deposition. Cores, bottom grab samples, and suspended sediments were collected in the Inside Passage of southeast Alaska and Canada between 56°0′N–59°30′ and 133°–137°W during the summers of 1965, 1966, and 1967 on the R/V ACONA (Fig. 5.1) to study the chemical

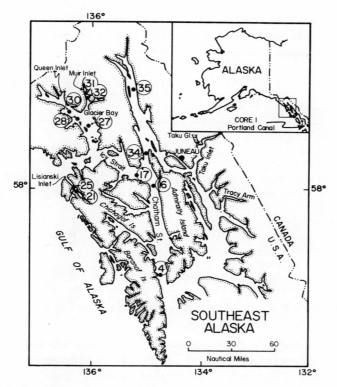

Figure 5.1 Location map, southeast Alaska.

changes in sediments and waters resulting from transfer from continental to marine environment.

Sediment-Seawater Interaction During Transport (Halmyrolysis)

Sediments obtained from melt water at the face of the Carrol Glacier in Queen Inlet were used in a study of rapid changes in interstitial waters. Replicate samples were thoroughly mixed with synthetic seawater. The pH of synthetic seawater dropped from 7.8 to 6.9 with a simultaneous release of calcium and removal of sodium from the supernatant solution (Table 5.1).

Suspended sediments from Queen River and Queen Inlet when mixed with synthetic seawater removed sodium, potassium and magnesium and released calcium and silica to the supernatant solution (Table 5.1). The

Table 5.1. Changes in the Composition of Synthetic Seawater[a] During Contact with Sediments

Sample	Loss (meq)			Gain (meq)	Soluble Silica (ppm)
	Na^+	K^+	Mg^{2+}	Ca^{2+}	
Carrol Glacier melt water sediments (7 days)	34.78	0.59	9.46	2.99	
Queen River suspended sediment (14 months)	34.78	1.66	8.22	2.49	8.7
Queen Inlet low tide suspended sediment from surface depth (14 months)	17.39	1.02	4.93	0.74	9.2
Queen Inlet low tide suspended sediment from 20-m depth (14 months)	17.39	0.89	4.11	0.74	9.2

[a] Synthetic seawater of pH 7.8 contained less than 0.5 ppm silica.

maximum removal of sodium, potassium, and magnesium and maximum release of calcium were observed for suspended river sediments, with lesser amounts of exchange for suspended sediments from the inlet. The amounts of sodium removed from the synthetic seawater by glacier sediments and by suspended river sediments were identical. The amount of sodium removed by river-suspended sediment is greater than the amounts removed by suspended sediments from Queen Inlet. Furthermore, the surface sediments removed slightly larger amounts of magnesium and potassium than suspended sediments collected from 20-m depth. The increase in exchangeable sodium and magnesium and decrease in calcium with prolonged seawater-sediment interaction were observed in Taku Inlet by Rae and Knowles (1965). In these results (Table 5.2) the total exchangeable cations in the sediments remain approximately the same; with passage of glacial sediment into the marine environment the exchangeable sodium increased by a factor of 16, potassium by 5, and magnesium by 2, and calcium decreased 2.2 times. Remarkably little exchangeable sodium in brackish water mud samples was also noted by McCrone (1967). Exchangeable magnesium increases with decrease in calcium from nonmarine to marine environment because of higher concentration of magnesium in seawater (Carroll and Starkey, 1960). Similar

Table 5.2. Exchangeable Cations Associated with
Suspended Sediments in the Taku System

Station	Distance from Taku Glacier	Exchangeable Cations (meq/100 g sediments)				
		Na	Mg	K	Ca	Total
1	15 km above glacier (river)	0.67	6.15	0.40	26.43	33.65
2	Face	0.87	2.96	0.46	25.95	30.24
3	50 m	1.39	8.22	1.33	19.96	30.91
4	100 m	3.15	7.84	1.72	17.75	30.45
5	200 m	3.96	8.64	3.04	19.46	35.09
6	300 m	5.48	9.95	3.71	14.37	33.51
7	400 m	8.70	11.35	2.10	11.96	34.10

^{a} From Rae and Knowles (1965).

explanation for exchange in terms of free energy of the cations resulting from the transfer of sediments from rivers to seawater has been given by Keller (1963).

In Muir Inlet noticeable differences were observed between the chemical composition of interstitial waters from bottom sediments and that of the overlying waters (Fig. 5.2). In general, the total ionic concentration in interstitial water is less than that in the overlying water, as has been observed in other southeast Alaskan inlets (Burrell, 1967). Sodium and magnesium ion concentrations are lower in interstitial water than in overlying waters. Conversely, interstitial waters are enriched with potassium and calcium ions. The differences in sodium and calcium ion concentrations between interstitial waters and overlying waters are greatest at the head of the inlet and decrease toward the mouth of the inlet. The maximum difference in magnesium ion concentration, however, is at the mouth and the minimum at the head of the inlet.

In view of both laboratory and field observation, it appears that sediments deposited near the source area, when transported from a glaciofluvial environment, do not have sufficient time to absorb sodium ions to reach seawater-sediment equilibrium and thus continue to remove large quantities of sodium ion from the interstitial fluid after deposition.

An attempt to explain the cation exchange on the basis of stoichiometry was futile, since there seems to be complex interaction between sediments

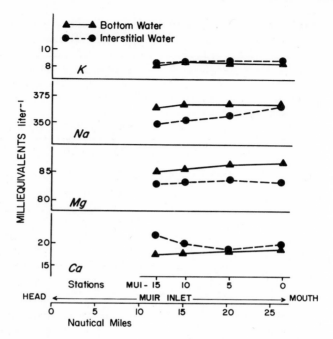

Figure 5.2 Major cationic concentrations in interstitial water from bottom sediments and overlying waters of Muir Inlet.

and seawater. The mixing of glacial sediments with distilled water of pH 7.3 in the laboratory did not change the final pH of the supernatant solution. Addition of the same sediments to synthetic seawater of pH 7.8, however, lowered the pH to 6.9. It appears that glacial and fluvial sediments are at equilibrium with freshwater environment and are saturated with hydrogen ion. These sediments when immersed in seawater exchange hydrogen ions with sodium, thus lowering the pH. After quasi-equilibrium between sodium ion concentration and hydrogen ions in sediment-seawater mixtures is attained, further transport and mixing comminute sediment particles and magnesium, calcium, and potassium ions play an important part in reaching the final equilibrium.

Mineralogy of various sand and silt fractions in glacial and adjacent saline environments were also studied in Taku Inlet. In the glacial environment, where periglacial weathering and the grinding of detritus result primarily in mechanical size reduction, the hornblende content increased with a decrease in grain size (Sharma, 1970a). Silt-sized hornblende is markedly less abundant in the adjacent saline environment, however, suggesting some chemical dissolution of silt-sized hornblende. It is suggested that the smaller hornblende particles are more readily taken into solution. Solution of various minerals in the seawater occurs during transportation, deposition, and burial and undoubtedly affects the composition of the interstitial waters. Furthermore, it has been shown that ions are released from the surface of minerals freshly ground in the laboratory (Laventure and Warkentin, 1965). The glacial sediments in southeast Alaskan inlets are mechanically ground but chemically unweathered. In a marine environment such material should tend to release potassium and magnesium ions from biotite and hornblende. The amounts of various cations released through the solution of minerals in nature are difficult to evaluate; however, in the laboratory solubility of minerals in seawater has been determined by various investigators (Garrels, 1960).

Sediment-Seawater Interaction After Deposition

Changes in cationic concentrations of the interstitial waters have been plotted within cores as well as between cores (Figs. 5.3, 5.4, 5.5, and 5.6). Although there are variations in cationic concentration from horizon to horizon, total ionic concentration increases with depth in a core. The sodium ion concentration shows most dramatic changes with depth. Changes in concentration of sodium relative to other cations signify that the variations in sodium ion concentration with depth are real rather than due to evaporation or contamination.

There are no systematic vertical variations in the sediment's grain size for the cores studied. In relation to the composition of the interstitial water, a trend between sodium and the percentage of silt and clay is observed with depth in cores. An increase in the percentage of silt is generally accompanied by an increased sodium ion concentration in the interstitial water (Figs. 5.7 and 5.8). To some extent sodium concentrations in the interstitial waters vary inversely with the percentage of clay in the samples. The relationship between sodium ion concentration and percentage of silt is shown in Fig. 5.9. The concentration of the major cations in the interstitial water was not influenced by the relative percentages of chlorite or illite (Figs. 5.7 snd 5.8). Furthermore, the relative amounts of magnesium-rich and iron-rich chlorites in sediments did not have any relationship to the major ions in interstitial waters (Fig. 5.10). The trace quantities of manganese generally decreased with depth in all cores. In Glacier Bay, the manganese content in the interstitial water from cores increased as the distance of the core from the sediment source increased (Fig. 5.4).

The cation exchange capacity of the sediments from cores is related to the percentage of clays—the greater the illite and chlorite percentage, the greater the cation exchange capacity (Burrell et al., 1968). From these observations it can be inferred that larger proportions of 4 μ (illite and chlorite) fraction in the sample have a higher proportion of reactive

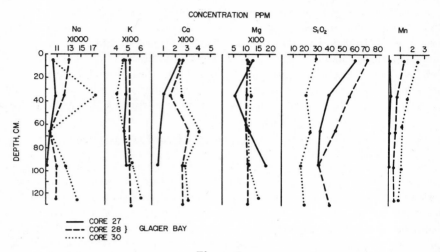

Figure 3

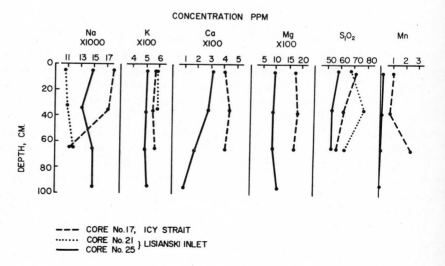

Figure 4

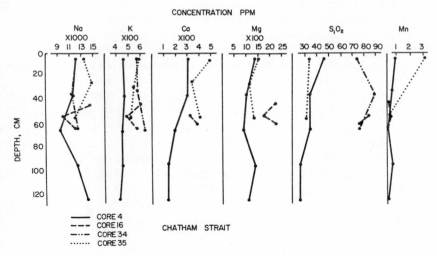

Figure 5

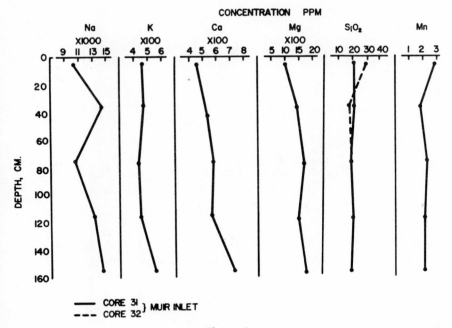

Figure 6

Figures 5.3 to 5.6 Depth profiles of major and minor elements in interstitial waters from core sediments.

area in broken surfaces and thus a higher exchange capacity. Higher exchange capacity with increased surface areas has been demonstrated by Laventure and Warkentin (1965). The mineralogy and the particle size of clays regulate the negative charge per unit surface area on which water and ions are adsorbed and on which the chemical factors can make their effects apparent. The depletion of sodium ions in the interstitial waters in sediments with larger proportions of clays bear out the effects of adsorption of sodium on clays. Thus it seems that the composition of interstitial waters is largely controlled by the mineralogy and the particle size of clays, which control the exchange capacity of the sediments. The removal and the enrichment of cations in interstitial water by sediments also depend on the distance transported and the length of sediment exposure to marine environment.

The chemical reaction between the clay particles and the fluid containing them occurs as isomorphous substitution or by ion exchange.

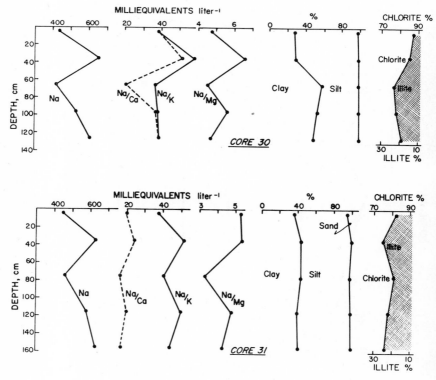

Figure 7

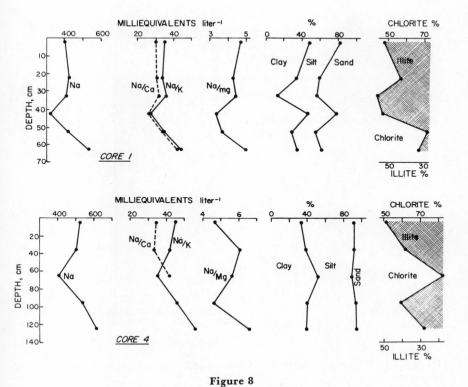

Figure 8

Figures 5.7 to 5.8 Depth profiles of sodium concentration and elemental ratios, percentages of sand, silt, and clays, and the relative percentages of illite and chlorite.

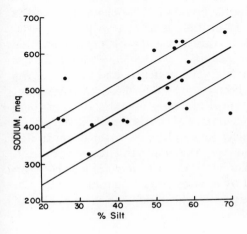

Figure 5.9 Relationship between percentage silt and sodium ion concentration in interstitial waters from core sediments. The data were fitted with a least-square regression line having a regression coefficient of 0.63 and standard error of ±78.3 meq Na/l.

181

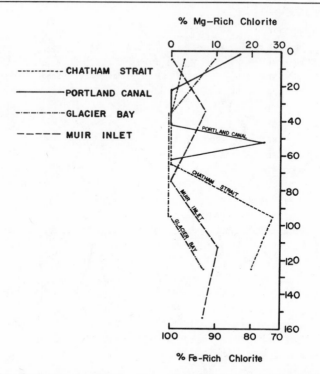

Figure 5.10 Percentage variations of magnesium- and iron-rich chlorites in core sediments.

The substitution can take place in the tetrahedral layer, the octahedral layer, or the interlayers. The ion exchange, which includes absorption phenomena, occurs at the solid-liquid interface. The chemical complex of water components controlling adsorption consists of several interdependent parameters. These include texture and structure, organic fraction of the particles, chemistry of the medium, ion-exchange equilibria, gas equilibria, and microbiological activity. The structure and texture of the sediment particles is important because it establishes the framework within which chemical and biochemical processes operate. Both of these phenomena are dependent on the chemistry of the fluid and on the nature of the sediments. The extent of exchange on clays is primarily controlled by the number of unbalanced charges, the structural framework, and the degree of hydration. On the other hand, the movement of ions between sediments and liquid is influenced by environmental factors such as pH, Eh, pressure, temperature, chemistry of the fluid, and availability of exchangeable ions in the mobile phase.

The incorporation of organic molecules in clays is complex. Some progress has been made during the past few years in the field of organo-clay complexes; recent studies have shed light on the type of bonding between organic molecule and clay minerals, the structural orientation, and the packing of the organic component in the interlayer of clays (Degens, 1965). The mode of preservation of organic matter from organisms in sediments during burial has been also elucidated by these studies.

The two basic ways for seaborne organic material to become incorporated into the sediments are by settling out of the detrital organic particles or by association with the minerals which are settling from the water column; the latter is classical adsorption. The organic molecules associated in this form are complex and the mechanism for association is not well understood.

Experimental data indicate that montmorillonite is capable of removing the greatest quantities of organic matter from solution. Illite, kaolinite, and quartz follow in that order. Exchange capacities, controlled by the mineral structure, are a primary factor in this system. The surface area of the mineral also exercises some control in this association, smaller particles remove more organic matter from solution. In some instances, as much as 70 mg of organic compounds was removed from solution by only 20 mg of mineral. The removal is dependent on the structure, functional groups, and molecular weight of the organic compound. Such mineral-organic associations are sensitive to environmental factors and are generally not reversible. Enzyme hydrolysis of proteins attached to clay surfaces depend on the particular clay involved. In the case of kaolinite, protein hydrolysis is not hindered significantly. On three-layered clays, such as montmorillonite, adsorption between the interlayers blocks hydrolysis and decomposition is retarded as a result of orientation and inaccessibility of the active group. Limited available data suggest that organic compounds adsorbed between montmorillonite interlayers become resistant to decomposition.

The mechanisms involved in biodegradation of organic matter associated with clays are no doubt complex and depend on the properties of both the clays and the adsorbate.

Post-Burial Sediment-Seawater Interaction

The influence of sediments in the oceans continues to some extent after the deposition of sediments. Postburial interaction between sediments and trapped pore water controls the chemical composition of the interstitial water. The ions in the interstitial water may migrate because of compaction of sediments, ionic gradation, and diffusion.

Diffusion and exchange in trapped pore water proceed upward toward free water overlying the sediments. The profile is primarily a chemical and biological adjustment between the oxygenated and mobile overlying water and the deoxygenated and immobile sediments that exist at depth. Compounds in solution in the interstitial water will diffuse along their respective concentration. This means that oxidized compounds will move down and reduce compounds up. Insoluble compounds such as elemental sulfur and ferrous sulfides will be held at the level where they are formed. It is therefore expected that H_2S will be diffusing upward and, after reaching the oxygenated zone, will be either oxidized spontaneously or used by bacteria. This diffusion causes stratification in the sediment column in the oceans. A brown oxidized layer is generally underlain by a layer of olive-gray sediments and by the very dark gray or black mud which represents the zone in which microbial activity reaches a peak. It is across this sedimentary boundary that nutrients and trace elements diffuse from the sediments into the overlying waters.

Sulfate, phosphate, and silicate may show slight or marked increases in concentration in a thin layer of water just above the sediments. Nitrate concentration varies in both directions depending on the basin, although total fixed nitrogen probably increases in the bottom water of the oceans in the southern California basins and the Black Sea. The changes at the bottom are due to biodiagenetic processes at the sediment-seawater interface and within the sediment column, resulting in the release of ions to the water. Changes may also be due to ion-exchange processes, but it cannot be taken for granted that an increase of ions in the overlying water invariably occurs. Koczy (1953), for instance, showed that in bottom water of the Atlantic Ocean there is a marked decrease not only in oxygen, but also in phosphate and silicate, which is explained by base exchange at the mud-water interface.

Interrelations of Sediment and Biota

The sea has always been considered a natural receptacle for human wastes. Grant M. Gross (personal communication) has reported that about 10×10^6 tons/yr of waste solids (excluding rubbish and floatable debris) are being delivered to the Atlantic by New York City, making this the largest single source of sediment from North America to the Atlantic Ocean. This waste consists of radioactive material, waste water composed of domestic effluents, and waste water from industry carrying chemical products and toxic substances in large quantities. The biological and chemical features of these wastes differ from those of the seawater into which they are discharged. The discharge of the wastes into the sea produces physical, chemical, or biological modifications in the environment and may result in pollution which spreads over great distances depending on the currents and the distribution of the marine superficial layers.

The man-induced changes in the form of siltation, changes in the current pattern by dredging and marine construction, also disturb the natural environment. The problem facing the biologist is to separate the effects of man's activities from natural environmental variations. The effects of silting and high turbidity resulting from excavations along the coast of Rumania have been studied by Bascescu (1966). He reported suffocation and respiratory difficulties in certain groups of the indigenous fauna, inhibition of photosynthesis, low numbers of planktonic organisms, and disappearance from the area of many species of mobile organisms. Some advantageous effects were also observed, particularly in the elimination of fouling species and in an increase in species preferring turbid waters. Quantitative studies of sessile forms showed that heavy silting resulted in the temporary disappearance of stenobiotic species of the fauna and flora from rocks, these being displaced by forms tolerant to turbid conditions.

Higher mortality rates of oysters in Matsushima Bay, Japan, were observed in areas where water temperatures were elevated and salinity depressed during the warmer season; this was also related to the area of higher turbidity and higher H_2S content due to organic pollution (Kan-No et al., 1965). The authors were unable to attribute the mass mortalities directly to external conditions, but noted that the total H_2S and organic matter in the bottom sediments had shown marked increase during the past 10 yr and the total yield of oyster per unit of rack or raft had correspondingly increased, indicating enrichment of the waters by organic pollution. They concluded that the cause of the mortalities was related to the physiological conditions of oysters' growth under the environmental conditions found in the bay. Furthermore, studies in the same area noted that death of oysters occurred during gonadal maturation and spawning which induced long-term organic pollution.

The number of species of phytoplankton in polluted areas and the chlorophyll content of the water decrease near sewer and pulpmill outfalls but increase where dilution is sufficient for the added nutrients to be used as fertilizer. Copeland (1966) reported on the measurements of phosphate-phosphorus, Ohle anomalies, redox potential, species diversity, salinity, alkalinity, and chlorophyll in St. Joseph's Bay, Florida. High phosphorus concentrations, reducing Ohle anomalies and redox potential, and low species diversity indicated some degree of organic pollution in the bay. Tibby (1965) related the production and the productivity of phytoplankton to measure the biological effects of sewage disposal in an open coastal marine environment. Survey of an area affected by sewage showed a pattern of low surface productivity near the point of discharge. An increase in productivity with values well above background was found downstream after about 6 hr. Perdriau (1964), on the basis of occurrence of carcinogenic hydrocarbons in marine sediments and edible mollusks in samples collected from the coasts of France and West Greenland, described the causes of pollution in these areas.

Summary

Human activity on land and at sea greatly influences the usefulness of the oceans to mankind. Fish and shellfish life, recreation, international trade, and even human habitation of the shore are affected by increasing role of man in the sea. Changes in drainage pattern, damming of rivers on land, and discharge of industrial wastes and human refuse in the ocean generally cause immediate as well as slower changes in the sedimentation, currents, chemistry, and biology of the area affected.

Rapid chemical and mineralogical changes occur when continental sediments come in contact with marine environment. An extreme example of such changes can be observed in southeast Alaska. Large amounts of glacial sediments with minimal chemical weathering are deposited in fiords, channels, and bays in this area. During early stages of sediment-seawater interaction sodium is removed from seawater by sediments while calcium is released to seawater. Removal of magnesium, accompanied by further removal of sodium from seawater, results from thorough mixing of sediments and seawater during transport. During deposition and early stages of burial the mineralogy and the particle size distribution, especially within the clay-sized range, control the chemical composition of the trapped pore waters. Deep-burial of sediments results in compaction, ionic gradation, and diffusion, hence in migration of interstitial fluids.

The nature and the rate of sediments deposited affects the density and the nature of biota in the area. Increase in sediments resulting from coastal structures and various industrial effluents may drastically alter the number and type of species dwelling in the region.

References

Bascescu, M. 1966. An instance of the effects of hydrotechnical work on the littoral marine life. *Studii Hidraul., Inst. Studii Cerc. Hidrotech.*, **9**:137; *Water Pollution Abstr. (Brit.)*, **39**:800.

Burrell, D. C. 1967. Major and minor cationic distribution at the sea-sediment interface. Preliminary data for Muir Inlet, Glacier Bay, pp. 59–96. In Clay-Inorganic and Organic-Inorganic Associations in Aquatic Environments, D. W. Hood and D. C. Burrell [EDS.]. Report to Atomic Energy Commission, contract AT-(04-3)-310 PA No. 3. Institute of Marine Science, University of Alaska.

Burrell, D. C., N. R. O'Brien, and G. D. Sharma. 1968. Geochemical and mineralogical investigations of sediments within an active glacial fiord. *Trans. Amer. Geophys. Union,* **49**:112.

Carroll, D., and J. C. Starkey. 1960. Effects of sea water on clay minerals, pp. 80–101. In *Clays and Clay Minerals, 7th Conference,* Pergamon Press, New York.

Copeland, B. J. 1966. Effects of industrial waste on the marine environment. *J. WPCF,* **38**:1000–1010.

Debyser, J. 1952. Variation du pH dans l'épaisseur d'une vase fluvio-marine. *Compt. Rend. Acad. Sci., Paris,* **234**:741–743.

Degens, E. T. 1965. *Introduction of geochemistry of sediments.* Prentice Hall, Englewood Cliffs, N.J., 342 pp.

Friedman, G. M., B. P. Fabricand, E. S. Imbimbo, M. E. Brey and J. E. Sanders. 1968. Chemical changes in interstitial waters from continental shelf sediments. *J. Sediment. Petrol.,* **38**:1313–1319.

Garrels, R. M. 1960. *Mineral Equilibria, at Low Temperature and Pressure,* Harper and Row, New York, 254 pp.

Garrels, R. M., and P. Howard. 1959. Reactions of feldspar and mica with water at low temperature and pressure. In *Proceedings of the 6th conference on clays and clay minerals,* Pergamon Press, London, pp. 68–88.

Grim, R. E., R. S. Dietz and W. F. Bradley. 1949. Clay mineral composition of some sediments from the Pacific Ocean off the California coast and the Gulf of California. *Bull. Geol. Soc. Amer.,* **60**:1785–1808.

Gross, Grant M. 1970. Personal Communication, State University of New York at Stony Brook.

Hummel, K. 1922. Die Entstehung eisenreicher Gesteine durch Halmyrolyse. *Geol. Rundschau,* **13**:40–81, 97–136.

Kan-No, H., M Sasaki, Y. Sakurai, T. Watanabe, and K. Suzuki. 1965. Studies on the mass mortality of the oyster in Matsushims Bay and its environmental conditions. *Bull. Tohoku Regional Fish Res. Lab. (Japan),* No. 25.

Keller, W. D. 1963. Diagenesis in clay minerals—a review: In *Clay and clay minerals. Proceedings of the 11th National Conference,* Pergamon Press, New York, pp. 36–157.

Koczy, F. F. 1953. Chemistry and hydrology of the water layer next to the ocean floor. *Statens Naturvetenskapliga Forskningsråd Årsbok. Sjatte Argangen,* **1951–1952**:97–102.

Kullenberg, G. 1952. On the salinity of the water contained in marine sediments. *Medd. Oceanog. Inst. Göteborg*, No. 21, Ser. B, **6**:38.

Laventure, R. S., and B. P. Warkentin. 1965. Chemical properties of Champlain Sea sediments. *Can. J. Earth Sci.*, **2**:299–308.

McCrone, A. W. 1967. The Hudson River Estuary. Sedimentary and geochemical properties between Kingston and Haverstraw, New York. *J. Sediment. Petrol.*, **37**:475–486.

McKinney, T. F., and G. M. Friedman. 1967. Geochemistry of seawater above and below the water-sediment interface on the New York and New Jersey continental shelves [abs.]. *Geol. Soc. Amer.*, Spec. Paper **101**:268–269.

Mikkelsen, V. 1956. The salinity of the water contained in brackish-water sediments compared with the content of diatoms and other organisms in the same sediments. *Medd. Dansk. Geol. For.*, **13**:104–112.

Perdriau, J. 1964. Marine pollution by carcinogenic hydrocarbon-type, 3,4-Benzopyrene-Biological incidences. II: *Cahiers Oceanog.*, **16**:205–219.

Powers, M. C. 1954. Clay diagenesis in the Chesapeake Bay area. In *Clays and clay minerals*. Natl. Acad. Sci.–Natl. Res. Council Publ. **327**:68–80.

Powers, M. C. 1957. Adjustment of land-derived clays to the marine environment. *J. Sediment. Petrol.*, **27**:355–372.

Rae, K. M., and L. I. Knowles. 1965. Ecological implications of clay-metal associations in aquatic environments. Report to Atomic Energy Commission, contract No. AT-(04-3)-310. Institute of Marine Science, University of Alaska.

Sharma, G. D. 1970a. Sediment-seawater interaction in glaciomarine sediments in Southeast Alaska. *Bull. Geol. Soc. Amer.*, **81**:1097–1106.

Sharma, G. D. and D. C. Burrell. 1970. The sedimentary environments and sediments of Cook Inlet, Alaska. *Bull. Amer. Assoc. Petrol. Geol.*, **54**:647–654.

Shishkina, O. V., and K. Poznaniyu. 1959. Metamorphism of the chemical composition of silt waters of Black Sea sediments. *Dokl. Akad. Nauk, SSSR*, Sbprnik Statei, 29–50.

Siever, R., R. M. Garrels, J. Kanwisher, and R. A. Berner. 1961. Interstitial waters of recent marine mud off Cape Cod. *Science*, **134**:1071–1072.

Siever, R., K. C. Beck, and R. A. Berner. 1965. Composition of interstitial waters of modern sediments. *J. Geol.*, **73**:39–73.

Tibby, R. B. 1965. Fate of wastes in the marine environment. Calif. State Water Quality Control Bd., Sacramento, California Publ. **29**.

Zhelezhova, A. A. 1958. The pH of sea sediments. *Trudy, Inst. Okeanol. Akad. Nauk, SSSR*, **1**:71–76.

Part II

Chemical Models of the Ocean

6

Oceanic Residence Time and Geobiochemical Interactions

DAYTON E. CARRITT, *Professor of Marine Sciences, University of Massachusetts, Amherst, Massachusetts*

The building of conceptual and mathematical models of the cosmos and its parts has engaged scientists and philosophers certainly since the beginning of scientific literature, and relics of past cultures suggest that ancient people were similarly occupied.

There is a thread of continuing questions and attempted answers in the literature of oceanography, geochemistry, and geophysics that today still are of major concern to workers in the fields. One series of these questions has to do with the evolution as well as the contemporary behavior of the ocean. Is the ocean becoming increasingly more saline? Has the volume of the ocean changed markedly during geological time? What are the phenomena that control the chemical composition of the ocean? How old is the ocean? How old is the earth? Of course we have answers to these questions today, but so did our predecessors 5, 50, and possibly 500 yr ago. It is likely that the changes in answers become what we refer to as our "advanced state of understanding" of our environment.

The purpose of this chapter is to outline briefly what appears to have been the manner in which some of the present-day conclusions have developed, to note the present-day state of the question-answer process with respect to the chemical composition of the ocean, and to suggest a few modifications that may become the basis for further study in this continuing process of question-answer-question.

Before the discovery of radioactive decay and the application of decay principles to the measurement of ages and time rates of change in natural systems, estimates of geological age depended largely on estimates of the rates of chemical denudation of exposed solid surfaces of the earth. As a consequence, much of the early literature concerning the age of the earth and the age and evolution of the ocean contained either speculations on or measurements of the extent and rates of what we would now call geochemical and biochemical reactions. For example, Halley (1715) suggested "the ocean itself is becoming salt, and we are there-by furnished with an argument for estimating the *duration of all things*,* from an observation of the increment of saltiness in their water" (italics mine).

Halley evidently considered the interactions between the earth and the ocean during all of geological time to be the same as he observed them to be in 1715. He regretted, for example, that the ancient Greeks and Latins had not measured the salinity of the sea so that "the increment of saltiness" over some 2000 yr could be used in an extrapolation technique to estimate the time at which the saltiness was zero–"the duration of all things"—the age of the earth.

Clearly Halley's model of the earth and the oceans while not verbalized must have been an extremely simplified revision of what we consider today to be "the real world." Because the various conceptual models that correspond to interpretations of experimental results are noted in the following discussion, it is desirable, if for no other reason than completeness, to attempt to reconstruct Halley's model. The main requirement is that "increments of saltiness" lead to estimates of "the duration of all things." Here we visualize that the state of a system is examined at two different times and the difference—increments of saltiness—provide the basis for extrapolation to that time when changes in the state of the system started—the duration of all things. In the simplest case we will make a linear extrapolation; Halley gives no indication for assuming nonlinear effects in the time rate of change of "getting salty."

Using modern values for saltiness and taking the annual input of dissolved solids by the world's rivers as a measure of increment of saltiness we find it would take 18×10^6 yr to accumulate the total ocean salt if

the total salt in the ocean = 48×10^{21} g

the annual input of salt by rivers = 2.7×10^{15} g/yr

---

* Presumably reference to the age of the earth.

What are the tacid assumptions concerning the model if 18×10^6 yr is truly "the duration of all things" or the age of the earth? It must be assumed:

1. The annual rate of introduction of 2.7×10^{15} g/yr, a value given by Clarke (1924) and obtained by summing analyses for all major rivers of the world, has either been constant over geologic time or represents an average value of a variable rate.
2. Dissolved solids, once introduced into the ocean, remain there.

Here we need not give a detailed critique of Halley's model: suffice it to say that the process of sedimentation suggests that the second assumption can not be valid.

Joly (1899) using *sodium* rather than *total salt* computed the age of the ocean to be 80 to 90×10^6 yr.

When this is corrected for cyclic sodium (sodium lifted from the surface of the ocean and deposited on the land) the computation yields approximately 200×10^6 yr. Becker (1910) further refined the computation by noting that river sodium could be assumed to be proportional to the mean sodium content of surface rocks, a quantity that would decrease with time. However, Conway (1943) noted that early in the history of the earth, surface rocks would have been largely igneous, and although the sodium content is higher in igneous than in sedimentary rocks, the rate with which sodium would be denuded from igneous rocks would be slower than from sedimentary. Combining these considerations gives an oceanic age of about 300×10^6 yr.

Finally, an age of 700 to 800×10^6 yr is obtained if the computations are modified to allow for a marked lowering in the concentration of dissolved solids in river waters during flood time as postulated by A. C. Lane in Holmes (1937).

All of the computations are essentially modifying details that do not change the basic Halley model or its underlying assumptions. None provide for sedimentation.

Goldschmidt (1954) as a part of his remarkable contribution to modern geochemistry summarized and gave quantitative expression to a description of the ocean that, at least partially, removed some of the present-day objections to Halley's model. Where Halley's model is of a static system, as far as chemical reactions are concerned, Goldschmidt attempted to put together a more dynamic system, one in which the ocean was modeled as being the result of both "put" and "take" processes.

Goldschmidt thought that both seawater and sediments were produced by the weathering and erosion of igneous rocks and assumed, like his predecessors, that sodium once dissolved from igneous rocks and carried to the ocean would forever stay in solution in the ocean:

$$\text{age of the ocean} = \frac{\text{total Na in the ocean}}{\text{rate of supply of Na}} = 120 \times 10^6 \text{ yr} \quad (1)$$

Here we see no formal change from Halley's model. Goldschmidt went on, however, to compute the amount of igneous rock that must have been weathered during geological time, together with the quantity of sediment formed:

weight of weathered rocks
$$= \frac{\text{total Na in the ocean}}{\% \text{ Na in igneous rocks} - \% \text{ Na in sedimentary rocks}} \quad (2)$$

and computed the amount of weathered igneous rock to be 160 kg/cm².

Barth (1961, 1962) pointed out that the validity of (1) and (2) depends on uniqueness of sodium that is not evident from our knowledge of the behavior of an aqueous solution of sodium salts. If (1) is to provide a reliable estimate of the age of the ocean, it follows that computations using data for other major dissolved constituents, magnesium, calcium, and potassium should yield the same age as that obtained from sodium computations, or the difference must be resolvable in terms of chemical or geochemical behavior. Available data give the following: "age of the ocean," (Barth, 1962, p. 374, Table V-3) for magnesium, calcium, sodium, and potassium is 23, 1.3, 120, and 10×10^6 yr, respectively.

The differences here are to be reconciled not by assuming a unique behavior for a single element as Goldschmidt did for sodium, but by noting that all elements may be unique in their geochemical behavior. This would now force a major revision in what we believe the overall behavior of the ocean to be—the need for a new model.

Barth (1952) introduced the next major change in model building. He pointed out that (1) stated for sodium or for *any other* element could not give a reliable estimate of the age of the ocean and that age computed from data for the other major dissolved cations show the effect of the inherent chemical reactivity of each of these elements. He noted that although computation using (1) will not give "the age of the oceans," the values obtained are nevertheless important properties of the marine environment. He further maintained that the values as computed give estimates of what he termed the residence time, that is, the mean length of time spent in the ocean by each element from the time of its introduction until its deposition in the sediments.

Barth's reconstruction of older models to incorporate known differences and similarities in the chemical behavior of elements is an "obvious" extension of our knowledge of basic solution chemistry to a geochemical problem. In retrospect it now seems out of character to find that Goldschmidt, who with justification is widely acclaimed as the founder of modern geochemistry, would have developed a model of one of the earth's major chemical systems in which sodium, at least by inference, is given unusual properties and other major constituents are neglected. It now seems likely that this is the result of a disruption in the continuity of Goldschmidt's thinking caused by World War II. Muir who edited Goldschmidt's *Geochemistry* (1954) noted that "the author felt that most attention had to be devoted to the geochemistry of the individual elements before the principles could be fully elucidated" and that Chapter IV, "Quantitative Treatment of Geochemical Processes: Some Fundamental Data," which is a major contribution in the "principles" Part I of *Geochemistry*, is a translation of a 1933 article, "Quantitative Geochemistry." Muir adds that "If the author had lived he would undoubtedly have modernized the presentation of much of what he had written in Part I . . . ." Certainly all the pieces for a satisfactory revision of his model can be found in his writings.

The present-day model of the igneous rock-seawater-sediment system is essentially that proposed by Barth (1952), with important additions by Goldberg and Arrhenius (1958). Focusing attention on the chemical composition of seawater, this system has the following important properties.

1. It is both physically and chemically a dynamic system. Chemical reactions that control the composition of seawater occur during weathering of the landmasses; in transit, as river water, to sea; while in the sea; and finally during the aging of sediments.

2. The system is not a closed system. It has at least one important and frequently neglected opening; the composition and quantity of seawater plus the composition and quantity of sediments cannot be accounted for by assuming the weathering of igneous rocks to be the sole source. The ocean contains massive quantities of volatile constituents in excess of the amounts that could have been produced by the weathering of igneous rocks. These are the so-called excess-volatiles. They include water, sulphur compounds, the halogens, boron, and carbon. These are accounted for by diffusion from the earth's interior, a process that is now going on and has been going on throughout geologic time. In addition, small quantities of matter are being added from outer space.

3. The processes of addition to and removal from the ocean are assumed to be in steady state, an assumption for which there is experimental verification. That is, the rate of addition of material as dissolved and suspended components of river water is the same as the rate with which that kind of material is removed during sedimentation.

4. Significant quantities of material taken to the sea by rivers is "cyclic salt," that is, salt driven from the surface of the sea as spray, which upon evaporation of the water forms aerosols, a part of which is then deposited on the land and finally back to the sea as river constituents. Most of the chloride in river water and significant fractions of the sodium and sulfate have been shown to be cyclic. Thus river composition is not a measure of erosion and weathering.

5. Chemically unreactive elements will remain in solution in the sea for a longer time than reactive elements. Reactive elements, which readily enter into reactions which produce solid phases, will be removed rapidly by sedimentation processes. This difference in chemical reactivity immediately accounts for (a) differences in residence time of different elements (difference in "age of ocean" in Goldschmidt's model), (b) difference in relative composition of river water and seawater, and (c) spacial distribution of some elements in sediments.

The present-day model of the oceans in which *residence time* has replaced *age of the ocean* is represented graphically by Fig. 6.1 and mathematically by (4) and (5).

$$\tau = \frac{A}{dA/dt} \qquad (3)$$

$$\tau_{1,\text{in}} = \frac{A_1}{R_{1,\text{in}}} \cdots \tau_{n,\text{in}} = \frac{A_n}{R_{n,\text{in}}} \qquad (4)$$

$$\|\leftarrow \text{at steady state} \rightarrow \|$$

$$\tau_{1,\text{out}} = \frac{A_1}{R_{1,\text{out}}} \cdots \tau_{n,\text{out}} = \frac{A_n}{R_{n,\text{out}}} \qquad (5)$$

where $A_1 \cdots A_n$ = the quantity of constituent 1 to n in the ocean
$R_{1,\text{in}}$ = rate of introduction of constituent 1 into the ocean
$R_{n,\text{out}}$ = rate of sedimentation of constituents n
τ = residence time computed for constituents 1 to n and from input (in) or sedimentation data (out)

This model, while superficially similar to previous ones, actually is very different, and some of the differences are quite subtle. For example,

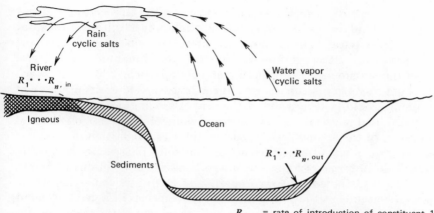

Figure 6.1 Model of the ocean based on residence time.

equations of the form of (5) occur in previous models where τ is referred to as "age of the earth" or "age of the ocean" and now "residence time." Strictly speaking, (3) cannot apply to the Barth model in which a steady state between introduction and removal is assumed. If the quantity of a given element A in the ocean is constant, then dA/dt must be zero.

A more precise statement of Barth's model is given in (4) and (5), which, while of the same form as (3), explicitly set out residence time of each element and provide the means of computing it from input and from removal data. In addition, the equality of residence times at steady state for a given element, computed from the two sources, is shown.

There are several features implicit in the presently accepted model that either have not been identified as important features of the real ocean or have not been tested in a quantitative way. One distinguishing feature of the present model is that it requires that the chemical systems in the sea be dynamic ones. That is, chemical reactions are believed to occur in which the order of chemical reactivity should be predictable from the behavior of the elements in isolated systems in the laboratory. This difference in reactivity is the basis on which we rationalize the observed difference in residence time from element to element.

There is an additional consequence of the residence time being a reflection of chemical reactivity; it follows that stable physical and chemical states and species of a single element may each be characterized by a residence time of its own. This simply expresses the fact that an element in different oxidation states of chemical combination will have different chemical reactivities. For example, aluminum may be carried by rivers to the sea in at least two forms—one in which the aluminum is an integral part of the alumino-silicate structure of a clay particle, and one in which aluminum exists in a soluble form. In the first case the residence time of aluminum would be a measure of the rate of settling of the solid phases concerned, and reactivity here has little meaning in the sense of chemical reactivity. In the second case, residence time as a reflection of chemical reactivity vis-á-vis adsorption reactions on solid phases can be readily rationalized in terms of known chemical behavior of solutions of aluminum salts, and the adsorptive properties of finely divided solids. If we accept the argument that different forms and species of a given element may have different residence times, what can be said of the measurement now reported in the literature? Certainly answers to this question must consider several features of the chemical analyses that produce the raw data for the computations. For example, some analytical methods "see" all of a given element regardless of form or species. Some methods, especially those using extraction and spectrophotometric techniques, give excellent results in simple inorganic systems but must suffer some deterioration in reliability when applied directly to natural waters that may contain significant quantities of complexing substances. Matson (1968), for example, notes that significant fractions of dissolved lead and copper in several natural freshwaters studied were complexed with high molecular weight organic ligands. It is reasonable to ask, were these complexes "seen" by the analytical methods used in the analyses of river water when obtaining data for "input" computations of residence time? Are the complexes stable enough or is a significant fraction of these elements always in a complex (steady state between formation and decomposition) in the sea to make it necessary to consider essentially two different "elements," that is, inorganic lead and complexed lead when estimating residence time? Certainly the chemical reactivities of the two will be different.

Goldberg (1965) noted that a tacit assumption of the Barth model is that the ocean be homogeneous with respect to elements for which we wish to compute residence times. This assumption extends the notion of constancy of relative proportions to all materials in the sea. While this can be shown to be invalid on theoretical grounds, as any *local* reaction which preferentially liberates or deposits an element must set up a concentration gradient with respect to that element, thus destroying the "constancy" of the system, it is nevertheless important to discover to what extent observable departures from constancy influence computed residence time in a practical sense.

The effects of the failure of the constancy notion should be magnified in the case of elements which have a computed residence time that is short compared with physical mixing times. In the case of aluminum, iron, and titanium for which residence times are of the order of 10^2 yr, is there a concentration gradient extending downstream from regions of introduction?

References

Barth, T. F. W. 1952. *Theoretical petrology*. 1st Ed. John Wiley & Sons, New York, 387 pp.

Barth, T. F. W. 1961. Abundance of the elements, area averages and geochemical cycles. *Geochim. Cosmochim. Acta*, **23**:1–8.

Barth, T. F. W. 1962. *Theoretical petrology*. 2nd Ed. John Wiley & Sons, New York, 416 pp.

Becker, G. F. 1910. The age of the earth. *Smithsonian Inst. Misc. Collections*, **56**:No. 6, 28 pp.

Clarke, F. W. 1924. The data of geochemistry. *U.S. Geol. Sur., Bull.* **770**.

Conway, E. J. 1943. The chemical evolution of the ocean. *Proc. Roy. Irish Acad.* **B48, B9**:119–159.

Goldberg, E. D., and G. O. S. Arrhenius. 1958. Chemistry of Pacific pelagic sediments. *Geochim. Cosmochim. Acta*, **13**:153–212.

Goldberg, E. D., 1965. Minor elements in the sea. Riley & Skirrow (eds.). In *Chemical Oceanography*, Vol. 1, Academic Press. New York, pp. 163–196.

Goldschmidt, V. M. 1954. *Geochemistry*. Oxford University Press, London.

Halley, E. 1715. A short account of the cause of the saltness of the ocean, and of the several lakes that emit no rivers, with a proposal by help theory, to discover the age of the world. *Phil. Trans. Roy. Soc.*, **29**:296–300.

Holmes, A. 1937. *Age of the earth*. Nelson, London, 287 pp.

Joly, J. 1899. An estimate of the geological age of the earth. *Sci. Trans. Roy. Dublin Soc.*, ser. 7 **3**:23–64.

Matson, W. R. 1968. Trace metals, equilibrium, and kinetics of trace metal complexes in natural media. Ph.D. Thesis, M.I.T., Cambridge, Mass., 258 pp.

7

Anoxic versus Oxic Environments

FRANCIS A. RICHARDS, *Professor of Oceanography, Department of Oceanography, University of Washington, Seattle, Washington*

The degradation of organic matter in aquatic systems proceeds via biological oxidations in which oxygen or oxygen-containing ions or compounds act as hydrogen acceptors. Organic matter can also be degraded by fermentative processes, which can be represented chemically as double decompositions rather than oxidation-reduction reactions. Unless the biota of a system is poisoned, it can accept organic matter which will be decomposed to inoffensive inorganic end products (and perhaps refractory organic compounds) as long as dissolved oxygen is present. We have thus learned to associate the presence of dissolved oxygen with an ecologically healthy system and to evaluate water quality, at least in part, in terms of its biochemical oxygen demand. Almost all of the ocean contains residues of dissolved oxygen and in this sense can be considered ecologically healthy.

There are relatively simple chemical models to describe the decomposition of organic matter in the ocean. These models are based on statistically average organic matter of planktonic origin (Redfield, 1934, 1942; Redfield et al., 1963). They permit one, within certain limitations, to relate quantitatively changes in concentrations of dissolved oxygen, carbon dioxide and carbonates, phosphate ions, and nitrogen compounds to the quantities of organic matter that have decomposed in the system. These models are useful in understanding the processes of organic decomposition, but their usefulness in assessing the balance between the supply of dissolved oxygen and organic matter to a particular part of the ocean system is limited. This limitation can be overcome only with rather extensive quantitative information on advective and diffusive processes and information on rates of organic production that is generally not available.

Certain generalities can be drawn from our models for decomposing organic matter that may be useful in evaluating the state of balance between oxygen supply and demand in ocean areas. It has become evident over the past 40 yr that the absolute quantities of organic matter that can be photosynthetically produced in the ocean are limited by the available amounts of phosphorus and nitrogen compounds. If one considers the oceans as a whole, this amount of organic matter is somewhat less than the amount required to consume all the oxygen that would be dissolved in the ocean if equilibrated, at its contemporary salinity and temperature, with air (at 1 atm pressure). The argument is that the maximum concentrations of phosphorus compounds in the world ocean are about 3 μg-atoms/l. If combined into organic matter having the composition of average plankton material, the organic matter would have an oxygen demand of about 850 μg-atoms/l of oxygen. Only about 750 μg-atoms/l of oxygen is present in air-saturated "average" seawater, so that if the situation prevailed in the "average" ocean, the eventual disappearance of dissolved oxygen from the ocean might be expected. However, as Redfield (1958) pointed out, the average situation in the world ocean would leave a small excess of dissolved oxygen after satisfaction of the potential oxygen demand. This overly simplified model ignores the dissolved oxygen produced during the photosynthetic formation of the organic matter, but this oxygen is produced in the upper tens of meters and apparently much of it diffuses into the atmosphere. Nonetheless, the fact remains that the balance between oxygen supply and demand in the world ocean is precarious. Vast areas of the eastern tropical Pacific Ocean contain very small quantities of dissolved oxygen, and it can be demonstrated that denitrification takes place in these areas. The introduction of small additions of organic matter into these regions could result in sulfate reduction with the concomitant formation of sulfides.

If the open ocean is in a precarious state of balance between oxygen supply and demand, the situation is even more tenuous in many near-shore systems where topography and vertical water structure limit the physical circulation of the water and therefore the circulatory replacement of dissolved oxygen to the deeper waters. These basins and fjords may receive increasing amounts of organic pollution, so that man's activities can foreseeably result in oxygen-bearing systems becoming anoxic. Pollution cannot be blamed for most of the anoxic systems that have been studied to date (reviewed by Broenkow, 1969), but the anoxic degradation of the Oslofjord is, or has been, threatened by pollution, and the anoxic regime in almost all of the known systems could be worsened by the introduction of organic pollutants.

Chemical Models

Oxic Conditions

Redfield (1934, 1942), Redfield et al. (1963), and Richards (1965a, b) have proposed models for the photosynthetic formation and respiratory oxidation of organic matter in the ocean based on the average elemental composition of plankton, in which the ratios of carbon, nitrogen, and phosphorus are 106:16:1, by atoms. We assume that the average oxidative state of carbon is that of carbohydrate, CH_2O, nitrogen is in the oxidative state of ammonia, NH_3, and the oxidation or reduction of phosphorus is quantitatively unimportant to the argument. Then photosynthesis can be represented by the equation

$$106CO_2 + 106H_2O + 16NH_3 + H_3PO_4 \rightarrow$$
$$(CH_2O)_{106}(NH_3)_{16}H_3PO_4 + 106O_2 \quad (1)$$

Respiration would be represented by the reverse of the equation. However, the ammonia so produced is transient and is biochemically oxidized to nitrate, in which form most of the combined nitrogen in the ocean occurs:

$$16NH_3 + 32O_2 \rightarrow 16HNO_3 + 16H_2O \quad (2)$$

The changes in concentrations of nitrogen, carbon, and phosphorus compounds and of dissolved oxygen during photosynthesis with ammonia as the nitrogen source would be, by atoms,

$$\Delta O : \Delta C : \Delta N : \Delta P = +212 : -106 : -16 : -1 \quad (3)$$

However, if nitrate is the nitrogen source, the ratios would be

$$\Delta O : \Delta C : \Delta N : \Delta P = +276 : -106 : -16 : -1 \quad (4)$$

The same ratios, with opposite signs, would hold during oxidative decomposition and respiration.

Many observations of the distribution of carbonate, phosphate, nitrate, and dissolved oxygen in the ocean tend to corroborate this model. Linear relationships between the variables were demonstrated to exist in all oceans by Redfield as early as 1934, and numerous workers have confirmed the matter. However, the relationships hold only for the ratios of changes brought about by biological processes in the body of the oceans and do not necessarily apply to ratios of absolute concentrations. This is because the nutrient content of a parcel of water may be made up of two fractions. One part has resulted from the oxidative mineralization of organic matter after the water left the sea surface, another part was already present when the parcel acquired its organic burden and dissolved oxygen content. Redfield (1942) has called the latter part the "preformed" nutrient content; Sugiura (1965) prefers to call it the "reserved" nutrient content. The other portion is referred to as being of oxidative origin. Thus

$$N = N_{preformed} + N_{oxidative} \tag{5}$$

where N is the nitrate or phosphate content, $N_{preformed}$ is the preformed nitrate or phosphate, and $N_{oxidative}$ is the nitrate or phosphate of oxidative origin. On the average, $NO_3^-{}_{(oxidative)} : PO_4^{3-}{}_{(oxidative)} = 16:1$, but the ratio of $NO_3^-{}_{(preformed)} : PO_4^{3-}{}_{(preformed)}$ and consequently of $NO_3^- : PO_4^{3-}$ may vary. Oxidative carbonates presumably occur at a ratio of 106 times the oxidative phosphate and 6.6 times the oxidative nitrate, but oxidative carbonates make up a maximum of 10 per cent of the total carbonate content of seawater, so most of the carbonate is preformed. Most or all of the nitrate and phosphate may be oxidative.

It follows that

$$PO_4^{3-}{}_{(oxidative)} = \frac{AOU}{106} \tag{6}$$

where AOU (apparent oxygen utilization) is the amount of dissolved oxygen that has been consumed from a parcel of water in oxidizing organic matter, and both phosphate and dissolved oxygen concentrations are expressed in μg-atoms/l. AOU is generally estimated as the difference between the observed oxygen content of a water parcel and the oxygen content the water would have if it were equilibrated with air at the sea surface at the observed values of salinity and temperature. This method of calculation tacitly assumes that the water has at some time been at the sea surface and has become saturated with air at that time.

Similarly,

$$NO_3^-{}_{(oxidative)} = \frac{AOU}{17.25} \tag{7}$$

and

$$(HCO_3^- + CO_3^{2-})_{oxidative} = \frac{AOU}{1.72} \tag{8}$$

The nutrients of oxidative origin are obviously controlled by biological processes, and conversely the preformed nutrients are conservative properties whose distributions, in the body of the oceans, are controlled by the physical processes of advection and diffusion. Thus Redfield (1942) was able to demonstrate a close relationship between the distributions of preformed phosphate and salinity in the Atlantic Ocean.

Departures from the model can be expected and occur under a variety of circumstances. The model is intended to apply only in the body of the ocean and departures can be expected at the boundaries. Oxygen can diffuse across the sea surface with relative ease, whereas combined nitrogen and phosphorus cannot. Thus biological changes in nitrogen and phosphorus contents at or near the sea surface may not be reflected in corresponding changes in the dissolved oxygen content. Redfield (1948) and Pytokowicz (1964) have used this fact to estimate rates of exchange of oxygen across the sea surface.

The model is concerned only with the beginning nutrient and oxygen concentrations and the final end products of the oxidative decomposition of organic matter. It tells us nothing of the intermediate formation and final disposition of nitrite and ammonium ions in the nitrogen cycle. Similarly, it cannot describe intermediate states that may result from nitrogen, carbon, and phosphorus in organic matter and organisms being oxidized and converted into inorganic end products at different rates. The model might lead one to expect, for example, that vertical distributions of nitrate and phosphate would have the same shape. In fact, they have similar but not identical shapes, possibly because differing water masses present under one oceanic area have differing preformed nutrient contents, possibly because phosphate is remineralized faster than nitrate, or the discrepancy may arise from a combination of causes.

Anoxic Conditions

Although oxic conditions prevail in almost all of the world's ocean, there are important systems in which the biochemical oxygen demand exceeds the oxygen supplies, and anoxic conditions arise. We refer to anoxic systems as those in which any part of the water column becomes completely devoid of oxygen. The condition may be permanent or intermittent. The intermittently anoxic systems and systems that have very small oxygen contents may be of particular interest in the context of the impingement of man on the oceans, because they may be in a critical state of balance and therefore particularly sensitive to man's activities.

In general, we can consider that oxygen-bearing water has at one time or another entered or even filled these systems. The decomposition of organic matter will proceed, so long as appreciable quantities of dissolved oxygen are present, according to the model for oxic conditions discussed above. However, as additional organic matter enters the system, all the dissolved oxygen will be consumed. When the dissolved oxygen disappears or reaches very low concentrations, denitrification and ultimately sulfate reduction will ensue. Equations representing these steps have been presented by Richards (1965a, b).

1. Denitrification:*

$$(CH_2O)_{106}(NH_3)_{16}(H_3PO_4) + 84.8HNO_3 = 106CO_2 \\ + 42.4N_2 + 148.4H_2O + 16NH_3 + H_3PO_4 \quad (9)$$

The ammonia released during this process may or may not be oxidized by some biochemical or inorganic process. If it is oxidized, the process might be represented by the equation

$$(CH_2O)_{106}(NH_3)_{16}(H_3PO_4) + 94.4HNO_3 = 106CO_2 \\ + 55.2N_2 + 177.2H_2O + H_3PO_4 \quad (10)$$

An intermediate step in this process is usually represented by accumulations of nitrite ions.

In seawater the reduction of sulfate ions can be expected to follow denitrification when all or nearly all of the nitrate and nitrite ions have been reduced to N_2.

2. Sulfate reduction:

$$(CH_2O)_{106}(NH_3)_{16}(H_3PO_4) + 53SO_4^{2-} = 106CO_2 + 53S^{2-} \\ + 16NH_3 + 106H_2O + H_3PO_4 \quad (11)$$

We know of no systems connected with the open ocean where other sources of oxygen appear to be used. All the systems studied still contain excess concentrations of sulfate ions, concentrations adequate to serve as hydrogen acceptors during the oxidative decomposition of any reasonable additions of organic matter. However, there are systems containing completely landlocked seawater in which all the sulfate ions have been reduced. Such systems have been reported from Norway by Strøm (1957, 1961) and from British Columbia by Williams et al. (1961). These systems

---

* Denitrification is used in this chapter, unless otherwise indicated, to mean the formation of free nitrogen from nitrate or nitrite ions, or both. Nitrate reduction to nitrite may take place as an intermediate step.

are characterized by large concentrations of carbonates and methane and by the absence of sulfates.

Methane could be formed in anoxic systems either by anaerobic fermentation or by carbonates serving as hydrogen acceptors in the oxidative decomposition of organic matter. However, the fermentative pathway is the most probable (Atkinson and Richards, 1967):

$$(CH_2O)_{106}(NH_3)_{16}(H_3PO_4) = 53CH_4 + 53CO_2 + 16NH_3 + H_3PO_4 \quad (12)$$

The fermentation can proceed at the expense of a variety of substrates, so that the equation may not be very realistic; acetate is a more probable substrate than carbohydrate, and carbohydrate is used in the formulation only to indicate the approximate oxidation state of the carbon, not the specific kinds of carbon compounds.

Under the extreme conditions that might arise in landlocked seawater, other ions, such as phosphate, might serve as hydrogen acceptors in the oxidative decomposition of organic matter:

$$(CH_2O)_{106}(NH_3)_{16}(H_3PO_4) + 53H_3PO_4 = 106CO_2 \\ + 106H_2O + 53PH_3 + 16NH_3 + H_3PO_4 \quad (13)$$

Phosphine and phosphonium salts are unstable and are unlikely to be detected in the natural environment. In the laboratory preparation of phosphine, bubbles of a mixture of phosphine and methane or hydrogen can be produced, which ignite spontaneously on exposure to air. Pure phosphine is not spontaneously combustible and is probably ignited by traces of P_2H_4. It is possible that this spontaneous combustion is responsible for ignition of the will-o'-the-wisp (*ignis fatuus*), but we know of no such occurrences over marine systems.

Properties of Anoxic Systems

1. By definition, anoxic systems are devoid of oxygen. However, the biological consequences of very low oxygen concentrations, such as those found in the intermediate waters of the eastern tropical Pacific region, are of considerable interest. Living fish have been taken from these waters (Wooster, 1952), and at least certain organisms can survive very low oxygen tensions. Sulfide-bearing waters are generally devoid of living organisms except bacteria.

2. Larger quantities of organic matter have decomposed in anoxic systems than in oxygen-bearing systems, so larger concentrations of certain products of organic decomposition accumulate in these environments than

ever occur in the open ocean. This includes phosphates, carbonates, ammonia, and silicates (Richards, 1965a, b).

3. The unusual accumulation of carbonates may result in increased concentrations of alkaline earth cations, which may be caused by the solution of carbonates:

$$MeCO_3 + H_2O + CO_2 = Me^{2+} + 2HCO_3^- \qquad (14)$$

in which Me represents an alkaline earth metal (Knull and Richards, 1969).

4. Presumably nitrogen is released on decomposition of organic matter in a reduced state, for example, ammonia, amino acids, urea, and uric acid. However, in the ocean most of this nitrogen is eventually biochemically oxidized to nitrate. This is an oxygen-consuming step, as indicated in (2). The oxidized nitrogen is then reduced to free N_2 by denitrification as anoxic conditions are approached or reached. This comprises a loss of biologically available combined nitrogen and consequently a loss of plant nutrient from the marine ecosystem. The region of denitrification in the eastern tropical Pacific Ocean comprises two large lobes extending westward, centered at about 12°N and 12°S latitude, and the loss of combined nitrogen in this region must be large enough to be biogeochemically significant. As yet we have no estimates of the magnitude of this loss to the marine ecosystem.

5. Redfield (1958) has pointed out that the process of sulfate reduction can ultimately serve to replenish oxygen (consumed by organic decomposition) to the atmosphere. During sulfate reduction, sulfide ions and CO_2 are formed. Upon subsequent transport of this mixture to the photic zone, the S^{2-} may be oxidized to free sulfur while the CO_2 becomes photosynthetically bound, concurrently releasing O_2 to the system. The simplified equations follow:

$$\text{Sulfate reduction: } 2CH_2O + SO_4^{2-} = 2CO_2 + 2H_2O + S^{2-} \qquad (15)$$

$$\text{Sulfide oxidation: } S^{2-} + \frac{1}{2}O_2 + 2H^+ = S^0 + H_2O \qquad (16)$$

$$\text{Photosynthesis: } CO_2 + H_2O = CH_2O + O_2 \qquad (17)$$

$$\text{Overall: } SO_4^{2-} + 2H^+ = S^0 + H_2O + \frac{3}{2}O_2 \qquad (18)$$

6. The formation of sulfides in anoxic systems is accompanied by a lowering of the redox potential of the system and by a change of oxidation state of certain metals, including iron and manganese. Both iron and

manganese tend to be solubilized by the process, because the sulfides of their lower oxidation states are more soluble than the oxides of their higher oxidation states, which apparently control the concentrations of these metals in solution in oxygen-bearing waters. The sulfide-bearing waters of Lake Nitinat, British Columbia, are apparently saturated with ferrous sulfide (Richards et al., 1965). Sulfides also form extremely insoluble compounds with many metals, such as copper and mercury, and Richards (1965a) reported that copper concentrations drop to below detection limits in the sulfide-bearing waters of the Cariaco Trench.

7. If the conjugate acids of weak acids or bases, such as sulfide or ammonium ions, are added to seawater, the alkalinity of the system will be altered, although the addition of neutral molecules, such as H_2S or CO_2, will leave the alkalinity unchanged (Knull and Richards, 1969). Knull and Richards found the specific alkalinity of Lake Nitinat to be measurably higher than at a nearby open-ocean station, and most of the excess alkalinity could be accounted for by the amounts of S^{2-} and HS^- ions present. Their analytical methods were not sufficiently precise to demonstrate whether such anions as acetate and other organic acid anions might be making a significant contribution to the alkalinity, as Gripenberg (1960) suggested is the case in the Baltic Sea. However, a relatively high specific alkalinity can be considered to be a usual characteristic of anoxic marine environments.

8. Carbon dioxide is a product of all the processes degrading organic matter, and its accumulation will tend to lower the pH, although this lowering is minimized by the buffering of seawater. In Lake Nitinat, pH values decrease from 7.3 or 7.5 in the oxygen-bearing layers to about 7.2 in the sulfide-bearing water (Knull and Richards, 1969). Skopintsev (1957) reported pH values in the Black Sea decreasing from surface values of around 8.3 to about 7.8 at the sulfide-oxygen interface (ca. 200 m), and then decreasing only a few hundredths of a pH unit as the bottom (2000 m) was approached.

9. The interface between oxygen- and sulfide-bearing waters is one of considerable chemical and biochemical interest. If the interface is in the photic zone, the activities of photosynthetic sulfur bacteria may become dominant with the production of free sulfur:

$$CO_2 + 2H_2S = CH_2O + 2S^0 + H_2O \tag{19}$$

The inorganic reaction between dissolved oxygen and sulfides has been studied by Cline and Richards (1969). The mechanics and kinetics of the reaction, which they investigated in a closed system filled with natural seawater from an anoxic fjord, are complicated and appear to depend, among other things, on the relative initial concentrations of sulfide and oxygen and on the presence or absence of a variety of catalysts, especially metallic cations. A variety of products is formed, the major ones being sulfate and thiosulfate, but sulfite appeared as a fairly long-lived unstable intermediate.

10. The sediments laid down under anoxic waters tend to be richer in organic matter than those laid down under oxygen-bearing waters. The latter generally contain 1 to 2 per cent of organic carbon, but sediments in Saanich Inlet contain up to 7.2 per cent organic carbon and Cariaco Trench sediments have up to 4 per cent organic carbon (Richards, 1970). The sediments of Saanich Inlet and apparently of other anoxic environments are rich in chlorophyll, chlorophyll degradation products, or both (Richards, 1965a).

The water in anoxic systems is apparently not enriched in either soluble or particulate organic matter. However, gross organic matter, such as bones and scales, tends to be preserved on the bottom under anoxic waters, presumably because of the absence of benthic organisms.

The bottom is probably the site of denitrification, sulfate reduction, and anaerobic fermentation in at least the shallower anoxic systems, with the products of these processes diffusing upward into the water column. These processes probably also take place in the water column itself, particularly in deeper systems such as the Black Sea.

Anoxic Systems in Nature

Table 7.1 (Broenkow, 1969) contains a listing and the main features of anoxic marine basins in communication with the open sea. The table thus excludes the landlocked fjords containing relict seawater described by Strøm (1957, 1961) and by Williams et al. (1961).

The Black Sea (Caspers, 1957) is by far the largest of the anoxic marine basins. It is characterized by rather low-salinity surface waters, diluted by major rivers entering the Sea of Azov and the northwestern part of the Black Sea. Dissolved oxygen disappears at depths of around 150 to 250 m, the shallower parts of the sulfide-oxygen interface lying in the central parts of the eastern and western lobes of the sea. Maximum sulfide concentrations of about 300 μM are reached near or at the bottom, around 2000 m. The vertical stability of the water column is maintained by an influx of saline water from the Mediterranean Sea through the Bosporus, and by the influx of freshwater, primarily from the rivers in the northern part of the Black Sea–Sea of Azov system.

Table 7.1. Anoxic Marine Basins in Communication with the Open Sea

Name and Location	Maximum Depth (m)	Sill Depth (m)	Maximum Hydrogen Sulfide (µg-atoms/l)	Bottom Salinity (‰)	Bottom Temperature (C)	Reference
Black Sea	2243	40	300	22.3	9.0	Caspers, 1957
Bolstadtfjord, Norway	140	1	47	21.1	4.1	Strøm, 1936
Cariaco Trench, North Coast of Venezuela	1390	150	25	36.2	17	Richards and Vaccaro, 1956
Darwin Bay, Isla Genovesa, Galapagos Islands	240	15	Present	34.9	18	Goering and Dugdale, 1966
Dramsfjord, Norway	117	8	215	29.9	5.3	Strøm, 1936
Fårö Deep, Baltic Sea	200	18	1	12.0	5.5	Fonselius, 1963
Fjellangervag, Norway	75	3	260	30.3	5.5	Strøm, 1936
Framvaren, Norway	175	2	1800	24.6	8.8	Strøm, 1936
Frierfjord, Norway	93	11	25	34.2	6.1	Strøm, 1936
Golfo Dulce, Costa Rica	200	50	9	34.7	16	Richards et al. (in press)
Gotland Deep, Baltic Sea	250	18	1	12.5	5.5	Fonselius, 1963
Gulf of Cariaco, Venezuela	80	54	33	36.6	21	Richards, 1960
Hellefjord, Norway	75	9	1800	33.3	5.9	Strøm, 1936
Helvigfjord, Norway	36	3	290	32.3	7.2	Strøm, 1936

Table 7.1. (*Continued*)

Name and Location

Maximum Depth (m)	Sill Depth (m)	Maximum Hydrogen Sulfide (μg-atoms/l)	Bottom Salinity (‰)	Bottom Temperature (C)	Reference
Indre Lyngdalsfjord, Norway					
116	3	300	32.3	7.2	Strøm, 1936
Isefjaerfjord, Norway					
27	1	350	31.1	6.0	Strøm, 1936
Kaoe Bay, Halmahera Island, Indonesia					
491	50	13	34.5	28	Van Riel, 1943
Lake Suigestsu, Japan					
34	2	3000	16	15	Shigematsu et al., 1961
Lake Nitinat, British Columbia, Canada					
205	4	350	30.2	10	Richards et al., 1965
Lake Varna, Bulgaria					
18		present	14	8	Caspers, 1957
Lenefjord, Norway					
240	3	175	34.2	5.3	Strøm, 1936
Lygrefjord, Norway					
180	6	3	33.8	6.7	Strøm, 1936
Nordasvatn, Norway					
87	3	340	31.0	6.3	Strøm, 1936
Saanich Inlet, British Columbia, Canada					
240	70	25	31.4	9.0	Herlinveaux, 1962
Skjoldafjord, Norway					
109	3	10	26.8	8.2	Strøm, 1936
Søndeledspoll, Norway					
17	2	240	33.1	6.8	Strøm, 1936
Tofino Inlet, British Columbia, Canada					
96	66	400	29.6	13	Coote, 1964
Vestrhusfjord, Norway					
55	2	440	32.4	5.0	Strøm, 1936

The fjords of Scandinavia and of British Columbia, Canada, offer a wide variety of anoxic systems, which have in common a seaward sill and generally diluted upper layers. They have a large range of sill depths and flushing characteristics. Some are not known to flush completely. An example is Lake Nitinat, Vancouver Island, British Columbia (Richards et al., 1965). This fjord has an effective sill depth of 3 to 4 m, and sulfide concentrations of up to 300 μM accumulate in its deepest water, around 200 m. About 20 per cent of this deep, sulfide-bearing water of Lake Nitinat was observed to be replaced by new water between June and August 1966, but the system is probably never free of sulfide-bearing waters.

An anoxic system close to and in many respects comparable to Lake Nitinat is Saanich Inlet, also on Vancouver Island (Richards, 1965a). In contrast to Lake Nitinat, Saanich Inlet has an effective sill depth of around 90 m and apparently flushes annually. Maximum sulfide concentrations are $\frac{1}{10}$ or less of those occurring in Lake Nitinat. One consequence of these differences is the relative roles of denitrification, which dominates in Saanich Inlet, and sulfate reduction, which dominates in Lake Nitinat, in the decomposition of organic matter. During the annual cycle in Saanich Inlet, about 30 μg-atoms/l of NO_3^--nitrogen and about an equal concentration of sulfate ions are reduced. In Lake Nitinat, the anoxic waters reflect the reduction of perhaps 30 μg-atoms/l of NO_3^--nitrogen and about 10 times this concentration of sulfate ions.

In contrast to the Black Sea and the fjords of the temperate latitudes, there are anoxic systems in the tropics that owe their vertical stability to their thermal stratification instead of salinity stratification. The largest known tropical anoxic marine system is the Cariaco Trench, off the Caribbean coast of Venezuela (Richards and Vaccaro, 1956; Richards, 1965a, b). This system has a limiting sill connecting it with the open Caribbean Sea at around 140 m and maximum sulfide concentrations of about 25 μM. Unlike the sulfide distribution in the Black Sea, the maximum sulfide concentrations in the Cariaco Trench probably do not occur on the bottom, at about 1400 m, but at some intermediate depth (Richards and Vaccaro, 1956; Richards and Benson, 1961; Atkinson and Richards, 1967). This and the much smaller sulfide concentrations in the Cariaco Trench suggest that its deep water is being replaced, continuously or intermittently, with new, oxygen-bearing water from the Caribbean Sea. Redfield et al. (1963) have estimated that the halflife of the water below sill depth in the Cariaco Trench is 71 yr. However, during many observations in the Trench since the first observation in December 1954, no oxygen-bearing water has been observed at depths greater than around 400 m, and the sulfide-bearing water is probably never completely replaced.

Other small tropical systems containing anoxic waters have been observed. The Gulf of Cariaco (Gade, 1961; Richards, 1960) is adjacent to the Cariaco Trench and intermittently contains sulfides. Sulfides were reported in Kaoe Bay in Indonesia during the *Snellius* Expedition (Van Riel et al., 1950). It was discovered during a cruise of the R.V. *Thomas G. Thompson* in March 1969 that Golfo Dulce, an embayment on the Pacific coast of Costa Rica, contained anoxic and sulfide-bearing water near the bottom (Richards et al., in press).

Rates of Oxygen-Consuming Processes

Although the processes leading to anoxic conditions appear to proceed in a relatively regular order, the rates of oxygen loss, denitrification, and sulfate reduction in anoxic systems are mostly unknown. We are beginning to get a time series of observations from Saanich Inlet that should yield some of these rates from the observation of the annual establishment of anoxic conditions. In similar studies, Barnes and Collias (1958) estimated oxygen consumption rates in several Puget Sound basins. These basins never became wholly anoxic, but average oxygen consumption rates of 0.005 to 0.02 ml/(l)(day) were observed. Richards and Broenkow (to be published) have recently made a direct observation of the rate of nitrate reduction in Darwin Bay, in the Galapagos Islands. The deep waters of the bay, the flooded caldera of a volcano, are prevented by the topography from exchanging horizontally with the open waters of the Pacific Ocean. The distributions of dissolved oxygen, nitrite, nitrate, and other chemical variables in the bay were observed in March and again in May 1969. Very little oxygen was present on either occasion, but between the two visits nitrate ions had disappeared and nitrite ions had appeared in a 1:1 correspondence. The observed loss of NO_3^- ions leads to a rate of nitrate reduction of about 70 μg-atoms/(l)(yr). This compares with the range of oxygen consumption rates in Puget Sound of 163 to 653 μg-atoms/(l)(yr) observed by Barnes and Collias (1958).

Natural and Polluted Systems

The processes and anoxic systems that have been discussed in this chapter arise as the result of natural conditions occurring in somewhat unusual marine environments. However, even these natural, essentially unpolluted systems can be detrimental. Occasionally stagnant, sulfide-bearing fjords have been observed to overturn, with resulting mass mortality of marine populations (Brongersma-Sanders, 1957). Sulfide-bearing seawater is corrosive to metal objects, and would present a difficult environment for man-made objects in the sea. These systems also represent accumulations of nutrients, which, if stirred into the photic zone, could result in overfertilization and the development of weed crops of algae. The introduction of additional organic wastes into these systems can be expected to result in the deterioration of the environment. Whether the deterioration results in conditions detrimental to man and nature depends on the balance of physical, chemical, and biological conditions. Physical alteration of marine systems, as by damming and dredging operations, may result in stagnation and anoxic conditions. In general, these are conditions to be avoided.

We have discussed systems that become anoxic through natural processes and have not attempted to cover anoxic conditions that might arise as a result of gross, or toxic, pollution. These cases represent widely variable individual situations, and the chemical models that help describe the naturally anoxic systems cannot be expected to apply to them.

Acknowledgments

Contribution No. 587 from the Department of Oceanography, University of Washington. This research was supported by the National Science Foundation, grant GA-644.

References

Atkinson, L. P., and F. A. Richards. 1967. The occurrence and distribution of methane in the marine environment. *Deep-Sea Res.*, **14**:673–684.

Barnes, C. A., and E. E. Collias. 1958. Some considerations of oxygen utilization rates in Puget Sound. *J. Marine Res.*, **17**:68–80.

Broenkow, W. W. 1969. The distributions of nonconservative solutes related to the decomposition of organic material in anoxic marine basins. Ph.D. Dissertation, University of Washington, Seattle, 207 pp.

Brongersma-Sanders, M. 1957. Mass mortality in the sea. In *Treatise on Marine Ecology and Paleoecology, Vol. I, Ecology*. J. Hedgpeth (ed.), *Mem. Geol. Soc. Amer.*, No. 67, 941–1010.

Caspers, H. 1957. Black Sea and Sea of Azov. In *Treatise on Marine Ecology and Paleoecology, Vol. I, Ecology*. J. Hedgpeth (ed.), *Mem. Geol. Soc. Amer.*, No. 67, 801–889.

Cline, J. D., and F. A. Richards. 1969. Oxidation of hydrogen sulfide in seawater at constant salinity, temperature, and pH. *Environ. Sci. Technol.*, **3**:838–843.

Coote, A. R. 1964. A physical and chemical study of Tofino Inlet, Vancouver Island, British Columbia. M.S. Thesis, University of British Columbia, Vancouver, 74 pp.

Fonselius, S. H., 1963. Hydrogen sulfide basins and a stagnant period in the Baltic Sea. *J. Geophys. Res.*, **68**:4009–4016.

Gade, H. G. 1961. Informe sobre las condiciones hidrograficas en el Golfo de Cariaco, para el periodo que empieza en Mayo y termina en Noviembre de 1960. *Bol. Inst. Oceanog. Univ. Oriente, Cumana, Venezuela*, **1**:21–47.

Goering, J. J., and R. C. Dugdale. 1966. Denitrification rates in an island bay in the Equatorial Pacific Ocean. *Science*, **154**:505–506.

Gripenberg, S. 1960. On the alkalinity of Baltic waters. *J. Cons. Perm. Int. Explor. Mer*, **26**:5–20.

Herlinveaux, R. H. 1962. Oceanography of Saanich Inlet in Vancouver Island, British Columbia. *J. Fish. Res. Bd. Canada*, **19**:1–37.

Knull, J. R., and F. A. Richards. 1969. A note on the sources of excess alkalinity in anoxic waters. *Deep-Sea Res.*, **16**:205–212.

Pytkowicz, R. M. 1964. Oxygen exchange rates off the Oregon coast. *Deep-Sea Res.*, **11**:381–389.

Redfield, A. C. 1934. On the proportions of organic derivatives in seawater and their relation to the composition of plankton, pp. 176–192. In *James Johnstone Memorial Volume*, University of Liverpool Press.

Redfield, A. C. 1942. The processes determining the concentration of oxygen, phosphate, and other organic derivatives within the depth of the Atlantic Ocean. *Papers Phys. Oceanog. Meteorol.*, **9**.

Redfield, A. C. 1948. The exchange of oxygen across the sea surface. *J. Mar. Res.*, **7**:347–361.

Redfield, A. C. 1958. The biological control of chemical factors in the environment. *Amer. Scientist*, **46**:205–221.

Redfield, A. C., B. H. Ketchum, and F. A. Richards. 1963. The influence of organisms on the composition of seawater. In *The Seas*, M. N. Hill (ed.), Interscience Publishers, New York, Vol. II, Chap. 2.

Richards, F. A., and R. F. Vaccaro. 1956. The Cariaco Trench, an anaerobic basin in the Caribbean Sea. *Deep-Sea Res.*, **3**:214–228.

Richards, F. A. 1960. Some chemical and hydrographic observations along the north coast of South America. I. Cabo Tres Puntas to Curaçao, including the Cariaco Trench and the Gulf of Cariaco. *Deep-Sea Res.*, **7**:163–182.

Richards, F. A., and B. B. Benson. 1961. Nitrogen/argon and nitrogen isotope ratios in two anaerobic environments, the Cariaco Trench in the Caribbean Sea and Dramsfjord, Norway. *Deep-Sea Res.*, **7**:254–264.

Richards, F. A., and W. W. Broenkow. Chemical changes, including nitrate reduction, in Darwin Bay, Galapagos Archipelago, over a two-month period, 1969. *Limnol. Oceanog.* (to be published).

Richards, F. A., J. D. Cline, W. W. Broenkow, and L. P. Atkinson. 1965. Some consequences of the decomposition of organic matter in Lake Nitinat, an anoxic fjord. *Limnol. Oceanog.*, **10** (Suppl.): R185–R201.

Richards, F. A. 1965a. Chemical observations in some anoxic, sulfide-bearing basins and fjords. *Proc. 2nd Int. Water Pollution Conf.*, Tokyo, 1964, pp. 215–232.

Richards, F. A. 1965b. Anoxic basins and fjords, pp. 611–645. In *Chemical Oceanography*, J. P. Riley and G. Skirrow (eds.). Academic Press, New York, Vol. I.

Richards, F. A., J. J. Anderson, and J. D. Cline. Chemical and physical observations in Golfo Dulce, an anoxic basin on the Pacific coast of Costa Rica. *Limnol. Oceanog.* (in press).

Richards, F. A. 1970. The enhanced preservation of organic matter in anoxic marine environments. In *Proc. Symp. organic matter in natural waters*, D. W. Hood (ed.). University of Alaska, College, Alaska.

Shigematsu, T., M. Tabushi, Y. Nishikawa, T. Muroga, and Y. Matsunaga. 1961. Geochemical study on Lakes Mikata. *Bull. Inst. Chem. Res., Kyoto Univ.*, **39**:43–56.

Skopintsev, B. A. 1957. The study of redox potential of the Black Sea waters. *Gidrokhim. Materialy*, **27**:21–36.

Strøm, K. 1936. Land-locked waters, hydrography and bottom deposits in badly-ventilated Norweigian fjords with remarks upon sedimentation under anaerobic conditions. *Skrifter Norske Videnskaps-Akad.*, **7**.

Strøm, K. 1957. A lake with trapped sea-water? *Nature*, **180**:982–983.

Strøm, K. 1961. A second lake with old sea-water at its bottom. *Nature*, **189**:913.

Sugiura, Y. 1965. On the reserved nutrient matters. *La Mer, Bull. Soc. Franco-Japanaise Oceanog.*, **2**:7–91.

Van Riel, P. M. 1943. Oceanographic results, the bottom waters. Introductory results and oxygen content. *Snellius Expedition Rept.* **2**.

Van Riel, P. M., H. C. Hamaker, and L. Van Eyck. 1950. Oceanographic results, tables, serial and bottom observations, temperature, salinity and density. *Snellius Expedition*, 1929–1930, **2**(6).

Williams, P. M., W. H. Mathews, and G. L. Pickard. 1961. A lake in British Columbia containing old sea-water. *Nature*, **191**:830–832.

Wooster, W. S. 1952. Shellback expedition, 17 May to 27 August 1952. University of California, Scripps Instit. Oceanogr., Unpublished Ms., Ref. 52–63.

8

Thermodynamics

PAUL W. SCHINDLER, *Professor of Chemistry, Institut für anorganische, analytische und physikalische Chemie, University of Berne, Berne, Switzerland*

The chemical reactions which are responsible for the formation of the ocean and the atmosphere can be summarized by the mass balance

primary igneous rock + volatile substances →
$$\text{sedimentary rocks + seawater + atmosphere} \quad (1)$$

Goldschmidt (1933) estimates that the formation of 1 liter of seawater is connected with weathering of 600 g of primary igneous rock. Recent computer-derived balances (Horn and Adams, 1966) for some major elements are summarized in Table 8.1. From the individual mass balances it becomes apparent that the distribution of the elements between the solid phases (sedimentary minerals) and the aqueous solution differs widely. Some elements such as sodium, potassium, calcium, and magnesium are fairly represented in the sea but other elements (silicon, aluminum, iron, and titanium) are preponderantly accumulated in the solids. To obtain an understanding of the factors that govern these distributions, consider (1) as representing a true equilibrium. Such equilibrium models were first introduced by Sillén (1961) whose pioneer work has stimulated much activity in this field. The first part of this chapter therefore outlines the thermodynamic basis of equilibrium models and illustrates their applicability by considering some particular systems.

Table 8.1 also gives indications for possible limitations of equilibrium considerations. The low residence time, that is, the rate of their removal from the aqueous phase, suggests that the ocean is strongly oversaturated with some elements. For these compounds the equilibrium model should be replaced by a steady state model.

Table 8.1. Geochemical Mass Balances for Some Major Elements[a]

Element	From Weathered Igneous Rock	Present in Sedimentary Rock[b]	Present in the Ocean	Residence Time in the Ocean[c] (yr)
Si	5820	5820	4.2×10^{-2}	8.0×10^3
Al	1620	1620	1.4×10^{-4}	1.0×10^2
Fe	862	862	1.4×10^{-4}	1.4×10^2
Ca	739	733	5.61	8.0×10^6
Na	572	425	147	2.6×10^8
K	524	519	5.33	1.1×10^7
Mg	358	340	18.2	4.5×10^7
Ti	98.6	98.6	1.4×10^{-5}	1.6×10^2

From Horn and Adams (1966).
[a] Unit: $G = 10^{20}$ g.
[b] Including connate water.
[c] E. D. Goldberg, in *Chemical Oceanography*, Vol. I, J. R. Riley and G. Skirow (eds.), Academic Press, London, 1965, p. 164.

Finally it is shown that interactions between equilibrium systems and steady state systems may—by feedback via the atmosphere—cause minor fluctuations in both seawater composition and general environmental conditions.

Thermodynamic Principles

The application of thermodynamic principles to geochemical problems requires, first, the isolation of the interesting part of the earth against its surroundings. The isolated part is called a system. There are three fundamental forms of isolation which in turn lead to three different types of systems.

1. Adiabatic systems are surrounded by walls, which prevent exchange of matter and thermal energy with their environment. Mechanical work, however, may be done or taken up by the system (Fig. 8.1a).

2. Closed systems can exchange energy with the surroundings whereas exchange of matter is prevented. An exchange of matter between different phases within the system is possible (Fig. 8.1b).

3. Open systems are enclosed by walls which allow exchange of energy as well as exchange of matter (Fig. 8-1c).

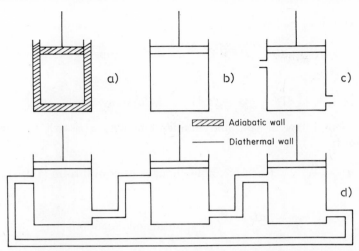

Figure 8.1 Thermodynamic systems: a) adiabatic system; b) closed system; c) open system; and d) open systems connected to a closed system.

For constructing equilibrium models, the consideration of closed systems will prove to be the most useful. Nevertheless, we start our discussion with open systems; later we may connect two or more open systems so as to make the total mass leaving one system equal to that entering the other system (Fig. 8.1d). Such an assembly of open systems then forms a closed system.

The first law of thermodynamics states that

$$dU = dQ - dW \qquad (2)$$

That is, the change in internal energy U of the system is calculated from dQ, the amount of thermal energy absorbed by the system minus dW, the amount of work done by the system. The W comprehends all kinds of work done by the system (or acted upon the system). Here we are concerned with mechanical work $P\,dV$ and chemical work $\mu_i\,dn_i$. The intensive factor μ_i was first introduced by Gibbs. It is called chemical potential and relates the chemical work done by the system with dn_i, the number of moles of compound i which are transported from the system to the environment. Summing up, we obtain

$$dW = P\,dV - \sum_{i=1}^{k} \mu_i\,dn_i \qquad (3)$$

Equation 2 now becomes

$$dU = dQ - P\,dV + \sum_{i=1}^{k} \mu_i\,dn_i \quad \text{(open system)} \tag{4}$$

or

$$dU = dQ - P\,dV \quad \text{(closed system)} \tag{4a}$$

The second law of thermodynamics connects dQ with the entropy change dS:

$$dS = \frac{dQ}{T} \quad \text{(for reversible changes or equilibrium states)} \tag{5}$$

or

$$dS > \frac{dQ}{T} \quad \text{(for spontaneous reactions)} \tag{5a}$$

From (4) and (5) we obtain Gibbs' equation:

$$dU = T\,dS - P\,dV + \sum_{i=1}^{k} \mu_i\,dn_i \tag{6}$$

Integration with the aid of Eulers' theorem gives

$$U = TS - PV + \sum_{i=1}^{k} \mu_i\,dn_i \tag{7}$$

Instead of the internal energy U, the Gibbs free energy G is often used:

$$G = U - TS + PV \tag{8}$$

The change of G is given by

$$dG = dU - T\,dS - S\,dT + P\,dV + V\,dP \tag{9}$$

or, substituting (6) for dU,

$$dG = -S\,dT + V\,dP + \sum_{i=1}^{k} \mu_i\,dn_i \tag{10}$$

Equation 10 has been derived for open systems. As mentioned earlier, we may now connect an appropriate assembly of open systems to form a closed system. Within this closed system there will be some mass transport summarized by the equation

$$\nu_A A + \nu_C C + \cdots \to \nu_B B + \nu_D D + \cdots \tag{11}$$

which turns out to be an ordinary chemical reaction. The advancement of this reaction is conveniently expressed by the progress variable ξ (which is also called the degree of advancement):

$$dn_i = \nu_i \, d\xi \tag{12}$$

Taking ν_i values positive for products and negative for reactants, we obtain

$$dG = -S \, dT + V \, dP + d\xi \sum_{i=1}^{k} \nu_i \mu_i \tag{13}$$

The term $\sum_{i=1}^{k} \nu_i \mu_i$ is called the free energy change ΔG of the reaction; the quantity $-\sum_{i=1}^{k} \nu_i \mu_i$ is the affinity A of the reaction

$$\sum_{i=1}^{k} \nu_i \mu_i = \Delta G = -A \tag{14}$$

For reactions at constant temperature and pressure we obtain

$$dG = \Delta G \, d\xi = -A \, d\xi \tag{15}$$

or

$$\left(\frac{\partial G}{\partial \xi}\right)_{P,T} = \Delta G = -A \tag{15a}$$

Furthermore, (4a), (6), (13), and (14) result in

$$dS = \frac{dQ}{T} - \frac{1}{T} \Delta G \, d\xi = \frac{dQ}{T} + \frac{1}{T} A \, d\xi \tag{16}$$

Inserting the condition (5) for *equilibrium states*, it is seen that

$$\frac{1}{T} \Delta G \, d\xi = 0 \tag{17}$$

Since T and $d\xi$ are different from zero, equilibrium states are characterized by

$$\Delta G = \left(\frac{\partial G}{\partial \xi}\right)_{P,T} = -A = 0 \tag{18}$$

Spontaneous reactions will occur if (5a) is

$$dS - \frac{dQ}{T} > 0$$

or

$$\left.\begin{array}{r}-\dfrac{1}{T}\Delta G\,d\xi > 0 \\[4pt] \dfrac{1}{T} A\,d\xi > 0\end{array}\right\} \qquad (19)$$

Hence the condition for any spontaneous reaction is given by

$$\left.\begin{array}{r}\Delta G < 0 \\[4pt] \left(\dfrac{\partial G}{\partial \xi}\right)_{P,T} < 0 \\[4pt] -A < 0\end{array}\right\} \qquad (20)$$

The change of Gibbs free energy G with ξ (in closed systems at constant temperature and pressure) is shown in Fig. 8.2. The equilibrium state is characterized by the minimum of the curve; the slope of the curve gives the negative affinity.

The preceding considerations provide a formal basis for deciding whether a given system is in its equilibrium state. For practical application a transformation of the rather abstract terms (A, ΔG, μ_i) into measurable quantities such as concentration or partial pressure is needed. We

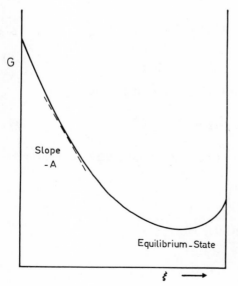

Figure 8.2 The change in Gibbs' free energy G with the degree of advancement of chemical reactions in closed systems at constant temperature and pressure.

start with (21) which relates the chemical potential μ_i of a substance i with its activity a_i:

$$\mu_i = \mu_i^\circ + RT \ln a_i \qquad (21)$$

The μ_i° is the standard chemical potential, that is, the chemical potential for the case when substance i is in its standard state. Evidently, the definition of this particular state is

$$a_i = 1 \qquad (22)$$

We point out later that the appropriate choice of the standard state is of great importance. For the moment we proceed by combining (21) and (14):

$$\Delta G = \sum_{i=1}^{k} \nu_i \mu_i^\circ + RT \sum_{i=1}^{k} \nu_i \ln a_i \qquad (23)$$

The sum

$$\sum_{i=1}^{k} \nu_i \mu_i^\circ = \Delta G^\circ \qquad (24)$$

is the standard (state) Gibbs free energy of the reaction. Equation 3 now becomes

$$\Delta G = \Delta G^\circ + RT \ln \prod_{i=1}^{k} a_i^{\nu_i} \qquad (25)$$

or, applied to reaction (11),

$$\Delta G = \Delta G^\circ + RT \ln \frac{a_B^{\nu_B} a_D^{\nu_D} \cdots}{a_A^{\nu_A} a_C^{\nu_C} \cdots} \qquad (26)$$

The quotient on the right-hand side of (26) is conveniently written as Q (for $\Delta G \neq 0$) or K (for $\Delta G = 0$):

$$Q = \left(\frac{a_B^{\nu_B} a_D^{\nu_D} \cdots}{a_A^{\nu_A} a_C^{\nu_C} \cdots} \right)_{\Delta G \neq 0} \qquad (27)$$

or

$$K = \left(\frac{a_B^{\nu_B} a_D^{\nu_D} \cdots}{a_A^{\nu_A} a_C^{\nu_C} \cdots} \right)_{\Delta G = 0} \qquad (28)$$

Equation 28, which relates the equilibrium activities in a closed system at constant temperature and pressure, states the law of mass action. The K is called the equilibrium constant. Hence

$$\Delta G^\circ = -RT \ln K \qquad (29)$$

and

$$\Delta G = \Delta G° + RT \ln Q = RT \ln \frac{Q}{K} \tag{30}$$

Provided that a_i values are experimentally available quantities, (30) turns out to be a most powerful tool for predicting chemical reactions in closed systems at constant temperature and pressure.

Suppose that we wish to learn whether a reaction such as

$$CaCO_{3(s)} \rightarrow Ca^{2+} + CO_3^{2-}$$

is proceeding in the ocean (a problem that is discussed in some detail in a subsequent section). We may first compute the equilibrium constant K from tabulated $\Delta G°$-values. A value for Q may then be calculated from information on the composition of seawater. Comparing Q and K we may find the following:

1. $K < Q \Rightarrow \Delta G > 0$: $CaCO_3$ will precipitate.
2. $K > Q \Rightarrow \Delta G < 0$: $CaCO_3$ will dissolve.
3. $K = Q \Rightarrow \Delta G = 0$: seawater is saturated with $CaCO_3$.

Activity Scales and Standard States

The problem of computing Q-values from analytical data is far from trivial, however. Let us therefore consider (21) in more detail:

$$\mu_i = \mu_i° + RT \ln a_i \tag{21}$$

The activity a_i is related to the molar concentration $[i]$ by

$$a_i = [i]\gamma_i \tag{31}$$

The quantity γ_i is the molar activity coefficient. (For the case when concentrations are expressed in molalities or mole fractions, γ_i is replaced by molal or rational activity coefficients. For gases (31) becomes

$$a_i = P_i \varphi_i$$

where P_i and φ_i are partial pressure and fugacity coefficient.) The conversion of concentrations into activities and vice versa requires the defini-

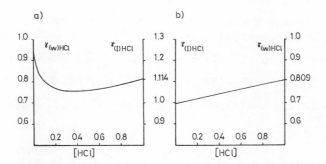

Figure 8.3 Molar activity coefficients of HCl: a) in pure water; b) in KCl–HCl mixtures of the constant ionic strength 1 M. $\gamma_{(w)}$ values refer to pure water whereas $\gamma_{(I)}$ values are related to 1 M KCl as a standard state. The given numbers are derived from Robinson and Stokes (1965) ommitting any corrections for conversion from molal into molar units.

tion of an activity scale or, similarly, the fixation of a standard state for component i where γ_i is unity. Activity scales (or standard states) for solute species are defined in such a way that γ_i approaches unity as the concentration approaches zero. Hence we may also say that the standard state is related to the pure solvent. We therefore refer to a particular standard state by indicating the solvent used.

The change of activity coefficients with concentration is largely dependent on the solvent chosen. In pure water the activity coefficients approach unity only in very dilute solutions. A plot of γ_{HCl} against [HCl] (Fig. 8.3a) shows clearly that the extrapolation [HCl] → 0 is rather cumbersome. Moreover, a theoretical prediction of the function $\gamma_i = f([i])$ is restricted to dilute solutions of 1:1, 1:2, or 2:1 electrolytes. It should be noted that at present there is no direct method for calculating activities related to pure water from analytical data of concentrated electrolyte solutions such as seawater. The course of γ_{HCl} in aqueous KCl-HCl mixtures of the ionic strength 1 (Fig. 8.3b) demonstrates that such extrapolations are much facilitated if pure water is replaced by an appropriate ionic medium of constant ionic strength. Experimental work has shown that activity coefficients in such media remain nearly constant as long as the concentrations of the reacting species are kept low (Biedermann and Sillén, 1953). Choosing ionic media as standard states, we therefore find that at low reactant concentration the difference between activity and concentration usually becomes smaller than the experimental errors, so that we can write

$$a_i = [i] \tag{32}$$

We can deduce the difference between the two activity scales from Fig. 8.3. The chemical potential of 1 M HCl can be written as

$$\mu = \mu_{(w)}{}^\circ + RT \ln \gamma_{(w)} = \mu_{(w)}{}^\circ + RT \ln 0.809$$

$$\mu = \mu_{(I)}{}^\circ + RT \ln \gamma_{(I)} = \mu_{(I)}{}^\circ + RT \ln 1.114$$

where indices (w) and (I) refer to pure water and to 1 M KCl as standard states.

Hence

$$\mu_{(I)}{}^\circ = \mu_{(w)}{}^\circ + RT \ln \frac{\gamma_{(w)}}{\gamma_{(I)}} = \mu_{(w)}{}^\circ + RT \ln \frac{0.809}{1.114}$$

This means that changing the activity scale is equivalent to a shift of μ° by a constant. In practice, most of the tabulated ΔG° and K values refer to pure water. There is, however, an increasing number of equilibrium constants related to particular ionic media such as 3 M NaClO$_4$ or 0.1 M KCl. Since ocean water is an ionic medium of practically constant composition, an activity scale related to a simplified model seawater (for instance, 0.7 M NaCl) would be of great value for chemical oceanography.

For solid solutions, the pure component solids are usually chosen as standard states.

Formulating Equilibrium Constants

The general expression for the equilibrium constant K is given by (28). Practical applications to some important reactions are illustrated by the following examples. Since all of the equilibrium constants used in this chapter are related to constant ionic media, (32) is used consistently.

$$\text{HCO}_3^- \leftrightarrows \text{H}^+ + \text{CO}_3^{2-} \tag{33}$$

$$K = \frac{[\text{H}^+][\text{CO}_3^{2-}]}{[\text{HCO}_3^-]}$$

$$\text{CO}_{2(g)} + \text{H}_2\text{O} \leftrightarrows \text{H}^+ + \text{HCO}_3^- \tag{34}$$

Since the water activity in a given ionic medium is constant, we can write

$$K = \frac{[\text{H}^+][\text{HCO}_3^-]}{P_{\text{CO}_2}}$$

$$\text{CaCO}_{3(s)} \leftrightarrows \text{Ca}^{2+} + \text{CO}_3^{2-} \tag{35}$$

$$K = \frac{[\text{Ca}^{2+}][\text{CO}_3^{2-}]}{a_{\text{CaCO}_3}}$$

or

$$[Ca^{2+}][CO_3^{2-}] = Ka_{CaCO_3}$$

In pure calcite, a_{CaCO_3} is unity by definition. In solid solutions such as magnesian calcites, a_{CaCO_3} is usually different from unity.

Equilibrium Models

As mentioned earlier, seawater is the product of a gigantic reaction between primary igneous rock and volatile substances. Assuming that this reaction has reached an equilibrium state, we might try to calculate the composition of seawater on the basis of the law of mass action (28). We begin by focusing our interests on a very limited number of major components, for instance H_2O, Ca^{2+}, Mg^{2+}, and CO_2. To calculate $[Ca^{2+}]$, $[Mg^{2+}]$, $[CO_3^{2-}]$, $[HCO_3^-]$, $[H^+]$, and P_{CO_2}, we must make assumptions concerning the nature and composition of the solid phases present. Suitable choices for these solid phases are $CaCO_3$ (calcite), $CaCO_3$ (aragonite), $MgCO_3$ (magnesite), $MgCO_3 \cdot 3H_2O$ (nesquehonite), $CaMg(CO_3)_2$ (dolomite), $Ca_xMg_{(1-x)}CO_3$ (magnesian calcite), $Mg_4(CO_3)_3(OH)_2 \cdot 3H_2O$ (hydromagnesite), and $Mg(OH)_2$ (brucite). A brief consideration shows that we can not have them all together. Since aragonite is unstable with respect to calcite,

$$CaCO_3 \text{ (aragonite)} \rightarrow CaCO_3 \text{ (calcite)} \quad \Delta G^0 = -0.25 \text{ kcal}$$

we can not have both solids in an equilibrated system. Similar restrictions apply for the coexistence of magnesite, brucite, and hydromagnesite. The number of coexisting solids can be calculated from Gibbs' phase rule which is conveniently written in the form

$$F = C - P + 2 \tag{36}$$

where F is the number of independent internal variables, C is the number of components, and P is the number of phases in the equilibrated system.

A phase is a homogeneous domain of the system, that is, a domain with uniform composition and properties. Examples of phases are gases, gaseous mixtures, liquid or solid solutions, and uniform solid substances. The number of components C is the minimum number of substances that are required to duplicate the system. The internal variables F are temperture T, pressure P, and the concentrations of the components in each phase. The system under consideration consists of four components: H_2O, CO_2, MgO, and CaO. The restrictions imposed on the system by the phase rule are shown in Table 8.2. The equations that relate independent

Table 8.2. Phase Rule Relationships in the System
H_2O–CO_2–MgO–CaO; Variables of Interest:
P_{CO_2}, T, $[Ca^{2+}]$, $[Mg^{2+}]$, $[HCO_3^-]$, and $[CO_3^{2-}]$

Phases		Independent Variables	
Number	Type	Number	Examples
3	Gaseous phase Aqueous solution $CaCO_3$	3	$T, P_{CO_2}{}^a, \dfrac{[Ca^{2+}]}{[Mg^{2+}]}$
4	Gaseous phase Aqueous solution $CaCO_3$ $CaMg(CO_3)_2$	2	$T, P_{CO_2}{}^a$
5	Gaseous phase Aqueous solution $CaCO_3$ $CaMg(CO_3)_2$ $Mg(OH)_2$	1	T

^{a} The total pressure P is fixed by $P = P_{H_2O} + P_{CO_2}$.

and dependent variables may now be derived using the pertinent equilibrium constants (Table 8.3). For coexistence of the stable phases we obtain for $CaCO_3$ (calcite) and $CaMg(CO_3)_2$ (dolomite):

$$\log \frac{[Ca^{+2}]}{[Mg^{2+}]} = \log \frac{([Ca^{2+}][CO_3^{2-}])^2}{([Ca^{2+}][Mg^{2+}][CO_3^{2-}]^2)} = 2.2 \tag{37}$$

For $CaMg(CO_3)_2$ (dolomite) and $MgCO_3$ (magnesite):

$$\log \frac{[Ca^{2+}]}{[Mg^{2+}]} = -0.8 \tag{38}$$

For $MgCO_3$ (magnesite) and $Mg(OH)_2$ (brucite):

$$\log P_{CO_2} = -6.7 \tag{39}$$

For $CaMg(CO_3)_2$ and $Mg(OH)_2$ (brucite):

$$\log \frac{[Ca^{2+}]}{[Mg^{2+}]} = -14.2 - 2 \log P_{CO_2} \tag{40}$$

Table 8.3. Equilibrium Constants in the System H_2O–CO_2–MgO–CaO

Reaction	Equilibrium Constant		T (C)	Standard State	Reference
	Symbol	log K			
Homogeneous reactions					
$H_2O \leftrightarrows H^+ + OH^-$	K_w	−13.7	25	Seawater	a
$CO_{2(g)} + H_2O \leftrightarrows H^+ + HCO_3^-$	K_{pa_1}	−7.5	25	Seawater	a
$HCO_3^- \leftrightarrows H^+ + CO_3^{2-}$	K_{a_2}	−9.0	25	Seawater	a
Involving stable solids					
$CaCO_3$ (calcite) $\leftrightarrows Ca^{2+} + CO_3^{2-}$	K_{s0}	−6.1	25	Seawater	b
$MgCO_3$ (magnesite) $\leftrightarrows Mg^{2+} + CO_3^{2-}$	K_{s0}	−6.8	25	Seawater	b
$CaMg(CO_3)_2$ (dolomite) $\leftrightarrows Ca^{2+} + Mg^{2+} + 2\,CO_3^{2-}$	K_{s0}	−14.4	25	Seawater	b
$Mg(OH)_2$ (brucite) $\leftrightarrows Mg^{2+} + 2\,OH^-$	K_{s0}	−11.0	25	Seawater	b
Involving metastable solids					
$MgCO_3 \cdot 3H_2O$ (nesquehonite) $\leftrightarrows Mg^{2+} + CO_3^{2-} (+3H_2O)$	K_{s0}	−3.2	25	Seawater	b
$xCa^{2+} + (1-x)Mg^{2+} + CO_3^{2-} \leftrightarrows$ $Ca_xMg_{(1-x)}CO_3$ (magnesian calcite)	$\log\left(\dfrac{[Mg]}{[Ca]}\right)_{solid}$ $= \log\dfrac{0.048([Mg^{2+}]/[Ca^{2+}])}{1 + 0.320([Mg^{2+}]/[Ca^{2+}])}$		25	3 m $NaClO_4$	b

Activities of $CaCO_3$ and $MgCO_3$ in magnesian calcites $Ca_xMg_{(1-x)}CO_3$ related to calcite and magnesite as standard states[b]

x	$(1-x)$	log a_{CaCO_3}	log a_{MgCO_3}
1.000	0.000	0.00	
0.990	0.010	0.04	0.14
0.980	0.020	0.21	0.65
0.970	0.030	0.34	1.00
0.960	0.040	0.39	1.22
0.950	0.050	0.42	1.40
0.930	0.070	0.42	1.67
(0.900)	(0.100)	(0.27)	(1.97)

[a] Schindler (1967).
[b] Calculated from Riesen (1969).

For $CaCO_3$ (calcite) and $Mg(OH)_2$ (brucite):

$$\log \frac{[Ca^{2+}]}{[Mg^{2+}]} = -6 - \log P_{CO_2} \qquad (41)$$

Equations 37 through 41 can now be plotted into a $\log P_{CO_2}$ vs $\log ([Ca^{2+}]/[Mg^{2+}])$ diagram (Fig. 8.4) which gives further illustration of the phase rule.

1. One solid phase present: $F = 3$. The temperature is implicitly fixed by the choice of numerical values for the equilibrium constants. The mutual independence of the variables results in areas that indicate the region where the particular solid is stable.
2. Two solid phases present: $F = 2$. At fixed temperature there is only one independent variable. (Note that the proper choice of this variable is essential: P_{CO_2} for (calcite + dolomite) and (magnesite + dolomite); $[Ca^{2+}]/[Mg^{2+}]$ for (brucite + magnesite). For coexisting (calcite + brucite) and (dolomite + brucite) both variables are equivalent.
3. Three solid phases present: $F = 1$. At a given temperature both $\log P_{CO_2}$ and $\log [Ca^{2+}]/[Mg^{2+}]$ are fixed.

A similar diagram may be constructed with $Mg(OH)_2$ and $Ca_xMg_{(1-x)}CO_3$ (magnesian calcite) as solid phases (Fig. 8.5). For each $[Ca^{2+}]/[Mg^{2+}]$ value the corresponding value for x can be evaluated from the distribution function

$$\frac{1-x}{x} = f \frac{[Ca^{2+}]}{[Mg^{2+}]}$$

given in Table 8.3. Once x is fixed, the interrelated numbers for a_{CaCO_3} and a_{MgCO_3} are read from Table 8.3.

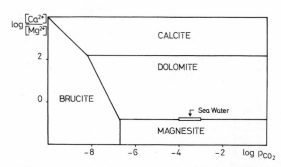

Figure 8.4 Stability area diagram for the system H_2O, CO_2, CaO, MgO including solids stable at 25 C and a total pressure of 1 atm.

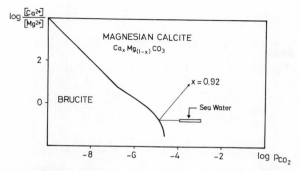

Figure 8.5 Stability area diagram for the system H_2O, CO_2, CaO, MgO including $Mg(OH)_2$ (brucite) and metastable magnesian calcites (25 C, 1 atm).

The equilibrium condition requires that

$$[Ca^{2+}][CO_3^{2-}] = Ks_{0(calcite)} a_{CaCO_3} \quad (42)$$

Hence for coexistence of $Ca_xMg_{(1-x)}CO_3$ and $Mg(OH)_2$ we obtain

$$\log \frac{[Ca^{2+}][CO_3^{2-}]}{[Mg^{2+}][OH^-]^2} = \log \frac{Ks_{0(calcite)} a_{CaCO_3}}{Ks_{0(brucite)}} \quad (43)$$

or, since

$$[CO_3^{2-}] = Kpa_1 Ka_2 P_{CO_2} \cdot [H^+]^{-2}$$

$$[OH^-] = K_w [H^+]^{-1}$$

we obtain

$$\log P_{CO_2} = \log \frac{Ks_{0(calcite)} a_{(calcite)}}{Ks_{0(brucite)}} + \log \frac{(Kw)^2}{Kpa_1 Ka_2} - \log \frac{[Ca^{2+}]}{[Mg^{2+}]} \quad (44)$$

Simple Models

Figures 8.4 and 8.5 show that the ratio $[Ca^{2+}]/[Mg^{2+}]$ in seawater could be controlled by (a) coexisting magnesite and dolomite or (b) magnesian calcite $Ca_{0.92}Mg_{0.08}CO_3$. For further testing we now compute the individual concentrations of Mg^{2+}, Ca^{2+}, H^+, HCO_3^-, and CO_3^{2-}, assuming $\log P_{CO_2} = -3.5$.

Model a

    4 components: H_2O, CO_2, CaO, MgO
    4 phases: gaseous, aqueous-solution, dolomite, magnesite
    2 independent variables: T, P_{CO_2}

The charge condition

$$2[Mg^{2+}] + 2[Ca^{2+}] + [H^+] = 2[CO_3^{2-}] + [HCO_3^-] + [OH^-] \quad (45)$$

can first be simplified by neglecting $[H^+]$ and $[OH^-]$. Considering the implication

$$\log \frac{[Ca^{2+}]}{[Mg^{2+}]} = -0.8; \quad \frac{[Ca^{2+}]}{[Mg^{2+}]} \simeq 0.2$$

we obtain

$$2.4[Mg^{2+}] = 2[CO_3^{2-}] + [HCO_3^-]$$

or

$$2.4 \frac{Ks_{0(\text{magnesite})}}{Kpa_1 Ka_2 P_{CO_2}} \cdot [H^+]^2 = 2Kpa_1 Ka_2 P_{CO_2}[H^+]^{-2} + Kpa_1 P_{CO_2}[H^+]^{-1} \quad (46)$$

Equation 46 can be solved by trial and error. The results are presented in Table 8.4.

Model b

4 components: H_2O, CO_2, CaO, MgO
3 phases: gaseous, aqueous-solution, magnesian calcite $Ca_xMg_{(1-x)}CO_3$
3 independent variables: T, P_{CO_2}, x $\left(\text{or } \dfrac{[Ca^{2+}]}{[Mg^{2+}]} \right)$

By similar arguments we obtain (47):

$$2.4 \frac{Ks_{0(\text{magnesite})} a_{MgCO_3}}{Kpa_1 Ka_2 P_{CO_2}} [H^+]^2 = 2Kpa_1 Ka_2 P_{CO_2}[H^+]^{-2} + Kpa_1 P_{CO_2}[H^+]^{-1} \quad (47)$$

As seen from Table 8.4, both model a and model b is in rather poor agreement with reality. Model a gives satisfactory numbers for pH and alkalinity, whereas model b is superior with respect to $[Mg^{2+}]$ and $[Ca^{2+}]$.

Improved Models

We can construct improved models by adding HCl as a new component. Gibbs phase rule now requires that the number of phases or the number of independent variables be increased by unity. Since there is little chemical evidence for the occurrence of sparingly soluble chlorides in the sediments, we keep the number of phases unchanged. Models c and d can

Table 8.4. Calculated and Observed Concentrations of $[Mg^{2+}]$, $[Ca^{2+}]$, $[H^+]$, $[HCO_3^-]$, and $[CO_3^{2-}]$. Units: $M = $ mole/l; $\log P_{CO_2} = -3.5$; $T = 25°$; $x = 0.92$

	Seawater	Models					
		a	b	c	d	e	f
$[Mg^{2+}]$	5×10^{-2}	7.5×10^{-4}	4×10^{-3}	5×10^{-2}	5×10^{-2}	1×10^{-3}	6.3×10^{-2}
$[Ca^{2+}]$	1×10^{-2}	1.5×10^{-4}	8×10^{-4}	1×10^{-2}	1×10^{-2}	2×10^{-4}	1.25×10^{-2}
$[HCO_3^-]$		1.4×10^{-3}	4.8×10^{-3}	1.8×10^{-4}	1.4×10^{-3}	1.26×10^{-3}	1.26×10^{-3}
$[CO_3^{2-}]$		2×10^{-4}	2.4×10^{-3}	3×10^{-6}	1.9×10^{-4}	1.6×10^{-4}	1.6×10^{-4}
Alkalinity[a]	2.36×10^{-3}	1.8×10^{-3}	9.6×10^{-3}	1.8×10^{-4}	1.8×10^{-3}	1.58×10^{-3}	1.58×10^{-3}
$-\log [H^+]$	8.1	8.16	8.69	7.25	8.14	$(8.1)^b$	$(8.1)^b$
$[Cl^-]$				$(1.2 \times 10^{-1})^b$	$(1.2 \times 10^{-1})^b$	1×10^{-3}	1.5×10^{-1}

[a] Alkalinity = $[HCO_3^-] + 2[CO_3^{2-}]$.
[b] Independent variable.

now be constructed by taking [Cl⁻] as an independent variable. The Cl⁻-concentration is chosen in such a way that the charge condition

$$2[Mg^{2+}] + 2[Ca^{2+}] + [H^+] = 2[(CO_3^{2-}] + [HCO_3^-] + [OH^-] + [Cl^-]$$

gives the appropriate values for [Mg²⁺] and [Ca²⁺].

Model c

    5 components: H_2O, CO_2, HCl, CaO, MgO
    4 phases: gaseous, aqueous-solution, dolomite, magnesite
    3 independent variables: T, P_{CO_2}, [Cl⁻]

Model d

    5 components: H_2O, CO_2, HCl, CaO, MgO
    3 phases: gaseous, aqueous-solution, $Ca_xMg_{(1-x)}CO_3$
    4 independent variables: T, P_{CO_2}, [Cl⁻], and x (or [Ca²⁺]/[Mg²⁺])

On the other hand, we can insert the hydrogen ion concentration as an independent variable. [Sillén (1961) has proposed that the pH value of seawater might be fixed by silicate equilibria and, within our models, we can select an appropriate number for pH.]

Model e

    5 components: H_2O, CO_2, HCl, CaO, MgO
    4 phases: gaseous, aqueous-solution, dolomite, magnesite
    3 independent variables: T, P_{CO_2}, $-\log[H^+] (\approx pH)$

Model f

    5 components: H_2O, CO_2, HCl, CaO, MgO
    3 phases: gaseous, aqueous-solution, $Ca_xMg_{(1-x)}CO_3$
    4 independent variables: T, P_{CO_2}, pH, x

The results of the numerical calculations (Table 8.4) show that models d and f represent a good approach to the real situation, whereas models c and e are less satisfactory. Thus by the criterion stated in (30), seawater seems to be equilibrated with magnesian calcite but oversaturated with magnesite and dolomite. Since there is little evidence for recent formation of magnesite and dolomite in marine environments, we conclude that the concentrations of Ca²⁺, Mg²⁺, and the carbonate species in seawater are controlled by the presence of magnesian calcites.

So far we have restricted our calculations to model systems at 25 C and a total pressure of about one atmosphere. Hence our conclusions apply merely to surface seawater. Lowering the temperature to 5 C and increasing the pressure to 500 atm generally enhance the solubilities of carbonate solids. This statement is (by the present lack of pertinent equilibrium data) preponderantly based on the observation that the amount of calcareous sediments decreases sharply at depths below 3000 to 4000 m. Nevertheless, from a purely inorganic point of view, we may portray the sea as an equilibrium system with respect to the components under consideration. Fluctuations within this system are caused by biological activity. It is well known that composition (and therefore solubilities) of carbonate skeletons varies widely. Moreover, the carbon dioxide partial pressure is subject to change by photosynthesis or respiration.

On the Way to a Complete Model

For the sake of simplicity we have thus far assumed P_{CO_2} to be an independent variable. This assumption is not justified, however, since most of the total carbon dioxide is fixed in some form in solids or in aqueous solutions. It is thus obvious to assume that P_{CO_2} is actually not independent but regulated by some processes between these solid and liquid phases. We shall therefore look for a model that turns P_{CO_2} into a dependent variable.

For this purpose, we consider a system consisting of the components H_2O, HCl, Al_2O_3, SiO_2, Na_2O, and K_2O. Model g is now made in such a way that all variables but T and $[Cl^-]$ become dependent. This requires the presence of six phases.

Model g

6 components: H_2O, HCl, Al_2O_3, SiO_2, Na_2O, K_2O

6 phases: gaseous, aqueous-solution, quartz, kaolinite, illite, and montmorillonite

2 independent variables: T, $[Cl^-]$

As shown by Sillén (1961), model system g can act as a pH-stat. The numerical elaboration of this model, however, turns out to be rather troublesome. The pertinent silicate equilibria such as

$$3.5 \text{ kaolinite} + 4H_4SiO_4 + Na^+$$
$$\leftrightarrows 3\text{Na-montmorillonite} + 11\tfrac{1}{2}H_2O + H^+$$

have been investigated by Hemley (1959; Hemley et al., 1961; Hemley and Jones, 1964) at temperatures between 200 and 400 C. The somewhat risky extrapolations to 25 C have been performed by several authors (Garrels and Christ, 1965; Holland, 1965; Stumm and Stumm-Zollinger, 1968), which in turn give somewhat different numbers. The figures derived for the pH value of this model lie between 7.3 and 8.1. This is in reasonable agreement with observation; it might, however, indicate that seawater, if completely equilibrated with silicate minerals, could have a pH that is somewhat lower than the present value. We return to this later.

For the moment, we proceed by combining models f and g and by inserting two more solid phases.

Model h

9 components: H_2O, CO_2, HCl, CaO, MgO, SiO_2, Al_2O_3, Na_2O, K_2O
9 phases: gaseous, aqueous-solution, magnesian calcite, quartz, kaolinite, illite, montmorillonite, Ca-feldspar, and chlorite
2 independent variables: T, $[Cl^-]$

This model controls the concentrations of nearly all major components of seawater. It is understood that within model h the silicate equilibria fix the pH value which in turn controls the distribution of carbonate species and thus regulates P_{CO_2}.

Steady State Models

So far we have discussed closed systems at equilibrium conditions. The ocean is an open system, however, with the peculiar property that the mass input from rivers is almost perfectly balanced by precipitation and evaporation. Hence for any element E we have the condition

$$\left(\frac{d[E]}{dt}\right)_{input} = \left(\frac{d[E]}{dt}\right)_{output} \tag{48}$$

or, within the ocean,

$$[E] = \text{constant}; \quad \left(\frac{d[E]}{dt}\right) = 0 \tag{49}$$

The particular state in which (48) and (49) apply is called a steady state. The mobility of a given element is characterised by its residence time

$$\tau = \frac{[E]}{\left(\dfrac{d[E]}{dt}\right)_{input\ or\ output}} \tag{50}$$

From the preceding section it should be clear that if equilibrium models apply, the residence time must be infinity. Actually, as mentioned earlier, the residence times vary from 10^2 to 10^8 yr.

The connections between residence time and deviation from equilibrium state have been elucidated by Morgan (1967). Following this author we may consider a reaction

$$A \underset{k'}{\overset{k}{\rightleftharpoons}} B \tag{51}$$

which takes place in an open system. The k and k' are first-order rate constants related to the equilibrium constant K by

$$K = \frac{k}{k'} \tag{52}$$

As pointed out by Morgan (1967), the steady state condition (49) results in

$$\frac{[B]}{[A]} = \frac{k \dfrac{V}{g}([A]o + [B]o) + [B]o}{k' \dfrac{V}{g}([A]o + [B]o) + [A]o} \tag{53}$$

Here V is the volume of the open system into which A and B are introduced at concentrations $[A]o$ and $[B]o$ (and from which A and B are removed) with the volume flow rate g. It is seen from (53) that the steady state $[B]/[A]$ approaches K as the flow g becomes small compared to the rate constants. Moreover, for the simple case $[B]o = 0$ one obtains

$$\frac{[A]}{[B]} = \frac{1}{K} + \frac{\tau_{1/2}/\ln 2}{\tau} \tag{54}$$

where $\tau_{1/2} = \ln 2/K$ is the half time of the first-order reaction.

Hence for the case when τ is large compared to $\tau_{1/2}$, reaction (51) approaches the equilibrium state. The applicability of (53) and (54) to the ocean has been demonstrated by Stumm and Stumm-Zollinger (1968) and in a more qualitative manner by the present author (1967). The general conclusion is that the residence time indicates whether the concentration of a given element is preponderantly controlled by an equilibrium state or by a steady state. Thus the concentrations of Ca^{2+} ($\tau = 8 \times 10^6$ yr) and Mg^{2+} ($\tau = 4.5 \times 10^7$ yr) are almost equilibrium concentrations, whereas the concentrations of aluminum and silicate species are probably maintained by steady state conditions.

A Dynamic Model

It has often been stated that the composition of seawater has been nearly constant during geological time. There are, however, indications for fluctuations, which are reflected by fluctuations in climatic conditions. It will be shown that such fluctuations might be caused by interactions of steady state systems with equilibrium systems including some feedback by the atmosphere. We have already considered model system f as an equilibrium system. Instead of P_{CO_2}, $[Cl^-]$ may now be taken as an independent variable. Hence, when a given pH value is imposed to model system f, P_{CO_2} will be fixed. On the other hand, model system g acts as a pH-stat. The short residence time of both aluminum and silicon implies now that model g could be a steady state system. As mentioned earlier, the equilibrium pH within model system g lies probably somewhere between 7 and 8. Hence, under conditions where the input of aluminum and silicon is reduced, this lower pH may be approached. This in turn shifts the equilibria in model f in such a way that P_{CO_2} increases. An enhancement of the CO_2 content of the atmosphere is (by virtue of its greenhouse effect) followed by an increase in temperature. These correlations are the basis for the following cycle.

Stage 1

Let us assume that the mean temperature at the earth's surface is reduced (for some unknown reason). As a consequence the glacial formation will increase and large areas of the continents will be covered with ice. Since weathering of rocks is greatly reduced under glacial conditions, the annual amount of aluminum and silicon introduced into the ocean will decrease. This lowers the pH value and increases P_{C_2O} and temperature.

Stage 2

Now, at enhanced temperature, the glacial period will end. Weathering conditions will improve and the input of silicon and aluminum will increase. As a consequence, model system g is shifted away from equilibrium state, pH will rise, and P_{CO_2} and temperature will decrease.

Thus far there is no proof that such a dynamic model is workable. It can, however, serve as a working hypothesis in looking for new facts and searching for the processes that regulate both the composition of seawater and the climatical conditions at the earth's surface.

Acknowledgments

The author is greatful to Werner Stumm (Harvard University) and Heinz Gamsjäger (University of Berne) for stimulating discussions and helpful advices.
The financial support by the Swiss National Foundation is acknowledged (Project No. 4951.2).

References

Biedermann, G., and L. G. Sillén. 1953. Studies on the Hydrolysis of Metal Ions. IV. Liquid Function Potentials and Constancy of Activity Factors in $NaClO_4$-$HClO_4$ Medium. *Archiv. Klusi*, **5**:425–440.

Garrels, R. M., and C. L. Christ. 1965. *Solutions, Minerals and Equilibria*. Harper & Row, New York, 450 pp.

Goldschmidt, V. M. 1933. Grundlagen der quantitativen Geochemie. *Fortschr. Mineral. Krist. Petr.*, **17**:112–156.

Hemley, J. J. 1959. Some mineralogical equilibria in the system K_2O-Al_2O_3-SiO_2-H_2O. *Amer. J. Sci.*, **257**:241–270.

Hemley, J. J., C. Meyer, and D. H. Richter. 1961. Some alteration reactions in the system Na_2O-Al_2O_3-SiO_2-H_2O. *U.S. Geol. Surv. Prof. Paper*, 424D: D338–D340.

Hemley, J. J., and R. W. Jones. 1964. Chemical aspects of hydrothermal alteration with emphasis on hydrogen metasomation. *Econ. Geol.*, **59**:538–569.

Holland, H. D. 1965. The history of ocean waters and its effect on the chemistry of the athmosphere. *Proc. Natl. Acad. Sci.*, **53**:1173–1183.

Horn, M. K. and J. A. S. Adams. 1966. Computer-derived geochemical balances and element abundances. *Geochim. Cosmochim. Acta*, **30**:279–297.

Morgan, J. J. 1967. Applications and limitations of chemical thermodynamics in natural water systems. In *Equilibrium Concepts in Natural Water Systems*. R. F. Gould (ed.) *Adv. Chem. Series*, **67**:1–29.

Riesen, W. F. 1969. Ph.D. Thesis. University of Bern, Switzerland.

Robinson, R. A., and R. H. Stokes. 1965. *Electrolyte Solutions*. Academic Press, New York. Butterworth & Co., London. 2nd Ed., Rev.

Schindler, P. W. 1967. Heterogenous equilibrium involving oxides, hydroxides, carbonates, and hydroxide carbonates. In *Equilibrium Concepts in Natural Water Systems*. R. F. Gould (ed.) *Adv. Chem. Series*, **67**:196–221.

Sillén, L. G. 1961. The physical chemistry of seawater. In *Oceanography*. M. Sears (ed.). *Amer. Assoc. Advan. Sci.* **67**:549–581.

Stumm, W., and E. Stumm-Zollinger. 1968. Chemische Prozene in natürlidien gewässern. *Arimig*, **22**:325–336.

Part III

Artifacts of Man

9

Lead

CLAIR PATTERSON, *Division of Geological and Planetary Sciences, California Institute of Technology, Pasadena, California*

Recent observations of the depth distribution of common lead in the oceans suggest dissimilarities of occurrence between it and chemically analogous metals, barium and radium. The latter tend to have relatively low concentrations in young surface waters and relatively high concentrations in old deep waters (Chow and Goldberg, 1960; Koczy, 1958; Turekian and Johnson, 1966; Chow and Patterson, 1966). Lead, however, appears to have high, variable concentrations in young surface waters and uniform, low concentrations in old deep waters (Tatsumoto and Patterson, 1963a, 1963b; Chow and Patterson, 1966). Lead distributions are similar to those of radioactive debris from nuclear explosives washed out of the atmosphere directly onto the oceans, although there are different opinions about the depth of penetration (Bowen et al., 1969).

It is likely that man now uses lead on such a large scale that the magnitude of industrial use overshadows and perturbs the natural flow of lead in the main sedimentary cycle. Lead was first used industrially about 2500 B.C. near the end of the Chalcolithic era in southwest Asia. By that time men were mining and smelting metal ores and were well acquainted with the remarkable utility of bronze and the value of gold. In their attempts to smelt useful bronzes from different, abundant ores, they had finally discovered that small amounts of silver could be parted (cupelled) from lead metal which had been first smelted from lead ores. Previously, native silver had been much rarer than gold, and perhaps more esteemed. Traces of silver occur commonly in galena, an abundant lead sulfide ore, and silver production from lead ores boomed after 2500 B.C. About 400 tons of lead in oxide form were produced for every ton of silver obtained by cupellation. Lead oxide was easily converted to metal again by reduction with charcoal. With the advent of coinage in 650 B.C. silver became a keystone in the founding and operation of successive civilizations and lead production rose sharply.

During the two millennia beginning with the cupellation of silver from lead ores and ending with the commencement of coinage, men found a multitude of uses for the huge piles of lead that began to accumulate. By Classic Greek times water was collected from lead-covered roofs and transported through lead metal gutters to lead-lined cisterns. Stone building blocks were held together with iron clamps embedded in lead metal that had been poured into holes drilled in the stone. Ships were sheathed with lead metal to repel woodworms. Salves, ointments, cosmetics, and paints were made of lead compounds. Grape sugars boiled down in lead pots were added to wines to keep them from souring (Gilfillan, 1965). Lead-tin alloys were widely used to line the insides of bronze utensils to keep copper out of foods and liquids.

The industrial use of lead in the world reached a monumental plateau during the first and second centuries of the Christian era. In Italy, the annual industrial use at the time amounted to nearly 0.004 ton of lead per person per year, which approximated the 0.01 ton of lead per person per year industrially used in the United States 2000 yr later. The Romans exhausted all the lead mines within their empire down to several hundred meters below the water table by about A.D. 300; and world lead production declined precipitously with the fall of Rome, remaining at a low ebb for the next six centuries. Production began to climb again slowly in Europe after the ninth century A.D. when most lead mining activities shifted from holdings in the declining Byzantine Empire to virgin deposits operated by free entrepreneurs in central Europe. During the late Renaissance, world lead production finally reached again the mighty levels achieved by the Romans after entire Indian nations became enslaved in the lead mines of Peru and Mexico. Beginning with the Industrial Revolution, world lead production climbed exponentially from 100,000 tons/yr in 1750 to 3,500,000 tons/yr in 1966 (Patterson, 1969). The rise in production after 1750 is shown in Table 9.1, where primary production from ores and secondary recovery from recycled scrap have been separated, as have operations in the northern and southern hemispheres. The rise in production of the important atmospheric contaminants, lead alkyls, has also been tabulated.

Industrial lead enters the oceans by way of rivers and atmospheric washout. Contributions from both routes have changed with time, gradually growing during past centuries, with pollution from the atmosphere

Table 9.1. 10^3 Metric Tons of Lead Smelted or
Burned as Alkyls per Year Since 1750 A.D.

Date	Northern Hemisphere Primary Smelting	Northern Hemisphere Secondary Smelting	Northern Hemisphere Burned Alkyls	Southern Hemisphere Primary Smelting
1966	2400	700	310	350
1960	1900	600	180 (1958)	360
1950	1300	550	110	240
1940	1300	400	36	230
1930	1200	400	4	170
1920	880	200	0	110
1910	940	60		100
1900	750	0		80
1890	520			40
1880	400			+
1860	220			+
1800	90			50
1750	60			40

From Murozumi et al. (1969).

increasing abruptly during the last two decades as a consequence of a sudden rise in the rate of burning of leaded automotive fuels.

It has been suggested (Chow and Patterson, 1962) that in prehistoric times the natural cycle of lead in the oceans was as follows. It entered the oceans mainly from rivers both as an insoluble form attached to solid particles and as a dissolved form chelated with water and organic acids. There was exchange of lead between the dissolved and insoluble forms, most of the insoluble lead settling out on the continental shelves, with only a minor fraction entering and settling in abyssal areas. Much of the soluble lead was ingested by organisms living in the upper mixed layers of the oceans where it traveled up the food chain until it began to leave the mixed zone in dead organisms large enough to settle rapidly. Near and at the ocean bottom lead again dissolved from disintegrating dead organisms, where it was chemically incorporated in bottom sediments, principally in an insoluble manganese-iron precipitate; a large fraction of the lead incorporated in silicate mineral particles that had settled in pelagic areas originated from airborne dusts.

The lead-silicon ratio is much higher in pelagic sediments than in soil and rocks, and organisms may yield this enrichment by sedimentation, disintegration, and dissolution. Tatsumoto and Patterson (1963a) have shown that observed oceanic lead concentrations are compatible with lead residence times of only a few years in surface waters of the oceans, which means they should respond relatively quickly to perturbations from industry.

Chow and Patterson (1962) evaluated the rate at which lead was removed from pelagic oceans by sediments during the Pleistocene by measuring lead concentrations in pelagic sediments that accumulated at various rates. They estimated that about 1.3×10^{10} g of dissolved lead was removed annually from pelagic seas in prehistoric times in the form of chemical manganese-iron precipitates, together with about 0.5×10^{10} g of lead mechanically removed in settled silicate minerals. They pointed out that of the approximately 50×10^{10} g of lead annually entering the seas in silicate particulates from rivers today about 99 per cent settles on the continental shelves and in adjacent trenches, thus failing to reach pelagic areas. On the other hand, they suggested the major fraction of dissolved lead in river waters reached the open seas, and today this input of dissolved river lead greatly exceeds the prehistoric output. From a total soluble denudation of 4.8×10^{15} g/yr (Poldervaart, 1955) containing a measured 50 ppm of lead (Durum et al., 1960) the input of soluble river lead to the oceans today is about 24×10^{10} g/yr. Part of this excess (about one seventh) originates in agricultural stimulation of soluble denudation (Conway, 1942).

It is not clear exactly how industrial pollution lead reaches the oceans via rivers. Rama and Goldberg (1961) showed that initially large concentrations of radioactive Pb^{210} in freshly fallen snow decrease as a result of adsorption and exchange with soil and settling suspended matter in streams. The removal of sludge in sewage treatment plants probably also removes a large fraction of the industrial lead in such wastes by similar mechanisms (Public Health Service, 1965a).

Runoff of storm waters from paved areas, buildings, and plant foliage in coastal cities does not undergo treatment that would remove much lead, however, and significant amounts of industrial lead may be washed into the oceans by this process. Present concentrations of industrial pollutant lead in rivers are so excessive that exchange between particulate and dissolved forms is significant in providing large amounts of dissolved lead even though the major fraction is adsorbed onto solids. Turekian and Scott (1967) and Kharkar et al. (1968) have shown that from one fifth to one half of such metals as silver and cobalt in river waters is typically attached to suspended material, but under conditions of high pollution a greater fraction of the metals is attached to solids. The size of the attached fraction seems to be independent of the exchange capacity or size of inorganic particles. These investigators have carried out experiments which indicate that when inorganic suspended material reaches the seas, the attached trace metals are displaced into solution by sodium and magnesium, which exist in much higher concentrations in the oceans than in rivers. The method of collecting, storing, and filtering water samples used by the National Water Pollution Surveillance System functions to reduce dissolved lead concentrations in the samples before they are analyzed, and makes their data unreliable. Unfiltered waters taken from rivers draining mainly industrialized regions show total lead concentrations of about 300 μg Pb/l (Ettinger, 1966). The average concentration of dissolved lead in rivers during prehistoric times was probably about 0.5 μg Pb/l (Patterson, 1965). Today concentrations average 5 μg Pb/l (Patterson, 1965).

It is claimed that the United States, which is by far the largest consumer of lead in the world, eventually recycles as scrap 60 per cent of the lead it uses (Merrill, 1952). This figure is excessive with respect to total United States industrial production over past decades, and it is certainly much too large for total world industrial production, since the United States is the principal user of lead scrap. More than one third of the lead production goes into automobile batteries, where the lead can be recycled without excessive loss, but the annual overall irrecoverable loss of world lead production is greater than 0.5 yr^{-1} today and was even greater in the past. Most of that irrecoverable loss has ended in soil within and adjacent to urban area complexes. Surface layers of soils from cities contain many hundreds of ppm of lead (Cholak et al., 1961) in contrast to a natural expected level of about 10 ppm, while street sweepings contain many thousands of ppm (Kaye and Reznikoff, 1947), but this lead does not enter the oceans today by means of streams and rivers that are well inside continents. It is only the excessively industrially polluted wastes and storm runoffs that are discharged into streams near coastal regions that are significant. It is likely that atmospheric industrial lead pollution arising from smelteries, incinerators, and automobiles and falling on and washing out on paved surfaces, buildings, and plant foliage is a major contributor. The amount of lead flowing through such routes to the seas from coastal cities could amount to about 1 per cent of the total annual world lead production. It is not known at present how the very large reservoirs of irrecoverable industrial lead wastes, now stored on unpaved land surfaces, will eventually affect the seas.

It is probable that the atmospheric route of entry of industrial lead to the seas is important today because very large amounts of lead are now burned in gasoline. This practice first began in 1924, and if it serves as an important means of introducing lead into the oceans, lead concentrations should have increased markedly in the atmosphere of the northern hemisphere after 1924. Murozumi et al. (1969) have demonstrated that this did occur by measuring increases in concentrations of lead in annual layers of snow in northern Greenland. Their observations are shown in Fig. 9.1. The observed concentrations of sea salts and silicate dusts did not change appreciably over time, and natural lead contributed by these materials amounted to about 0.0004 µg Pb/kg snow. At 800 B.C. the analytical lead blank of 0.003 µg Pb/kg snow exceeded lead levels in the snow. At the beginning of the Industrial Revolution in 1750 A.D., however, industrial

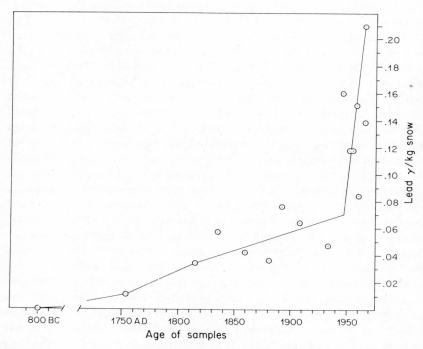

Figure 9.1 Increase of industrial lead pollution in northern Greenland snow with time since 800 B.C. Natural concentration of lead is 0.0004 γ Pb/kg snow. Concentrations of salts and dusts have changed systematically with time by less than a factor of 2 since 1750 A.D. (dusts probably increasing, salts probably decreasing). From Murozumi et al. (1969).

lead concentrations had risen to about 25 times the natural levels. By 1940 industrial lead concentrations had climbed to 175 times the natural levels. After 1940 industrial lead rose abruptly over a short interval to 500 times the natural levels. Scatter in data is due mainly to the samples being comprised of about one and one-half annual layers at random, some containing two winters and one summer and others the reverse. Lead concentrations were markedly higher in winter and lower in summer layers of snow.

Industrial lead aerosols which were removed from the atmosphere by polar snowflakes originated mainly from smelteries before 1940 and from burned gasoline after that date. In Table 9.2 atmospheric contributions from these two sources at different times are compared with levels of lead deposited in northern Greenland snow. The burning of coal has never been a significant contributor of lead aerosols, even compared to the output from lead smelteries in the early thirties. Jaworowski (1967) suggests that burning coal is important, even today. The concentrations of lead pollutant aerosols rose 300 per cent in Greenland snows after 1940, when the burning of large amounts of lead alkyls rose 900 per cent, but the rate of coal burning increased only 40 per cent during this same period. Material balance considerations also suggest that coal smoke is a minor contributor of lead aerosols compared to other sources. Lead concentrations were found to be much less in South Polar snows than in North Polar snows (Murozumi et al., 1969), confirming their industrial origin, since most of the smelting and burning of lead has been confined to the northern hemisphere, and meridional circulating cells in the troposphere would hinder the migration of lead aerosols from the northern hemisphere to Antarctica.

In regions closer to the sources of industrial lead aerosols, atmospheric precipitations contain much higher concentrations of lead. Recent firn samples from glaciers in Poland were found to contain 80 to 150 μg Pb/kg ice (Jaworowski, 1967). Schroeder and Balassa (1961) could not find lead (< 1 μg Pb/kg snow) in fresh snow from rural Vermont, however, and Tatsumoto and Patterson (1963b) found only 2 μg Pb/kg snow in fresh snow from a wilderness area in California. It is likely that Jaworowski's samples reflect the effects of intense, local sources of atmospheric lead pollution. Extensive investigations of lead concentrations in urban rain in the United States show very high concentrations of lead. Chow and Earl (1969), and Lazrus et al. (1970) found an average of about 40 μg Pb/kg rain in cities.

These increases of lead concentrations in recent snow and rain, in proceeding from remote polar locations to rural continental areas and then

Table 9.2. Lead Aerosol Production in the Northern Hemisphere Compared with Lead Concentrations in Northern Greenland Snow at Different Times

Date	10^5 Tons of Lead Smelted per Year	Fraction Converted to Aerosols (%)	Tons of Lead Aerosols Produced per Year from Smelteries ($\times 10^3$)	10^5 Tons of Lead Burned as Alkyls per Year	Fraction Converted to Aerosols (%)	Tons of Lead Aerosols Produced per Year from Alkyls ($\times 10^3$)	Total Tons of Lead Aerosols Produced per Year ($\times 10^3$)	Lead per Kilogram Snow in Greenland
1753	1	2	2	—	—	—	2	0.01
1815	2	2	4	—	—	—	4	0.03
1933	16	0.5	8	0.1	40	4	10	0.07
1966	31	0.06	2	3	40	100	100	0.2

to cities, corresponds closely with observed trends in concentrations of atmospheric lead among these same locations. Lead is <0.0005 µg Pb/m^3 at Camp Century in northern Greenland (Murozumi et al., 1969), 0.001 µg Pb/m^3 in the central Pacific Ocean (Chow et al., 1969), 0.1 to 0.4 µg Pb/m^3 in rural areas near highways (Robinson and Ludwig, 1964), and 2 to 30 µg Pb/m^3 in cities (U.S. Public Health Service, 1965b). In all observed situations of lead in precipitation and in air, concentrations are hundreds and thousands of times above the levels expected from natural materials. It can be anticipated, therefore, that atmospheric precipitation over the oceans would contain higher concentrations of lead near industrial coasts and lesser concentrations over remote open seas.

Meteorological factors greatly affect lead concentrations in the atmosphere and in precipitation. The residence time in the atmosphere of lead aerosols in the 5–0.05µ range is probably of the order of several weeks (Burton and Stewart, 1960), so that a multitude of meteorological factors have sufficient time to operate and perturb lead concentrations, but changes in rates of industrial lead aerosol productions are quickly reflected in atmospheric lead concentrations because residence times are sufficiently short. From their extensive observations on atmospheric lead concentrations in the San Diego area, Chow and Earl (1969) noted an annual cycle for urban atmospheres which is clearly related to the average height of the inversion layer—the lower (winter) the inversion ceiling, the higher the lead concentrations, and vice versa. The total variation of lead concentrations during a single annual cycle is about a factor of 3. Observations in other cities by other investigators (U.S. Public Health Service, 1965b) show similar effects. Ventilation under this inversion ceiling by horizontal winds regulates urban atmospheric lead concentrations on a daily basis, and such mixing rapidly dilutes pollutants to lower levels outside the immediate boundaries of source areas (Junge, 1963). Lead concentrations in urban atmospheres follow diurnal cycles which are regulated by the twice-daily rise and fall in the intensity of automobile traffic (U.S. Public Health Service, 1965b). In polar regions, Murozumi et al. (1969) found marked seasonal variations in lead concentrations in snow, those in winter layers being about three times higher than those in summer layers. There may be similar seasonal variations of lead in rain over the oceans, since the salt content varies systematically between winter and summer (De Bary and Junge, 1963), and similar variations of salt have been found in polar snows (Murozumi et al., 1969).

The observed distribution of lead in the oceans is illustrated in Fig. 9.2. Lead concentrations in waters below about 500 m cluster between 0.02 and 0.04 µg Pb/kg seawater. In shallow water layers lead concentrations

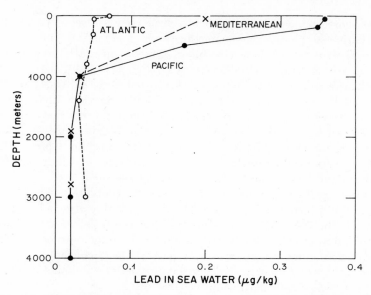

Figure 9.2 Profiles of lead concentrations in ocean waters of the Atlantic (Bermuda, 32°10′N, 64°30′E), Mediterranean (West End, 40°39′N, 05°40′E), and Pacific (California Coast, 29°13′N, 117°37′W). Data from Chow and Patterson (1966).

vary according to the nearness of the sampling site to intense sources of industrial lead aerosols, ranging from 0.07 to 0.35 µg Pb/kg seawater. These values refer to dissolved and exchangeable lead, but do not include lead bound up inactively in suspended particles. Many investigators have reported lead concentrations in surface seawaters that range from 5 to 10 µg Pb/kg water, and it is possible that some of these figures refer to the sum of dissolved, exchangeable, and particulate lead present. Chow (1968) reviewed these data and pointed out that such high values probably result from lead contamination by the analyst, which is extremely difficult to control in ships because of the widespread use of lead aboard them.

Barium and radium concentrations in surface waters are about half those in deep waters, and on a similar basis the prehistoric concentrations of lead in surface waters may have been about half those now observed in deep waters, or about 0.01 to 0.02 µg Pb/kg water. It is estimated that the average value of lead concentrations in surface waters in the northern hemisphere is now about 0.07 µg Pb/kg water (Chow and Patterson, 1966). This industrial pollution increment of 0.05 µg Pb/kg water in the upper several hundred meters of the oceans of the northern hemisphere can readily be accounted for by a few years' rain, polluted with lead to only one tenth the concentrations common in urban rain. Since such pollution effects have been in operation for more than two decades and additional excesses of industrial lead have been entering the oceans through rivers, the question becomes one of explaining where all the excess industrial lead went, not whether there is a sufficient supply.

There is evidence, from variations in the isotopic compositions of lead in pelagic sediments, that the residence time of lead in the upper layers of the oceans is very brief (Chow and Patterson, 1962). The isotopic compositions of pelagic sediment leads varies systematically with location in the oceans and generally reflects the isotopic character of lead on the surfaces of drainages in nearby continents. This can happen only if lead entering the oceans is removed again very quickly before it can be isotopically homogenized. Tatsumoto and Patterson (1963a) evaluated a biological removal constant for the simple case of equilibrium and no mixing with deep waters, expressed by

$$I = BN$$

where I is the lead input, N is the concentration in surface waters, and B is the removal constant. They found B to be 0.1 yr (a 7-yr half-life—which is actually too long for the isotopic effect) and used this biological removal mechanism to predict what the lead concentrations in surface waters should be, employing generous estimates for the introductions of industrial lead via rivers and the atmosphere over the last four decades. They calculated that one would expect to find lead concentrations of 0.25 μg Pb/kg water in the 100-m surface layer of northern hemisphere oceans. For 200 m this would reduce to 0.12 μg Pb/kg water. It seems probable that some sort of removal mechanism such as this is operating to keep industrial lead concentrations down in ocean surface waters.

It is clear that much of the present information concerning lead in the oceans is tentative and that additional work is needed to understand the situation with assurance. Perhaps the greatest single future advance in this field will come with an understanding of the speciation of oceanic lead, because that is the major unknown at present which casts serious doubt on most aspects of what is believed about oceanic lead.

It seems likely that man has polluted the mixed zone of the oceans in the northern hemisphere with industrial lead to such a degree that industrial lead has displaced most of the natural lead originally there and has elevated average lead concentrations by a factor of 2 or 3. It is probable that lead body burdens in higher organisms near the ends of food chains have been elevated as a consequence. Lead is highly toxic, and since there is no known threshold for physiological damage, such organisms may be adversely affected. The significance of industrial lead pollution in the oceans does not reside in this possibility, however. More importantly, it characterizes the enormous lengths to which man has gone in constructing an inhospitable environment for himself.

Unfortunately, learning of these matters does not automatically lead to their cure because of the conflict of public and private interests that are involved. The pollution of man's environment on a worldwide scale by industrial lead has been proceeding virtually unnoticed and unchallenged for thousands of years and with an intensity that has been increasing exponentially during the last two centuries. The major factor that allowed this to happen was the simultaneous development in medicine, during the nineteenth century, of the useful and humane concept of occupational health hazard together with knowledge of the toxicity of lead. This prevented the recognition of lead as one of the first great environmental poisons because lead poisoning was relegated to the category of an occupational health hazard which seemed amenable to cure by proper care during the industrial production and manufacture of leaded materials. The seriousness of the problem of environmental pollution by industrial lead was made known recently (Patterson, 1965), and since that time the subject has received wide recognition (Douglas-Wilson and Munro, 1966; U.S. Public Health Service, 1966; U.S. Senate, 1966), but not much has been accomplished because of the massive opposition by lead-producing and lead-consuming industries to any attempts to curb the industrial production and use of lead.

Acknowledgments

Principal support of the listed investigations which involve this author and support for writing of this study originated from AEC Contract AT(04-3)-767, CALT. 767-44. Contribution number 1982 of the Division of Geological and Planetary Sciences.

References

Bowen, V. T., V. E. Noshkin, N. L. Volchok, and T. T. Sugihara. 1969. Strontium-90: Concentrations in surface waters of the Atlantic Ocean. *Science*, **164**:825–827.

Burton, W. M., and N. E. Stewart. 1960. Use of long-lived natural radioactivity as an atmospheric tracer. *Nature*, **186**:584–589.

Cholak, J., L. J. Schafer, and T. O. Sterling. 1961. The lead content of the atmosphere. *J. Air Pollution Control Assoc.*, **11**:281–288.

Chow, T. J. 1968. Isotope analysis of sea water by mass spectrometry. *J. Water Pollution Control Federation*, **40**:399–411.

Chow, T. J., and J. Earl. 1969. Trends of increasing lead aerosols in the atmosphere. Presented at American Chemical Society Meeting, Houston, Texas. February 1970.

Chow, T. J., J. Earl, and C. F. Bennett. 1969. Lead aerosols in the marine atmosphere. *Environ. Sci. Tech.*, **3**:737–740.

Chow, T. J., and E. D. Goldberg. 1960. On the marine geochemistry of barium. *Geochim. Cosmochim. Acta*, **20**:192–198.

Chow, T. J., and C. Patterson. 1966. Concentration profiles of barium and lead in Atlantic waters off Bermuda. *Earth Planetary Sci. Lett.*, **1**:397–400.

Chow, T. J., and C. Patterson. 1962. The occurrence and significance of lead isotopes in pelagic sediments. *Geochim. Cosmochim. Acta*, **26**:263–308.

Conway, E. J. 1942. Mean geochemical data in relation to oceanic evolution. *Proc. Roy. Irish Acad.*, **48**:119–127.

DeBary, E., and C. E. Junge. 1963. Distribution of sulfur and chlorine over Europe. *Tellus*, **15**:370–381.

Douglass-Wilson, I., and I. A. H. Munro. 1966. Lead absorption and lead poisoning. *Lancet*, **1**:1307–1308.

Durum, W. H., S. G. Heidel, and L. J. Rison. 1960. World-wide runoff of dissolved solids. *Intern. Assoc. Sci. Hydrol., Publ.* **51**:618–621.

Ettinger, M. B. 1966. Lead in drinking water. *U.S. Public Health Ser. Publ.* **1440**:21–29.

Gilfillan, S. C. 1965. Lead poisoning and the fall of Rome. *J. Occupational Med.*, **7**:53–60.

Jaworowski, Z. 1967. Stable and radioactive lead in environment and human body. *Rev. Rept. Nucl. Energy Inform. Center*, **29**. (Warsaw, Poland.)

Junge, C. E. 1963. *Air Chemistry and Radioactivity*, Academic Press, New York.

Kaye, S., and P. Reznikoff. 1947. A comparative study of the lead content of street dirts of New York City in 1924 and 1934. *J. Ind. Hyg. Toxicol.* **29**: 178–179.

Kharkar, D. P., K. K. Turekian, and K. K. Bertine. 1968. Stream supply of dissolved silver, molybdenum, antimony, selenium, chromium, cobalt, rubidium, and cesium to the oceans. *Geochim. Cosmochim. Acta*, **32**:285–298.

Koczy, F. F. 1958. Natural radium as a tracer in the ocean. Proc. 2nd Intern. Conf. Peaceful Uses Atomic Energy, Geneva, **18**:351–370.

Lazrus, A. L., E. Lorange, and J. P. Lodge. 1970. Lead and other metal ions in the United States precipitation. *Environ. Sci. Tech.* **44**: 55–58.

Merrill, C. W. 1952. Accumulation and conservation of metals in use. Quoted in Report of the President's Materials Policy Commission, Pressures for freedom, **2**:42.

Murozumi, M., T. J. Chow, and C. Patterson. 1969. Chemical concentrations of pollutant lead aerosols, terrestrial dusts, and sea salts in Greenland and Antarctic snow strata. *Geochim. Cosmochim. Acta,* **33**:1247–1294.

Patterson, C. 1965. Contaminated and natural lead environments of man. *Arch. Environ. Health,* **11**:344–360.

Patterson, C. 1969. Data on the history of lead production are taken from an unpublished manuscript in preparation by the author.

Poldervaart, A. 1955. Chemistry of the earth's crust. *Geol. Soc. Amer.,* Spec. Paper **62**:119–144.

Rama, M. Koide, and E. D. Goldberg. 1961. Lead-210 in natural waters. *Science,* **134**:98–99.

Robinson, E., and F. L. Ludwig. 1964. Size distributions of atmospheric lead aerosols. Stanford Research Institute Project No. **PA-4788**.

Schroeder, H. A., and J. J. Balassa. 1961. Abnormal trace metals in man: lead. *J. Chronic Diseases,* **14**:408–425.

Tatsumoto, M., and C. Patterson. 1963a. The concentration of common lead in sea water. In *Earth Science and Meteoritics,* J. Geiss and E. Goldberg (eds.) North Holland Publishing Co., Amsterdam. Chap. 4, p. 74.

Tatsumoto, M., and C. Patterson. 1963b. Concentrations of common lead in some Atlantic and Mediterranean waters and in snow. *Nature,* **199**:350–352.

Turekian, K. K., and D. G. Johnson. 1966. The barium distribution in seawater. *Geoch. Cosmochim. Acta,* **30**:1153–1174.

Turekian, K. K., and M. R. Scott. 1967. Concentrations of chromium, silver, molybdenum, nickel, cobalt, and manganese in suspended material in streams. *Environ. Sci. Tech.,* **1**:940–942.

U.S. Public Health Service. 1965a. Interaction of heavy metals and biological sewerage treatment processes. *U.S. Public Health Ser. Publ.* **999-WP-22**.

U.S. Public Health Service. 1965b. Survey of lead in the atmosphere of three urban communities. *U.S. Public Health Ser. Publ.* **999-AP-12**.

U.S. Public Health Service. 1966. Symposium on Environmental Lead Contamination. *U.S.Public Health Ser. Publ.* **1440**.

U.S. Senate. 1966. Committee on Public Works, Subcommittee on Air and Water Pollution. Hearings. 89th Congress, 2nd Session, 311–453.

10

Chlorinated Hydrocarbons

R. W. RISEBROUGH, *Institute of Marine Resources, Department of Nutritional Sciences, University of California, Berkeley, California*

This chapter was originally entitled "Artifacts of Man: Pesticides," but the word "pesticide" has become obsolete in pollutant ecology and it is time to discard it. Since political and economic considerations, aesthetics, ethics, and values frequently intrude into what has become the scientific field of pollutant ecology, it is imperative that the scientist use only language that is exact and as free as possible of emotional and commercial connotations. "Chlorinated hydrocarbon" conveys the meaning of the compounds encountered in the sea, which might also be referred to as "biocides," "insecticides," or simply as "pollutants."

United States production of biocidal materials, including herbicides, fungicides, and insecticides, is now over a billion pounds per year (United States Tariff Commission Reports). They comprise several hundred different products, the vast majority of which are degraded readily into water-soluble derivatives within a short time interval after their application. At present there is no evidence to suggest that any of these water-soluble degradation products are accumulating as persistent pollutants in the marine environment. Only a very few "pesticides," all of them chlorinated hydrocarbons, have ever been detected in the sea, and it is therefore misleading to speak of these as if they encompassed all biocides now in use. Moreover, the most abundant of the synthetic marine pollutants usually considered under the heading "pesticide" are not biocides at all. The DDT compound p,p'-DDE(DDE) is a derivative of the insecticidal DDT compound p,p'-DDT, but has a very low toxicity to insects, vertebrates, and almost all other forms of life, and has consequently never been used as a biocide. DDE appears to be the most abundant of the synthetic pollutants in the sea (Risebrough, 1969; Risebrough et al., 1968). The polychlorinated biphenyls (PCB) are also abundant and widespread pollutants in marine ecosystems. Like the DDT compounds they are chlorinated hydrocarbons, but they are industrial products and not used as "pesticides" (Risebrough et al., 1968; Jensen et al., 1969, Table I). It is therefore inaccurate to refer to these pollutants as "pesticides"; in scientific writing the term "chlorinated hydrocarbon" is much more exact.

The current political debate about the continued use of chlorinated hydrocarbons or similar materials which yield persistent waste products that are transferred to the sea and that have biological activity has been characterized by lack of clarity of language. At issue is not whether insecticides or biocides should be used, but whether persistent waste products derived from either industrial or agricultural uses are compatible with the long-term technological development of our planet. The industry spokesmen continue to insist that an "attack on DDT is part of a general attack on pesticides" and that the choice is between "pesticides or pests" (Jukes, 1970). These arguments acquire a different perspective when the language becomes more exact. Moreover, DDT is a mixture of compounds. The composition of the commercial mixture used as the insecticide is very different from the composition of the DDT compounds that are found in the sea. It is therefore important to specify which compound is meant and to restrict the use of the term "DDT" to the commercial formulation.

The term "pesticide" also implies a value judgment about the organisms killed. Several of the biocides that have been recently developed are relatively specific in their action and affect only a few species within an ecosystem. Other biocides, such as the Shell Company's product, Azodrin, are extremely toxic not only to both harmful and beneficial insects, but to many other classes of organisms as well (Van Den Bosch, 1969). An ecologist attempting to assess the impact of such an economic poison is hardly justified in referring to all of the species affected as "pests" if his approach is scientifically neutral, and if he wishes to use language that describes the situation in scientifically neutral terms.

Understanding the pollutant ecology of the sea and determining measures to be taken to preserve its integrity encompass a broad scope of activity. The characteristics that cause certain technological waste products to become persistent pollutants must be defined. The amounts of these compounds that are manufactured and the amounts that are eventually released into the global environment must be calculated. The rates and routes of entry into the marine environment must be established. Studies of their distribution among the components of marine communities must be carried out, and most important of all, the effects of the pollutants upon the marine ecology must be determined. Only a limited amount of work has so far been undertaken and few data are available.

Evidently a pollutant that has the capacity to damage a marine ecosystem must be present in sufficient amounts. A useful measure for purposes of comparison is the amount of inorganic carbon annually incorporated into organic material by phytoplankton, since this sets an upper limit to the total biological activity of the sea. This figure has been determined to be 2.0×10^{16} g/yr (Ryther, 1969). A pollutant that is accumulated by organisms and that is introduced into the ocean at rates of 10^{10} g/yr or higher has evidently a potential capacity to inflict damage, since biological activity of pollutants is expressed frequently at concentrations in the order of 1 ppm. Pollutants that may be accumulated by marine organisms include several of the heavy metals as well as nonpolar organic compounds that are poorly soluble in water but highly soluble in lipids. Mobility is also an essential characteristic since pollutants that are released on land but that are nonmobile will evidently never reach the sea in significant quantities. Persistence is also a necessary characteristic. Most of the waste products of contemporary technology are the same waste products that are produced by primitive societies or by plant and animal communities that do not include man. The most abundant are carbon dioxide, water, and the mineral salts. Like the biblical ashes and dust, they are recycled indefinitely through the organisms of successive generations. The persistent pollutants that are now causing concern to environmental scientists evidently do not obey this ecological and biblical law, partly because of the unprecedented scale of technological activity in a planet with finite dimensions and partly because of the inherent stability of the molecules that renders them comparatively resistant to breakdown by microorganisms within a meaningful time span.

In Table 10.1 the results of analyses of several petrels and shearwaters are listed. These are strictly oceanic species that do not approach land except to breed on islands. They feed at the surface of the sea upon planktonic organisms and do not, like pelicans, dive for fish beneath the surface. The concentrations of the DDT compounds and of the polychlorinated biphenyls in these marine birds are higher than the concentrations recorded in terrestrial birds in either North America (Risebrough et al., 1968; Keith and Hunt, 1966) or Great Britain (Reports of the Royal Society for the Protection of Birds, London).

The fish-eating marine birds inhabiting the southern California coast are heavily contaminated with both DDE and PCB. The brown pelicans (*Pelecanus occidentalis*) and double-crested cormorants (*Phalacrocorax*

Table 10.1. DDT and PCB Residues in Petrels and Shearwaters; DDT Residues Consist of p,p'-DDE plus p,p'-DDT plus p,p'-DDD

Species, Locality (Breeding Area) Tissue	Number	DDT (wet weight, ppm)	PCB (wet weight, ppm)
Fulmar, Fulmarus glacialis, California, whole birds (Alaska)	3	7.1	2.3
Pink-footed Shearwater, Mexico, whole bird (Chile) *Puffinus creatopus*	1	3.0	0.4
Sooty Shearwater (New Zealand and Chile) *Puffinus griseus*			
California, whole bird	2	11.3	1.1
New Brunswick, fat	3	40.9	52.6
Slender-billed Shearwater (*Australia*), *Puffinus tenuirostris*, California, whole bird	1	32.0	2.1
Greater Shearwater, *Puffinus gravis*, New Brunswick (Southern Atlantic), fat	3	70.9	104.3
Bermuda Petrel, *Pterodroma cahow* (Bermuda), eggs and chicks	5	6.4	—
Ashy Petrel, *Oceanodroma homochroa* (California), whole birds	12	66.0	24.0
Black Petrel, *Loomelania melania* (Mexico), whole birds	8	9.2	1.0
Least Petrel, *Helocyptena microsoma* (Mexico), whole birds	3	3.2	0.35
Leach's Petrel, *Oceanodroma leuchorhoa*			
(Baja California), egg fat	1	953	351
(New Brunswick), body fat	3	164	192
Wilson's Petrel, *Oceanites oceanicus* New Brunswick (Antarctica), body fat	3	199	697

From Risebrough et al. (1968); Risebrough (unpublished results); Wurster and Wingate (1968); Hickey and Grubb (personal communication).

auritus) contain a thousand ppm or more of DDE and several hundred ppm of PCB in their body lipids (Risebrough et al., 1970; Gress, Risebrough, Jehl, and Kiff, unpublished results, 1970). The DDE residues are occasionally as high as 2500 ppm in the lipid. Residues in this case have been expressed on a lipid rather than a fresh weight basis since the lipid concentrations permit better comparisons with abnormal physiological symptoms (Risebrough et al., 1970).

San Francisco Bay receives the drainage waters from the principal agricultural areas of California, and fish from the Bay might therefore be assumed to have much higher residue concentrations than fish of the same species in the open ocean. Analyses of collections of the northern anchovy (*Engraulis mordax*) from San Francisco Bay and from the coastal waters showed, however, that residue concentrations were higher in the oceanic fish than in the fish of San Francisco Bay and that residues increased from north to south with highest concentrations in the vicinity of Los Angeles (Table 10.2). Analyses of collections of other marine fish from the coastal waters of California, San Francisco Bay, and the eastern Pacific Ocean show that the majority of residues are within the range of 0.2 to 2.0 ppm (Table 10.2). Also included in Table 10.2 are residues reported in fish from the Baltic Sea (Jensen et al., 1969). Recent surveys carried out to determine the concentrations of chlorinated hydrocarbon insecticides in freshwater fish of California and of major United States river systems have found that the majority of residues also fall within the range of 0.2 to 2.0 ppm in whole fish (Keith and Hunt, 1966; Henderson et al., 1969). Residues in interior freshwater lakes are frequently lower than 0.2 ppm (Kleinert et al., 1968). No data are as yet available from the Atlantic and Indian Oceans but the limited information collected to date indicates that the fish of the oceans may be as contaminated and in at least one area of the sea are more contaminated with the DDT compounds than are freshwater fish.

DDE almost always comprises at least 80 per cent of the DDT residues found in marine organisms. The remainder consists principally of p,p'-DDT and p,p'-DDD (Risebrough et al., 1971). Most studies with "DDT" have usually been carried out with p,p'-DDT and when "DDT" is reported to be degraded, p,p'-DDT is usually converted to DDE. DDE is, however, the compound of greatest concern to environmental scientists since it is much more persistent than is p,p'-DDT. Apparently so far only one paper has reported the degradation of DDE to another unknown compound by organisms found in cheese (Ledford and Chen, 1969). Other bacteria will undoubtedly be found that are capable of using DDE or PCB as an energy source. The high concentrations of DDE and

Table 10.2. Concentrations of the Total DDT Compounds and of PCB in Marine Fish from California, the Eastern Pacific, and the Baltic Sea[a]

Species, Locality	Number	DDT (wet weight, ppm)	PCB (wet weight, ppm)
Northern Anchovy, *Engraulis mordax*			
San Francisco Bay, July 29, 1965	17	0.59 ± 0.11	NM[b]
San Francisco Bay, November 4, 1965	29	0.33 ± 0.04	NM
Monterey, November 30, 1965	30	0.90 ± 0.22	NM
Morro Bay, June 16, 1965	29	0.74 ± 0.22	NM
Port Hueneme, February 24, 1966	15	3.04 ± 1.00	NM
Terminal Island, Los Angeles, June 25, 1966	44	14.0 ± 1.9	1.0
English Sole, *Parophrys vetulus*			
San Francisco Bay, July 29, 1965	18	0.55 ± 0.07	0.11
San Francisco Bay, November 4, 1965	33	0.55 ± 0.12	0.11
San Francisco lightship, December 1, 1965	15	0.19 ± 0.04	0.05
Monterey, February 15, 1966	15	0.76 ± 0.16	0.04
Shiner Perch, *Cymatogaster aggregata*			
San Francisco Bay, October 20, 1965	14	1.0 ± 0.1	1.2
San Francisco Bay, October 20, 1965	10	1.4 ± 0.3	0.4
San Francisco Bay, November 4, 1965	15	1.1 ± 0.1	0.9
Jack Mackerel, *Trachurus symmetricus*			
Channel Islands, November 22, 1965	31	0.56 ± 1.0	0.02
Hake, *Merluccius productus*			
Puget Sound, January 29, 1966	22	0.18 ± 0.05	0.16
Channel Islands, February 24, 1966	6	1.8 ± 1.1	0.12
Bluefin Tuna, *Thunnus thynnus*			
Isla Geronimo, Mexico, August 29, 1965			
Body muscle	7	0.56 ± 0.24	0.04
Liver	9	0.22 ± 0.13	0.04
Yellowfin Tuna, *Thunnus albacares*			
Galapagos Islands, November 1965			
Liver	13	0.07 ± 0.02	NM
Central America (90°W, 8°N) August 1965			
Liver	13	0.62 ± 0.19	0.04
Herring, *Clupea harengus*			
Baltic Sea, 1966–1968	18	0.68	0.27
Fresh tissue			
Plaice, *Pleuronectes platessa*			
Baltic Sea, 1967	6	0.018	0.017
Fresh tissue			
Cod, *Gadus morhua*			
Baltic Sea, 1967	5	0.063	0.033
Fresh tissue			
Salmon, *Salmo salar*			
Baltic Sea, 1968	11	3.4	0.30

[a] From Risebrough (1969); Risebrough et al. (1971); Jensen et al. (1969).
[b] NM: not measured.

PCB in marine organisms so far analyzed indicate, however, that marine bacteria are not degrading a significant part of the DDE and PCB entering the sea. It is possible that in a system where compounds such as DDE and PCB are widespread but nowhere sufficiently concentrated to stimulate bacteria to synthesize the enzymes that might degrade them, accumulation may occur that gives concentrations that have ecological significance.

Of the other chlorinated hydrocarbon insecticides, dieldrin and heptachlor epoxide have been detected in resident wildlife on Signy Island in Antarctica in an area remote from any of the scientific stations (Tatton and Ruzicka, 1967). Dieldrin is also present in sea birds of Great Britain (Robinson et al., 1967) and of North America (Risebrough et al., 1968b). The distribution of endrin is more localized since it was not reported from Antarctic wildlife and is frequently not detected in California marine wildlife. It has been found in brown pelicans from Florida (Schreiber, 1970), Panama, and the Gulf of California (Risebrough et al., 1968). The concentration reported in one egg from the Gulf of California was 1 ppm in the yolk lipid. Because the Gulf of California is a relatively closed area of the sea with a valuable shrimp and fishing industry, the presence of endrin, a very poisonous compound, in these amounts is cause for concern for the future of these industries, indicating the need for additional investigations. Most likely the source is the agricultural lands of western Mexico.

The polychlorinated biphenyls were first discovered in Sweden (Jensen et al., 1969). They were found subsequently in wildlife from Great Britain (Holmes et al., 1967), the Netherlands (Koeman et al., 1969), and North America (Risebrough et al., 1968). PCB has been extensively used in industry in various paints, plastics, adhesives, coating compounds, electrical equipment, and many other products (Monsanto Company Bulletins). The usefulness of the chlorinated biphenyl molecule is derived mainly from its chemical stability which imparts fire and chemical resistance and resistance to hydrolysis. PCB is compatible with many natural and synthetic polymers, and is therefore useful in adhesive and coating compounds. Because of its dielectric properties, fire resistance, inertness, and stability at higher temperatures, PCB is also used in large volumes in transformers, capacitors, and heat transfer systems and for various hydraulic applications. These same properties, combined with the very low solubility of PCB in water, ensure that the molecules will persist after the manufactured products become waste products. Once in the environment PCB appears to move through food chains in ways identical to the DDT compounds (Risebrough et al., 1968).

Confirmation of the identity of these environmental pollutants as PCB has been based upon mass spectrometric determinations in Sweden (Widmark, 1967; Jensen et al., 1969), the Netherlands (Koeman et al., 1969), and North America (Bagley et al., 1970).

In 1969 fish from the Pacific Ocean were confiscated and condemned as unfit for consumption by the U.S. Food and Drug Administration. Jack mackerel (*Trachurus symmetricus*) from the southern California coastal waters were found to contain 10 ppm of the DDT compounds in their tissues (Food Chemical News, June 23, 1969, p. 28). Other fish from the same area of the sea are similarly contaminated and it is likely that other fisheries will be closed and that fisheries in other areas will be affected if accumulation continues. No tolerance limits have as yet been set for PCB in human food, but since PCB has been measured in human milk (Risebrough et al., unpublished results) it is likely that tolerance limits will be set in the near future by the Food and Drug Administration. Some fisheries can be expected to become unavailable for human use because of excessive PCB contamination. Residues of PCB in estuarine organisms in Escambia Bay, Florida, ranged up to 184 ppm in fish tissue, 1.5 to 2.5 ppm in shrimp, and 1.0 to 7.0 ppm in blue crabs (Duke et al., 1970). The source in this area was the effluent of a factory.

United States production of DDT averaged about 150 million lb between 1961 and 1966 (U.S. Tariff Commission Reports) or 6×10^{10} g. Figures for total world production are unknown but it is unlikely that they are greater than twice the United States amount. World production for this time interval might therefore be estimated as 10^{11} g/yr. This figure is only five orders of magnitude lower than the primary productivity of the ocean as measured by the incorporation of carbon into phytoplankton (Ryther, 1969). United States production of the aldrin-dieldrin-endrin-toxaphene group averaged about 110 million lb in the same time interval (U.S. Department of the Interior Special Scientific Report No. 119). Figures for PCB production are not available because United States laws enable manufacturers to regard such information as privileged if fewer than three companies are engaged in the manufacture of a chemical. As a result the Monsanto Company is not required by law to release these figures even when requested to do so by congressional leaders (Congressional Record H 3055-H3056, April 14, 1970). A knowledge of these production figures and a list of all products containing PCB is essential to environmental scientists attempting to determine the pathways of entry into environment, human food, and people. This custom is clearly at odds with both the scientific need and the public interest.

In the past it was usually assumed that the chlorinated hydrocarbons in the sea must have arrived by water transport in rivers. The failure to correlate concentrations of the DDT compounds in fish from the California coastal waters with input to the sea from agricultural drainage waters (Risebrough et al., 1971), the detection of DDT compounds in Antarctic organisms (Sladen et al., 1966; Tatton and Ruzicka, 1967) and in Antarctic snow (Peterle, 1969); the presence of chlorinated hydrocarbons in marine airborne particulate material (Risebrough et al., 1968) and in rain water (Tarrant and Tatton, 1968); and the high concentrations of both DDE and PCB in the surface-feeding petrels and shearwaters from the open sea (Table 10.1) suggest that air transport with subsequent fallout is also an important dispersal mechanism and route of entry into the marine environment. Determination of the future rates of accumulation of the chlorinated hydrocarbons in the sea evidently requires an accurate assessment of the relative importance of river and air transport and of the total amounts entering by each route.

Direct discharge of chlorinated hydrocarbon insecticides or of PCB into drainage waters by manufacturing plants might be a source of contamination in local marine environments, although industry spokesmen have denied that this source is significant. In 1963 at least 5 million fish were killed in the lower Mississippi River. In an investigation that followed (U.S. Department of Health, Education, and Welfare, Conference in the matter of Pollution of the Interstate Waters of the Lower Mississippi River Proceedings, 1964) lethal amounts of endrin were found in fish by investigators of the United States Public Health Service (Hearings, Department of Agriculture Appropriations for 1966, p. 185). Endrin is very poisonous to fish and birds and concentrations of only 2.6 ppb in water are sufficient to kill juvenile fish of certain species (U.S. Department of Health, Education, and Welfare, *Proceedings,* op cit., p. 369). In his testimony, Dr. Bernard Lorant, Vice-president in charge of research for the Velsicol Chemical Corporation, which manufactures endrin in Memphis, Tennessee, stated that although the endrin concentrations may reach 116,000 ppb in the creek that receives the effluents from the factory 45,000 times higher than the concentration that will kill some juvenile fish, no endrin is carried downstream (Hearings, op cit. pp. 486–487). A recent monitoring program has measured very high concentrations of endrin in the sediments below the Velsicol plant in Memphis, and a compound which is formed during the manufacture of endrin had been detected as far as 500 miles downstream (Barthel et al., 1969). Formulating plants for insecticides in other areas of the Mississippi drainage were also contributing chlorinated hydrocarbon insecticides to the river waters, but only relatively small amounts were entering drainage waters from agricultural lands where the insecticides were intensively used (Barthel et al., 1969). The Shell Chemical Company, which manufactures aldrin, dieldrin, and endrin, subsequently denied that leakage from manufacturing plants was responsible for the fish kills or that materials from the factories could enter the sea, stating that "the implication of endrin in the 1963 Mississippi River fish kill has not been verified by recent studies" citing measurements made in 1964 and 1965 that "showed no significant levels of endrin (greater than 0.02 ppm) in the fish, shrimp, oysters, mud, or water analyzed" (Aldrin, Dieldrin, Endrin: A Status Report, the Shell Chemical Company, 1967, p. 44). It is not clear how residues found in 1964 and 1965 disprove lethal levels found in 1963, but 0.02 ppm in water is 8 times the concentration required to kill some juvenile fish. It is unfortunate that no analyses were made of brown pelican tissues since the species suddenly disappeared from the Louisiana coast in the early sixties. Previously it had been the state bird of Louisiana. Endrin is present in sublethal amounts in the brown pelicans that remain in Florida (Schreiber, 1970) and it is possible that the endrin in the fish eaten by this species was the cause of its disappearance on the Gulf Coast. On the west coast a large fraction of the DDT compounds in the southern California marine ecosystem has come from the discharge at the Montrose Chemical Corporation factory in Los Angeles, the largest American producer of DDT (Risebrough et al., 1971). Leakage and spillage from manufacturing plants could therefore be significant in local areas, but no figures are available that indicate yearly outputs from these sources.

In addition to the Mississippi River studies quoted above, other work has also shown that rivers with low silt loads passing through agricultural areas may be almost free of residues. The fish in a Wisconsin stream draining orchards where chlorinated hydrocarbons had been used extensively were found to contain lower residues than the fish in the marine waters of California (Moubry et al., 1968). An initial report of the monitoring program for residues in river waters in the western United States concluded that over half of the streams where residues were detected contained only 0.005 ppb of the DDT compounds (Brown and Nishioka, 1967), considerably lower than an average of 0.08 ppb measured in British rainfall (Tarrant and Tatton, 1968). Subsequent work showed that residues increased with silt content and that the highest concentration recorded was only 0.12 ppb. Eighty-two per cent of all residues found in this study consisted of the DDT compounds (Manigold and Schulze, 1969). The average concentration of the DDT compounds in waters entering San

Francisco Bay were in the order of 0.1 ppb, about one tenth the solubility of p,p'-DDT in water (Bailey and Hannum, 1967; Bowman et al., 1960).

By multiplying the average yearly flow and the average concentration of the DDT compounds it is possible to calculate the input into the sea from each river system. In this way Risebrough et al. (1968) calculated that the drainage waters of the interior valleys of California were bringing about 1900 kg/yr into San Francisco Bay and that the Mississippi was bringing about 10,000 kg/yr into the Gulf of Mexico. These amounts included the DDT adsorbed to silt particles since the analyses did not separate this fraction. A reasonable estimate for the total amount of DDT compounds brought by all of the rivers of the United States into the oceans might therefore be about 20,000 kg. This amount, although large, is only a small fraction of the DDT used each year in the United States. Between 1961 and 1966 the average United States domestic consumption of DDT was 56 million lb, or 2.5×10^{10} g (United States Tariff Commission Publications). Only about 0.1 per cent of this amount is therefore carried by rivers into the sea.

The total world output of river water into the sea is 37,000 km³/yr (Sverdrup et al., 1942). If all of the rivers of the world carried the maximum amount of DDT found in United States rivers, the maximum contribution of the rivers would be about 4×10^9 g, or about 4 per cent of total yearly production. The true value is probably closer to 0.1 per cent of total production, or 10^8 g, on the assumption that the world runoff patterns are similar to United States runoff.

No measurements have yet been made of polychlorinated biphenyls in river water that might permit estimates of the amounts transferred by rivers to the sea.

The relatively low amounts of the DDT compounds that enter the sea from rivers are a result of the very low solubility of the DDT compounds in water and a high affinity for the surfaces of soil and silt particles. The solubility of p,p'-DDT in fresh water is 1.2 ppb (Bowman et al., 1960). The solubility of p,p'-DDE or of the polychlorinated biphenyls in either freshwater or seawater has not yet been determined. These are values that must be obtained in order to understand the movements of these pollutants through marine ecosystems and to determine the partition coefficients between water and the various components of marine communities.

Chlorinated hydrocarbons therefore will tend to remain in soil rather than be leached into drainage waters. After 20 yr, 35 to 50 per cent of the DDT used for subterranean termite control in Mississippi was still present, and there was no significant migration to adjacent soil in spite of intensive rainfall (Smith, 1968). Residues in river water were found to increase with increasing silt content (Manigold and Schulze, 1969) and most of the DDT compounds transported by rivers may be assumed to be adsorbed to the silt particles.

The Shell Company has concluded that since there is negligible leaching of the chlorinated hydrocarbons from soils, little or no environmental contamination will result from the application of DDT, aldrin, dieldrin, endrin, or other chlorinated hydrocarbon insecticides to soils (Shell Chemical Company; Aldrin, Dieldrin, Endrin: A Status Report, pp. 28–30). In this report, as in most reports from the Shell Company or other members of the National Agricultural Chemicals Association (Sobleman, 1970), the facts documenting detrimental effects are usually left out. In this instance Shell does not mention the very considerable loss of the chlorinated hydrocarbon insecticides into the atmosphere through volatilization. The volatility of these biocides and their loss to the atmosphere is considered in a recent review of the persistence of insecticides in soils (Edwards, 1966). Many factors may influence the residence time of the insecticides in soils and the rates at which they volatilize. Retention tends to be highest in soils with high organic content (Edwards, 1966). There is no evidence that DDE, which is frequently formed from p,p'-DDT in soils is in turn converted to other compounds. Greater persistence in organic soils with higher bacterial populations suggests that there is no appreciable bacterial degradation of DDE in soils and that disappearance most likely proceeds via the atmosphere.

The volatility of the chlorinated hydrocarbon insecticides may be demonstrated in several ways. In a laboratory study, half of the aldrin applied to a Petri plate disappeared within 24 hr. Loss by vaporization of dieldrin proceeded at a rate of 4 to 8 per cent/day, slower than that of aldrin but rapid enough to be very significant ecologically (Lichtenstein et al., 1968). The annual loss of dieldrin, aldrin, and endrin from soils is thus explained. DDT has also a tendency to "disappear" when it is applied to an ecosystem (Keith and Hunt, 1966). Codistillation with water, a phenomenon that greatly increases the rate of volatilization of DDT when the DDT is applied to water surfaces (Acree et al., 1963), enhances the rate of entry of the DDT compounds into the atmosphere. As a result of aerial transport the DDT compounds are found in areas where they have never been applied. There is negligible transfer from soil to the upper parts of an alfalfa plant, but nevertheless significant residues are frequently found on alfalfa, a result of the adsorption of DDT compounds present in the air in the vapor phase to leaves and stems (Ware et al., 1968). DDT compounds in frogs of the high Sierra in California where DDT has never been applied must have come from aerial fallout (Cory et al., 1970). The presence of DDT compounds in forest ecosystems of Pennsylvania in regions which had never been treated is yet another example of aerial transfer of DDT compounds to terrestrial ecosystems (Cole et al., 1967).

Atmospheric transfer and subsequent fallout into the sea could therefore explain the observed distribution and amounts of both DDT and PCB compounds in the marine environment.

In an attempt to collect extraterrestrial dust of meteorite origin, the Science Research Council of the United Kingdom mounted a nylon mesh screen on the eastern tip of Barbados (Delaney et al., 1967). The Northeast Trades that reach Barbados have blown over 5000 km of the tropical Atlantic, and it was first considered very unlikely that any dust accumulating on the screen could be of continental origin. This assumption was, however, in considerable error because significant quantities of airborne particulate material were collected on the screen which upon mineralogical and biological examination proved to be most likely from Africa and Europe. A century earlier Darwin had noticed dust blowing from Africa during the voyage of the Beagle and had speculated upon its contribution to the sedimentary deposits in the sea (Darwin, 1846). From the fallout rates of the dust over the tropical Atlantic and from analyses of the dust collected in Barbados for chlorinated hydrocarbons it was possible to calculate that 600 kg/yr of DDT and dieldrin were entering the Atlantic between the equator and 30°N (Risebrough et al., 1968). This was a minimal value since the efficiency of collecting very small particles was low and since chlorinated hydrocarbons entering the sea directly from the vapor phase were not included. Nevertheless this figure is of the same order of magnitude as the amount of DDT compounds and dieldrin which annually enter San Francisco Bay from the San Joaquin and Sacramento Rivers.

Airborne dust collected at the Scripps pier in La Jolla, California, was also analyzed for the chlorinated hydrocarbons (Risebrough et al., 1968). The winds here are landward, with an unknown admixture of air from agricultural areas of California. Concentrations of chlorinated hydrocarbons were approximately 1000 times higher than those recorded in Barbados dust and ranged from 1 to 81 ppm with an average value of 18.2 ppm. Although measurements of the fallout rate of airborne particulates into the sea off southern California have not been made, it is clear that air transport brings large amounts of DDT compounds into the local marine ecosystem.

Estimates of the total amount of aerial fallout of the DDT compounds or of any other pollutants into the sea will require determinations of the total fallout per unit area per unit of time. Permanent snowfields are an obvious source of much valuable information since they preserve the historical record. It is possible, however, to make a preliminary estimate from the determinations of chlorinated hydrocarbon insecticides in rainfall. Seven stations were maintained in the United Kingdom between August 1966 and July 1967 and concentrations of chlorinated hydrocarbons were measured throughout the year. The average total concentrations for all stations of the three DDT compounds was approximately 0.08 ppb (Tarrant and Tatton, 1968). If this is multiplied by the total yearly precipitation over the ocean, 297,000 km^3 (Sverdrup et al., 1942), a value of 2.4×10^{10} g is obtained, approximately one quarter of the world's annual production. This figure does not include dry fallout of the kind reported in the tropical Atlantic (Risebrough et al., 1968) nor does it consider direct impingement of the DDT compounds in the vapor phase upon the sea surface. Loss by re-evaporation is not considered, and the average concentration in total precipitation might be considerably lower than 0.08 ppb. Nevertheless it is an order of magnitude estimate that might stimulate other work designed to answer the basic questions. This estimate of the amount of DDT compounds reaching the sea via aerial fallout is two orders of magnitude greater than the amount estimated to enter from the rivers.

Polychlorinated biphenyls were also present in the British rain samples. Quantitative estimates of fallout of these compounds are evidently needed.

If all of the DDT ever produced, estimated to be in the order of 10^{12} g, were uniformly dissolved in the ocean's total volume of 10^{21} l, the average concentration would be in the order of 0.001 ppb. This is not a trivial concentration, since the amount in Lake Michigan, where the fish contain in the order of 3 ppm in body tissues and where ecological degradation is evident, is in the order of 0.001 to 0.003 ppb in the water (Reinert, 1970). Since the range in concentrations in ocean fish was between 0.2 and 2.0 ppm, a reasonable value for mean concentration would be 1 ppm. If the ratio between concentrations in fish and concentrations in water were the same in freshwater and seawater, the concentration to be expected in the upper layers of the sea would be about 0.001 ppb. No measurements, however, of concentrations of chlorinated hydrocarbons in seawater have yet been made, nor is it known whether they have yet reached the deeper waters. The total amount of marine fish has been estimated to be in the order of 2.4×10^{14} g (Ryther, 1969). The total amount of DDT compounds in fish would therefore be in the order of 2×10^8 g, only 1 per cent of the amount that could reach the sea each year in precipitation. This speculation is intended to be nothing more than speculation, pointing out the evident research needed. Quantitative data on fallout are needed, as is the determination of the partition coefficients among the various components of the marine community, residence time in the mixed layer, and rate of deposition in the sediments.

Effects upon fish reproduction have been noted by Burdick et al. (1964); Macek (1968a and b). Concentrations of a combination of the DDT compounds necessary to impair reproduction were of the same order as those found in Lake Michigan eggs of the coho salmon (*Oncorhynchus kisutch*) and in the kiyi (*Coregonus kiyi*), a species formerly abundant in Lake Michigan but which has become rare (Reinert, 1970). They are also equivalent to the concentrations now found in the anchovies and jack mackerel of the Southern California coastal waters. There is no proof that these or other marine species are or will be affected. On the other hand there are very reasonable grounds for concern, especially if accumulation continues.

The evidence that the chlorinated hydrocarbons in the sea are affecting bird populations is considerably more concrete and is accumulating rapidly. Conclusive experimental proof that one or more chlorinated hydrocarbons would affect the reproductive success of birds at levels now present in the marine environment was not published until 1969, but naturalists had long suspected that chlorinated hydrocarbons in the environment of insecticide origin were the cause of the reproductive failures and population declines of a number of fish-eating and raptorial birds that had begun in the 1950's. The most notable example was the peregrine falcon (*Falco peregrinus*), a species characterized by its population stability, its ability to coexist with man in cities, and its position at the very top of food chains. Following the discoveries that a population crash had occurred in Great Britain and that no more peregrines were breeding in the United States east of the Mississippi, J. J. Hickey convened a conference in Madison, Wisconsin, in 1965 that considered all conceivable causes of the decline of the peregrine and that reviewed the status of other species that were then experiencing unexplainable reproductive failures (Hickey, 1969). Evidence that environmental residues of chlorinated hydrocarbons, accumulated by the peregrine and by such fish-eating species as the osprey (*Pandion haliaetus*) and the bald eagle (*Haliaeetus leucocephalus*) via the food chains, were the contributing cause was convincing at that time to several scientists who were familiar with the natural history of the species affected, but it was largely circumstantial and was not yet sufficient to convince the scientific community at large. It certainly was not sufficient, but probably never will be, to convince the spokesmen of the biocide industries. A lively and at times bitter debate began with the publication of Rachel Carson's book *Silent Spring* and has continued to the present day between the industry spokesmen and scientists associated with industry on one side, and environmental scientists, conservationists, naturalists, and concerned citizens on the other. The scope of the debate has been very broad and has included questioning the role of industry in the decision-making process and the ability of the government bureaucracies to cope with the problem of environmental pollution, but the beneficial aspects of insecticide use have never been at issue. The strictly scientific debate has focused on the question whether the persistent insecticides, as opposed to the nonpersistent degradable compounds, were accumulating in the environment and whether they were contributing to environmental degradation. The killing of song birds by direct poisoning, such as frequently occurred during the Dutch elm disease program, was a separate issue, based on aesthetic and private values.

T. H. Jukes has been the most persistent and vociferous of the industry spokesmen, beginning with an article published in 1963 in which the individuals questioning insecticide use were categorized as organic farmers, emotional bird-watchers, or members of an undefined group that supported "the anti-scientific measures introduced by Lysenko and fostered by Stalin," themes Dr. Jukes has continued to use through 1970 (Jukes, 1963; Jukes, 1970). Throughout the intervening years countless letters went to the editors of newspapers, magazines, and journals in which it was vigorously denied that DDT had any deleterious effect upon wildlife or upon the environment, including a letter to *Science* published in 1968, the theme of which was that environmental pollution poses no problems as long as there are robins on the lawn (Jukes, 1968). Dr. Jukes has yet to publish, however, a scientific paper about any of the chlorinated hydrocarbons or a topic in the fields of environmental science or pollution ecology. The task of assembling and documenting the evidence has been left to selected scientists concerned about environmental pollution and specific effects upon populations.

By far the most significant paper that has appeared in the field of environmental science was published by Ratcliffe (1967). For the first time it was shown that the population declines of several birds of prey have a physiological basis caused by an environmental factor that appeared about 1947. Eggs of the peregrine, golden eagle (*Aquila chrysaetos*), and British sparrow hawk (*Accipiter nisus*) from declining populations had deficient amounts of calcium carbonate.

The debate continued in 1968 at a conference held in Rochester, New

York, and the papers presented at that conference have been published in book form under the title, *Chemical Fallout*, Miller and Berg, Eds. (1969). Robinson (1969) of the Shell Chemical Company and Spencer (1969) of the National Agricultural Chemicals Association presented the industry position. Environmental effects were discussed in papers by Herman et al. (1969), Butler (1969), Wurster (1969), and Risebrough (1969), and in addition a number of papers were presented by Swedish investigators on the effects of mercury pollution. Robinson's paper considered an absence of effects in British birds that are low in the food chains, and by means of selected quotations concluded that raptorial populations did not appear to be affected by environmental chlorinated hydrocarbons. The theme of Spencer's paper at that meeting (Spencer, 1969) was that the machinery of the government agencies provides "extensive provision for protection of wildlife and the environment from adverse effects of pesticides." The crucial decisions about registration and use and about enforcement, however, have been made only by the Department of Agriculture. Spencer wrote (Spencer, 1969, p. 506) the following: "Despite the fact that the Department of Agriculture alone has the final responsibility for registration of pesticides, seldom, if ever, have they disregarded the advice of the members of the interdepartmental review." Congressional investigation has revealed, however, that in 1969 alone there were apparently 185 registrations of biocide formulations by USDA over the objections of the Department of Health, Education, and Welfare; and that none of these objections had been referred to the Secretary of Agriculture in accordance with the provisions of the Interdepartmental Agreement; that despite evidence of repeated violations by some shippers, Pesticides Regulation Division of the Department of Agriculture (PRD) had not initiated a single criminal prosecution for 13 yr; and that employees of the Shell Chemical Company were serving as consultants to the Department of Agriculture on matters relating to the registration of Shell products (Committee on Government Operations, Eleventh Report to the 91st Congress: Deficiencies in Administration of Federal Insecticide, Fungicide, and Rodenticide Act, U.S. Government Printing Office, 1969). Spencer's comment at the Rochester conference, "Does it make any difference as to where a man works," (Spencer, 1969, p. 470), illustrates that "scientific" conclusions are very frequently affected not only by the interests of the employer and the employee but also by prevailing attitudes in segments of industry, the government, or the military.

The paper published by Hickey and Anderson later in 1968 showed that the shell thinning phenomenon was also occurring in North America, that populations that showed the thinning were in decline but populations of the same species in other areas that showed no thinning were still stable, and that thinning in eggs of the herring gull (*Larus argentatus*) was significantly correlated with residues of DDE. These findings greatly strengthened the scientific argument on the environmentalist side.

In late 1968 several Wisconsin communities were preparing to begin the annual spraying with DDT in an attempt to control the Dutch elm disease. Sufficient evidence was available at that time to indicate that environmental DDT contamination was becoming a significant pollution problem, and under Wisconsin procedure that permits public hearings on the interpretation of laws, hearings were held by the Department of Natural Resources to determine whether continued DDT use in Wisconsin would be in violation of a law that prohibits the use of any substance that becomes a pollutant in the waters of the state and that is harmful to fish and wildlife. The transcript of the hearings has been published by the Wisconsin Department of Natural Resources, Madison, Wisconsin (1969). The initial evidence presented included work showing that DDT may harm fish populations at levels present in Lake Michigan (Macek, 1968a, b), that any DDT use could lead to contamination of the world environment (Risebrough et al., 1967; Risebrough et al., 1968; Risebrough, 1969) and that the declines of several bird species were linked with environmental DDT residues. However, this evidence was still largely circumstantial. The participants were aware of work carried out at the Patuxent center of the Bureau of Fish and Wildlife that would greatly strengthen their case, but the results were not yet published. The Department of the Interior then consented to release these data in advance of publication, and in a brief but brilliant testimony Dr. Lucille Stickel provided the experimental proof that until that time was lacking. Dietary levels of p,p'-DDT and dieldrin at levels that might be found in the environment significantly reduced shell thickness and reproductive success of the American kestrel (*Falco sparverius*), a species in the same genus as the peregrine (Porter and Wiemeyer, 1969). Even more significant was the finding that small concentrations of DDE in the diet could induce shell thinning in eggs of mallard ducks (Heath et al., 1969). The case for the environmental side, which was being conducted by the Environmental Defense Fund, was therefore immeasurably strengthened with Dr. Stickel's testimony. The lawyer for the National Agricultural Chemicals Association subsequently attempted to show that even though DDE might induce the effect of shell thinning under experimental conditions, polychlorinated biphenyls were also abundant as widespread pollutants in the environment and that the damage might in fact be caused by PCB rather than by DDE. In support of this argument, Mr. Coon of the Wisconsin Alumni Research Foundation, the commercial laboratory that has analyzed the samples of many environmental samples, including those of Dr. Hickey, testified that his laboratory had misidentified p,p'-DDT and p,p'-DDD in many samples, and that what had been reported to be these compounds was in fact PCB. Mr. Coon also suggested that the DDE values reported by his laboratory might also be suspect. During that time the methodology for distinguishing the DDT compounds from PCB was being worked out (Anderson et al., 1969; Risebrough et al., 1969) and the data were therefore presented by the environmental side to show that the DDE values reported in Wisconsin wildlife were not affected by the presence of PCB in the samples. A more formidable difficulty, however, remained.

The mechanism originally proposed to explain the shell thinning was the degradation of estrogen by the microsomal mixed function oxidase enzymes in the liver induced by the chlorinated hydrocarbons (Welch et al., 1969; Wurster, 1969). These enzymes are induced in birds and mammals by foreign lipid-soluble compounds that are frequently poisonous or otherwise deleterious, and since they are poorly soluble in water, they can be excreted by the kidney only with difficulty. The detoxification process includes hydroxylation by the liver enzymes and the inducers usually induce the enzymes that metabolize them. Induced enzyme activity is usually of short-term duration. Estrogen, testosterone, and other steroids are also degraded by these enzymes (Conney, 1967; Welch et al., 1969, Peakall, 1967) as is thyroxine (Schwartz et al., 1969). Estrogen induces the formation of medullary bone, the source of much of the calcium subsequently deposited in the eggshell (Wurster, 1969). PCB, however, is a stronger inducer than is DDE (Risebrough et al., 1968), and if the induction phenomenon were the only one involved with thin eggshells PCB could be expected to have a greater effect than DDE. The Environmental Defense Fund then introduced evidence from a paper by Anderson et al. (1969) showing that no correlation between shell thinning and PCB residues could be found in eggs of the white pelican (*Pelecanus erythrorynchos*) but that a correlation did exist between thinning and the amounts of the DDT compounds. Correlations between shell thinning and both DDE and PCB in eggs of the double-crested cormorant were reported in the same paper. Whether PCB did influence shell thinning in some way was not clear.

While the DDT Hearings in Wisconsin were still underway, findings were made on the west coast that documented the impact of chlorinated hydrocarbons in the marine environment. An earlier analysis of brown pelican tissue had revealed concentrations of 80 ppm in the breast muscle, levels that might be suspected of causing damage (Risebrough et al., 1967). Unsuccessful attempts were made to visit the nesting areas on the Channel Islands in California in 1968, but in March 1969 a party was able to make a preliminary survey. The pelicans were breeding on Anacapa Island, but the initial attempt by approximately 300 pairs failed when all of the eggs broke during incubation because of insufficient amounts of calcium carbonate in the shells. The pelicans continued their nesting attempts throughout the spring and one young pelican finally hatched in late June. At that time four additional eggs were present. Of approximately 1200 nesting attempts therefore made by the pelicans in 1969 on Anacapa Island, at most five young pelicans were produced (Risebrough et al., 1970). The story was much the same on the Coronados Islands, further to the south, where all eggs also collapsed during incubation. On the islands off western Baja California shell thinning was observed, but eggs were sufficiently thick to permit hatching in most cases (Jehl, 1970). Shell thinning was also evident in pelican eggs from Florida (Schreiber, 1970). In 1970 approximately 500 nesting attempts on Anacapa have produced no hatchlings up to the time this chapter was written in early July (Gress, 1970), and there has been no successful breeding on the Coronados (Jehl, 1970). Analyses of approximately 180 egg fragments and of whole eggs from Anacapa, the Islas Coronados, the Islas San Martin and San Benitos in western Baja California, the Gulf of California, four Florida localities, Jamaica, Venezuela, Panama, and Peru, for DDE, PCB, and other chlorinated hydrocarbons show that DDE is the principal cause of shell thinning in the brown pelican (Risebrough and Anderson, 1970). When all colonies are combined there is a highly significant correlation between thinning and DDE concentration, with the eggs from Florida and the Gulf of California which contain lower amounts of DDE showing lesser amounts of thinning. The eggs with the thinnest shells on Anacapa and the Coronados, consisting of little more than a membrane with almost no calcium carbonate, contained about 2500 ppm of DDE in the lipid fraction (Risebrough and Anderson, 1970).

The double-crested cormorants of southern California also experienced complete reproductive failures in 1969 and 1970, the cause also being the collapse of eggs during incubation. Like the pelicans this species in this ecosystem preys only upon marine fish. Correlation between thinning and DDE concentrations was also found (Gress et al., unpublished observations). Following the brown pelican discoveries, other marine species were also investigated. The common murres (*Uria aalge*) and the ashy petrels (*Oceanodroma homochroa*) of the Farallon Islands at the latitude of San

Francisco, an area less contaminated than the marine ecosystem of southern California, also show shell thinning to a degree approaching the critical level but which nevertheless permits most young birds to hatch (Gress, Risebrough, and Sibley, unpublished results; Risebrough and Coulter, unpublished results). The American egrets (*Casmerodius albus*) breeding at a coastal site north of San Francisco were found to be laying thin-shelled eggs in 1969 (Pratt and Risebrough, unpublished observations). Many marine species, therefore, now appear threatened and can be expected to disappear if contamination levels continue to increase.

The curves relating shell thickness to DDE concentrations in the eggs of the brown pelican, double crested cormorant, and herring gull all show a very sharp decrease with relatively small amounts of DDE. Moreover the effect is linear to zero and a range of DDE concentrations that produces no effect is not observed. This finding is an apparent contradiction with one of the fundamental laws of classical toxicology: "For every chemical there is a level of exposure that produces no effect. The body is capable of handling this amount by metabolizing it, excreting it unchanged, or by isolating it in some depot site. As the level of exposure increases, the body's ability to deal with the chemical is exceeded, and effects begin to occur." (Ottoboni, 1969). "Unlike 'natural' chemicals such as lead and arsenic which have been used in large quantities for many years, organic pesticides are not permanent. Their biological action follows well-established principles of toxicology. The key concept is that of acute versus chronic exposure" (Crosby, 1970). These statements are clearly not compatible with the conclusions emerging from the field that might be called "environmental" toxicology. The optimum thickness of an eggshell has been determined in countless evolutionary processes. Any deviation might be considered a selective disadvantage.

The observed effects at low DDE concentrations suggest that hormone metabolism is not the immediate cause of thin eggshells. Enzyme induction may lower the concentrations of estrogen (Peakall, 1970), thereby delaying the breeding cycle (Jefferies, 1966), but the pelican eggs with almost no eggshell were nevertheless laid by birds that had gone through the breeding cycle and which had functional oviducts, both of which would require adequate amounts of circulating estrogen. Although estrogen is broken down by induced enzymes, feedback mechanisms can be expected to maintain concentrations at levels adequate to stimulate oviduct formation. Since estrogen stimulates the deposition of medullary bone, the source of some but not all of eggshell calcium, it is not possible to explain the complete absence of calcium in some eggs, therefore, by the enzyme induction theory.

DDE and p,p'-DDT may also inhibit the enzyme carbonic anhydrase which is essential for the deposition of calcium carbonate and for the maintenance of pH gradients across membranes such as those found in the shell gland (Risebrough et al., 1970; Peakall, 1970; Bitman et al., 1970). Because this enzyme is almost always found in excess of physiological requirements, it is doubtful whether carbonic anhydrase inhibition could explain the DDE effect at low concentrations (Risebrough et al., 1970) but could exert an effect at higher concentrations of DDE. Since chlorinated hydrocarbons, including DDE, are known to inhibit ion transport across membranes through inhibition of membrane ATPases, and since calcium transport across membranes is also an energy-requiring process in which the energy is released with the participation of membrane proteins functioning as ATPases, it is likely that the principal DDE effect in the shell gland is upon the calcium transport process (reviewed by Risebrough et al., 1970). Since PCB is not a nerve poison as are the chlorinated hydrocarbon insecticides, this mechanism would account for the failure to find a PCB effect upon shell thinning.

PCB, as well as DDE and other chlorinated hydrocarbons, may nevertheless exert an effect upon the reproductive success of marine birds through enzyme induction and the consequent upsetting of hormone balances. The abnormal behavior of the contaminated brown pelicans on Anacapa Island observed in 1970 could be a result of unbalanced metabolism of the steroid hormones or of thyroxine (Gress, 1970). Birds fed a chlorinated hydrocarbon may show symptoms of hyperthyroid activity (Jefferies, 1969; Jefferies and French, 1969) or significant decreases in estrogen concentration (Peakall, 1970).

Marine organisms other than birds are also dependent upon carbonic anhydrase and ion transport phenomena and it is still an open question whether any other effects upon marine communities caused by chlorinated hydrocarbons will be observed. The thin-shelled pelican eggs were found only when investigators went out to look for them. The future of many populations of marine birds and of the fisheries is clearly dependent upon the rates of accumulation of the chlorinated hydrocarbons in local marine ecosystems. Brown pelicans on Anacapa were still producing young in 1963 and 1964 (Banks, 1966) but eggs showing a critical amount of shell thinning had been collected in 1962 (Anderson and Hickey, 1970). Between 1962 and 1969 therefore a level of contamination critical to the brown pelican was reached and exceeded in the marine ecosystem of southern California.

The DDT producers of the United States have recently published an extensive list of statements and papers by industry spokesmen and by scientists sympathetic to the industry position (Sobleman, 1970). There are a number of statements about the brown pelican but none are based upon personal research, nor are they based upon data presented in scientific publications since the relevant papers were still in press and had not been published, nor had any of the industrially oriented scientists asked for or obtained unpublished data that might have corrected the somewhat extravagant statements compiled in this publication. The other statements about the environmental effects of the DDT-derived compounds in this publication have no firmer foundation than do those about the brown pelican. The DDT producers and their sympathizers have clearly shown a lack of public responsibility, but in producing this publication they have provided an invaluable source of material for the study of the motivations, psychology and values of the industrially oriented scientist. The Monsanto Company, by withholding information about production figures and PCB use, is also clearly acting against the public interest. The deficiencies uncovered in the Pesticides Regulation Division of the Department of Agriculture indicate that government agencies will not always put public interest before private interest. The continuing need is for solid data that will answer the pertinent questions and for impartial, independent interpretations of these data by scientists who employ a variety of methods and approaches and whose conclusions are not influenced by political or commercial considerations. Scientists in a university or other private capacity are essential for the fulfillment of this responsibility and they, by default, must assume the major part of this responsibility.

References

Acree, F., M. Beroza, and M. C. Bowman. 1963. Codistillation of DDT with water. *Agric. Food Chem.*, **11**:278–280.

Anderson, D. W., and J. J. Hickey. 1970. Oological data on egg and breeding characteristics of Brown Pelicans. *Wilson Bull.*, **82**:14–28.

Anderson, D. W., J. J. Hickey, R. W. Risebrough, D. F. Hughes, and R. E. Christensen. 1969. Significance of chlorinated hydrocarbon residues to breeding pelicans and cormorants. *Can. Field-Nat.*, **83**:91–112.

Bagley, G. E., W. L. Reichel, and E. Cromartie. 1970. Identification of polychlorinated biphenyls in two bald eagles by combined gas-liquid chromatography-mass spectrometry. *JAOAC*, **53**:251–261.

Bailey, T. E., and J. R. Hannum. 1967. Distribution of pesticides in California. *J. Sanit. Engin. Div.*, **5**:27–43.

Banks, R. C. 1966. Terrestrial Vertebrates of Anacapa Island, California. *Trans. San Diego Soc. Nat. Hist.*, **14**:173–188.

Barthel, W. F., J. C. Hawthorne, J. H. Ford, G. C. Bolton, L. L. McDowell, E. H. Grissinger, and D. A. Parsons. 1969. Pesticides in water: Pesticide residues in sediments of the lower Mississippi River and its tributaries. *Pesticides Monitoring J.*, 3:8–66.

Bitman, J., H. C. Cecil, and G. F. Fries. 1970. DDT-induced inhibition of avian shell gland carbonic anhydrase: a mechanism for thin eggshells. *Science*, **168**:594–596.

Bowman, M. C., F. Acree, and M. K. Corbett. 1960. Solubility of carbon-14 DDT in water. *Agric. Food Chem.*, **8**:406–408.

Brown, E. and Y. A. Nishioka. 1967. Pesticides in water: Pesticides in selected western streams. A contribution to the national program. *Pesticides Monitoring J.*, **1**:38–46.

Burdick, G. E., E. J. Harris, H. J. Dean, T. M. Walker, J. Skea, and D. Colby. 1964. The accumulation of DDT in lake trout and the effect on reproduction. *Trans. Amer. Fish. Soc.*, **93**:127–136.

Butler, P. A. 1969. The significance of DDT residues in estuarine fauna. In *Chemical Fallout* M. W. Miller and G. G. Berg (eds.), Charles C Thomas, Springfield, pp. 205–218.

Cole, H., D. Barry, D. E. H. Frear, and A. Bradford. 1967. DDT levels in fish, stream sediments, and soil before and after DDT aerial spray application for fall cankerworm in northern Pennsylvania. *Bull. Environ. Cont. Toxicol.*, **2**:127–146.

Conney, A. H. 1967. Pharmacological implications of microsomal enzyme induction. *Pharmacol. Rev.*, **19**:317–366.

Cory, L., P. Fjeld, and W. Serat. 1970. Distribution patterns of DDT residues in the Sierra Nevada Mountains. *Pesticides Monitoring J.*, **4**:204–211.

Crosby, D. G. 1970. Statement before the State of Washington DDT hearings, Seattle Wash., October, 1969, in *DDT: selected statements from state of Washington DDT hearings and other related papers* (compiler M. Sobleman), Montrose Chemical Corp., Torrance, Calif.

Darwin, C. 1846. An account of the fine dust which falls on vessels in the Atlantic Ocean. *Quart. J. Geol. Soc. (London)*, **2**:26.

Delaney, A. C., A. C. Delaney, D. W. Parkin, J. J. Griffin, E. D. Goldberg, and B. E. F. Reimann. 1967. Airborne dust collected at Barbados., *Geochim. Cosmochim. Acta*, **31**:885–909.

Duke, T. W., J. I. Lowe, and A. J. Wilson. 1970. A polychlorinated biphenyl (Aroclor 1254) in the water, sediment, and biota of Escambia Bay, Florida. *Bull. Environ. Cont. Toxicol.*, **5**:171–180.

Edwards, C. A. 1966. Insecticide residues in soils. *Residue Rev.*, **13**:83–132.

Gress, F. 1970. Reproductive failures of the brown pelicans on Anacapa Island in 1970. Unpublished manuscript.

Heath, R. G., J. W. Spann and J. F. Kreitzer. 1969. Marked DDE impairment of mallard reproduction in controlled studies. *Nature*, **224**:47–48.

Henderson, C., W. L. Johnson, and A. Inglis. 1969. Organochlorine insecticide residues in fish. *Pesticides Monitoring J.*, **3**:145–171.

Herman, S. G., R. L. Garrett and R. L. Rudd. 1969. Pesticides and the western grebe. In *Chemical fallout* M. W. Miller and G. G. Berg, (eds.), Charles C Thomas, Springfield, pp. 24–51.

Hickey, J. J. (ed.) 1969. *Peregrine Falcon Populations: Their Biology and Decline*, University of Wisconsin Press, Madison, Wis.

Hickey, J. J., and D. W. Anderson. 1968. Chlorinated hydrocarbons and eggshell changes in raptorial and fish-eating birds. *Science*, **162**:271–273.

Holmes, D. C., J. H. Simmons, and J. O'G. Tatton. 1967. Chlorinated hydrocarbons in British wildlife. *Nature*, **216**:227–229.

Jefferies, D. J. 1966. The delay in ovulation produced by p,p'-DDT and its possible significance in the field. *Ibis*, **109**:266–272.

Jefferies, D. J. 1969. Induction of apparent hyperthyroidism in birds fed DDT. *Nature*, **222**:578–579.

Jefferies, D. J., and M. C. French. 1969. Avian thyroid: Effect of p,p'-DDT on size and activity. *Science*, **166**:1278–1280.

Jehl, J. 1970. Reproductive failures of the brown pelicans of Baja California. Unpublished manuscript.

Jensen, S., A. G. Johnels, M. Olsson, and G. Otterlind. 1969. DDT and PCB in marine animals from Swedish waters. *Nature*, **224**:247–250.

Jukes, T. H. 1963. People and pesticides. *Amer. Scientist*, **51**:355–361.

Jukes, T. H. 1968. The naivete of science. *Science*, **159**:695.

Jukes, T. H. 1970. Statement before the State of Washington DDT Hearings, Seattle, Wash., October, 1969, in *DDT: Selected Statements from State of Washington DDT Hearings and Other Related Papers* (compiler M. Sobleman), Montrose chemical Corp, Torrance, Calif.

Keith, J. O., and E. G. Hunt. 1966. Levels of Insecticide Residues in Fish and Wildlife in California. *Trans. 31st Amer. Wildlife and Natural Resources Conference*, Wildlife Management Institute, Washington, D.C.

Kleinert, S. J., P. E. Degurse and T. L. Wirth. 1968. Occurrence and significance of DDT and dieldrin residues in Wisconsin fish. Department of Natural Resources, Madison, Wis. *Tech. Bull.*, **41**.

Koeman, J. H., M. C. Ten Noever de Brauw, and R. H. de Vos. 1969. Chlorinated biphenyls in fish, mussels and birds from the river Rhine and the Netherlands coastal area. *Nature*, **221**:1126–1128.

Ledford, R. A., and J. H. Chen. 1969. Degradation of DDT and DDE by cheese organisms. *J. Food Sci.*, **34**:386–388.

Lichtenstein, E. P., J. P. Anderson, T. W. Fuhremann, and K. R. Schulz. 1968. Aldrin and Dieldrin: Loss under sterile conditions, *Science*, **159**:1110–1111.

Macek, K. J. 1968a. Reproduction in brook trout, *Salvelinus fontinalis*, fed sublethal concentrations of DDT. *J. Fish. Res. Board Can.*, **25**:1787–1796.

Macek, K. J. 1968b. Growth and resistance to stress in brook trout fed sublethal levels of DDT. *J. Fish. Res. Board Can.*, **25**:2443–2451.

Manigold, D. B., and J. A. Schulze. 1969. Pesticides in water: Pesticides in selected western streams. A progress report. *Pesticides Monitoring J.*, **3**:124–135.

Monsanto Company, Technical Bulletins O/PL-311A, O/PL-306, and O-FF/1. Monsanto Company, St. Louis.

Moubry, R. J., J. M. Helm, and G. R. Myrdal. 1968. Chlorinated pesticide residues in an aquatic environment adjacent to a commercial orchard. *Pesticides Monitoring J.*, **1**:27–29.

Ottoboni, A. 1969. DDT: The world has been doused with it for 25 years. With what results? *Calif. Health*, **27**:1–2, 15.

Peakall, D. B. 1967. Pesticide-induced enzyme breakdown of steroids in birds. *Nature*, **216**:505–506.

Peákall, D. B. 1970. p,p'-DDT: Effect on calcium metabolism and concentration of estradiol in the blood. *Science*, **168**:592–594.

Peterle, T. J. 1969. DDT in Antarctic snow. *Nature*, **224**:620.

Porter, R. D., and S. N. Wiemeyer. 1969. Dieldrin and DDT: Effects on sparrow hawk eggshells and reproduction. *Science*, **165**:199–200.

Ratcliffe, D. A. 1967. Decrease in eggshell weight in certain birds of prey. *Nature*, **215**:208–210.

Reinert, R. E. 1970. Pesticide concentrations in Great Lakes Fish. *Pesticides Monitoring J.*, **3**:233–240.

Risebrough, R. W. 1969. Chlorinated hydrocarbons in marine ecosystems. In *Chemical fallout*, M. W. Miller and G. G. Berg (eds.), Charles C Thomas, Springfield, pp. 5–23.

Risebrough, R. W., and D. W. Anderson. 1970. Pollutants and changes in shell thickness of eggs of the brown pelican. Unpublished manuscript.

Risebrough, R. W., J. D. Davis, and D. W. Anderson. 1970. Effects of various chlorinated hydrocarbons on birds. In *Biological impact of pesticides in the environment*. J. W. Gillette (ed.). In press.

Risebrough, R. W., J. D. Davis, R. Anastasia, F. A. Beland, and J. H. Enderson. Polychlorinated biphenyls: Industrial pollutants in human milk. Unpublished manuscript.

Risebrough, R. W., R. J. Huggett, J. J. Griffin, and E. D. Goldberg. 1968. Pesticides: Transatlantic movements in the northeast trades. *Science*, **159**:1233–1236.

Risebrough, R. W., D. B. Menzel, D. J. Martin, and H. S. Olcott. 1967. DDT Residues in Pacific sea birds: a persistent insecticide in marine food chains. *Nature*, 216:589–591.

Risebrough, R. W., D. B. Menzel, D. J. Martin, and H. S. Olcott. 1971. DDT residues in Pacific marine fish. *Pesticides Monitoring J.* In press.

Risebrough, R. W., P. Reiche, and H. S. Olcott. 1969. Current progress in the determination of the polychlorinated biphenyls. *Bull. Environ. Cont. Toxicol.*, 4:192–201.

Risebrough, R. W., P. Reiche, D. B. Peakall, S. G. Herman, and M. N. Kirven. 1968b. Polychlorinated biphenyls in the global ecosystem. *Nature*, 220: 1098–1102.

Risebrough, R. W., F. C. Sibley, and M. N. Kirven. 1971. Reproductive failures of the brown pelican on Anacapa Island in 1969. *American Birds*, 25:8–9.

Robinson, J. 1969. Organochlorine Insecticides and bird populations in Britain. In *Chemical Fallout*, M. W. Miller and G. G. Berg (eds.). Charles C Thomas, Springfield, pp. 113–173.

Robinson, J., A. Richardson, A. N. Crabtree, J. C. Coulson, and G. R. Potts. 1967. Organochlorine residues in marine organisms. *Nature*, 214:1307–1311.

Ryther, J. H. 1969. Photosynthesis and fish production in the sea. *Science*, 166:72–76.

Schreiber, R. W. 1970. Pollutants and shell thinning in the Florida brown pelicans. Unpublished manuscript.

Shell Chemical Company. 1967. *Aldrin, Dieldrin, Endrin (A Status Report)*.

Schwartz, H. L., V. Kosyreff, M. I. Surks, and J. H. Oppenheimer. 1969. Increased deiodination of L-thyroxine and L-triiodothyronine by liver microsomes from rats treated with phenobarbital. *Nature*, 221:1262–1263.

Sladen, W. J. L., C. M. Menzie, and W. L. Reichel. 1966. DDT residues in Adelie penguins and a crabeater seal from Antarctica. *Nature*, 210:670–673.

Smith, V. K. 1968. Pesticides in soil: Long-term movement of DDT applied to soil for termite control. *Pesticides Monitoring J.*, 2:55–57.

Sobleman, M. 1970. *DDT: Selected Statements from State of Washington DDT Hearings and Other Related Papers*. Montrose Chemical Corp. Torrance, Calif.

Spencer, D. A. 1969. The procedure for processing information on pesticides. In *Chemical fallout*, M. W. Miller and G. G. Berg (eds.). Charles C Thomas, Springfield, pp. 501–507.

Sverdrup, H. U., M. W. Johnson, and R. H. Fleming. 1942. *The oceans: their physics, chemistry and general biology*, Prentice-Hall, Englewood Cliffs, N.J.

Tarrant, K. R., and J. O'G. Tatton. 1968. Organochlorine pesticides in rainwater in the British Isles. *Nature*, 219:725–727.

Tatton, J. O'G., and J. H. A. Ruzicka. 1967. Organochlorine pesticides in Antarctica. *Nature*, 215:346–348.

Van Den Bosch, R. 1969. The toxicity problem-comments by an applied insect ecologist. In *Chemical Fallout*, M. W. Miller and G. G. Berg (eds.). Charles C Thomas, Springfield, pp. 97–112.

Ware, G. W., B. J. Estesen, and W. P. Cahill. 1968. Pesticides in soil: an ecological study of DDT residues in Arizona soils and alfalfa. *Pesticides Monitoring J.*, **2**:129–132.

Welch, R. M., W. Levin, and A. H. Conney. 1969. Effect of chlorinated insecticides on steroid metabolism. In *Chemical fallout*, M. W. Miller and G. G. Berg, (eds.), Charles C Thomas, Springfield, pp. 390–404.

Widmark, G. 1967. Possible interference by chlorinated biphenyls. *JAOAC*, **50**:1069.

Wurster, C. F. 1969. Chlorinated hydrocarvon insecticides and avian reproduction: How are they related? In *Chemical fallout*, M. W. Miller and G. G. Berg, (eds.), Charles C Thomas, Springfield, pp. 368–387.

Wurster, C. F. and D. B. Wingate. 1968. DDT residues and declining reproduction in the Bermuda petrel. *Science*, **159**:979–981.

11

Carbon Dioxide—Man's Unseen Artifact

WALLACE S. BROECKER, YUAN-HUI LI, and TSUNG-HUNG PENG, *Lamont-Doherty Geological Observatory of Columbia University, Palisades, New York*

During the past century man has recovered 10^{11} tons of coal, 10^{11} barrels of petroleum, and 10^{12} m^3 of natural gas from the sedimentary rocks which mantle the earth's surface. The carbon in these fuels has been delivered via combustion to the atmosphere as CO_2 gas. During the coming century an order of magnitude greater of fuel will be required to meet our energy demands. As a consequence, sometime during the first half of the next century the CO_2 content of our air will reach a level twice as high as it is today.

Fortunately CO_2 is not toxic. Our medical well-being is not endangered. Instead we must worry about the effect on our climate. The CO_2 and H_2O in our atmosphere trap outgoing infrared light. Because of this impediment to loss of energy our planet maintains a temperature considerably warmer than would exist in the absence of these gases. By adding more CO_2 to the atmosphere man will shift the balance in the earth's radiation budget. In order to maintain equilibrium between incoming and outgoing energy some change in cloudiness, water vapor content, and temperature will take place. As a result climate will gradually change. Whether these changes will be beneficial, harmful, or of little consequence to the various nations of the earth is a matter of immense concern to man.

In order to evaluate the situation we must know (*a*) the amounts of CO_2 which have been and will be produced, (*b*) the fraction of this CO_2 which has and will remain in the atmosphere, and (*c*) the manner in which climate changes with increasing CO_2 content of the air. As the information needed is not complete, in addition to summarizing the state of our knowledge, we point out areas needing further research.

CO_2 Production: Past and Future

A compilation of world fossil fuel production made by the United Nations, Department of Economic and Social Affairs (1956) and the 1950–1959 World Energy Supplies Statistical Papers, Series J, has generally been adopted as the basis for CO_2 generation estimates (Revelle and Suess, 1957; Revelle, 1965). The weights of coal, lignite, liquid hydrocarbon, and natural gas are given in Table 11.1 for each decade from 1860 to 1960. Assuming the mean carbon content of coal to be 70 per cent, of lignite 42 per cent, of liquid hydrocarbon 85 per cent, and of natural gas 70 per cent, and that all the carbon in these fuels has been converted to CO_2 gas, the decade by decade production estimates for CO_2

Table 11.1. World Production of Fossil Fuel by Decade[a]

Decade	Coal	Lignite	Liquid Hydrocarbon	Natural Gas	Reference
1860–9	1,660	90	4	—	1[b]
1870–9	2,560	180	16	—	1[b]
1880–9	3,850	290	56	12	1[b]
1890–9	5,400	490	142	39	1[b]
1900–9	8,560	880	300	82	1[b]
1910–9	11,270	1,260	590	154	1[b]
1920–9	11,850	1,870	1,510	285	1[b]
1930–9	11,500	2,130	2,340	484	1[b]
1940–9	13,570	3,000	3,610	970	1[b]
1950–9	14,960	6,470	7,710	2400	2[c]
Σ	85,200	16,700	16,300	4400	

[a] Units: Millions of Metric Tons—10^{12} g.
[b] World energy requirements in 1975 and 2000, United Nations Proceedings of the International Conference on the Peaceful Uses of Atomic Energy (1956).
[c] World energy supplies statistical papers, United Nations, Series J, New York.

Table 11.2. Carbon Dioxide Produced by Fossil Fuel Combustion by Decade[a]

Decade	Coal	Lignite	Liquid Hydro-carbon	Natural Gas	Decade Total	Cumulative Total
1860–9	0.44	0.01	0.00	0.00	0.45	0.45
1870–9	0.68	0.03	0.01	0.00	0.71	1.16
1880–9	1.02	0.04	0.02	0.00	1.08	2.24
1890–9	1.43	0.08	0.04	0.01	1.56	3.80
1900–9	2.26	0.14	0.09	0.02	2.51	6.31
1910–9	2.98	0.17	0.18	0.04	3.37	9.68
1920–9	3.13	0.29	0.47	0.07	3.96	13.64
1930–9	3.04	0.33	0.73	0.12	4.22	17.86
1940–9	3.58	0.46	1.12	0.24	5.41	23.27
1950–9	3.95	1.00	2.40	0.62	7.96	31.23

[a] Units: 10^{16} g of CO_2.

are given in Table 11.2. As of 1960, 31×10^{16} g of CO_2 (0.71×10^{16} moles) had been generated.

Prediction of future CO_2 production requires not only an extrapolation of man's energy needs but also an estimate as to how rapidly nuclear power will replace chemical power. In the absence of any adequate means of assessing these trends we have assumed that the 4.5 per cent/yr increase in fuel consumption for the period 1958 to 1966 (see Table 11.3) will continue indefinitely. This means of extrapolation yields the CO_2 production estimates given in Table 11.4. By the year 2010 the total CO_2 produced will reach 210×10^{16} g (5.3×10^{16} moles) or about 7 times that produced up to 1960. The rate of CO_2 production 50 yr hence would be four times the current rate.

Distribution of Excess CO_2

Although initially injected into the atmosphere as CO_2 gas, the carbon produced by man will gradually become dispersed throughout the oceans and terrestrial biosphere. Two questions must be answered in this connection. By how much will this dispersal reduce the excess CO_2 burden of the atmosphere, and how long will the dispersal take? In Table 11.5 the carbon contents of the available reservoirs are summarized. It is clear from a comparison of their sizes that the atmosphere contains only a small fraction of the total carbon in the dynamic system. The ocean contains 55 times as much carbon and the terrestrial biosphere about two and one half times as much. Once the fossil fuel carbon has been dispersed through these reservoirs only about one atom in 60 will remain in the

Table 11.3. Annual Rate of Increase of CO_2 Production[a]

Year	Annual CO_2 Production	Increase in CO_2 Production	Per Cent Increase
1958	0.93	—	—
1959	0.98	0.05	5.4
1960	1.04	0.06	6.1
1961	1.06	0.02	1.9
1962	1.11	0.05	4.7
1963	1.14	0.03	2.7
1964	1.21	0.07	6.1
1965	1.26	0.05	4.1
1966	1.33	0.06	4.8
		Average	4.5

[a] Units: 10^{16} g of CO_2.
From World energy supplies statistical papers, Series J, United Nations, New York.

Table 11.4. Extrapolated CO_2 Production Based on an Annual Increase Rate of 4.5 Per Cent[a]

Decade	Decade Total	Cumulative Total
1960–9	12.4	44
1970–9	19.2	63
1980–9	29.8	93
1990–9	46.3	139
2000–9	71.9	211

[a] Units: 10^{16} g of CO_2.

Table 11.5. Carbon Contents and Approximate Response Times for the Dynamic Carbon Reservoirs on the Earth's Surface

	Total Carbon (10^{16} moles)	Carbon (moles/m^2)[a]	Total Carbon (per cent)	Response Time (yr)	Reference
Atmosphere	5	140	1.5	—	[b]
Ocean	325	9050	94.5	—	[c]
Warm surface	5	140	1.5	~2	
Main thermocline	80	2250	23.3	~100	
Deep ocean	240	6660	69.7	~1000	
Biosphere	14	370	4.0	—	
Living organics	3	70	0.8	~5	[d]
Soil organics	11	300	3.2	~300	[e]
Total	343	9570	100.0	—	

[a] Area of ocean surface (3.6×10^{14} m).
[b] Pales and Keeling (1965).
[c] Li et al. (1969).
[d] Goldschmit (1958).
[e] Craig (1958).

atmosphere. This does not mean, however, that upon complete dispersal the increase in CO_2 content of the atmosphere will be only $\frac{1}{60}$th the amount of carbon oxidized by man. The increase in CO_2 content proves to be much greater than this because the uptake of fossil fuel carbon by the biosphere and oceans displaces normal carbon from these reservoirs back to the atmosphere.

The input of fossil fuel CO_2 has been a complex function of time (see Fig. 11.1). For this reason dispersal calculations taking into account the actual temporal sequence of inputs are tedious. These complexities can be avoided and a good approximation obtained by assuming that all the CO_2 was added at a single time. The time selected is either the mean time elapsed since combustion (as of 1960, 30 yr) or the mean square root time (as of 1960, 25 yr) depending on the dispersal model used.

The time constant for mixing between atmospheric and surface ocean carbon is about 10 yr (Craig, 1957). The time required for these two reservoirs in turn to mix with the deep sea is about 1000 yr (Broecker, 1963a). The living biosphere replaces its carbon at the rate of about once every 5 yr and soil organics at the rate of once each several hundred

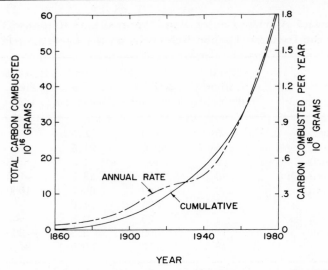

Figure 11.1 Production rate and cumulative total of CO_2 generated by the combustion of fossil fuel (see text). The curve is extrapolated beyond 1967 by assuming that consumption will rise at the annual rate of 4.5 per cent.

years (Broecker and Olson, 1960). Thus in the 30 yr available for mixing only the living biosphere and surface ocean have had an adequate opportunity to absorb their full component of man-made CO_2. Only a relatively small portion of the soil organic and subsurface ocean reservoirs can have responded in this short time. Our next step is to analyze as precisely as possible the uptake of CO_2 by these reservoirs.

CO_2 Uptake by the Biosphere

A mere trading of carbon atoms between the atmosphere and terrestrial biosphere will not lead to a reduction in the amount of excess CO_2 in the atmosphere. Only if the size of terrestrial biosphere increases in response to the increased atmospheric CO_2 content will this be the case. Is there any reason why this should happen? There seems to be only one reasonable link between the sizes of the two reservoirs. If the availability of atmospheric CO_2 allows photosynthetic fixation to proceed at a more rapid pace, this in turn could lead to a larger steady-state size for the terrestrial biosphere. However, because the major factors limiting photosynthesis by terrestrial plants are light, water, and nutrients, it is unlikely that a greater CO_2 concentration in air would cause a substantial rise in photosynthetic activity. In the absence of such a rise it is hard to see why the size of the carbon reservoir would change. We therefore conclude that interaction with the biosphere only mixes the carbon in the growing atmospheric reservoir with that in a biosphere of static size. There would be no consequent reduction in the atmospheric CO_2 excess.

An example shows why this conclusion is plausible. Let us assume that for each 10 per cent rise in atmospheric CO_2 content the rate of terrestrial photosynthesis and subsequently the steady state size of the terrestrial biosphere will rise by 1 per cent. Taking the amount of carbon in the terrestrial biosphere to be 2.5 times that in the atmosphere, complete equilibration of this would lead to a transfer of only 20 per cent of the excess CO_2 to the biosphere (i.e., $100 \times [0.1 \times 2.5/0.1 \times 2.5 + 1.0]$). However, since 80 per cent of the biospheric carbon is stored in the slowly responding soil organic reservoir we would expect that in the average time of 30 yr available for equilibration only a small fraction of ultimate steady state increase would have occurred. Taking the residence time of carbon in soil organics to be 300 yr only one tenth the steady state increase could have taken place and only 6 per cent of the excess CO_2 would be removed from the atmosphere (i.e., $100 \times [0.1(0.5 + 0.1 \times 2.0)/0.1(0.5 + 0.1 + 2.0) + 1.0]$). Although this example makes reasonable the conclusion that the biosphere is not an important sink for excess CO_2, our understanding of the response of this reservoir is sufficiently poor that its role must be reviewed as we improve our knowledge of the relationship between CO_2 pressure, rates of photosynthesis, and the standing crop of organic compounds.

CO_2 Uptake by the Ocean

The ocean turns out to be an important sink for excess CO_2. This is the case because CO_2 chemically reacts with water to form bicarbonate (HCO_3^-) and carbonate (CO_3^{2-}) ions. Were CO_2 relatively inert like N_2 or O_2, the excess CO_2 would become distributed between the ocean and atmosphere as these gases are (i.e., $\sim 98\%$ in the atmosphere and only $\sim 2\%$ in the oceans).

Three steps must be considered in the uptake of CO_2 by the oceans. First, the relatively thin layer of well-mixed water lying above the seasonal thermocline at the ocean surface will quickly establish equilibrium with the CO_2 in the atmosphere. Next, downward mixing and the sinking of the particles generated by organisms will slowly carry this carbon to the deep sea. Finally, as the CO_2 content of the ocean rises, water masses currently supersaturated with respect to the calcium carbonate minerals calcite and aragonite will become undersaturated and the calcium carbonate in sediments bathed in these waters will begin to dissolve. Such solution will enhance the ocean's capacity for fossil fuel CO_2. We consider each of these three steps in turn.

The Surface Ocean

CO_2 gas is exchanged between the air and sea by molecular transfer across the interface between these media. Upon entering the sea, the CO_2 is mixed quite rapidly through a layer which averages roughly 70 m in thickness (large seasonal and geographic variations in this depth make difficult the precise definition of its thickness). The base of this layer is defined as the point where the temperature of the water commences its decrease toward the 1 to 3 C range characteristic of the deep sea. The enhanced stability of the water column resulting from the temperature decrease proves to be a major barrier to further mixing.

We must first determine the relationship between the total dissolved carbon content of this 70-m-thick layer of surface water and the CO_2 content of the overlying air. We can express this mathematically.

$$\frac{\Delta \Sigma CO_2}{\Sigma CO_2} = \alpha \frac{\Delta p_{CO_2}}{p_{CO_2}}$$

where ΣCO_2 is the concentration of dissolved inorganic carbon in the surface sea and P_{CO_2} is the partial pressure of CO_2 in the air. The deltas represent changes in these values resulting from fossil fuel combustion and α is the constant of proportionality relating the two fractional changes. It is the value of α we seek.

As mentioned above, when CO_2 is added to seawater, it not only dissolves as ordinary gases but also undergoes chemical reaction with the carbonate and borate ions present in the water. It is the chemical reactions which permit the CO_2 to be taken up by the sea in sizable quantities.

The reactions taking place can be written as follows:

$$CO_2 + H_2O + CO_3^{2-} = 2HCO_3^-$$

and

$$CO_2 + H_2O + H_2BO_3^- = H_3BO_3 + HCO_3^-$$

As the increase in the amount of CO_2 dissolved as a gas is inconsequential,

$$\Delta \Sigma CO_2 = -\Delta CO_3^{2-} - \Delta H_2BO_3^-$$

It can be shown that for small increases in the CO_2 content of seawater the proportions of CO_2 reacting with borate and carbonate ions are in proportion to the concentrations of these species in surface seawater. Hence

$$\frac{\Delta H_2BO_3^-}{\Delta CO_3^{2-}} \simeq \frac{H_2BO_3^-}{CO_3^{2-}}$$

Thus

$$\Delta \Sigma CO_2 = -\left(1 + \frac{[H_2BO_3^-]}{[CO_3^{2-}]}\right) \Delta CO_3^{2-}$$

or

$$\Delta CO_3^{2-} = \frac{-[CO_3^{2-}]}{[CO_3^{2-}] + [H_2BO_3^-]} \Delta[\Sigma CO_2]$$

If carbon is to be conserved

$$\Delta HCO_3^- = \Delta \Sigma CO_2 - \Delta CO_3^{2-}$$

Hence

$$\Delta HCO_3^- = \frac{2[CO_3^{2-}] + [H_2BO_3^-]}{[CO_3^{2-}] + [H_2BO_3^-]} \Delta[\Sigma CO_2]$$

Prior to the production of fossil fuel CO_2 equilibrium between surface ocean and atmosphere required that

$$p_{CO_2} = \frac{[HCO_3^-]^2}{[CO_3^{2-}]} \frac{1}{K}$$

Once the added CO_2 has distributed itself between the surface ocean and atmosphere

$$p_{CO_2} + \Delta p_{CO_2} = \frac{([HCO_3^-] + \Delta HCO_3^-)^2}{([CO_3^{2-}] + \Delta CO_3^{2-})} \frac{1}{K}$$

Substituting the relationships for ΔHCO_3^- and ΔCO_3^{2-} and dividing the first of these equations by the second

$$1 + \frac{\Delta p_{CO_2}}{p_{CO_2}} = \frac{\left(1 + \frac{2[CO_3^{2-}] + [H_2BO_3^-]}{[CO_3^{2-}] + [H_2BO_3^-]} \frac{\Delta \Sigma CO_2}{[HCO_3^-]}\right)^2}{\left(1 - \frac{\Delta \Sigma CO_2}{[CO_3^{2-}] + [H_2BO_3^-]}\right)}$$

If the change in ΣCO_2 is sufficiently small

$$\frac{\Delta p_{CO_2}}{p_{CO_2}} \simeq \left(1 + \frac{4[CO_3^{2-}] + 2[H_2BO_3^-]}{[HCO_3^-]}\right) \frac{\Delta \Sigma CO_2}{[CO_3^{2-}] + [H_2BO_3^-]}$$

or

$$\frac{\Delta \Sigma CO_2}{\Sigma CO_2} = \frac{\frac{[CO_3^{2-}] + [H_2BO_3^-]}{[\Sigma CO_2]}}{\left(1 + \frac{4[CO_3^{2-}] + 2[H_2BO_3^-]}{[HCO_3^-]}\right)} \frac{\Delta p_{CO_2}}{p_{CO_2}}$$

The sought for constant α is then

$$\alpha = \left(\frac{[CO_3^{2-}] + [H_2BO_3^-]}{[HCO_3^-] + 4[CO_3^{2-}] + 2[H_2BO_3^-]}\right)\left(\frac{[HCO_3^-]}{[HCO_3^-] + [CO_3^{2-}]}\right)$$

The concentrations of these species in the warm portions of the surface ocean average

$[CO_3^{2-}] = 0.25$ moles/m³   (Li et al., 1969)

$[HCO_3^-] = 1.80$ moles/m³   (Li et al., 1969)

$[H_2BO_3^-] = 0.10$ moles/m³   (Sverdrup et al., 1942)

Hence

$$\alpha \cong 0.10$$

and

$$\frac{\Delta \Sigma CO_2}{\Sigma CO_2} = 0.10 \frac{\Delta p_{CO_2}}{p_{CO_2}}$$

An increase of 10 per cent in the CO_2 content of the atmosphere will result in a 1 per cent increase in the total dissolved carbon content of warm surface ocean water.

The carbonate and borate ion concentrations in deep seawater are considerably lower and the bicarbonate ion content somewhat higher than that in warm surface water. For such water α is even smaller than 0.1. Since surface waters in high latitude regions are mixtures of warm surface and deep seawaters their α values will also be somewhat lower. However, as the area of the ocean covered by these waters is relatively small, the error resulting from the use of $\alpha = 0.1$ for the entire surface ocean layer is unimportant.

As shown in Table 11.5 a layer of surface ocean water 70 m thick contains nearly the same amount of carbon as the air. Since if a small amount of excess CO_2 gas is added to the surface ocean-atmosphere system, the fractional increase in the CO_2 content of the surface ocean will be only one tenth that for the air, the excess CO_2 will distribute itself at equilibrium such that 9 per cent enters the surface ocean and 91 per cent remains in the air. Thus in the absence of transfer to the deep sea the ocean does not constitute a significant sink for fossil fuel CO_2.

Before turning our attention to the deep sea, it is necessary to establish the time required for the surface ocean and atmosphere to reach equilibrium. This requires a knowledge of the rate at which CO_2 molecules are traded across the air-sea interface.

The average rate at which this process takes place can be established from the distribution of natural radiocarbon between the air and surface ocean. This is done as follows. Prior to man's interruption of steady state the rate at which CO_2 atoms left the air to enter the sea must have just balanced the rate they left the sea to enter the air. In order to balance the radioactive decay of ^{14}C in the sea the incoming CO_2 must have carried more radiocarbon than the outgoing CO_2. For steady state

$$I C_A A - I C_S A = k_0 C_0 V \lambda$$

where I is the average CO_2 exchange rate per unit area, and A, of the ocean surface; C_A, C_S, and C_0 are, respectively, the mean $^{14}C/C$ ratios in the atmosphere, surface ocean, and mean ocean; k_0 and V are the mean ΣCO_2 concentration and volume for the entire ocean; and λ is the fraction of ^{14}C atoms decaying each year. The difference between the two terms on the left is the net influx of ^{14}C to the ocean and that on the right, the rate at which ^{14}C is disappearing from the sea by radio decay. Solving for I and setting V/A equal to the mean depth, $\hbar$, we have

$$I = \frac{C_0/C_A}{1 - C_S/C_A} k_0 \hbar \lambda$$

Using the following measured values, I turns out to be 15 ± 5 moles/$(m^2 \text{ yr})$.

$\hbar = 3800$ m

$\lambda = 1/8000/\text{yr}$

$k_0 = 2.4$ moles/m^3 (Li et al., 1969)

$C_S/C_A = 0.94 \pm 0.02$ (Broecker, 1963b)

$C_0/C_A = 0.82 \pm 0.04$ (Broecker, 1963a)

As the upper 70 m of the ocean contain close to 150 moles of dissolved inorganic carbon per square meter of ocean surface, the mean residence time for carbon atoms in the surface ocean with respect to transfer to the atmosphere is 10 yr [150 moles/m^2/15 moles/$(m^2$ yr$)$]. However, the time required for the surface ocean to adjust its carbon content to a new atmospheric CO_2 pressure is a factor α shorter than the mean residence time (hence 1 yr rather than 10 yr). The reason for this difference can be most easily seen by considering a hypothetical example. Suppose the air and surface sea are at equilibrium and suddenly enough CO_2 is added to the air to increase its CO_2 content by 11 per cent. We have shown that when equilibrium is re-established the CO_2 content of the water will rise 1 per cent and that of the air will drop from an 11 to a 10 per cent excess. Thus to establish equilibrium we will have to transfer about 1.5 moles of CO_2/m^2 from air to sea. Before the injection of the CO_2 the fluxes into and out of the ocean were equal. After the 11 per cent increase in air CO_2 content the influx will rise by 11 per cent. If the CO_2 exchange rate was 15 moles/m^2 yr this increase would yield a net inflow of 1.65 moles/m^2 yr. The ratio of the ultimate uptake (1.5 moles/m^2) to the initial influx rate (1.65 moles/m^2 yr) yields a response time of just under 1 yr.

As the CO_2 content of the ocean builds toward its new equilibrium value, the rate at which CO_2 leaves the ocean will rise and the rate at which CO_2 enters will fall. When equilibrium is reestablished the two rates will become equal and will be 10 per cent higher than their initial value. The net rate of CO_2 influx will fall from its initial value of 1.65 moles/m^2 to zero when equilibrium is reestablished. The reduction will proceed in an exponential manner such that half its initial value is reached in 0.693 times the response time (i.e., $\sim$0.7 yr), to one quarter its value in twice this time (i.e., $\sim$1.4 yr), and so forth. Correspondingly the CO_2 content of the ocean will rise half way to its new equilibrium value in 0.7 yr, $\frac{3}{4}$ of the way in 1.4 yr, and so forth (this relationship is shown graphically in Fig. 11.2).

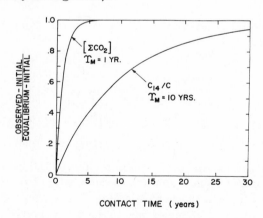

Figure 11.2 Response of the total dissolved inorganic carbon content and the ratio of radiocarbon to stable carbon in a seawater reservoir 70 m thick to a sudden injection of C^{14}-free CO_2 to the overlying atmosphere. Because the constant α has a value of 0.1, the concentration adjustment will approach its new equilibrium value 10 times more rapidly than the isotope ratio.

As we have stated, the average fossil fuel CO_2 molecule has been in existence for 30 yr. As we shall see, surface water in the ocean is replaced by deep water no more frequently than once every 10 yr. Thus it is safe to assume that because of its rapid equilibration with the atmosphere the surface ocean always has a ΣCO_2 content very close to equilibrium with that in the atmosphere. Bolin and Eriksson (1959) made this point in their treatment of this problem.

The Deep Sea

Carbon is carried from the surface to the deep sea not only by the complex advective and diffusive mixing processes but also by the sinking remains of organisms. These processes will slowly mix the man-made CO_2 into the most inaccessible portions of the sea. The question we must answer is how rapidly this dispersal proceeds.

Unfortunately our knowledge of oceanic mixing processes is far from complete and we must resort to highly simplified models of the ocean in order to estimate the rates of vertical transport. We view the ocean as shown in Fig. 11.3. The thin skin of warm water which covers much of the ocean is separated from a much larger reservoir of cold deep water by a gradational layer averaging 1000 m in thickness which we refer to as the main thermocline. This layer rises to the surface in the high latitude regions separating that portion of the ocean surface covered with warm

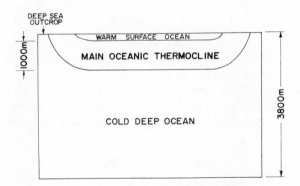

Figure 11.3 Simplified model of vertical mixing in the ocean. To reach the deep sea, CO_2 added to the system must either enter the warm surface layer and pass down through the main thermocline which constitutes the major barrier to oceanic mixing or pass directly through the high latitude outcrop of the cold ocean. The rates of transfer are chosen to be in accord with the distribution of natural radiocarbon.

water ($\sim$90%) from that portion where the cold deep waters are in direct contact with the atmosphere. The excess CO_2 in the atmosphere and the warm surface ocean can reach the deep sea either by transfer through the main thermocline or by transfer through the small region where the atmosphere and deep sea are in direct contact.

The distribution of natural radiocarbon between the surface ocean and deep sea allows limits to be placed on the rates at which these transfers occur. Just as the average air-sea exchange rate was fixed by the requirement that radiocarbon atoms be added to the sea at a rate balancing their decay within the sea, radiocarbon atoms must be added to the deep sea at the rate they are disintegrating in the deep sea.

We assume the transfer of radiocarbon across the main thermocline can be approximated by diffusive mixing. The rate at which ^{14}C enters the deep sea is given by

$$\frac{C_S k_S - C_D k_D}{\hbar_T} DA$$

where C_S and C_D are, respectively, the $^{14}C/C$ ratios in warm surface and deep seawater, k_S and k_D are the ΣCO_2 concentrations in these water types, $\hbar_T$ is the mean thickness of the thermocline layer, A is the area of the ocean, and D is the coefficient of eddy mixing.

If the assumption is made that nearly equal amounts of CO_2 are transferred into and out of the deep sea by direct exchange with the atmosphere through the high latitude region outcrop, then designating this exchange as IA we can write this component of the ^{14}C input as follows:

$$(C_A - C_D)IA$$

where C_S and C_D are, respectively, the $^{14}C/C$ ratios in the atmosphere and deep sea. For convenience, A is taken to be the area of the ocean and I the total exchange divided by this area.

In order for the total carbon content of the deep sea to remain constant, the net upward diffusion of carbon through the thermocline must be just balanced by the downward transport of carbon by particulate matter, B. Hence

$$B = \frac{k_D - k_S}{\hbar_T} DA$$

Taking the $^{14}C/C$ ratio in these particles to be equal to that in the surface ocean the ^{14}C transport by this mechanism is given by

$$\frac{C_S k_D - C_S k_S}{\hbar_T} DA$$

These three inputs must just balance the decay of ^{14}C in the deep sea which may be written

$$C_D\, k_D\, \hbar_D\, A\, \lambda$$

where C_D, k_D, and $\hbar_D$ are, respectively, the $^{14}C/C$, total dissolved inorganic carbon concentration, and mean thickness of the deep sea reservoir, A is the area of the ocean, and λ is the fraction of ^{14}C decaying per year.

Setting the sum of the ^{14}C inputs equal to the decay rate we have

$$(C_S - C_D)\frac{k_D D}{\hbar_T} + (C_A - C_D)\, I = C_D\, k_D\, \hbar_D\, \lambda$$

If we define the fraction of ^{14}C entering the deep sea by transfer through the main thermocline as f, then

$$(C_S - C_D)\frac{k_D D}{\hbar_T} = C_D k_D \hbar_D\, \lambda f$$

or

$$D = \frac{\hbar_T \hbar_D\, \lambda}{\left(\dfrac{C_S}{C_D} - 1\right)} f$$

and

$$(C_A - C_D)\, I = C_D k_D \hbar_D\, \lambda(1-f)$$

or

$$I = \frac{k_D \hbar_D\, \lambda}{\left(\dfrac{C_A}{C_D} - 1\right)}(1-f)$$

Although most of the ^{14}C supplied to the deep sea must proceed down through the main thermocline rather than via the deep sea outcrop, there is presently no way to determine the exact value of f. Despite the fact that they constitute no more than 10 per cent of the ocean area, the greater ^{14}C contrast and the more rapid-than-average CO_2 exchange rates in the outcrop areas might permit as much as 30 per cent of the radiocarbon to enter by this route. We assume that f lies between 0.7 and 1.0.

Taking $\hbar_D$ to be 2800 m, $\hbar_T$ to be 1000 m, $1/\lambda$ to be 8000 yr, and $(C_S - C_D)/C_D$ to be 0.14 (Broecker, 1963a), the diffusion coefficient turns out to be 2.5×10^3 m²/yr for $f = 1.0$ and 1.8×10^3 m²/yr for $f = 0.7$. Taking k_D to be 2.4 moles/m³ and $(C_A - C_D)/C_D$ to be 0.20 (Broecker, 1963a), the value of I turns out to be 1.3 moles/(m² yr) for $f = 0.7$ and, of course, zero for $f = 1.0$ (remember I and D are calculated for the area of the whole ocean, and not for the outcrop and thermocline areas).

With this model for oceanic mixing we can estimate the amount of man-made CO_2 which has and will be transported to the deep sea. Let us first consider transport via the main thermocline. Transport by diffusion proceeds as the square root of time. A rough estimate of the mean penetration distance is given by $\sqrt{Dt}$. Taking t to be 25 yr and D to be 2.5×10^3 m²/yr, we get 250 m for the mean thickness of an equivalent layer of oceanwide extent into which CO_2 has entered. Since the main thermocline has a thickness of 1000 m, this means that the amount of the man-made CO_2 which has penetrated to the base of the main thermocline and into the deep sea is as yet very small. This simplifies the calculation by permitting the main thermocline to be considered a semi-infinite half space. Thus we will assume that excess CO_2 is diffusing from a well-mixed reservoir (the warm surface ocean plus atmosphere) into a semi-infinite half space (the main thermocline). To avoid the complications introduced by simulating the actual time distribution of inputs we assume that the CO_2 produced up to 1960 was all added to the atmosphere in the year 1935. Crank (1956) gives the solution to this problem.

Considering first the case where transfer through the deep sea outcrop is negligible (i.e., that for $f = 1.0$), we find that 28 per cent of the CO_2 will have entered the thermocline. Since one tenth of the remainder (or 7%) resides in the warm surface ocean, the total fraction in the ocean turns out to be 35 per cent and that remaining in the atmosphere 65 per cent (see Fig. 11.4).

This result can be compared with that obtained using the upper limit for transfer through the outcrop of the deep sea. Upon addition to the atmosphere CO_2 will be transferred into the cold ocean at the rate

$$\frac{\Delta p_{CO_2}}{p_{CO_2}} IA$$

The amount of excess CO_2 initially in the air is

$$\frac{\Delta p_{CO_2}}{p_{CO_2}} M$$

where M is the normal atmospheric CO_2 content. The fraction, q, of the CO_2 excess transferred to the ocean in this way during the first year is then

$$q = \frac{IA}{M} = \frac{I}{M/A}$$

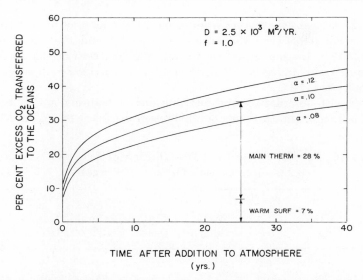

Figure 11.4 Fraction of CO_2 reaching the ocean as a function of the time elapsed since addition to the atmosphere. The assumption is made that all the CO_2 enters via the warm surface ocean (rather than via the deep water outcrop). The warm surface ocean layer is assumed at all times to be at equilibrium with the atmosphere. The amount of fossil fuel CO_2 in the ocean as a whole at time zero is then that contained by this layer. The mean age of the fossil fuel CO_2 is taken to be 25 yr. (This result is calculated by multiplying the amount of CO_2 added in each decade by the square root of the time elapsed since that decade. These products are summed and divided by the total CO_2 production. The result is squared to give the mean age.)

For $f = 0.7$, I is 1.3 moles/(m² yr). M/A is about 140 moles/m². Thus about 1 per cent of the total CO_2 excess is initially transferred per year directly to the deep ocean. Were this rate to apply for the entire 30-yr period, as much as 30 per cent could be disposed of in this manner. The rate will, however, decrease with time. First, it will fall because of the decrease in the atmospheric excess CO_2 content due to transfer to the warm surface ocean, to the thermocline, and directly to the deep sea. Second, it will drop because mixing between the outcrop region and the remainder of the deep sea reservoir may not be fast enough to prevent a rise in the CO_2 partial pressure in the cold surface water toward that in the atmosphere. As the importance of the latter effect cannot now be evaluated, we will be able to obtain only an upper limit on the uptake by assuming it to be negligible. By simultaneously considering the loss to the thermocline and transfer to the deep sea through its outcrop we find that about 20 per cent of the excess CO_2 will enter the deep sea and about 25 per cent the main thermocline. The remaining 55 per cent will be distributed between the atmosphere (90%) and the warm surface ocean (10%). In this case then a total of 50 per cent of the man-made CO_2 must reside in the sea. This value should provide an upper limit for uptake by the sea.

Table 11.6 summarizes the distributions of man-made CO_2 between the atmosphere and the three oceanic reservoirs for the two limits on the

Table 11.6. Predicted Per Cent Distribution of Excess CO_2 Among the Atmospheric and Oceanic Reservoirs[a]

Year	Mean Age	f	Atmosphere	Warm Ocean	Main Thermocline	Deep Ocean	Total Ocean
1960	25[b]	1.0	65	7	28	0	35
	30[c]	0.7	50	5	25	20	50
1980	24[b]	1.0	66	7	27	0	34
	29[c]	0.7	52	5	24	19	48
2000	22[b]	1.0	67	7	26	0	33
	27[c]	0.7	54	5	23	18	46

[a] See text.
[b] Square root mean.
[c] Linear mean.

mode of radiocarbon transfer to the deep sea. The results suggest that 40 ± 7 per cent of the CO_2 generated by man as of 1960 reside in the sea. Similar calculations are given for the years 1980 and 2010 (assuming a 4.5 per cent/yr production increase). Since the mean age of the man-made CO_2 molecules does not change very much, the proportions of CO_2 in the various parts of the system will remain nearly the same with time.

Sedimentary Calcium Carbonate

In the absence of any interaction with the sediments on the ocean floor, complete mixing between the atmosphere and ocean would (assuming $\alpha = 0.10$) lead to an oceanic uptake of only 85 per cent of the excess CO_2. The rest would remain in the atmosphere. However, eventually the acidity of the ocean will rise to the point where $CaCO_3$ stored in the sediments will begin to dissolve. As can be seen from the following reaction, solution of 1 mole of $CaCO_3$ effectively enhances the oceans capacity for CO_2 by 1 mole.

$$CO_2 + CaCO_3 + H_2O \Rightarrow Ca^{2+} + 2HCO_3^-$$

To neutralize the 0.7×10^{16} moles of CO_2 already generated by man would require the solution of only 0.2 g of $CaCO_3/cm^2$ of sea floor. This amount is present in the upper centimeter of sediment.

Currently the calcium carbonate-bearing sediments are bathed in water supersaturated with $CaCO_3$ (Li et al., 1969). Before solution could take place, enough CO_2 would have to be added to these waters to bring them to undersaturation. Once this is accomplished the capacity of the ocean for CO_2 will begin to rise.

As shown in Fig. 11.5 the degree of supersaturation is highest at the ocean surface and decreases with depth. Depending on the mineral (calcitic forams and coccoliths are less soluble than aragonitic pteropods, mollusks, and corals) and location in the ocean (the degree of supersaturation decreases more rapidly with depth in the Pacific than in the Atlantic) saturation is achieved somewhere between a few hundred and 5000 m depth in the sea. Below the point where saturation is achieved the waters are currently undersaturated and the sediments correspondingly free of calcium carbonate.

Man-made CO_2 is currently restricted largely to the warm surface ocean and the upper regions of the main thermocline. These regions are the most highly supersaturated in the ocean. As shown in Fig. 11.5 the warm surface ocean is currently about sevenfold supersaturated with calcite and fourfold with aragonite. Thus for carbonate solution to begin, the carbonate ion concentration would have to be cut from its current value of 0.24 moles/m³ to 0.06 moles/m³ (for aragonite) and to 0.035 moles/m³ (for calcite). If, as suggested in Table 11.6, 6 ± 1 per cent of the CO_2 generated by man as of 1960 (i.e., 4×10^{14} moles) resides in the warm surface ocean then since this layer has a volume of 2.5×10^{16} m³ (70 m $\times$ 10^{14} m²) the CO_2 increase has been only 0.016 moles/m³. Since part of this CO_2 reacts with borate, the decrease in carbonate ion has been only about 0.012 moles/m³. A decrease of 15 times this amount is needed before the surface sea becomes undersaturated. We will be well into the next century before this is accomplished.

Considerably smaller amounts of CO_2 are needed to bring deeper waters to undersaturation. However, considerably smaller amounts of CO_2 have been made available at these depths. The question as to whether solution will commence first in the deep sea or in the surface water thus involves a detailed knowledge as to show excess CO_2 penetrates into the sea. As we are not currently in a position to do any more than guess, this question is best left undiscussed.

Another problem that arises is how fast the neutralization will proceed once undersaturation is achieved at any given depth. Very little is known regarding the rates of solution of calcium carbonate in seawater. The problem is further complicated by the fact that the carbonate which will undergo solution is generally surrounded by a matrix of silicate detritus. If we are to make long-range predictions, these kinetic questions must be studied.

Inventory Summary

In order to predict the fate of the CO_2 being generated by man, a detailed knowledge of the environment is needed. We have seen that the dependence of photosynthesis rates on atmospheric CO_2 content, the

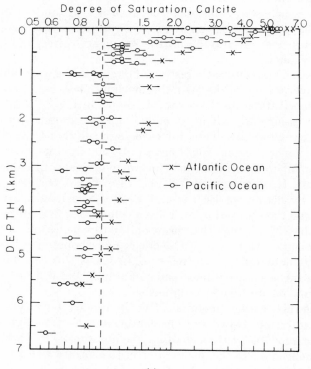

(a)

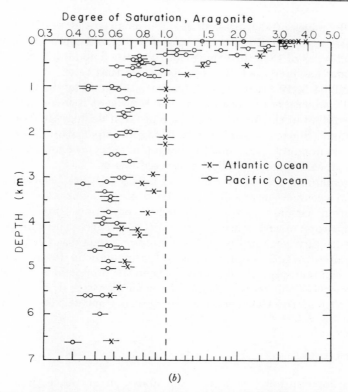

Figure 11.5 Degree of supersaturation of (a) calcite and (b) aragonite as a function of depth in the Atlantic and Pacific Oceans. From Li et al., 1969.

reservoir size, and turnover time of soil organics, air-sea CO_2 exchange rates, ocean circulation rates and patterns, and the marine chemistry of CO_2 and $CaCO_3$ must be thoroughly understood. Unfortunately we are just beginning to understand most of these processes.

From what we know it appears that the terrestrial biosphere will not aid in the reduction of the atmospheric CO_2 excess. The sea, on the other hand, has and will continue to remove about 40 per cent of the excess. Sometime within the next 100 to 200 yr the acidity of the sea will reach the point where $CaCO_3$ begins to dissolve from the sediments. At this point an even greater proportion of the excess CO_2 will be taken up by the sea.

As of 1960, 0.71×10^{16} moles of CO_2 had been produced. If 40 per cent (0.28×10^{16} moles) has been taken up by the sea, then of the

5.7 × 10^{16} moles of CO_2 present in the atmosphere at that time 0.43 × 10^{16} moles represent the increase caused by man's combustion of fuels. Presumably, as of 1850 the atmosphere contained about 5.3 × 10^{16} moles of CO_2 and has increased by 8 per cent since that time.

About 0.04 × 10^{16} moles of CO_2 has been taken up by the warm surface ocean. This increase has resulted in a 1.4 per cent increase in the bicarbonate content of this reservoir and a 5 per cent decrease in its carbonate ion content. Since the surface sea is highly supersaturated with both calcite and aragonite, this small reduction in carbonate ion has not led to solution of carbonate in shallow water sediments.

The remaining 0.24 × 10^{16} moles of excess CO_2 has been carried into the main oceanic thermocline and the deep sea. Our knowledge of its distribution within this vast reservoir is very poor. Only by following the penetration of such tracers as ^{90}Sr, ^{137}Cs, ^{3}H, and so on will we be able to say with any certainty how this mixing proceeds (see Chap. 22).

If man maintains the 4.5 per cent/yr increase in fuel consumption that has characterized the last decade, then the proportions of CO_2 in any given reservoir will remain nearly the same in years to come. Roughly 60 per cent of the CO_2 excess at any given time will be found in the atmosphere.

Crosschecks

Many assumptions went into the above inventory estimates. How much confidence can we have in their validity? We have two quantitative observations which can be used as checks. First, Pales and Keeling (1965) showed that during the interval March 1959 to June 1963, the CO_2 content of the air at Mauna Loa Observatory on the island of Hawaii was increased from 313.0 to 315.8 ppm (by volume; see Fig. 11.6). This corresponds to an annual increase of 0.68 ppm. These workers have also shown that the global atmosphere is sufficiently uniform in CO_2 content that measurements at one station should give a meaningful world average. Our computer calculations show that instantaneous partitioning of CO_2 between the atmosphere and the oceans in this interval of time is very nearly equal to the cumulative partitioning. Thus if about 60 per cent of the excess CO_2 generated up to the year 1960 is currently in the atmosphere, the rate at which the CO_2 content of the atmosphere was rising during the time interval monitored by Keeling and his co-workers should be 60 ± 10 per cent the rate at which CO_2 was being generated by fuel consumption. From the United Nations summaries this would correspond to a rate of 0.77 ± 0.13 ppm/yr. The agreement

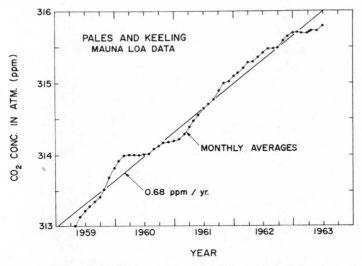

Figure 11.6 Twelve-month running means of the CO_2 content of the air on top of the extinct Mauna Loa volcano on the Island of Hawaii. From Pales and Keeling, 1965. The results suggest an average rise of 0.68 ppm/yr (i.e., ∼0.2 per cent) over this 4-yr period.

between the observed and predicted values is encouraging. The fact that the observed increase is more nearly equal to that predicted by the extreme case of maximum transfer through the deep sea outcrop could be interpreted in a number of ways. One could conclude that the CO_2 production estimates based on the United Nations summary are 15 per cent or so too great. One could conclude that the mixing model adopted is not adequate. One could conclude that the Pales and Keeling (1965) CO_2 increase is biased due to some natural phenomena or due to a small drift in their standardization. As none of these possibilities can be easily dismissed, we conclude only that the observations are not in serious conflict with the inventories presented above.

A second crosscheck comes from the observation that the $^{14}C/C$ ratio in atmospheric CO_2 fell by 2.0 ± 0.3 per cent between 1850 and 1950. This result is based on radiocarbon measurements on tree rings of known age (see Fig. 11.7). As first suggested by Suess (1955), the CO_2 generated by man comes from fuels too old to contain any ^{14}C. Thus if all the CO_2 generated by man had remained unexchanged in the atmosphere the $^{14}C/C$ ratio would have fallen 10 per cent between 1850 and 1950. The observation that the decrease was much smaller clearly demonstrated

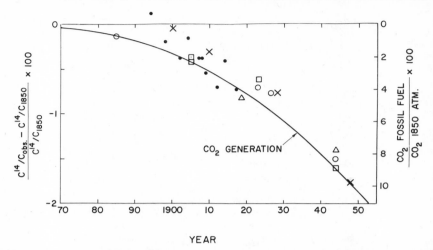

Figure 11.7 Comparison of C^{14}/C data on tree rings over the period 1880 to 1950 with the CO_2 production curve. The small black dots are for tree ring samples from Trondheim, Norway (Lerman et al., 1967). The squares are for samples from Wisconsin; the triangles, Peru; the circles, Wellington, New Zealand; and the crosses, Christchurch, New Zealand (Fergusson, 1958). Had all the CO_2 produced remained in the atmosphere the C^{14}/C ratio should have dropped by 10 percent over this period. The fact that the ratio dropped by only 2 per cent suggests that the fossil fuel CO_2 has mixed with an amount of oceanic and biospheric carbon 4 times greater in amount than that in the atmosphere.

that during the time available the atmosphere had exchanged its carbon with a reservoir about 5 times as large (distributed among 5 times the carbon the resulting $^{14}C/C$ decrease would be only 2.0% rather than 10%). Would our model predict such a decrease?

Let us first consider the terrestrial biosphere. Although this reservoir has not changed its size appreciably as a result of fuel consumption, it has exchanged carbon with the atmosphere. As exchange between the atmosphere and living biosphere requires less than a decade we can assume that the carbon from fossil fuels is distributed uniformly between these two reservoirs. As the amount of carbon in the living biosphere is one half that in the atmosphere, the expected ^{14}C, reduction would be

$$\frac{1.0\ M_{atm}}{1.0\ M_{atm} + 0.5\ M_{atm}} \times -10 = -6.7 \text{ per cent}$$

There would be a small additional reduction resulting from mixing with carbon in the soil humic reservoir. This reservoir has twice as much carbon as the atmosphere but has equilibrated only about 10 per cent of this with the atmosphere in the 30 yr available. Taking this into account the expected reduction would be

$$\frac{1.0\, M_{atm}}{1.0\, M_{atm} + 0.5\, M_{atm} + 0.2\, M_{atm}} \times -10 = -6.0 \text{ per cent}$$

Clearly, considerable mixing with carbon in the ocean is required to explain the small magnitude of the observed decrease.

It was shown above that surface waters had adequate time to achieve the equilibrium total CO_2 content dictated by the atmosphere. This is not true in the case of isotopic equilibrium. As mentioned previously the time constant for CO_2 content adjustment and that for isotopic adjustment differ by a factor α. Hence isotopic equilibration requires 10 times longer than CO_2 content equilibration. That this is important can be seen from the fact that prior to man's impingement on the system the $^{14}C/C$ ratio (corrected for isotopic fractionation) in the warm surface ocean was about 6 per cent lower than that in the atmosphere and 14 per cent higher than that in the deep sea (Broecker, 1963a). Thus deep water brought to the surface achieved only 0.70 isotopic equilibration with the atmosphere before returning to the deep sea. We will assume that water brought to the surface today will achieve the same extent of equilibration with man-made carbon in the atmosphere.

Now for the case where outcrop exchange is negligible (i.e., $f = 1.0$) we can quite easily estimate the amount of oceanic carbon (in the warm surface ocean and main thermocline) with which the atmospheric carbon has mixed. It is 0.70 the amount of carbon in the water required to carry away the amount of excess CO_2 given in Table 11.6 (i.e., 35% of the added CO_2). This amount turns out to be 3.8 times the carbon content of the atmosphere. Thus the expected reduction is

$$\frac{M_{atm}}{M_{atm} + 0.5\, M_{atm} + 0.2\, M_{atm} + 3.8\, M_{atm}} \times -10 = -1.8 \text{ per cent}$$

a value in good agreement with the tree ring observation.

For the maximum outcrop exchange (i.e., $f = 0.7$) the amount of warm surface and main thermocline carbon would be reduced to 3.2 times that in the atmosphere. However, added to this would have to be the amount of deep ocean carbon isotopically equilibrated with the atmosphere. In this case the amount would be equal to the exchange rate [1.3 moles/(m^2 yr)] times the mean lifetime of the fossil fuel CO_2 molecules (30 yr) or about 40 moles of CO_2/m^2 of ocean surface. This is 0.3 times the amount of CO_2 in the atmosphere. The expected reduction in this case is

$$\frac{M_{atm}}{M_{atm} + 0.5\ M_{atm} + 0.2\ M_{atm} + 3.2\ M_{atm} + 0.3\ M_{atm}} \times -10 = -1.9 \text{ per cent}$$

Again, the prediction is close to the observed value.

Thus both the CO_2 rise measured by Pales and Keeling (1965) and the $^{14}C/C$ ratio decrease measured by Suess (1955) and by Fergusson (1958) point to degrees of mixing between the ocean and atmosphere consistent with the model presented here. The conclusion that 40 ± 10 per cent of the excess CO_2 generated by man to date has reached the ocean seems to be fairly well founded. The prediction that this proportionation will apply during the coming 50 yr also appears valid.

Climate Changes Resulting from CO_2 Production

For each 100 units of solar energy received by the earth roughly 30 are reflected back into space. The remaining 70 are absorbed either by the atmosphere or the ocean and land surface. In order to maintain energy balance infrared light exactly equal in energy to that absorbed is emitted from the top of the atmosphere. The amount of infrared light leaving the top of the atmosphere depends both on its temperature and its transparency to infrared light. CO_2 and H_2O are the only gases in the atmosphere capable of absorbing important quantities of infrared light. The higher the content of these gases the greater the opacity of the atmosphere to infrared light and the higher its temperature must be to radiate away the necessary amount of energy.

A change in the CO_2 content of the atmosphere upsets the earth's radiation balance by holding back departing infrared light. If no other change were to occur in the system, the net amount of energy accumulated by the earth would raise its surface temperature until the enhanced infrared emission reestablished balance between incoming and outgoing radiation. The problem is greatly complicated by the fact that other changes will certainly take place. For example, if the earth's temperature rises, the water vapor content of the atmosphere is likely to rise. More water will have the same effect as more CO_2 creating positive feedback in the system and hence forcing the temperature to rise even higher. A rise in water vapor would quite likely increase the fraction of the globe covered by clouds. Such an increase would cause the amount of primary solar radiation absorbed by the earth to fall. Some combination of increased temperature and cloudiness will balance the enhanced absorption of infrared light by the added CO_2 and H_2O vapor. The problem is to establish the relative magnitudes of these effects. In order to determine how temperature, water vapor content, and cloudiness will change with rising CO_2 content we consider the effect of each of these variables on the radiation balance.

Temperature

Changing the earth's surface and atmospheric temperature (while holding cloudiness, H_2O, CO_2, etc. constant) has little effect on the system's ability to absorb solar radiation. Since the ability of any object to emit infrared light varies as the fourth power of temperature for a small fractional change in temperature $\Delta T/T$ to the first approximation, a four-times-larger fractional change in escaping infrared energy will result, that is,

$$\frac{\Delta E}{E} \cong 4 \frac{\Delta T}{T}$$

The mean temperature of the earth's surface is 15 C, or 288 K. Other parameters held constant, for each degree rise in temperature the rate of energy loss should rise by 1.4 per cent ($100 \times 4 \times 1/288$). This corresponds to 1.8 mcal/(cm² min deg).

CO_2 Content

CO_2 absorbs infrared light in the wave length region 12 to 18 μ. As shown in Fig. 11.8 the current amount of CO_2 in the atmosphere absorbs almost all the radiation from 14.5 to 15.5 μ and almost none below 12.0 or above 18.0 μ. The critical regions in the spectrum are then 12.0 to 14.5 and 15.8 to 18.0 μ. Infrared light currently escaping in these ranges would be partially absorbed were the CO_2 content of the air to rise. Taking into account the competition of water vapor in absorption of infrared light in these spectral ranges and also the influence of cloudiness, Möller (1963) estimates that, other factors held constant, an increase in CO_2 content of 10 per cent would reduce the outgoing energy by 0.43 mcal/(cm² min). Doubling the CO_2 content would reduce it by 3.0 mcal/(cm² min). If only temperature changed as a result of these increases they would necessitate, respectively, a 0.15 C and a 1.0 C rise in mean global temperature. Manabe and Wetherald (1967), using a model which includes heat balance within the atmosphere as well as at the earth's surface, obtain 1.33 C

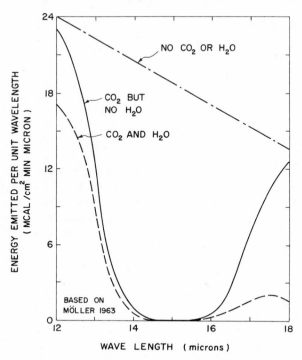

Figure 11.8 Opacity of the atmosphere to infrared light in the wavelength range 12 to 18μ. The energy loss from a CO_2- and H_2O-free atmosphere is compared with that for a water-free atmosphere containing an amount of CO_2 equal to that in our atmosphere, and with an atmosphere containing CO_2 and H_2O in their normal atmospheric abundance. From Möller, 1963.

for the temperature increase associated with doubling the CO_2 content with fixed *absolute* humidity and cloudiness.

H_2O Vapor Content

Water vapor plays a dual role in the atmosphere. Water vapor scatters incoming light and absorbs outgoing light. Möller (1963) estimates that a 10 per cent increase in water vapor content of the air would reduce the incoming radiation absorbed by 1.0 mcal/(cm² min) and reduce the outgoing infrared by 3.5 mcal/(cm² min). The net effect would be for the earth to store energy at the rate of 2.5 mcal/(cm² min) (compared with 0.43 mcal/(cm² min) for a 10% rise in CO_2 content). According to Manabe and Wetherald (1967), Möller's failure to take into account the changes in vertical turbulence which would accompany changes in carbon dioxide content led to a serious over-estimate of the sensitivity of earth temperature to variations in water vapor content.

Cloudiness

Cloud cover also affects both the amount of solar energy absorbed by the earth and the amount of infrared energy lost from the atmosphere. Using a value of 50 per cent for the current cloud cover, Möller estimates that for each 1 per cent cloud cover increase the earth will absorb 1.6 mcal/(cm² min) less solar energy and will absorb 0.8 mcal/(cm² min) more outgoing infrared radiation. The effect of an instantaneous 1 per cent cut in cloudiness would be to cause a net loss of 0.8 mcal/(cm² min) from the earth.

Snowcover

Changes in temperature, water vapor content, and cloudiness will almost certainly lead to changes in snowcover (see Budyko, 1969). These in turn will change the earth's albedo. As areas which are snow-covered reflect perhaps 90 per cent of the radiant energy received while those covered by soil and vegetation reflect only 20 to 50 per cent of the radiation received, snowcover may prove to be an important component in the calculation.

Interdependences

As the manner in which cloudiness and snowcover change with the temperature and water vapor content of the air is not known, we first confine our attention to the relationship between water vapor content, CO_2 content, and temperature. Assuming that at all times the earth must lose as much solar energy as it receives

$$\Delta E = \frac{dE}{dW}\frac{dW}{dT}\Delta T + \frac{dE}{dT}\Delta T$$

where ΔE is the amount of energy held back per unit time by man-made CO_2, ΔT is the resulting temperature change, dE/dW and dE/dT are, respectively, the change in rate in energy loss with water vapor content and temperature, and dW/dT is the change in water vapor content with temperature. Solving for ΔT

$$\Delta T = \frac{\Delta E}{\dfrac{dE}{dW}\dfrac{dW}{dT} + \dfrac{dE}{dT}}$$

Although the manner in which water vapor content changes with temperature is not well understood, the assumption that the relative humidity remains constant should provide a first approximation. If this is the case the water vapor content of the atmosphere should rise 6.5 per cent/deg. Taking the dependence of energy loss on water vapor content given by Möller (1963)

$$\frac{dE}{dW}\frac{dW}{dT} = -1.5 \text{ mcal}/(\text{cm}^2)(\text{min})(\text{deg})$$

The minus sign indicates that added water vapor reduces the rate at which energy is lost. This change nearly cancels the enhanced loss generated by the temperature rise itself, that is,

$$\frac{dE}{dT} = 1.8 \text{ mcal}/(\text{cm}^2)(\text{min})(\text{deg})$$

The sum of these numbers is only 20 per cent of their separate magnitudes, that is

$$\frac{dE}{dW}\frac{dW}{dT} + \frac{dE}{dT} = 0.3 \text{ mcal}/(\text{cm}^2 \text{ min deg})$$

This creates a serious problem. The magnitude of ΔT and even its sign is extremely dependent on the values chosen for dE/dT, dE/dW, and dW/dT. Small changes in any one of these parameters would completely change the conclusions drawn. Were we to accept Möller's values for dE/dW and dE/dT and the assumption of constant relative humidity, the 8 per cent increase in atmospheric CO_2 content [$\Delta E = 0.8 \times 0.43 = 0.34$ mcal/(cm² min)] that has already taken place should have yielded a temperature rise of 1.5 C. Manabe and Wetherald (1967) concluded that the great amplification yielded by water vapor in Möller's model is an artifact of his failure to take into account the dependence of the intensity of vertical turbulence upon the carbon dioxide content in the atmosphere. Computing the states of radiative convective equilibrium of atmosphere for various concentration of carbon dioxide, they suggest that if relative humidity is held constant, the increase in the temperature resulting from doubling CO_2 content of the atmosphere will be 2.36 C (compared to 1.33 C for constant absolute humidity).

When ocean water is heated in the laboratory, its CO_2 partial pressure rises about 3 per cent/C (Takahashi, 1961). The rate of increase of atmospheric CO_2 content with rising surface temperature would not be nearly this high because that portion of the ocean equilibrating its temperature and CO_2 content with the air in the time available is quite limited. We estimate that rate of CO_2 rise would be more like 0.5 per cent/deg. If so, the self-induced CO_2 rise has not played an important role in the radiation balance adjustment.

These considerations allow two conclusions to be drawn.

1. The effects of CO_2 on climate can by no means be passed off as negligible.

2. Simple mean global calculations are inadequate for the precise evaluation of these effects. Only when a sufficiently rigorous computer analogue of the global system becomes available will this be possible. Cloudiness, snowcover, and water vapor content are key factors in these calculations.

Geographic Variations in Climate Change

Any climate change resulting from CO_2 addition will certainly effect different regions in different ways. Ultimately, man must concern himself with those areas where the climate is currently marginal. For example, precipitation changes in the arid regions or temperature changes in high latitude regions would be exceedingly significant. Although computer analog methods are absolutely essential to these questions, one point is worth making. Möller (1963) has shown that the amount of radiation held back by the man-made CO_2 in a water-free atmosphere would be twice that for an atmosphere with a water vapor content equal to the mean for the earth's mean water vapor content and cloudiness. This means that the geographic change in radiation balance will not be uniform but rather larger in areas of low water vapor content and cloudiness (i.e., deserts and polar regions).

Observed Climate Change Over the Last Century

Temperature and precipitation records made at many points on the globe allow fairly good estimates of the change in global climate to be made. Although these changes reflect not only those due to man-made CO_2 but also those resulting from other natural factors, their magnitude allows some limits to be placed on climate changes caused by man-made CO_2.

Let us first consider the mean global temperature. Mitchell's (1963) summary of the available temperature record (see Fig. 11.9) indicates that mean global temperature rose 0.5 C between 1880 and 1945. Since 1945 the temperature has steadily decreased. Between 1880 and 1960 the net change is only 0.35 C. Adopting Manabe and Wetherwald's (1967) model for constant relative humidity the predicted increase in tempera-

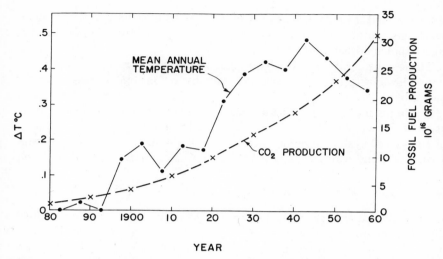

Figure 11.9 Changes in mean annual global temperatures calculated for 5-yr intervals from 1880 to 1960 compared with the cumulative production of fossil fuel CO_2. From Mitchell, 1963.

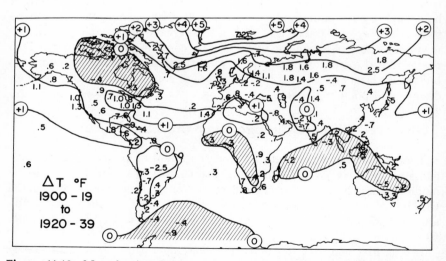

Figure 11.10 Map showing the geographic pattern temperature differences between the 20-yr intervals 1900 to 1920 and 1920 to 1940. Numbers show values in °F for individual stations. Shaded areas are those showing a drop in temperature. From Mitchell, 1963.

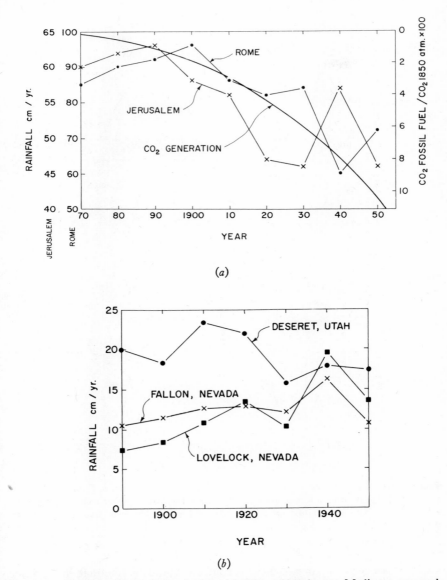

Figure 11.11 Rainfall records for the interval 1870 to 1960 for two Mediterranean and three Great Basin cities (based on data from the U. S. Weather Bureau, the Israel Meteorological Service, and on data published by H. H. Clayton et al.).

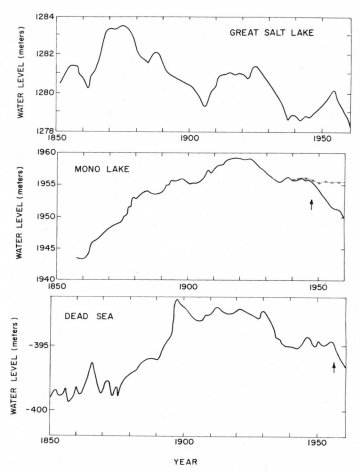

Figure 11.12 Water-level records for three desert closed basin lakes. From Neev, 1964; Harding, 1965. Arrow indicates onset of large-scale diversion of water from rivers supplying the lakes. The dotted line for Mono Lake represents the lake level were there no diversions.

ture for an 8 per cent rise in CO_2 content would be about 0.25 C. Since the observed change almost certainly (see below) involves a sizable component of natural (as opposed to man-induced) change the theoretical result is not incompatible with observation.

Mitchell (1963) also gives maps showing the geographical distribution of the temperature change (see Fig. 11.10). It is interesting to note that the largest changes occur in the regions adjacent to the Arctic Ocean.

OBSERVED CLIMATE CHANGE OVER THE LAST CENTURY   321

In Fig. 11.11 are summarized the precipitation records for two Mediterranean cities: Jerusalem and Rome, and for three Great Basin cities, Lovelock, Nevada, Fallon, Nevada, and Deseret, Utah. In Fig. 11.12 the levels of the desert lakes Mono and Great Salt Lake and the Dead Sea are shown. Although changes occur, these moisture records do not suggest any consistent trends in the degree of aridity in either the Mediterranean or Great Basin regions which follow the pattern of CO_2 addition. If the CO_2 addition is changing the aridity, these changes are being masked by the natural fluctuations.

In a recent publication Dansgaard and his co-workers (1969) presented $^{18}O/^{16}O$ ratios for ice from many levels in a core drilled at Camp Century on the Greenland ice cap. In Fig. 11.13 the last 12,000 yr of this record are reproduced. As shown by these investigators the variations observed reflect changes in the average temperature of the air masses from which the snow was derived. (The more deficient the ice is in ^{18}O

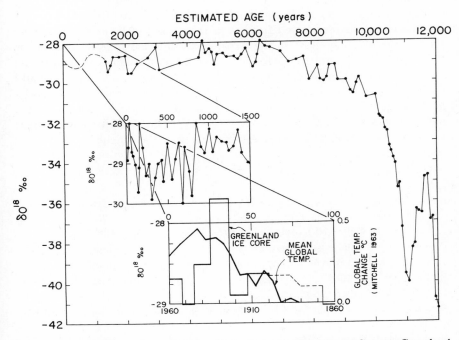

Figure 11.13 Variations of $^{18}O/^{16}O$ ratio in the ice from the Camp Century Greenland core. The time scale is based on the present accumulation rate of snow and a model for glacial flow. The global temperature variation curve of Mitchell (1963) is given for comparison purposes.

content the lower the mean air mass temperature.) This record allows the changes observed in the last century to be placed in perspective. First, they are extremely small compared to those which accompanied the end of the last glacial period ($\sim$11,000 yr ago). On the other hand, they are not unlike those which have characterized the last 1000 yr. It seems safe to say that the major source of the changes observed over the last 100 yr is quite likely natural. While a man-induced component may be present it has not as yet clearly emerged from the normal climatic "noise."

Conclusions

Man has, through the combustion of fossil fuels, increased the CO_2 content of the atmosphere by 8 per cent (as of 1960). It appears that this amount will rise in proportion to further production. At any given time 60 ± 10 per cent of the total will be in the atmosphere. Meteorological theory is just reaching the level where meaningful estimates of the climate change expected from such increases can be calculated. Historical records suggest that the 8 per cent increase produced no more than a 0.5 C temperature increase. No prominent changes in aridity attributable to CO_2 have as yet been observed in desert regions. Despite the absence of any alarming trends to date we must be cognizant of the much larger CO_2 rise to come in the next 50 yr. It is high time that a more conserted effort be made to understand more fully what Roger Revelle has termed man's greatest geophysical experiment.

References

Bolin, B., and E. Eriksson. 1959. Changes in the carbon dioxide content of the atmosphere and sea due to fossil fuel combustion. In *Rossby memorial volume*, B. Bolin (ed.), Rockefeller Press, New York, pp. 130–142.

Broecker, W. S. 1963a. Radioisotopes and large scale oceanic mixing. In *The sea*, Hill (ed.), Vol. II, John Wiley and Sons, Ltd., London, pp. 88–108.

Broecker, W. S. 1963b. $^{14}C/^{12}C$ ratios in surface ocean water. *Natl. Acad. Sci.—Natl. Res. Council Publ.* **1075**:138–149.

Broecker, W. S., and E. A. Olson. 1960. Radiocarbon from nuclear tests II. *Science*, **132**:712–721.

Budyko, M. I. 1969. The effect of solar radiation variations on the climate of the earth. *Tellus*, **21**:611–619.

Clayton, H. H., et al. (ed.) World Weather Records. Washington Smithsonian Inst. and U.S. Weather Bur.

Craig, H. 1957. The natural distribution of radiocarbon and the exchange time of carbon dioxide between atmosphere and sea. *Tellus*, **9**:1–17.

Craig, H. 1958. Distribution of radiocarbon and tritium; cosmological and geological implications of isotope ratio variations. *Natl. Acad. Sci.—Natl. Res. Council Publ.* **572**: 135–147.

Crank, J. 1956. *The mathematics of diffusion*, Clarendon Press, Oxford.

Dansgaard, W., S. J. Johnsen, J. Møller, and C. C. Langway, Jr. 1969. One thousand centuries of climatic record from Camp Century on the Greenland ice sheet. *Science*, **166**:377–381.

Fergusson, G. J. 1958. Reduction of atmospheric radiocarbon concentration by fossil fuel carbon dioxide and the mean life of carbon dioxide in the atmosphere. *Proc. Roy. Soc. London*, Ser. A **243**:561–574.

Goldschmit, V. M. 1958. *Geochemistry*. Oxford University Press, London.

Harding, S. T. 1965. Recent variations in the water supply of the western Great Basin. *Archives Series Report*, **16**:159. Water Resources Center Archives, University of California.

Israel Meteorological Service. Annual Weather Report; Annual Rainfall Summary. Tel Aviv.

Lerman, J. C., M. G. Mook, and J. C. Vogel. 1967. Effect of the Tunguska Meteor and sunspots on radiocarbon in tree rings. *Nature*, **216**:990–991.

Li, Y. H., T. Takahashi, and W. S. Broecker. 1969. The degree of saturation of $CaCO_3$ in the oceans. *J. Geophys Res.*, **74**:5507–5525.

Manabe, S., and R. T. Wetherald. 1967. Thermal equilibrium of the atmosphere with a given distribution of relative humidity. *J. Atmospheric Sci.*, **26**: 786–789.

Mitchell, J. M. 1963. On the world-wide pattern of secular temperature changes. Changes of climate, *Arid Zone Monograph*, **20**:161–181, UNESCO, Paris.

Möller, F. 1963. On the influence of changes in the CO_2 concentration in air on the radiation balance of the earth's surface and on the climate. *J. Geophys. Res.*, **68**:3877–3886.

Neev, D. 1964. Ph.D. Thesis, Hebrew University, Israel.

Pales, J. C., and C. D. Keeling. 1965. The concentration of atmospheric carbon dioxide in Hawaii. *J. Geophys. Res.*, **70**:6053–6076.

Revelle, R. 1965. In *Restoring the quality of our environment*, Report of Environmental Pollution Panel, President's Science Advisory Committee, White House, pp. 111–113.

Revelle, R., and H. E. Suess. 1957. Carbon dioxide exchange between atmosphere and ocean and the question of an increase of atmospheric CO_2 during the past decades. *Tellus*, **9**:18–27.

Suess, H. E. 1955. Radiocarbon concentration in modern wood. *Science*, **122**: 415–417.

Sverdrup, H. U., M. W. Johnson, and R. H. Fleming. 1942. *The oceans.* Prentice-Hall, Englewood Cliffs, N.J.

Takahashi, T. 1961. Carbon dioxide in the atmosphere and in Atlantic Ocean water. *J. Geophys. Res.*, **66**:477–494.

United Nations, Department of Economic and Social Affairs: World Energy Requirements in 1975 and 2000. *1956 Proc. of the Int. Conf. on the Peaceful Uses of Atomic Energy*, 3–33.

United Nations, World Energy Supplies Statistical Papers, Ser. J, United Nations, N.Y., 1950–1959.

U.S. Weather Bureau: Climatic Summary of the United States to 1930, Secs. 19 and 20, Suppl. for 1931 through 1952.

12

Radioactivity—Chemical and Biological Aspects

THEODORE R. RICE and DOUGLAS A. WOLFE, *National Marine Fisheries Service, Center for Estuarine and Menhaden Research, Beaufort, North Carolina*

General History

Man first initiated a controlled, self-sustaining nuclear chain reaction on December 2, 1942. He exploded the first destructive nuclear device at Alamogordo, New Mexico, on July 16, 1945. Since these historic events, many nuclear explosives have been tested in the air, underground, and underwater, and nuclear energy is producing electrical power for our rapidly expanding industrial civilization. Radioactivity was first detected in the marine environment in 1906 (Strutt, 1906), only 10 yr after Becquerel's discovery of the first radionuclide—uranium. But with the arrival of the nuclear age came large quantities of radioactive waste materials requiring disposal such that man would not be exposed to damaging radiation. Because of their vast volume and depths, the oceans have been considered a suitable repository for low-level wastes, which are usually diluted and dispersed, and for high-level wastes, which are usually concentrated to a minimal volume and packaged in semi-permanent shielded containers. The continued expansion of the use of nuclear energy, coupled with expectable occasional surface explosions for excavation or for weapons testing, will bring a gradual increase in oceanic radioactivity during the next several centuries. In the ensuing sections, we discuss the potential sources, cycling, and effects of oceanic radioactivity and the ways in which man can evaluate and control radioactive contamination.

Kinds of Radionuclides

Radioactive elements are classified according to their origin as those independent of man's activities (natural) and those produced by man (artificial). The principal categories of natural radionuclides are as follows: (a) primordial radionuclides and their decay products (daughters) and (b) radionuclides resulting from interaction between cosmic rays and elements in the atmosphere or in the earth. Primordial radionuclides have existed since the earth was formed and have produced daughter radionuclides through their decay. To persist through geological time, radionuclides must have relatively long half-lives; some have half-lives measured in millions and trillions of years. In contrast, daughter radionuclides have half-lives ranging from several thousand years to microseconds. Thus the existence of natural daughter radionuclides in the sea depends on the continuous decay of the parent radionuclides. Naturally occurring radionuclides arose in the oceans, at least in part, from the weathering of rock, and several have been identified (Table 12.1). Most of the natural radia-

Table 12.1. Principal Naturally Occurring Radionuclides in the Sea

Nuclide	Half-Life (yr)	Concentration (g/l)	Concentration ($\mu\mu C_i/l$)
^{3}H	1.2×10^1	3.2×10^{-18}	3.0
^{10}Be	2.7×10^6	1.0×10^{-13}	2.0×10^{-3}
^{14}C	5.5×10^3	3.1×10^{-14}	1.5×10^{-1}
^{32}Si	7.1×10^2		1.2×10^{-5}
^{40}K	1.3×10^9	4.5×10^{-5}	3.0×10^2
^{87}Rb	5.0×10^{10}	3.4×10^{-5}	2.8
^{226}Ra	1.6×10^3	8.0×10^{-14}	$(3.6-25) \times 10^{-2}$
^{228}Th (RdTh)	1.9	4.0×10^{-18}	$(0.25-1.4) \times 10^{-2}$
^{228}Ra (MsTh)	6.7	1.4×10^{-17}	2.3×10^{-3}
^{230}Th (Io)	8.0×10^4	6.0×10^{-13}	$(0.4-1.2) \times 10^{-2}$
^{231}Pa	3.2×10^4	5.0×10^{-14}	$(1.4-2.4) \times 10^{-3}$
^{232}Th	1.4×10^{10}	2.0×10^{-8}	$(0.02-1.0) \times 10^{-2}$
^{235}U	7.1×10^8	1.4×10^{-8}	5.2×10^{-2}
^{238}U	4.5×10^9	2.0×10^{-6}	1.15

tion in the sea originates from three of these: ^{40}K, ^{232}Th, and ^{238}U, and ^{40}K alone accounts for more than 90 per cent of it. In the atmosphere, nuclei of nitrogen, oxygen, and other gases react with cosmic rays to produce radionuclides. Two of the most interesting and abundant of these are ^{14}C and tritium;* others are ^{7}Be, ^{10}Be, ^{22}Na, ^{32}Si, ^{129}I, and ^{187}Re-^{187}Os. The distributions of several of these cosmic-ray produced radionuclides in the sea have been used in studies of currents, mixing processes, and geochronology of the ocean.

Artificial radionuclides are produced by (a) splitting the atomic nuclei of heavy elements such as plutonium or uranium (fission), (b) bombarding stable nuclei with neutrons (neutron induction), and (c) combining the nuclei of light isotopes such as tritium and deuterium (fusion). The first artificial radionuclides were produced by Frédéric Joliot and Irène Joliot-Curie in 1932, but it was not until 1945 that man-made radionuclides occurred in the oceans. Some of the artificial radionuclides that have been detected in the sea are shown in Table 12.2.

Sources of Artificial Radioactivity in the Oceans

The artificial radionuclides now in the oceans have arisen mainly from atmospheric testing of nuclear weapons. If a world-wide moratorium on such testing becomes accepted, this source soon after will cease to be significant. Peaceful uses of nuclear energy, however, are becoming more and more diverse, and, thus several modes may become established for the introduction of radionuclides into the oceans.

Fallout from Nuclear Explosions

Distribution and Classification of Fallout. Nuclear explosions produce large quantities of radioactive debris, much of which is swept skyward by the force of the blast. The radioactive debris that returns to the earth's surface is known as fallout. The process of fallout deposition on the earth's surface is extremely complex; the rate and distribution of the deposition depend on many factors, including the types and yields of the explosive devices, the latitude and altitude of the explosion, the season when the blast occurs, and meteorological conditions in general.

More than 200 fission-product radioisotopes may emerge from the detonation of a device involving fission or fission-fusion reactions (Table 12.2). In addition, a large number of radioisotopes are produced by neutron-bombardment of stable atoms near the center of the explosion. The composition of these neutron-activation products depends on the conditions of the detonation, that is, whether the explosion occurs in the air, on the ground, or underwater, and on the composition of the bomb casing. The

* Carbon 14 and tritium are formed also by nuclear explosions.

Table 12.2. Artificial Radionuclides that Have Been Identified in the Sea; concentrations Have Varied with Time and Space due to Man's Activities

Nuclide and Daughter	Half-Life
Fission products	
^{89}Sr	50.4 days
^{90}Sr; ^{90}Y	28 yr; 64.4 hr
^{91}Y	58 days
^{95}Zr; ^{95}Nb	63.3 days; 35 days
^{103}Ru; ^{103m}Rh	41.0 days; 54 min
^{106}Ru; ^{106}Rh	1.0 yr; 30 sec
^{129m}Te; ^{129}Te	33 days; 74 min
^{129}I	1.6×10^7 yr
^{137}Cs; ^{137m}Ba	30 yr; 2.6 min
^{141}Ce	32.5 days
^{144}Ce; ^{144}Pr	290 days; 17.5 min
^{144}Nd	2.0×10^{15} yr
^{147}Pm; ^{147}Sm	2.5 yr; 1.3×10^{11} yr
^{131}I	8.05 days
^{140}Ba	12.8 days
Induced nuclides	
^{32}P	14.3 days
^{35}S	87.1 days
^{51}Cr	27.8 days
^{54}Mn	300 days
^{55}Fe	2.94 yr
^{59}Fe	45.1 days
^{57}Co	270 days
^{58}Co	72 days
^{60}Co	5.27 yr
^{65}Zn	245 days
^{110m}Ag	249 days
^{113m}Cd	14 yr
^{185}W	74 days
^{187}W	24 hr

nuclides of major importance in worldwide fallout include ^{55}Fe, ^{90}Sr, and ^{137}Cs because greater amounts are produced, their physical half-lives are long, and all are biologically active. Of lesser significance are ^{54}Mn, ^{65}Zn, ^{95}Zr-^{95}Nb, ^{103}Ru-^{103m}Rh, ^{106}Ru-^{106}Rh, ^{141}Ce, and ^{144}Ce. Short-lived nuclides may also be important for brief intervals in the immediate vicinity of nuclear tests if the isotope is strongly adsorbed on biological surfaces (^{140}Ba-^{140}La) or actively metabolized by organisms (^{131}I).

Fallout is classified into three categories: close-in or local fallout, tropospheric fallout, and stratospheric fallout. The heaviest particles of the debris, which quickly settle in the immediate vicinity of the explosion, are called local fallout. Local fallout may be very dense if the explosion is low enough to involve soil or water from the earth's surface. The lighter radioactive debris is carried by the blast into the troposphere or the stratosphere. Tropospheric radioactivity is carried in a generally easterly direction around the world and settles to the earth within a few months at about the same latitude where the explosion occurred. In contrast, radioactive debris injected into the stratosphere enters circulation patterns in the upper atmosphere and falls out over a long period.

The pattern of fallout deposition has been cited as substantiating different theories of atmospheric circulation (Hagemann et al., 1959; Burton and Stewart, 1960; Feely and Spar, 1960). The generally accepted conception is that stratospheric fallout moves laterally from equatorial to polar regions and descends through the stratosphere into the troposphere over the higher latitudes of the temperate zones. Thus the areas of greatest deposition are at about 50° N and 50° S latitude (Fig. 12.1). Since most of the tests have been conducted in the U.S.S.R. and the United States, and the amount of transport across the equator is small, about 80 per cent of the fallout has been deposited in the northern hemisphere. Vertical mixing in the tropopause is accelerated in late winter and early spring, giving rise to a cyclic variation in fallout deposition. Tropospheric and stratospheric fallout both are increased also during periods of heavy rain or snow. Radionuclides from nuclear explosions thus enter the world oceans mainly from three sources: deposition of tropospheric and stratospheric fallout, close-in fallout from explosions conducted underwater or on small islands or atolls, and rain water runoff carried by rivers and underground aquifers.

Fallout deposition per unit area is generally believed greater over the oceans than over the land masses. Bowen and Sugihara (1960) integrated concentrations of ^{90}Sr in seawater through depth profiles, and found the total to be several times higher than the levels of ^{90}Sr in the same area on land. Volchok (1965) and Broecker et al. (1966b), however, developed

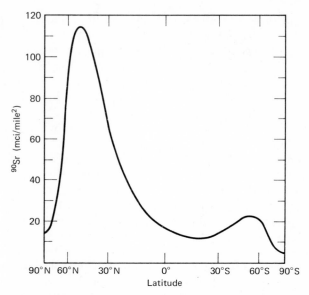

Figure 12.1 Latitudinal distribution of ^{90}Sr in soil, 1963–64. After List et al. 1965.

independent estimates which suggested that fallout is only about 50 per cent greater over the oceans than over land. Volchok computed total oceanic fallout by subtracting the increment of terrestrial fallout from the total stratospheric release, as estimated from direct stratospheric measurements, and Broecker et al. (1966b) measured ^{90}Sr accumulated in waters over the Bahama Banks during residence times of 12 to 180 days and compared this value to the average increment in ^{90}Sr at land stations at the same latitude during the same time. Kosourov (1966) found that fallout over the oceans on a "dry day" was 30 to 50 per cent of that on rainy days, whereas the bulk of fallout over land occurs during "precipitation," and the rate on dry days is only a few per cent of that on rainy days. It was hypothesized that sea spray may lead to the higher deposition of fallout over or near the ocean. Results of recent sampling off the coasts of Norway and Iceland, however, contradict the earlier studies by showing that ^{90}Sr deposition and rainfall are actually heavier over land masses than over the oceans (U.S. Atomic Energy Commission 1968e). Additional study is required to resolve the apparent discrepancy.

When nuclear explosions occur underwater, the resulting neutrons can react with the constituents of seawater, and a large number of radionuclides may be induced (Klement, 1959). Most of them, however, have very short half-lives and decay to insignificant levels within a few days after the explosion. The dominant nuclides induced in seawater represent those elements most abundant in seawater, and the composition of induced radionuclides changes considerably with time (Table 12.3). From the standpoint of potential significance to marine organisms and man, the more important radionuclides include fission products and those induced by the capture of neutrons in weapons casing, earth, and air (Table 12.2).

Table 12.3. Per Cent Abundance of Dominant Radionuclides Induced in Seawater at Different Times After an Underwater Explosion

Nuclide	$T_{1/2}$	Elapsed Time After Detonation			
		1 Hr	1 Day	1 Month	1 Yr
^{38}Cl	0.74 sec	68.2	—	—	—
^{24}Na	15.0 hr	28.2	96.8	—	—
^{80}Br	18 min	2.2	—	—	—
^{42}K	12.4 hr	1.0	2.8	—	—
^{27}Mg	9.5 min	0.2	—	—	—
^{82}Br	35.7 hr	—	0.39	—	—
^{35}S	86.7 days	—	—	78.1	48.5
^{45}Ca	165 days	—	—	21.3	47.2
^{36}Cl	3×10^5 yr	—	—	0.46	4.3

From data of Klement (1959).

Leaching of Fallout Radioactivity from Land Masses. Runoff of fallout radioactivity to the oceans appears to be very slow. It has been estimated that the annual runoff of ^{90}Sr is 1 to 10 per cent of the yearly deposition over land (Libby, 1956; Menzel, 1960), and the amount of ^{90}Sr in ground water is usually too small to measure (Yamagata, 1961). Cesium 137 is also very strongly bound to sediments (Davis, 1963); only about 2 to 6 per cent of annual terrestrial deposition is found in runoff (Yamagata, 1961). During 1954–1959 the tritium concentration in ground water from a depth of 40–50 m also was appreciably lower than in seawater and very much lower than in rain water (Brown, 1961). Thus it appears that radioisotopes are likely to be taken up by surface and subsurface layers of soil through ion-exchange processes and that the contribution of radioactivity to the oceans from runoff in surface streams or underground aquifers is very small.

Deposition of Fallout to Present. Because of their high fission yields, long half-lives, and relative ease of determination, ^{90}Sr and ^{137}Cs are used most frequently as indicators of fallout deposition, despite the probability that more ^{55}Fe has been deposited on the earth's surface to date. The concentration of ^{90}Sr and ^{137}Cs from fallout increased in surface waters of the oceans until about 1965, when the rate of introduction of additional fallout became less than the combined rates of removal of the isotopes into deeper layers of the oceans and of physical decay (Table 12.4).

Table 12.4. Worldwide Deposition of ^{90}Sr

Year	Annual Deposition[a] (MCi)	Cumulative Deposition[a] (MCi)	Average Concentration in North Atlantic Surface Waters[b] ($\mu\mu$Ci/1)
1957	—	2.3	—
1958	1.05	3.22	—
1959	1.29	4.40	—
1960	0.43	4.73	—
1961	0.54	5.12	0.08
1962	1.76	6.11	0.12
1963	2.95	9.46	0.18
1964	2.10	11.27	0.23
1965	1.14	12.12	0.16
1966	0.54	12.35	0.16
1967	0.28	12.31	0.12[c]
1968	0.28	12.28	—
1969	0.29	12.25	—

[a] Volchok, 1970.
[b] Adapted from Bowen et al., 1968, Fig. 6.
[c] For first 6 months of 1967 only.

Levels of ^{90}Sr and ^{137}Cs reportedly have been much higher in surface waters of the northwest Pacific than elsewhere (Miyake et al., 1961). Miyake and Saruhashi (1958) attributed this distribution to a rapid transfer of radioactivity from Pacific test sites. By the end of 1969,

SOURCES OF ARTIFICIAL RADIOACTIVITY IN THE OCEANS   333

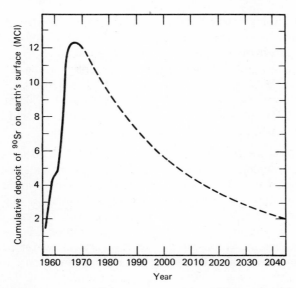

Figure 12.2 The amount of ^{90}Sr on the earth's surface reached a maximum during 1966–1967 and will decline exponentially with a half-life of 28 yr unless atmospheric testing is resumed. The level of ^{137}Cs deposited is approximately 1.7 times that shown for ^{90}Sr.

12.25 MCi of ^{90}Sr had accumulated on the earth's surface (Table 12.4 and Fig. 12.2), and about 0.34 MCi remained in the atmosphere (Krey et al., 1970). Atmospheric tests conducted by France and China in 1964–68 had a combined yield of less than 10 Mtons, and their contribution to worldwide fallout was negligible in comparison with that from the previous Russian and American tests. The deposition of ^{90}Sr during 1963–65 indicated that the stratosphere is cleared of ^{90}Sr according to an exponential function with a half-removal time of just less than a year. Unless intensive atmospheric testing is resumed, therefore, the stratosphere will have been effectively cleared by 1975, and the total cumulative amount of ^{90}Sr on the surface will decay with a half-life of 28 yr (Fig. 12.2). Since the half-life of ^{137}Cs is similar (30 yr), its disappearance should approximately parallel that of ^{90}Sr. The levels of ^{137}Cs deposited are about 1.6 times the activity of ^{90}Sr (United Nations, 1969).

The level of ^{55}Fe on the earth's surface has been much higher than has either ^{90}Sr or ^{137}Cs. Palmer and Beasley (1967) report the decay-corrected, cumulative deposition of ^{55}Fe at Westwood, New Jersey (41°N latitude), to be 0.4 Ci/mi² at the end of 1965. If the mean density of deposition in the northern hemisphere is assumed to have been only 0.3 Ci/mi², the total cumulative deposition would have been about 30 MCi in the northern hemisphere and about 37 MCi on the entire globe, as compared to 10.7 MCi of ^{90}Sr at the same time (Table 12.4 and Fig. 12.2). Since about 1965, the cumulative deposit of ^{55}Fe has decreased (Palmer and Beasley, 1965, 1967; Palmer et al., 1968) and the level of ^{55}Fe will become insignificant long before ^{90}Sr and ^{137}Cs have decayed away.

Fallout from Nuclear Excavation. Feasibility studies have suggested that nuclear explosives are ideally suited for large-scale excavation projects. Proposed uses include the construction of deepwater harbors and canals. Probably the most familiar proposal concerns the construction of a sea-level canal across the isthmus of Central America. Nuclear excavation can be engineered so that very little radioactivity is injected into the stratosphere or troposphere, but the earth or water near the center of the fission reaction would be bombarded by neutrons, with the resultant production of many significant radioisotopes. Most of the radioactivity produced (about 80 to 85 per cent) would remain in the bottom of the cavity, with the remainder constituting the local fallout. Radioactivity introduced into the ocean in the intense local fallout would include those isotopes induced in seawater (Table 12.3) and those induced in the soil (whose composition would depend on the nature of the local strata), the fission products from the explosion, thermonuclear reaction products such as ^{3}H and ^{7}Be, and residual fissionable or fusionable materials. The time required for local fallout to be deposited after an underground explosion is approximately proportional to the height of the resultant radioactive cloud; about 83 per cent of total local fallout from a cloud 12,000 ft high is deposited within 1 hr. The characteristics of the radioactive cloud depend on the total nuclear yield of the device, the yields of fission products and induced isotopes, the depth of emplacement, and the type of substratum. The composition and relative hazard of the resultant radionuclides have been the object of considerable discussion (Martin, 1969; Vogt, 1969; Lowman, 1969). The following isotopes may be of potential long-term concern in the marine environment: ^{3}H, ^{7}Be, ^{32}P, ^{46}Sc, ^{45}Ca, ^{54,56}Mn, ^{55,59}Fe, ^{65}Zn, ^{89,90}Sr, ^{95}Zr-^{95}Nb, ^{103,106}Ru, ^{134,137}Cs, ^{141,144}Ce, ^{185,187}W, ^{203}Pb, and ^{239}Pu.

Low-Level Wastes from Nuclear Power Reactors

The production of electrical power by nuclear plants in noncommunist countries is predicted to expand from about 5,000 MW at the end of

1968 to over 230,000 MW by 1980, with more than 60 per cent of the total produced within the United States (U.S. Atomic Energy Commission, 1968a). As of September 1970, electric utilities in the United States were operating 17 nuclear plants, had 54 under construction, and had 38 planned (Blakely, 1970). Most of these reactors routinely will release effluents containing small quantities of radioactivity into the aquatic environment, and there is always some possibility, however remote, of accidental uncontrolled releases of radioactivity from any of these reactors. Although the treatment and disposal of radioactive wastes are carefully planned at all reactor sites, the actual accepted practices vary, depending on the location and type of the reactor. The amount of radioactive waste disposed of annually by each power station is federally regulated, but the rates of release vary locally within the regulations.

Design of Nuclear Power Plants. A typical power plant, equipped with a pressurized-water reactor, is shown in schematic form in Fig. 12.3. Other reactor types used in power plants include boiling water, heavy water, gas-cooled, and (in breeder reactors) sodium-cooled. In all these types, heat from the fission process is transferred via a primary coolant,

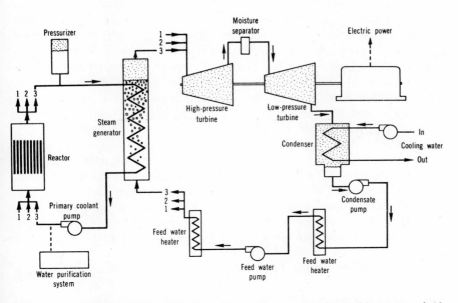

Figure 12.3 Schematic representation of a nuclear power plant with a pressurized-water reactor. Low-level radioactive wastes arise in the water-purification system and are released into the cooling water effluent. From Lyerly and Mitchell (1967).

sometimes through a secondary coolant loop, to a steam generator. After passing through a turbine which drives the generators, the steam is cooled and condensed by water from an external source, usually a river, lake, or estuary. Power plants are designed so that the accidental release of radioactivity to the environment is highly improbable, but significant quantities of radioactive wastes are produced during routine operation of any nuclear power plant. Radioactive materials in the primary coolant originate from the intense nuclear irradiation in the reactor. Fission products formed from the uranium fuel are effectively contained within the cladding of the fuel elements, but if the cladding develops a leak, radioactive halogens, fission product gases, and their decay products may escape into the primary reactor coolant. Some additional fission products arise in the coolant from traces of uranium dioxide naturally present in the water or on the outside surface of the fuel elements. The largest source of radionuclides in the primary coolant, however, is from neutron activation of impurities in the water and of the water itself. Impurities include trace materials not removed by the previous purification procedures and corrosion products from the insides of the reactor and the associated plumbing. Radioactive materials are removed from the primary coolant water continuously or intermittently by ion-exchange resins in the coolant loop of the reactor. Radioactive wastes are processed within the power plant to concentrate almost all the radioactivity into an easily packaged solid form which eventually leaves the plant for permanent burial underground. The relatively small amounts of radioactive materials that cannot be recovered for disposal as solid wastes are released into the environment in either gaseous or liquid form. For most water-cooled reactors, noble gases and tritium constitute by far the greatest proportion of the radioactivity released (Table 12.5). Krypton 85, because of its relatively long half-life (10.4 yr), may build up to substantial quantities in the atmosphere and concomitantly as dissolved gas in the oceans. Tritium results from ternary fission at very low yield (Albenesius and Ondrejcin, 1960), as well as from neutron activation of hydrogen, deuterium, and lithium, and becomes a part of water molecules in the primary coolant. Since tritium is not easily concentrated by present waste treatment procedures, it is released into the environment as either liquid water or water vapor.

Fusion Reactors. Fusion reactors are still far from being operational, but the long-term future for power generation is probably in breeder-fission or fusion reactors (Fowler and Post, 1966). The major advantages of the fusion process over that of fission for power production are the abundant and cheap supply of fuel and the absence of fission product

Table 12.5. Annual Release of Radionuclides Estimated for a Pressurized-Water Power Reactor of 1050 MW Electric Capacity

			Liquid Wastes			
Isotope	Half-Life	μCi/yr		Isotope	Half-Life	μCi/yr
^{3}H	12.26 yr	4×10^9		^{131}I	8 days	6.61×10^3
^{54}Mn	314 days	9.7×10^{-1}		^{132}Te	78 hr	6.99×10^2
^{56}Mn	2.58 hr	2.64×10^1		^{132}I	2.3 hr	2.8×10^2
^{58}Co	71 days	2.95×10^1		^{133}I	21 hr	5.13×10^3
^{60}Co	5.26 hr	3.48		^{134}I	53 min	2.16×10^1
^{89}Sr	50.4 days	9.1		^{135}I	6.7 hr	2.6×10^3
^{90}Sr	28 yr	5.76		^{134}Cs	2.1 yr	8.69×10^2
^{90}Y	64 hr	1.06		^{136}Cs	13 days	8.36×10^1
^{91}Sr	9.7 hr	2.49		^{137}Cs	30 yr	4.58×10^3
^{91}Y	59 days	2.11×10^1		^{140}Ba	12.8 days	2.28
^{92}Y	3.5 hr	5.13		^{140}La	40.2 hr	2.35
^{99}Mo	66 hr	1.25×10^4		^{144}Ce	285 days	7.82

	Gaseous Wastes		
Isotope	Half-Life	Ci/yr	
^{85}Kr	10.4 yr	5.62×10^3	
^{133}Xe	5.27 days	1.58×10^3	

From Preliminary Facility Description and Safety Analysis Report, Salem Nuclear Generating Station, Burlington Co., N.J. Docket No. 50-272.

waste materials. Some 3×10^6 Ci of tritium would be produced daily, however, in the moderator of a deuterium-tritium nuclear fusion reactor with a capacity of 1000 MW$_e$, and about 1.1×10^6 Ci/day would be released to the environment (Parker, 1968; Parker and Rose, 1968). Calculations have shown that if all nuclear power were produced by the fusion process, release of the resultant tritium would produce unacceptably high radiation dose by the year 2000. Methods for removing tritium from fusion effluents must therefore be improved before power can be produced safely by nuclear fusion (Parker, 1968).

Nuclear-Powered Ships. Insofar as the nature of radioactive wastes is concerned, nuclear ships are in the same general category as fixed power plants. Although shipboard reactors are much smaller than electrical power plants, there are more ship reactors presently operational. As of June 1970, the United States had about 106 nuclear-powered submarines and 8 surface vessels in operation or under construction (U.S. Atomic Energy Commission, 1970).

Radioactive wastes might be discharged to the marine environment from the following sources on nuclear-powered vessels: (a) The expansion volume of primary coolant during warm-up of a pressurized-water reactor; (b) leakage during routine operation, wastes from the laboratory and from equipment decontamination, and shower and laundry wastes associated with the reactor; (c) ion-exchange resins which remove activated corrosion products from the primary coolant; and (d) other contaminated solid materials. The major potential source of liquid waste is the primary coolant. During reactor warm-up, the coolant expands and a certain amount of the volume is released directly into the environment. This waste contains traces of fission products and also activated corrosion products, which are maintained at very low levels by ion exchange on a resin. It was estimated for the *USS Savannah* that the expansion volume during reactor warm-up is about 290 ft^3, which would contain about 0.68 Ci, consisting mainly of ^{51}Cr, ^{60}Co, ^{55}Fe, and ^{182}Ta. This amount might be supplemented by an additional 0.03 Ci/day from leakage and other minor sources of liquid waste (National Academy of Sciences—National Research Council, 1959b). After 50 days of operation, however, ion-exchange resins used in the bypass cleanup of the primary coolant would contain a total activity of about 400 Ci, consisting of the same isotopes dominant in the primary coolant. Most nuclear-powered vessels probably release considerably less radioactivity than the *USS Savannah*.

Hanford and Windscale-Production Reactors. Highly fissionable ^{239}Pu for enriched atomic fuels and for weapons is produced in nuclear reactors by neutron bombardment of naturally occurring ^{238}U. Except for the precautions necessary to contain the highly toxic plutonium, operation of a production reactor poses no unique problems of waste disposal. With two important exceptions—the Hanford Plant in the United States and Windscale Plant in England—production reactors have wastes similar to those produced by power reactors. The Hanford Plant at Richland, Washington, however, utilizes a single pass of Columbia River water through the reactor core for a primary coolant, and the effluent contains about 60 different radionuclides from leakage of fuel elements and from neutron activation within the reactor. The major isotopes released (Table 12.6) are ^{56}Mn, ^{64}Cu, ^{24}Na, ^{51}Cr, and ^{239}Np (Junkins et al., 1960). Nearly 25,000 Ci/month entered the Pacific Ocean from the mouth of the Columbia River during full operation of the plant (Mauchline and Templeton, 1964; Osterberg, 1965). Predominant isotopes reaching the mouth of the Columbia River some 2 weeks after release are ^{65}Zn and ^{51}Cr, although several others also have been detected in estuarine biota (Watson et al., 1963) and sediments there (Perkins et al., 1966).

Table 12.6. Major Isotopes Released to the Columbia River by the Hanford Production Reactors[a]

Isotope	Half-Life	Composition From Physical Decay (per cent)		
		at 4 hr	at 72 hr	at 2 wk
^{56}Mn	2.58 hr	27.5	—	—
^{64}Cu	12.9 hr	18.8	3.34	—
^{24}Na	14.97 hr	13.7	4.04	—
^{51}Cr	27.8 days	8.3	53.14	89.72
^{239}Np	2.35 days	8.2	24.55	2.13
^{76}As	26.8 hr	7.3	8.68	—
^{31}Si	2.62 hr	4.9	—	—
^{69}Zn	55 min	2.4	—	—
^{72}Ga	14 hr	1.3	0.30	—
^{92}Sr	2.7 hr	0.8	—	—
^{239}U	23.5 min	0.8	—	—
^{133}I	21 hr	0.7	0.51	—
^{92}Y	3.53 hr	0.6	—	—
^{97}Nb	72 min	0.6	—	—
^{91}Sr	9.7 hr	0.5	0.03	—
^{65}Zn	245 days	0.4	2.67	5.75
^{32}P	14.22 days	0.3	1.76	2.29
^{90}Y	64.4 hr	0.3	0.97	0.11
^{135}I	6.7 hr	0.3	—	—
^{93}Y	10.1 hr	0.3	0.02	—

[a] Modified and expanded from Junkins et al., (1960). The listed isotopes actually constitute only 98 per cent of the total 4 hr after release. The remaining 2 per cent may include as many as 80 other isotopes.

At the Windscale facility, near Sellafield, England, about 7500 Ci of fission products are discharged monthly about 2 miles offshore into the Irish Sea (Mauchline and Templeton, 1964). These wastes arise from the chemical processing of nuclear fuel and consist mainly of longer-lived fission products (^{89}Sr, ^{90}Sr-^{90}Y, ^{95}Zr-^{95}Nb, ^{144}Cr, ^{137}Cs, ^{103}Ru, ^{106}Ru), along with ^{238}U and ^{239}Pu (Longley and Templeton, 1965).

Reprocessing of Nuclear Fuel. In most reactors, the fuel elements must be replaced when only 1 or 2 per cent of the uranium fuel has fissioned. "Spent" fuel elements are therefore processed to reclaim the remaining uranium for reuse. This reprocessing of spent fuel is done at the Hanford and Windscale Plants and by government installations at Savannah River and Idaho Falls. The Savannah River Plant in Aiken, South Carolina, which also operates production reactors, has maintained releases of radioactive wastes at levels well below the Federal Radiation Council's Radiation Protection Guides (Evans et al., 1968). Civilian power reactors are increasing at such a fast rate that civilian reprocessing plants are required. The first such plant in the United States began operation during 1967 and 5 to 15 more are expected to be built within the next 10 yr (Eisenbud, 1968). Because of the very high levels of radioactivity in spent fuel, the reprocessing plants may encounter disposal problems of greater magnitude than those of the reactors themselves. The composition of radioactive wastes from new fuel reprocessing plants should be similar to that from power plants, in which tritium, xenon, and krypton contribute most of the radioactivity (Table 12.5). During 1966, the lower Savannah River contained 6 to 10 nCi/l of tritium from the Savannah River Plant. Residents of Beaufort, South Carolina, which derives its drinking water primarily from the Savannah River, received from this source an estimated whole body dose of 1.0 to 1.7 mrem/yr, or about 1 per cent of the recommended maximum dose to the general population (Moghissi and Porter, 1968).

Contained High-Level Solid Wastes

The ocean depths were once considered an appropriate burial site for solid wastes from nuclear power plants, production reactors, and fuel reprocessing plants. Wastes were contained in semi-permanent packages which (*a*) could not be easily damaged and which reached the bottom without appreciable loss of contents; (*b*) were free of voids; (*c*) had a minimum average density of 1.2 g/cm³; (*d*) were shielded for safe storage, shipment, and handling; and (*e*) could be handled quickly and conveniently. All packaged wastes had to be buried at a minimum depth of 1000 fathoms (National Committee on Radiation Protection, 1954).

The amount of radioactive material discharged into ocean depths is somewhat uncertain, partly because of the difficulty in determining the average concentration of radionuclides in large packages and partly because some of the users have been reluctant to identify themselves or the magnitude of their disposals (Parker, 1967). Up to 1960, approximately 10,000 Ci of radioactivity had been dumped off the California coast; about 6000 Ci were deposited in the Atlantic Ocean and Gulf of Mexico during 1951–58 (National Academy of Sciences—National Research Council, 1959a); and a total of some 61,000 Ci in solid wastes were discharged to the oceans during 1946–63 (Parker, 1967). Although the containers were sometimes damaged, no radioactivity was detectable in samples of mud and water taken from ocean burial sites (Brown et al., 1962, Parker, 1967). Nearly all radioactivity in solid form in the United States is now buried underground because this disposal method is less expensive than sea burial. An indeterminate amount of very low-level solid waste, however, is discharged, unpackaged, directly into the oceans. Other nations, excluding the U.S.S.R. (Polikarpov, 1966a), also continue to dump solid wastes at sea.

Other Sources

Aerospace Applications. Nuclear energy is now being tested as a source of power for space travel, not only to propel the vehicle but also to provide heat for the cabin and electricity for the purification and regeneration of water and air and operation of communications and control instruments. Extensive experimentation on nuclear rocketry has been conducted at a Nevada test site. The reactor at the heart of the nuclear rocket engine generates more than 5000 MW of thermal power and reaches temperatures near 2500 K (Corliss, 1967). Significant quantities of radioactive fission products could be released from such an engine only if the fuel elements disintegrated as a result of a malfunction. If such an accident occurred during lift-off over the ocean, very large quantities of fission products would fall onto the water.

Nuclear power is also appropriate to provide electrical power for manned or unmanned space missions, including such functions as radio and television relay satellites. These systems for nuclear auxiliary power (SNAP) would generate from 500 W to 1000 kW of electrical power. Designs include the use of heat from the decay of ^{144}Ce to generate 500 W of electrical power and the use of small reactors coupled to thermoelectric generators. As with space propulsion reactors, contamination of our environment might occur only from an accident during transportation, launch, or operation in space of the nuclear power plant. Potential types of accidents include possible criticality of the fuel as a result of a transportation accident before launch, poorly aimed or malfunctioning launch vehicle, and ablation of the reactor during reentry into the atmosphere.

Radioisotope Applications. Laboratories, hospitals, and industries are steadily increasing amounts of radioactive materials. During 1967, about 3.4 million Ci of radioisotopes were supplied by the U.S. Atomic Energy Commission to users (U.S. Atomic Energy Commission, 1968d), and commercial outlets supplied additional quantities. Since most of this radioactivity is in sealed sources, reasonable care and proper use will ensure that oceanic contamination from research and industrial applications will be negligible compared to that from other sources.

Accidents. In contrast to the small amounts of radioactive wastes released during routine operation of nuclear-powered vessels, the amount of radioactivity contained in the spent fuel elements is very large. More than 10^7 Ci of mixed fission products might result from a 1-yr operation of a 60-MW reactor (National Academy of Sciences—National Research Council, 1959b). It is therefore very difficult to assess the radioactive contamination in the oceans from the loss of nuclear submarines such as the *USS Thresher* in 1963 and the *USS Scorpion* in 1968. Although the integrity of the reactor could be maintained for some time on the ocean floor, it seems probable that radioactive materials were released to the ocean from broken fuel elements. Other significant sources of oceanic contamination could include accidental loss in the oceans of SNAP-reactors, or of nuclear weapons aboard submarines or aircraft. Although the possibility of nuclear explosion is extremely remote, the shock of a crash can cause rupture of the weapons casing and thus introduce ^{239}Pu into the environment as happened at Palomares, Spain, January 17, 1966 (Lewis, 1967; Szulc, 1967). Such occurrences are completely unpredictable and the consequences of the resultant contamination depend on the extent of the damage in each instance.

Distribution and Cycling of Artificial Radioactivity in the Oceans

In the marine environment, radionuclides may (*a*) remain in solution, (*b*) be adsorbed on suspended particulate matter or on bottom sediments,

(c) flocculate and precipitate to the bottom, or (d) be accumulated by plants and animals. Radionuclides are diluted and dispersed by currents, turbulent diffusion, isotopic dilution, and movement of animals. Simultaneous concentrating processes include bioaccumulation and passage through food webs, and physicochemical adsorption, ion-exchange, flocculation, and precipitation. The routes and rates followed by a particular isotope within the marine ecosystem depend largely on the chemical characteristics and biological utility of the isotope. Study of the cycling of radionuclides in natural environments constitutes the field of radioecology, which has received considerable attention in recent symposia (Radioecology, Fort Collins, Colorado, 1961; Radioecological Concentration Processes, Stockholm, Sweden, 1966; Disposal of Radioactive Wastes into Seas, Oceans, and Surface Waters, Vienna, Austria, 1966; Second National Symposium on Radioecology, Ann Arbor, Michigan, 1967; Sealevel Canal Bioenvironmental Studies, A.I.B.S., Columbus, Ohio, 1968), reviews (Mauchline and Templeton, 1964; Templeton, 1965; Burton, 1965; Miyake, 1963; Chipman, 1966) and books (Polikarpov, 1966a; and Seymour, in press).

Nonbiological Cycling of Radionuclides

Physical and Chemical States of Radionuclides in Seawater. The state of a radionuclide in seawater depends very much on its mode of introduction into the oceans. Greendale and Ballou (1954) determined the physical states of several fission product elements in seawater after simulating the conditions of vaporization for an underwater nuclear detonation (Table 12.7). The major sources of radioactivity in the oceans, however, are from atmospheric fallout and from reactor effluents. Most fallout enters the ocean as finely divided debris and the nuclides probably behave in a manner similar to that shown in Table 12.7. Certain rare-earth isotopes also tend to exhibit colloidal properties in water (Schweitzer, 1956). Highly radioactive particles are, however, occasionally encountered in the environment (Cutshall and Osterberg, 1964; Mamuro et al., 1963; Edvarson et al., 1959), and Baptist (1966) demonstrated the formation of particles when mixed fission products were added to flasks of seawater. It has been postulated that ingestion of particulate fallout could account for the high specific activities of ^{54}Mn observed in the estuarine scallop, *Aequipecten irradians* (Schelske et al., 1966). Fallout nuclides that enter the oceans with freshwater drainage, as well as isotopes in reactor effluents, probably have already entered normal biogeochemical cycles for the elements, and occur in seawater in the physical and

Table 12.7. Composition, by Physical State, of Some Fission Product Elements after Vaporization in Seawater

Element	Physical State (per cent)		
	Ionic	Colloidal	Particulate
Ru	0.35	4.65	95
Ce	1.5	4	94.5
Cs	70.5	7	22.5
Zr	1.5	2.5	96
Sr	87	2.9	10.1
I	89.5	8.5	2
Y	0.2	4.3	95.5
Nb	0.2	0.3	99.5
Mo	30	10	60
Te	45	43	12
Sb	73	14.5	12.5

From Greendale and Ballou (1954) and Freiling and Ballou (1962).

chemical states characteristic of each element. The most probable chemical species for many elements in seawater have been described by Sillén (1961, 1967), Goldberg (1963), and Goldberg et al. (1963), based on theoretical data, however experimental evaluation of the forms present is very limited and must be accomplished before adequate understanding is possible.

Vertical Transport of Radionuclides. Radionuclides may be transported downward by flocculation and precipitation, by adsorption on sedimenting particulate matter, by diffusion, and by convection in moving masses of water. The rate of vertical dispersion above the thermocline is much greater than molecular diffusion, but much lower than the rate of horizontal movement. Vertical stirring within the mixed layer depends on surface winds and on the range of the density gradient (Revelle and Shaffer, 1957). After nuclear tests in the Pacific, particulate radioactivity accumulated rapidly at the upper edge of the thermocline; and the radioactivity was not uniformly distributed within the mixed layer, probably because of concurrent settling of particulate radioactivity and rapid dispersion of surface waters by currents and wind (Lowman, 1960).

Determinations of artificial radionuclides in seawater can obviously be valuable in tracing transfer processes in the oceans (see Miyake, Chap. 22). Since ^{90}Sr and ^{137}Cs are largely ionic in seawater (Table 12.7), their movement probably represents the movement of large water masses. Early measurements of these radionuclides in surface waters of the oceans were summarized by Burton (1965) and Polikarpov (1966a). Studies of vertical penetration have produced discrepant results and interpretations. A significant fraction of the total depth profile of ^{90}Sr has been found in uniformly low concentrations at mid-depths (1000–2500 m) in the Atlantic Ocean (Bowen and Sugihara, 1960, 1963; Bowen et al., 1966) and in the northwest Pacific Ocean (Miyake et al., 1962; Folsom and Sreekumaran, 1966). The uniformity of ^{90}Sr concentrations in the mid-depth layers, coupled with the increase in ^{90}Sr during 1961–63, substantiates a rapid mixing of intermediate layers and the flow of intermediate-depth currents northward and southward toward the equator in the Atlantic Ocean (Bowen et al., 1966). Rocco and Broecker (1963) and Broecker et al., (1966a) refuted the concept of rapid penetration to intermediate ocean depths, however, when they did not detect significant quantities of either ^{90}Sr or ^{137}Cs at depths greater than 200 m in the Pacific or 500 m in the Atlantic. The extent and pattern of penetration of ^{137}Cs into the oceans may be clarified in the future by more intensive use of the highly selective (for cesium) ion-exchanger, potassium cobalt ferrocyanide (KCFC) (Folsom and Sreekumaran, 1966).

The rare-earth isotopes, ^{144}Ce and ^{147}Pm, are depleted in surface waters relative to ^{90}Sr, probably because of their particulate nature in seawater or because they adsorb onto sedimenting particulate matter (Sugihara and Bowen, 1962; Bowen and Sugihara, 1963; Carpenter and Grant, 1967). In surface waters, near the mouth of the Columbia River and off the Oregon coast, ^{95}Zr and ^{95}Nb were present but in surface waters in the upwelling region within 30 to 60 km of the coast, they were greatly reduced. This water upwells from offshore at depths of 100 to 200 m and probably did not receive any relatively short-lived fallout nuclides (Gross et al., 1965). The vertical distributions for 18 trace elements in the oceans were determined by Schutz and Turekian (1965).

Lateral Transport of Radionuclides. Horizontal dispersion of radioactivity depends on ocean currents, surface winds, and vertical and horizontal density gradients. Above the thermocline, and especially near the surface, winds may influence the lateral movement of radioactivity, although large-scale oceanic currents are probably the dominant force. If the wind is moving surface waters in a different direction from the prevailing current, it may forestall homogeneous mixing above the thermocline. Within the thermocline, most of the motion of soluble material is along surfaces of equal density; thus dispersion in the lateral direction would be much greater than in the vertical direction. At the Eniwetok test site, the radioactivity drifted westward usually with about the speed of the current (Lowman, 1960).

Adsorption of Radionuclides on Suspended Particulates and Sediments. Sediments have a high affinity for many radionuclides and must be considered in any discussion of distribution of marine radioactivity. When radioactive zinc was added to an enclosed estuarine pond community, 99.8 per cent of it became associated with the sediments within 60 days. In a second experimental pond having a restricted tidal connection to the estuary, 18 to 33 per cent of the ^{65}Zn added to the water was particulate during the first 96 hr, and 99.3 per cent of the isotope remaining in the pond after 100 days was adsorbed on the sediment (Duke et al., 1966). In the Columbia River and its estuary, concentrations of ^{65}Zn, ^{60}Co, and ^{54}Mn are higher in sediments than in overlying water (Osterberg et al., 1966; Jennings and Osterberg, 1969); and ^{65}Zn becomes adsorbed on particles in the river, which subsequently settle into the bottom layers (Nelson et al., 1964). Adsorption on fine particulate matter was favored; for example, more than 6.5 per cent of the ^{65}Zn was adsorbed on particles less than 0.074 mm in diameter, which constituted only 0.1 per cent of the total sediment sample, and more than 50 per cent of the ^{65}Zn was adsorbed on particles less than 0.84 mm diameter, or 13.7 per cent of the sample (Nelson et al., 1966). Elution with artificial seawater was ineffective in removing ^{65}Zn, ^{51}Cr, and ^{46}Sc from natural Columbia River sediment; a highly variable proportion (0 to 73 per cent) of ^{54}Mn was displaced from the sediment by seawater elution. Substantial portions of ^{65}Zn, though not of ^{51}Cr or ^{56}Sc, were eluted by solutions (in order of effectiveness) of divalent copper, cobalt, and manganese (Johnson et al., 1967). In the Columbia River ^{51}Cr (III) is also adsorbed on particulate matter and may remain bound to particles in seawater. Although chromium (VI) is the favored oxidation state of chromium in seawater, chromium (III) is actively adsorbed on living and nonliving particles at all pH values normal for seawater (Curl et al., 1965).

Fission products released by the Windscale Plant into the Irish Sea are much more strongly bound on mud-flat sediments than on sandy beaches. The concentration of radioactivity in the sediment decreases exponentially both with distance from the release point and with depth from the surface layer of sediment. Isotopes most highly concentrated in surface sediments are ^{95}Zr-^{95}Nb and ^{106}Ru-^{106}Rh, with accumulation factors of approximately 1.5×10^4 (Jefferies, 1968). The prominent fallout isotopes found in marine sediments from the Ligurian Sea were ^{144}Ce-^{144}Pr and ^{147}Pm, which together accounted for about 95 per cent of the fallout radioactivity in the top 4 cm of sediment; ^{155}Eu contributed 4 per cent and ^{90}Sr-^{90}Y made up the remaining 1 per cent. Nearly all the fallout radioactivity was in the top 8 cm of sediment; of the four reported isotopes, ^{90}Sr-^{90}Y seemed to have penetrated to the greatest depth (Cerrai et al., 1967; Schreiber, 1966).

The adsorption of cesium on marine clays decreases as the concentration of soluble Ca^{2+}, Mg^{2+}, Na^+, and K^+ is increased in the overlying seawater, and the effectiveness of desorption increases in the order shown (Cheng and Hamaguchi, 1968). This phenomenon would account for the steep concentration gradient of ^{137}Cs in water as the salinity increases in shallow estuaries. The existence of such a relationship in the environment has been deduced from the concentration factors for ^{137}Cs in estuarine clams and from the inverse logarithmic correlation between ^{137}Cs accumulated and salinity (Wolfe et al., 1969).

Quantification of the dynamic transfer of materials from seawater or biota to marine sediments, or vice versa, is extremely problematical. Elements usually exist in sediments in several different forms, which are not equally available for exchange. Estimates of exchangeable element in sediments depend therefore on the technique used for analysis, which usually include an acid extraction (Müller, 1967; Fukai, 1965). The total reservoir of an exchangeable element in sediments can be estimated accurately, however, from the amount of radioisotope in the sediment after accumulation has ceased and from the specific activity of the same isotope in the overlying water at that time. If conditions of exchange equilibrium are assumed, the rate of elemental exchange across the sediment-water interface can also be estimated (Duke et al., 1968).

Biological Cycling of Radionuclides

Marine organisms accumulate radioactive materials (a) directly from seawater by highly selective ion transport mechanisms or by nonspecific physical adsorption processes, and (b) by ingestion of other organisms or particulate detritus which contain previously accumulated radionuclides. Aquatic organisms normally obtain a radionuclide or element from a combination of sources, including water, food, suspended particulate matter, and sediment. The extent to which an organism in steady state enriches an element or isotope over its concentration in the medium is called the concentration factor. In many organisms, the concentration factors for certain elements or radioisotopes are of limited practical utility because of the difficulty in selecting a meaningful environmental basis for comparing the amounts of one element in the organism. Although seawater in the open oceans may be considered to exhibit a more or less constant composition, elemental concentration may be very different in the layers of water just above a bed of sediment or in the interstitial waters within the sediment, and may undergo wide seasonal or tidal fluctuations in shallow estuarine systems. Trace elements such as iron, manganese, cobalt, copper, zinc, and nickel exhibit a log-normal distribution in marine organisms, such that the maximum concentrations may be several times higher than the mean value (Ting and Roman deVega, 1969). Thus concentration factors for radioisotopes may exhibit considerable natural variation. Much of the utility of concentration factors is based on the assumption that the concentration factor is independent of the elemental concentration in the medium; that is, that increasing the concentration of an element in the medium causes a proportional increase in the content in the organism. Such changes have been observed in many studies (Corcoran and Kimball, 1963; Rice and Willis, 1959; Mauchline and Templeton, 1964), but Preston (1966a) has noted that the concentration factor for zinc in oysters is inversely related to the reported concentration of zinc in the medium. Polikarpov (1966b) concluded that the concentration of a chemical element in aquatic organisms in directly proportional to the amount of the element in the water up to the concentrations of 10^{-6} to 10^{-4} moles/l. At higher concentrations of the same element or of certain nonisotopic carriers, concentration factors may be reduced. Concentration factors may be changed also if the physicochemical state of the element is not consistent over its entire range of concentration. For most elements in seawater, the discharge of radionuclides at acceptable levels does not change significantly the concentration of the total element. A picocurie of radioactivity represents only 10^{-9} or 10^{-10} μg of most elements; and most trace elements occur in seawater at levels above 10^{-3} μg/l. The concentrations of only very rare elements, such as plutonium, might be influenced significantly by the introduction of artificial radioisotopes into the oceans; other changes are likely to occur in shallow nearshore areas which receive seasonally dependent freshwater inflow and industrial wastes containing relatively high concentrations of certain elements.

Accumulation of Radioisotopes. When an organism is placed in seawater containing a radioisotope, it accumulates the isotope at a rate which decreases until a steady state between the tissues and water is attained.

The accumulation can be described approximately by a generalized exponential relationship of the following type:

$$C = a - be^{-kt}$$

where C is the concentration of isotope in the organism at time t, and a, b, and k are constants. Probably both uptake and loss processes obey an exponential law only in the simplest cases, that is, where only two compartments are involved. Actual biological accumulation results from the interplay of many such simple systems, so that uptake and loss processes could better be described by more complex functions, including logarithmic, power, and power-logarithmic terms (Polikarpov, 1966b). The rate of accumulation at any time probably depends at least in part on the difference in specific activity* of the isotope in the medium and in the organism. Since the isotope enters the organism at some point or tissue and is then transported within the organism to other tissues where it may reside for some time before elimination, physical decay of the isotope occurs during the biological residence time; therefore, the specific activity of the isotope within the entire organism can never reach that of the medium. In like manner, the specific activity of any isotope decreases during passage of the element through a food chain; consequently the specific activity of each component in the chain is restricted to some value lower than the specific activity at the previous trophic level.

Radioactive and stable isotopes of the same element generally are accumulated by organisms in the same ratio as is present in the medium. Organisms fractionate isotopes of lighter elements such as hydrogen, lithium, carbon, oxygen, and sulfur (Bowen, 1960), and according to Bernhard and Zattera (1969), phytoplankton discriminates between radioactive and stable zinc under certain conditions, but this latter discrimination could result from a difference in chemical or physical form of the isotopes. The accumulation of some radioisotopes depends less on the concentration of the stable element than on the concentration of another more abundant chemically related element. Thus, ^{90}Sr and ^{137}Cs are accumulated and distributed by marine organisms much as are the stable elements of calcium and potassium, respectively, although organisms generally discriminate in favor of the calcium and potassium. In brackish-water invertebrates, accumulation was slower for cesium than for potassium, but the greater binding affinity of the cesium atom resulted in slightly higher steady-state concentration factors for cesium than for

* Specific activity of a radioisotope is the amount of radioisotope per unit of the corresponding element, frequently expressed as $\mu Ci/g$.

potassium (Bryan and Ward, 1962; Bryan, 1961, 1963a, 1963b, 1965). When strontium is accumulated in the presence of calcium, however, resultant strontium/calcium ratios in freshwater fish and molluscs are generally significantly lower (0.4–0.6) than the strontium/calcium ratio in the medium (Beninson et al., 1966; Templeton and Brown, 1963, 1964; Ophel and Judd, 1967). Since aquatic plants do not discriminate against strontium, higher strontium/calcium ratios might be expected in herbivorous or omnivorous fish than in carnivores (Ophel and Judd, 1967).

The extensive literature on concentration factors for the various elements and isotopes in marine organisms has been thoroughly reviewed up to 1965 (Mauchline and Templeton, 1964; Polikarpov, 1966a) and will not be reviewed here. The multiple applications of a nuclear technology will produce a wide variety of radionuclides, which may be introduced into the marine environment by many different routes. The nuclides most likely to be significant in the marine biosphere are listed in Table 12.8, along with typical concentration factors observed for the elements in marine algae, crustaceans, molluscs, and fish.

Role of Organisms in the Translocation of Radionuclides. In spite of the high degree of concentration exhibited for certain radionuclides in marine organisms (Table 12.8), living organisms probably would contain only an insignificant portion of the total amount of radioactivity present in the oceans at any time (Polikarpov, 1966b; Lowman et al., in press). Compared with the circulation by currents in the open ocean, horizontal migration by oceanic nekton is also an unimportant factor in the distribution of radionuclides. Some radionuclides are carried across the halocline by the diurnal vertical migrations of zooplankton (Pearcy and Osterberg, 1964), although this transport may be of little significance simply because of the small biomass involved (a maximum of <15 mg/m^3 in the reference cited and an estimated average of 2.94 mg/m^3 in the tropical eastern Pacific—see Blackburn, 1966). For those biologically important elements which are relatively rare in sea water, such as cobalt, the vertical movement of zooplankton may be more significant (Kuenzler, 1969).

Zooplankton might excrete radionuclides in fecal pellets below the halocline, after feeding on the surface the previous night. The adsorption of ^{95}Zr-^{95}Nb onto rapidly sinking fecal pellets was postulated to explain the high concentrations of the isotope in benthic sea-cucumbers off the Oregon coast (Osterberg et al., 1963). Comparison of ratios of ^{95}Zr-^{95}Nb/^{40}K in animals from deep and shallow water indicated that ^{95}Zr-^{95}Nb sank 2800 m in only 7 days. A similar situation did not exist for ^{65}Zn, probably because of rapid loss from fecal pellets back to the water. Sinking exo-

Table 12.8. Approximate Concentration Factors for Radioisotopes of Probable Significance in the Marine Biosphere

Radionuclide(s)	Algae	Crustacea	Molluscs	Fish
^{3}H	0.90	0.97	0.95	0.97
^{7}Be	250	—	—	—
^{14}C	4000	3600	4700	5400
^{24}Na	1	0.2	0.3	0.13
^{32}P	10^4	2×10^4	6×10^3	3.7×10^4
^{45}Ca	2	120	0.4	1.2
^{46}Sc	1200	300	—	750
^{51}Cr	2000	100	400	100
^{54,56}Mn	3000	2000	10^4	200
^{55,59}Fe	2×10^4	2500	10^4	1500
^{57,58,60}Co	500	500	500	80
^{65}Zn	10^3	2000	1.5×10^4	1000
^{85}Kr	~1	~1	~1	~1
^{89,90}Sr	50	2	1	0.2
^{90,91}Y	500	100	15	10
^{95}Zr-^{95}Nb	1500	100	5	1
^{103}Ru, ^{106}Ru-^{106}Rh	400	100	10	1
^{110}Ag	—	7	10^4	—
^{132}Te-^{132}I	—	—	—	—
^{131}I	5000	30	50	10
^{133}Xe	~1	~1	~1	~1
^{137}Cs	15	20	10	10
^{140}Ba-^{140}La	25	—	—	8
^{141,144}Ce	700	20	400	3
^{185,187}W	5	2	20	3
^{203,210}Pb	700	—	200	—
^{210}Po	1000	—	—	—
^{226}Ra	1000	100	1000	130
^{239}Pu	1300	3	200	5

skeletons, molted by crustacea at the surface, might carry radionuclides such as ^{65}Zn downward (Fowler and Small, 1967), especially since ^{65}Zn is intimately bound in the various layers of the exoskeleton (Cross et al., 1968). The continuous settling of cast exoskeletons and fecal pellets thus may slowly but uninterruptedly pump radionuclides from surface layers down through the halocline where vertical mixing by physical processes is normally impeded.

For man the most significant role of marine and estuarine organisms in the redistribution of radioactive materials results from our exploitation of fishery resources. At the Windscale Plant, release of radioactive materials is limited by the concentration of radionuclides (mainly ^{106}Ru) in the edible seaweed, *Porphyra umbilicalis*, growing on intertidal rocks south of the point of discharge (Howells, 1966). The maximum possible radiation dose estimated for people living in the vicinity of the Hanford Plant would be derived mainly from consumption of fish containing ^{32}P and ^{65}Zn (Foster and Soldat, 1966; Foster and Honstead, 1967; Honstead and Brady, 1967). Accumulation of ^{65}Zn by prolonged consumption of oysters from the mouth of the Columbia River was much less significant than the accumulation of ^{32}P from freshwater fish (Honstead and Hildebrandt, 1967). Between March and December 1956, almost 500 tons of fish caught by the Japanese tuna fleet had to be destroyed because of the content of radionuclides accumulated after the nuclear tests on Bikini Atoll in March, 1956 (Matsuda and Hayashi, 1959). Though not exactly a marine form, the edible land crabs on Bikini Atoll contained such high levels of ^{90}Sr more than 10 yr after the last nuclear explosion there that consumption of the crabs had to be restricted for the people who returned to live on the islands (U.S. Atomic Energy Commission, 1968c). Considerations for the routine protection of humans from radioactivity contained in seafood products are discussed in a later section.

Man's major point of contact with the oceans has been and will probably continue to be in nearshore areas, especially estuaries. About 29 per cent of the United States population lives within 50 miles of the coastline, and nuclear power plants will be constructed on or near the coast to meet the expanding power needs. Coastal areas will receive radioactive effluents directly from discharge pipes and also from rivers. Thus much waste radioactivity will never reach the open ocean but will accumulate instead in the sediments of estuaries, where flocculation and precipitation occur continually. Estuaries are nursery grounds for the larval and immature forms of many commercially important marine organisms, several of which are harvested directly from estuaries (e.g., oysters, clams, shrimp, blue crabs, and flounders). About 66 per cent of the total catch of commercial marine species comes from nearshore waters less than 25 fathoms deep. It is in estuaries, therefore, that the impact of man's nuclear technology is most likely to be felt.

It is in shallow coastal waters and estuaries also that the biota probably exert the greatest influence on the distribution of radionuclides. The concentrations of biomass for phytoplankton and zooplankton are much higher there than in the open ocean (Table 12.9), and the benthic fauna

Table 12.9. Phytoplankton and Zooplankton in Estuaries and Surface Waters in the Ocean

	Mean Standing Crop (mg C/m^3)	
	Phytoplankton	Zooplankton
Estuaries, North Carolina[a]	187	9.66
Open ocean, tropical eastern Pacific[b]	5.9	0.29

[a] Williams et al., (1968), Williams (1966), and Williams (personal communication).
[b] Blackburn (1966). Standing crop of phytoplankton determined from mean value of 9.8 mg/m^2 chlorophyll a over 100-m depth and the relation mg C = 60 × mg chlorophyll a; that for zooplankton determined from mean value of 8.26 mg/m^3 total micronekton, assuming a carbon content of 3.5 per cent.

constitutes a more significant part of the total biomass in the shallower water. In addition, many epibenthic and pelagic species, including shrimp, crabs, pinfish, flounders, and menhaden, migrate seasonally to and from estuarine areas, and a significant amount of biomass is removed annually by commercial fishing. Estimates of the relative importance of these biota in determining the distribution of radionuclides in estuaries are not currently available.

Estuarine organisms, especially bivalve molluscs, can be particularly sensitive indicators of radioactivity in the environment (Schelske et al., 1965, 1966, 1967). The fallout radionuclides, ^{65}Zn, ^{54}Mn, ^{137}Cs, ^{106}Ru-^{106}Rh, and ^{144}Ce, decreased somewhat in estuarine clams from North Carolina during 1965–67. Sudden influxes of short-lived nuclides (^{140}Ba-^{140}La, ^{95}Zr-^{95}Nb, ^{103}Ru-^{103}Rh, and ^{141}Ce) occurred in the clams, however, in less than 8 to 10 days after two atmospheric nuclear tests were conducted in China. Isotopes of ruthenium were about 10 times more concentrated in clams collected from salinities of 6 to 15‰ than in those from 0 to 3‰, perhaps because ruthenium becomes particulate at higher salinities and the clams can remove it from the water more effectively (Wolfe and Schelske, 1969). Cesium 137 was concentrated to higher levels by clams in relatively fresh water, and the ^{137}Cs content underwent seasonal fluctuations which corresponded directly to the seasonal variation of fallout deposition and inversely to that of salinity in the estuary (Wolfe, 1967; Wolfe and Schelske, 1969). The possible mechanisms for accumulation of radionuclides by bivalves, as reviewed by Brooks and Rumsby (1965) include (*a*) ingestion of suspended particulate material from seawater, (*b*) ingestion of food organisms, (*c*) complexing of metals by coordinate linkages with organic molecules or adsorption onto membranous surfaces, and (*d*) incorporation of ions into physiologically important systems. In oysters, which are well-known for their capacity to accumulate zinc (Chipman et al., 1958), zinc is required for the enzymatic activity of alkaline phosphatase, but about 95 per cent of the total zinc in oyster homogenates can be removed readily by dialysis without affecting the activity of alkaline phosphatase (Wolfe, 1970). Thus although zinc may be involved in several metabolic functions in oysters, much of the zinc is probably accumulated incidentally during feeding.

Effects of Waste Radioactivity in the Oceans

Biological effects of the increased radiation resulting from radioactive materials discarded into the oceans might be manifested as somatic or genetic changes in the organisms irradiated. People living and working in coastal areas will also be subjected to increased radiation from external exposure to radioactive seawater and from internal exposure from eating seafood organisms. It is generally assumed that the protection of man from excessive radiation from these sources will simultaneously protect communities of marine organisms. It must be emphasized, however, that the genetic effects of chronic exposure to low levels of radiation are not well understood either for marine organisms or for humans. Organisms in the sea have been exposed to less radiation during their evolution than have terrestrial organisms because of the shielding effects of water; marine populations may therefore be more susceptible to radiation-induced genetic mutation. In the following sections we discuss our present knowledge of the effects of radiation on marine species.

Environmental Observations

The effects of radiation on marine organisms have been very difficult to evaluate in the natural environment, for example, near sites of nuclear bomb tests. Dead fish were observed after the detonation of nuclear devices at Bikini-Eniwetok, but it was believed that they died from the shock of the blast (Donaldson, 1963). Some fish in the immediate vicinity of the detonation of a large nuclear device, however, can be expected to receive lethal doses of radiation. The dose of radiation received by fish in the vicinity of a nuclear explosion would of course depend on their distance from the explosion and the depth at which they were swimming. The gradient of radiation would be expected to extend the deaths over a long time and over a considerable area of the ocean. During the period between initial exposure to radiation and death, fish could migrate various distances. Although dead fish have not been seen after a nuclear test (except near the bomb site), most deaths could have been unobserved, especially since fish weakened by excessive radiation might be devoured by predators or succumb to other environmental stresses, seemingly unrelated to radioactivity in the ocean.

One instance of adverse effects on organisms from the accumulation of radioactive isotopes by marine organisms occurred after the detonation of a nuclear bomb in the Marshall Islands. In 1958, Gorbman and James (1963) found that herbivorous fish concentrated radioactive iodine from seaweeds which had previously accumulated this radioisotope from seawater. Carnivorous fish then further concentrated the iodine to such high levels that their thyroid glands were destroyed. Although radioactive iodine is accumulated by marine organisms, it is usually not of serious concern except in areas receiving close-in fallout. Since the radioactive decay rate of ^{131}I is relatively rapid (half-life, 8 days) and since ^{131}I is diluted with stable iodine in seawater, marine organisms seldom would accumulate enough of it to be detectably injured.

Somatic Effects of Radiation on Marine Organisms

Comparative Sensitivities. Although more primitive organisms are usually more resistant to radiation than the more complex animals (Fig. 12.4), variation among related species is considerable. For example, when snails of two genera were exposed to 10,000 roentgens (R), 50 per cent of the *Radix* died after about 1 month, whereas *Thais* lived for about 6 months (Bonham and Palumbo, 1951). Irradiation with about 93,000 R killed 50 per cent of a group of oysters, *Crassostrea virginica*, in 34 days and 50 per cent of a similar group of clams, *Mercenaria mercenaria*, in 38.5 days (Price, 1965). Although this similarity in survival time would appear to indicate a similar response to radiation for the two species, calculation of LD-50's earlier or later than 30 days after irradiation shows that the tolerances of the two species are not similar (Fig. 12.5). Oysters have an LD-50 almost twice that of clams 20 days after irradiation and

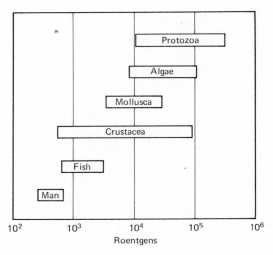

Figure 12.4 Lethal doses of radiation for different phylogenetic groups. Sensitivity to radiation generally increases with increasing complexity of organisms. After Donaldson (1963).

only one-fourth that of clams 40 days after irradiation (White and Angelovic, 1966).

Different stages in the life cycle of a species also differ in sensitivity to radiation (Table 12.10). The sensitivity of rainbow trout, *Salmo gairdneri*, decreases with increasing age (Welander, 1954; Welander et al., 1949),

Table 12.10. Amounts of Radiation Required to Kill 50 Per Cent of the Rainbow Trout, *Salmo gairdneri*, **Irradiated at Various States in Their Life Cycle**

Stage in Life Cycle	LD-50 (R)
Gametes	50–100
1 cell	58
32 cell	313
Germ ring	454–461
Eye	415–904
Adult	1500

From Welander (1954).

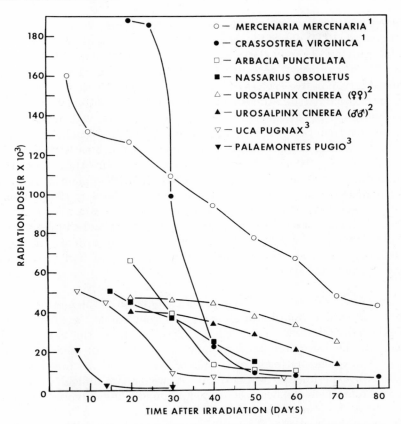

Figure 12.5 Radiation sensitivities of several marine invertebrates, showing the mean dose-time combinations at which 50 per cent of the experimental animals died. From White and Angelovic (1966).

and this relation appears to be the general rule for other species. Gametes and eggs through the one-cell stage are very sensitive to radiation; for example, during mitosis of the single cell of coho salmon, *Oncorhynchus kisutch*, the LD-50 is only 16 R (Donaldson and Foster, 1957).

The dormant eggs of aquatic invertebrates are especially resistant to radiation. Eggs of the brine shrimp, *Artemia salina*, that have been placed in water so that embryonic development is resumed are more than twice as sensitive to radiation as the dry dormant eggs (Bonham and Palumbo, 1951). Experiments on the effects of radiation from ^{90}Sr-^{90}Y on the eggs of the plaice have given inconsistent results; Fedorov et al. (1964) reported very high sensitivity to low concentrations, whereas Brown and Templeton (1964) detected no effect at concentrations as high as 10^{-4} Ci/l. Polikarpov and Ivanov (1961) found that a concentration of 10^{-10} Ci/l of ^{90}Sr-^{90}Y was sufficient to cause a significantly increased incidence of abnormal larvae in five species of marine fish. Concentrations of radioactivity in the sea have sometimes reached or exceeded this level. Strontium–90 has been reported at concentrations of 10^{-12} and 10^{-11} Ci/l in the Pacific Ocean and Irish Sea (Miyake and Saruhashi, 1960; Mauchline, 1963). In 1959 the waters of the Irish Sea contained as much as 10^{-11} Ci/l ^{137}Cs, 10^{-10} Ci/l ^{144}Ce, 10^{-10} Ci/l ^{95}Zr, and 10^{-9} Ci/l ^{106}Ru (Mauchline, 1963). Irradiation of rainbow trout eggs retarded the development of the young and increased the number of abnormalities (Welander, 1954). Somatic damage was proportional to the amount of radiation received, and the greatest damage was in tissues that were growing and dividing rapidly (Welander et al., 1948). Also, the growth of young rainbow trout from irradiated parent stock was considerably slower than that of the young from unirradiated parents (Foster et al., 1949).

Population Effects. Radioactive contamination of the marine environment will subject marine organisms to a continuous elevated exposure to external radiation. Thus all life stages, including the most sensitive ones, will be irradiated, and populations may be affected by somatic changes in immature forms or by genetic changes in the breeding stock. Results of experimental work to date, however, are difficult to relate to these potential effects, because most experiments have involved relatively short-term doses much higher than will occur in the environment. Any increase in the level of background radiation might produce slow, gradual changes in the population structure of marine ecosystems and in the genetic makeup of the species involved. Such changes may be undetected and it probably will be difficult or impossible to establish a relation between increased radiation and observed changes in natural populations.

Radiation should be considered a normal ecological factor which may affect a species' ability to tolerate stresses imposed by other ecological variables, such as temperature and salinity. Thus the estuarine fish, *Fundulus heteroclitus*, had an LD-50 (50 days) of nearly 2000 rads at temperature-salinity combinations of 12 C, 15‰ and 22 C, 5‰; 920 rads at 22 C, 25‰; and only 300–350 rads at 27 C, 5–25‰ (Angelovic et al., 1969).

Since the eggs are frequently the most sensitive life stage (Table 12.10), Zaystev and Polikarpov (1964) calculated the effect on certain fish populations of the annual removal of a constant proportion of the fertilized eggs. If fertility and post-hatch survival remain constant and more than

Table 12.11. Time (Years) Required to Reduce Populations of Fish 50 Per Cent, on the Basis of the Percentage of Eggs Destroyed Each Year

Eggs Destroyed (per cent)	Species		
	Mullet	Horse Mackerel	Anchovy
5	55	68	100
10	29	34	50
20	15	18	25
30	11	13	17
40	8	10	13
50	7	8	10

From Zaystev and Polikarpov (1964).

10 per cent of the eggs are killed each year, the population will be depleted noticeably within a few years (Table 12.11).

Chronic exposure to low-level radiation caused decreases in experimental populations of *Daphnia* (Marshall, 1962, 1966) and *Artemia* (Grosch, 1962). Relative to control populations of *Artemia*, those cultured in the presence of radiophosphorus had a shorter life span and deposited fewer zygotes per brood, but deposited a higher proportion of developed zygotes as viable cysts (Grosch, 1966). In *Daphnia*, however, external ^{60}Co radiation caused a slight increase in the mean brood size accompanied by decreased individual fertility and survival time (Marshall, 1966).

Genetic Effects

Although definite evidence of radiation-induced genetic changes in marine organisms is presently lacking, the potential effects of heritable recessive mutations in human populations can be estimated from experience with several offspring generations of irradiated *Drosophila* and mice. Such experiments suggest that for chronic low-level exposure, a total dose of 100 rems or more would be required to double the natural mutation rate in man (Little, 1966). As will be seen later, background radiation gives a dose of about 5 rems per generation; the doubling dose rate therefore would be about 20 times our present background. If humans are exposed to higher levels of radiation in a single acute dose instead of in a chronic dose, the doubling dose rate might be only 30 rems per generation. Since geneticists generally agree that doubling the natural mutation rate would have deleterious effects on human populations (Alexander, 1965), it is important that environmental contamination be restricted to levels well below the amount giving dose rates of 30 to 100 rems/generation. Individual genes in marine organisms are probably much less sensitive to radiation than those in humans, and higher dose rates are likely to be genetically safe for aquatic populations. It is generally assumed that protection of man from excessive radiation will also protect other natural populations in the environment.

Control of Environmental Contamination

Modes of Exposure

Human radiation exposure from disposed waste in the aquatic environment is from both external and internal sources. External exposure arises from wading or swimming in contaminated water, from handling fishing gear on contaminated water, from surf-fishing, just relaxing on beaches sprayed or occasionally covered with contaminated water, and so forth. External exposure from radioactivity in the open ocean could be significant only for people who spent very large amounts of time at sea, such as sailors and fishermen. Doses received by swimmers submerged in the water, water skiers and boaters on the water, and fishermen at the edge of the water should also be considered. Fishermen on the bank of the Columbia River receive most of their external dose from ^{65}Zn accumulated by algae growing at the river's edge; an external dose of 15 mR was estimated for fishermen who spent 500 hr on the river bank during 1965. Submersion in the Columbia River for 240 hr could deliver a dose of 20 mR (Foster and Soldat, 1966). When nuclear power plants are placed near regions of heavy sedimentation, such as estuaries, the external dose to fishermen from radionuclides on the silt may present the most restrictive limitation on the rate of discharge of radioactive wastes, as it does for certain English power stations (Preston, 1966b).

For the general population, the external dose received in this manner will probably be less than the internal dose obtained from eating radioactive seafood and from drinking radioactive water. Additional internal exposure will arise from food made radioactive from fertilizers or irrigation water. Disposal of radioactive wastes must be restricted and regulated to protect humans from excessive radiation doses from the above sources. How do we transform an acceptable dose level into specific limitations on radioactive contaminants released in any particular situation? The basic philosophy has been to restrict the concentrations of total radioactivity and of individual known radioisotopes in the environment, and moer especially, in drinking water (International Commission on Radiological Protection, 1959; National Committee on Radiation Protection, 1959).

Maximum Permissible Concentrations

For any given radioisotope, the level which must not be exceeded to avoid potentially harmful doses is called the Maximum Permissible Concentration (MPC). The International Commission on Radiological Protection (ICRP) and the National Committee for Radiation Protection (NCRP) have recommended MPC's for about 240 radionuclides for occupational exposure from either air or water during a 168-hr week. The recommended MPC for any single isotope must be diminished whenever exposure is from a known mixture of radioisotopes, from air and water simultaneously, or from some external source of radiation. The MPC for any completely unknown mixture of radionuclides is set at 10^{-7} μCi/cm^3 for continuous occupational exposure.

The calculation of MPC's depends entirely on biological data—data on elemental distributions and concentrations in various human organs, data on the average size of human organs and of humans themselves, data on the efficiency of accumulation of an ingested radioisotope by different organs, and data on the rate of elimination of elements from the various organs. For many nuclides and organs, these data not only are lacking for humans, but are extremely scarce for animals. This required information has been derived from experiments wherever possible, or when data are completely unavailable, estimates are made by comparison with chemically similar elements, or by assuming steady-state conditions for an element in a critical organ. Elemental composition and distribution for a "standard man" were first stipulated during the 1940's, but these values have since undergone many revisions and additions. The composition of the standard man is set forth by the ICRP (1959). Nearly always, the critical body organ is that organ which accumulates the greatest concentration of the radioisotope under consideration. The body burden is the amount of the radionuclide in the total body that produces the maximum permissible dose to the critical organ. The discussion of MPC's, up to this point, has been limited mainly to radionuclides in drinking water. When we consider waste disposal in the marine environment, however, we must be concerned primarily with radionuclides in seafoods. Since certain elements are concentrated to very high degrees by marine organisms, we must reconsider MPC's in terms of human consumption of fishery products.

Evaluation of marine contamination by radioactivity is complicated by the diverse environmental and biological variables that affect the disposition of a radionuclide introduced into the environment. Since the greatest potential hazard to man is almost certainly from the consumption of seafood containing radioactivity, waste disposal into the oceans must be sufficiently controlled to prevent the accumulation of unsafe concentrations in edible organisms. Two divergent approaches have been proposed for effecting this control: the first is merely an extension of the widely accepted MPC concept, introduced briefly above; and the second is based on the premise that man cannot exceed his allowable body burden of any radioisotope so long as the specific activity of that isotope in the environment is maintained below the allowable specific activity in man.

Conventional Derivation of MPC's for Marine Environment. Permissible concentrations of radionuclides in the marine environment are usually calculated directly from the previously recommended MPC's in drinking water for continuous occupational exposure. This further derivation involves additional information for which supporting data may be unavailable, and certain assumptions must therefore be made. Although the limiting concentration of any radionuclide will be found in edible organisms, the approach generally followed has been to establish MPC's for radionuclides in seawater (National Academy of Sciences—National Research Council, 1959a, 1959b; Aten, 1961; Yamagata, 1965; Freke, 1967). Any industrial company can thus readily relate the levels of radioactivity in its effluent to the MPC.

The first step in computing an MPC for seawater is to define the relation of the MPC for drinking water to an MPC for seafood, as follows:

$$(MPC)_{sf} = \frac{(MPC)_w 2200}{100 I} \quad \mu Ci/g$$

where $(MPC)_w$ is the MPC in drinking water for continuous occupational exposure ($\mu Ci/cm^3$); $(MPC)_{sf}$ is the MPC in seafood ($\mu Ci/g$); I is the rate of ingestion of seafood (g/day); 2200 is the rate of intake of water assumed for calculation of $(MPC)_w$ (cm^3/day); and 100 is the factor for converting $(MPC)_w$ for continuous occupational exposure to that for exposure of general population.*

* This factor is set by the ICRP at 100 if the critical organ is gonads or total body, and at 30 for other radionuclides.

The $(MPC)_{sf}$ calculated by means of the above equation refers to the radioactivity in the food when it is consumed, not when it is caught. Applying this $(MPC)_{sf}$ to organisms while still in the environment would result in a safety factor for humans due to physical decay of the radioactivity during the interim between capture and consumption. This factor is of little consequence for very long-lived nuclides, such as ^{137}Cs and ^{90}Sr, but could permit elimination by decay of such short-lived isotopes as ^{131}I. The $(MPC)_{sf}$ obviously refers also to the average concentration of radioactivity in any seafood eaten. Since many species of fish move around considerably during their life spans, the variation of radioactivity among individuals must be high. Individual fish containing high levels of radioactivity will be diluted (in the natural population, and during commercial processing and marketing) by fish from other geographical areas that contain less radioactivity. "Market dilution" can be accounted for by permitting organisms in the immediate area of waste disposal to contain more radioactivity than is allowable for the average seafood consumed. A factor of 10 for this dilution has been suggested (Dunster, 1958).

Maximum permissible concentrations for seawater are derived directly from $(MPC)_{sf}$ by dividing by the ratio of the elemental concentration in the seafood to that in the water (concentration factor).

$$(MPC)_{sw} = \frac{(MPC)_{sf}}{C_f}$$

It is evident that the value of $(MPC)_{sw}$ for any particular radioisotope depends entirely on which organism is considered in the calculation. Widely different values for C_f have been reported for the same element or isotope among different groups of edible marine organisms. Since metabolic uptake and physical adsorption are both affected by physical conditions such as temperature, pH, and salinity, concentration factors for a given isotope in any single species will also be subject to wide variation. Concentration factors for different elements in aquatic organisms have recently been compiled from the literature by Polikarpov (1966a). Since varying patterns of consumption of any given seafood species also affect $(MPC)_{sf}$, the $(MPC)_{sw}$ is a particularly vulnerable standard. Freke (1967), in his derivation of $(MPC)_{sw}$, selected a single reasonable (or accepted) published value of C_f for each of a long series of radionuclides in four groups of seafood organisms—fish, crustacea, molluscs, and seaweed. A similar, though less restrictive approach (National Academy of Sciences—National Research Council, 1959a), discriminated only between vertebrates and invertebrates. The group of organisms exhibiting the greatest C_f was considered limiting, and the $(MPC)_{sw}$ was computed on that basis.

The foregoing estimation of $(MPC)_{sw}$ is fraught with uncertainty. The factors I and C_f are subject to wide variability, and uptake of an isotope from seafood is not equatable to uptake from drinking water. Perhaps the greatest uncertainty, however, arises from the more or less arbitrary selection of a single species (necessarily one for which information is available) as the basis of the entire calculation. No single species is consumed at a uniform rate by the entire population. Using an "average" value of C_f for a large group of similar organisms, for example, molluscs, is also unacceptable, because certain species may exhibit much higher concentration factors than others, and may also constitute far different proportions of the diets in different localities. Many of these drawbacks are eliminated from consideration in the application of the specific activity approach.

Specific Activity Approach to MPC's for Marine Environment. The use of specific activity for establishing permissible levels of environmental radioactivity was first recommended for the control of radioactive waste disposal into United States Pacific coastal waters (National Academy of Sciences—National Research Council, 1962). Specific activity of any radioisotope is defined as the ratio of radioactivity from that isotope to the total amount of the element present (μCi/g, or equivalent). The application of this approach is based on only one major assumption: that a radioisotope introduced into the environment readily equilibrates with the stable isotope(s) of the same element, so that biological concentrating mechanisms will be unable to discriminate between the radioactive and stable forms of the element. The distribution of radionuclides will then correspond to the distribution of stable elements in the environment. The maximum permissible specific activity for any radioisotope is readily available from data on the standard man:

$$(MPSA) = \frac{q}{m \times C}$$

where (MPSA) is the maximum permissible specific activity (μCi/g); q is the maximum permissible burden of radioisotope in critical organ (μCi); m is the mass of the critical organ (g); and C is the concentration of stable element in critical organ (g/g). The (MPSA) is then converted to $(MPC)_{sw}$—or, if desired, $(MPC)_{sf}$—simply by multiplying by the concentration of stable element in seawater (or in seafood):

$$(MPC)_{sw} = K \times (MPSA)$$

where K is the grams of stable element per cm³ seawater. The $(MPC)_{sw}$ so derived is independent of concentration factors and of rate of consumption of seafoods. If the MPC is set for seawater instead of for the marketed seafood, physical decay of the radioisotope will introduce a safety factor. Radioactivity that is accumulated by primary producers and then passed up a food chain to man will undergo physical decay during its residence in each trophic level. This process will be partially offset by continuous bioaccumulation from the water surrounding the organism, but for most elements the biological turnover probably is slow enough that physical decay will reduce the specific activity of a radioisotope before the organism is consumed by an organism of the next higher trophic level.

The specific activity approach, as calculated in the above equations, is not applicable for any situation where the gastrointestinal tract is a critical organ. For radioisotopes that are not assimilated efficiently, the primary concern must be for the gastrointestinal tract, and the $(MPC)_{sw}$ must be derived from the allowable radioactivity estimated for seafood organisms.

Certain instances of bioaccumulation have led investigators to question the premise that radioisotopes behave in the natural environment just as does the stable isotope of the element. For example, Schelske et al., (1966) postulated that bay scallops, *Aequipecten irradians* Lamarck, accumulated ^{54}Mn from particulate fallout in seawater because the specific activities of scallop tissues were higher than that of the water.

The results of calculating $(MPC)_{sw}$ for several isotopes by the specific activity and the conventional approach are compared in Table 12.12 (Isaacs, 1964).

The specific activity approach has special merit for those radioisotopes which, because of long physical half-life or of continuous introduction into the environment, are chronic contaminants. Such radioisotopes will have had ample time to equilibrate with the stable isotopes already present in the environment, so that after the initial period of introduction, most of the radioisotope will be distributed in the same way as a stable carrier. This distribution probably holds for ^{137}Cs and ^{90}Sr from fallout in the biosphere, although neither isotope has yet equilibrated with stable isotopes in the deeper parts of the ocean (Broecker et al., 1966a). In any instance of acute contamination, however, the radioisotopes may be introduced in a physical or chemical form different from that of the stable isotopes in the environment, and bioaccumulation of the radioactivity might be independent of the distribution of the stable element. At present, nearly all radioactivity introduced into the environment is comparable to

Table 12.12. Comparison of $(MPC)_{sw}$ Values Derived by Conventional and Specific Activity Approaches

(1) Isotope	(2) PSC Conventional Approach $(\mu Ci/ml)$[a]	(3) MPCC Specific Activity Approach $(\mu Ci/g)$[b]	(4) (2)/(3)
^{60}Co	5×10^{-9}	5.0×10^{-9}	1.0
^{59}Fe	3×10^{-9}	3.0×10^{-9}	1.0
^{64}Cu	6×10^{-9}	6.0×10^{-9}	1.0
^{65}Zn	2×10^{-8}	1.2×10^{-8}	1.67
^{137}Cs	4×10^{-6}	1.5×10^{-4}	0.0267
^{90}Sr	5×10^{-8}	3.4×10^{-5}	0.00147
^{131}I	2×10^{-9}	1.6×10^{-6}	0.00125
^{32}P	1×10^{-9}	2.8×10^{-9}	0.357
^{35}S	1.2×10^{-5}	8.4×10^{-3}	0.00143
^{45}Ca	4.5×10^{-6}	1.2×10^{-4}	0.0375

Modified from Isaacs (1964).

[a] Permissible seawater concentrations as calculated in *Natl. Acad. Sci.-Natl. Res. Council* Publ. 655, but revised according to *NRCP Handbook 69* (1959) values for $(MPC)_w$, and according to the more recent biological concentration factors and the appropriate reduction factors for dose to the general population as recommended by *Natl. Acad. Sci.-Natl. Res. Council* Publ. 985.

[b] Maximum permissible seawater concentrations recommended by *Natl. Acad. Sci.-Natl. Res. Council* Publ. 985.

acute contamination, because significant concentrations of most radioisotopes are not yet widespread, but are localized in the immediate areas of contamination. As radioactive contamination becomes widespread, that is, if portions of the environment should approach their "maximum permissible burdens" of radioactivity, the specific activity approach will become increasingly applicable. In the meantime, isolated "acute" radioactive contaminations might be controlled more effectively on the basis of the nature of the contaminating effluent and of the immediate environmental circumstances than on the basis of the distribution of stable elements in the receiving environment. The conventional approach necessarily involves many conservative approximations but nevertheless is based on sound principles of bioaccumulation, and will adequately protect the population from excessive radiation.

International Viewpoints on Waste Disposal. The number of installations that purposely discharge radioactive wastes directly into the ocean or estuaries or into river systems flowing into the ocean is now increasing. Wastes presently are released to aquatic environments by installations in the United States, Canada, Great Britain, France, Sweden, Denmark, Netherlands, U.S.S.R., India, and Australia (Parker, 1967). The recommendations of the International Commission on Radiological Protection (ICRP) (1959) are universally used as guides for regulation of waste disposal, but various interpretations of the recommendations have arisen and the levels of radioactivity permitted in effluent water vary somewhat from nation to nation. The source of this discrepancy is usually in the point of application of the standards recommended by the ICRP. The British impose a restriction on the dose actually received by people living near the nuclear installation, the United States imposes a restriction on the content of radioactivity in the effluent that leaves the restricted area around the installation, and the U.S.S.R. imposes a more rigid restriction on the amount of radioactivity released from inside the installation. The British policy thus permits the greatest release of radioactivity and requires more intensive site evaluations and environmental surveillance to ensure that the dose limitation is not exceeded (Preston, 1966b). The United States permits the release of liquid wastes containing 3 to 10 times the maximum concentration of radionuclides that can be legally released to the environment within the U.S.S.R. (Parker, 1965). The Soviet policy is generally to enforce the ICRP standards for the general population, which were recommended as limits for human consumption, at the point of release inside a nuclear facility. This policy does not consider the dilution capacity of rivers, estuaries, or the ocean, or the rate of physical decay relative to the release of radioactive wastes. The United States law, on the other hand, limits the releases to concentrations less than 10 per cent of the MPC values recommended by the ICRP for 168-hr occupational exposure (Code of Federal Regulations), and this limit is placed at the point where the wastes leave the restricted area. Thus United States law allows concentrations 10 times as high as those allowed by the U.S.S.R. for nuclides with a critical organ of either gonads or total body, and 3 times as high for other nuclides. Although the United States policy more nearly reflects the international standards set by the ICRP for the general population, it necessitates that radiation doses to individuals living in the vicinity of points of discharge be evaluated more frequently (Honstead, 1968).

References

Albenesius, E. L., and R. S. Ondrejcin. 1960. Nuclear fission produces tritium. *Nucleonics*, **18**:100.

Alexander, P. 1965. *Atomic radiation and life*, 2nd ed. Penguin Books, Baltimore. 396 pp.

Angelovic, J. W., J. C. White, Jr., and E. M. Davis. 1969. Interactions of ionizing radiation, salinity, and temperature on the estuarine fish, *Fundulus heteroclitus*. In *Symposium on radioecology*, D. J. Nelson and F. C. Evans (eds.), USAEC CONF-670503, Oak Ridge, Tenn., pp. 131–141.

Aten, A. H. W., Jr. 1961. Permissible concentrations of radionuclides in sea water. *Health Phys.*, **6**:114–125.

Baptist, J. P. 1966. Uptake of mixed fission products by marine fishes. *Trans. Amer. Fish. Soc.*, **95**:145–152.

Beninson, D., E. Vander Elst, and D. Cancio. 1966. Biological aspects in the disposal of fission products in surface waters. In *Disposal of radioactive wastes into seas, oceans and surface waters*, IAEA Proc. Ser., IAEA, Vienna, pp. 337–354.

Bernhard, M., and A. Zattera. 1969. A comparison between the uptake of radioactive and stable zinc by a marine unicellular alga. In *Symposium on radioecology*, D. J. Nelson and F. C. Evans (eds.), USAEC CONF-670503, Oak Ridge, Tenn., pp. 389–398.

Blackburn, M. 1966. Relationships between standing crops at three successive trophic levels in the eastern tropical Pacific. *Pacific Sci.*, **20**:36–59.

Blakely, J. P. 1970. Action on reactor and other projects undergoing regulatory review or consideration. *Nucl. Safety*, **12**(1):59–71.

Bonham, K., and R. F. Palumbo. 1951. Effects of X-rays on snails, crustacea, and algae. *Growth*, **15**:155–188.

Bowen, H. J. M. 1960. Biological fractionation of isotopes. *Int. J. Appl. Radiat. Isotop.*, **7**:261–272.

Bowen, V. T., and T. T. Sugihara. 1960. Strontium-90 in the "mixed layer" of the Atlantic Ocean. *Nature*, **186**:71–72.

Bowen, V. T., and T. T. Sugihara. 1963. Cycling and levels of strontium-90, cerium-144, and promethium-147 in the Atlantic Ocean. In *Radioecology*, V. Schultz and A. W. Klement (eds.), Reinhold, New York, pp. 135–139.

Bowen, V. T., V. E. Noshkin, and T. T. Sugihara. 1966. Transport of strontium-90 toward the equator at mid-depths in the Atlantic Ocean. *Nature*, **212**:383–384.

Bowen, V. T., V. E. Noshkin, H. L. Volchok, and T. T. Sugihara. 1968. Fallout strontium 90 in Atlantic Ocean surface waters, pp. I-2-65. In Health and Safety Laboratory Fallout Program Quarterly Summary Report July 1, 1968. USAEC HASL-197.

Broecker, W. S., E. R. Bonebakker, and G. G. Rocco. 1966a. The vertical distribution of cesium 137 and strontium 90 in the oceans. *J. Geophys. Res.*, **71**:1999–2003.

Broecker, W. S., G. G. Rocco, and H. L. Volchok. 1966b. Strontium 90 fallout: Comparison of rates over ocean and land. *Science*, **152**:639–640.

Brooks, R. R., and M. G. Rumsby. 1965. The biogeochemistry of trace element uptake by some New Zealand bivalves. *Limnol. Oceanogr.*, **10**:521–527.

Brown, J. M., J. F. Thompson, and H. L. Andrews. 1962. Survival of waste containers at ocean depths. *Health Phys.*, **7**:227–228.

Brown, R. M. 1961. Hydrology of tritium in the Ottawa Valley. *Geochim. Cosmochim. Acta*, **21**:199–216.

Brown, V. M., and W. L. Templeton. 1964. Resistance of fish embryos to chronic irradiation. *Nature*, **203**:1257–1259.

Bryan, G. W. 1961. The accumulation of radioactive caesium in crabs. *J. Mar. Biol. Ass. U.K.*, **41**:551–575.

Bryan, G. W. 1963a. The accumulation of radioactive caesium by marine invertebrates. *J. Mar. Biol. Ass. U.K.*, **43**:519–539.

Bryan, G. W. 1963b. The accumulation of ^{137}Cs by brackish water invertebrates and its relation to the regulation of potassium and sodium. *J. Mar. Biol. Ass. U.K.*, **43**:541–565.

Bryan, G. W. 1965. Ionic regulation in the squat lobster, *Galathea squamifera*, with special reference to the relationship between potassium metabolism and the accumulation of radioactive caesium. *J. Mar. Biol. Ass. U.K.*, **45**:97–113.

Bryan, G. W., and E. Ward. 1962. Potassium metabolism and the accumulation of ^{137}caesium by decapod crustacea. *J. Mar. Biol. Ass. U.K.*, **42**:199–241.

Burton, J. D. 1965. Radioactive nuclides in sea water, marine sediments and marine organisms. In *Chemical oceanography*, J. P. Riley and G. Skirrow (eds.), Academic Press, New York, Vol. II, pp. 425–475.

Burton, W. M., and N. G. Stewart. 1960. Use of long-lived natural radioactivity as an atmospheric tracer. *Nature*, **186**:584–589.

Carpenter, J. H., and V. E. Grant. 1967. Concentration and state of cerium in coastal waters. *J. Mar. Res.*, **25**:228–238.

Cerrai, E., B. Schreiber, and C. Triulzi. 1967. Vertical distribution of Sr^{90}, Ce^{144}, Pm^{147}, and Eu^{155} in coastal marine sediments. *Energia Nucleare*, **14**:586–592.

Cheng, H., and H. Hamaguchi. 1968. Studies on the adsorption of radioisotopes on marine sediments. 1. Role of exchangeable cations on the adsorption of cesium and exchange equilibria. *Health Phys.*, **14**:353–363.

Chipman, W. A. 1966. Food chains in the sea. In *Radioactivity and human diet*, R. S. Russell (ed.), Pergamon Press, Oxford, pp. 421–453.

Chipman, W. A., T. R. Rice, and T. J. Price. 1958. Uptake and accumulation of radioactive zinc by marine plankton, fish and shellfish. *U.S. Fish Wildl. Serv., Fish. Bull.*, **58**:279–292.

Code of Federal Regulations, Title 10, Chap. 1, Part 20. Standards for protection against radiation (USGPO).

Corcoran, E. F., and J. F. Kimball, Jr. 1963. The uptake, accumulation and exchange of strontium-90 by open sea phytoplankton. In *Radioecology*, V. Schultz and A. W. Klement (eds.), Reinhold, New York, pp. 187–191.

Corliss, W. R. 1967. Nuclear propulsion for space. USAEC Div. Tech. Inform., Oak Ridge, Tenn., May 1967.

Cross, F. A., S. W. Fowler, J. M. Dean, L. F. Small, and C. L. Osterberg. 1968. Distribution of ^{65}Zn in tissues of two marine crustaceans determined by autoradiography. *J. Fish. Res. Bd. Can.*, **25**:2461–2466.

Curl, H., Jr., N. Cutshall, and C. Osterberg. 1965. Uptake of chromium (III) by particles in sea-water. *Nature*, **205**:275–276.

Cutshall, N., and C. Osterberg. 1964. Radioactive particle in sediment from the Columbia River. *Science*, **144**:536–537.

Davis, J. J. 1963. Cesium and its relationships to potassium in ecology. In *Radioecology*, V. Schultz and A. W. Klement (eds.), Reinhold, New York, pp. 537–556.

Donaldson, L. R. 1963. Evaluation of radioactivity in the marine environment of the Pacific Proving Ground. In *Nuclear detonations and marine radioactivity*, S. H. Small (ed.), Kjeller, Norway, pp. 73–83.

Donaldson, L. R., and R. F. Foster. 1957. Effects of radiation on aquatic organisms. In *The effects of atomic radiation on oceanography and fisheries. Natl. Acad. Sci.—Natl. Res. Council* Publ. **551**. Washington, D.C., pp. 96–192.

Duke, T. W., J. N. Willis, and T. J. Price. 1966. Cycling of trace elements in the estuarine environment. I. Movement and distribution of zinc-65 and stable zinc in experimental ponds. *Chesapeake Sci.*, **7**:1–10.

Duke, T. W., J. N. Willis, and D. A. Wolfe. 1968. A technique for studying the exchange of trace elements between estuarine sediments and water. *Limnol. Oceanogr.*, **13**:541–545.

Dunster, H. J. 1958. The disposal of radioactive liquid wastes into coastal waters. *Proc. 2nd U.N. Int. Conf. Peaceful Uses At. Energy*, Geneva, **18**:390–399.

Edvarson, L., K. Low, and J. Sisefsky. 1959. Fractionation phenomena in nuclear weapons debris. *Nature*, **184**:1771–1774.

Eisenbud, M. 1968. Report on environmental contamination by radioactive substances by the environmental radiation exposure advisory committee (Eisenbud, Chairman). USDHEW, PHS. December 1, 1967. iii + 24 pp. (USGPO).

Evans, A. G., W. L. Marter, and W. C. Reinig. 1968. Guides limiting the release of radionuclides by the Savannah River Plant. *Health Phys.*, **15**:57–65.

Fedorov, A. F., V. N. Podymakhin, V. P. Kilezhenko, N. I. Buyanov, and E. M. Goloskova. 1964. Radiation conditions in the fishery regions of the North Atlantic (June–August 1961). In *Radiochemical and ecological studies of the sea*. U.S. Dept. Commerce, JPRS: 25, 966. Washington, D.C., pp. 42–52.

Feely, H. W., and J. Spar. 1960. Tungsten-185 from nuclear bomb tests as a tracer for stratospheric meteorology. *Nature*, **188**:1062–1064.

Folsom, T. R., and C. Sreekumaran. 1966. Survey of downward penetration of fallout in the ocean by in situ absorption. Scripps Institute of Oceanography Ref. Ser. 67-21.

Foster, R. F., and J. F. Honstead. 1967. Accumulation of zinc-65 from prolonged consumption of Columbia River fish. *Health Phys.*, **13**:39–43.

Foster, R. F., and J. K. Soldat. 1966. Evaluation of the exposure resulting from the disposal of radioactive wastes into the Columbia River. In *Disposal of radioactive wastes into seas, oceans and surface waters*. IAEA Proc. Ser., IAEA, Vienna, pp. 683–696.

Foster, R. F., L. R. Donaldson, A. D. Welander, K. Bonham, and A. H. Seymour. 1949. The effect on embryos and young of rainbow trout from exposing the parent fish to X-rays. *Growth*, **13**:119–142.

Fowler, S. W., and L. F. Small. 1967. Moulting of *Euphausia pacifica* as a possible mechanism for vertical transport of zinc-65 in the sea. *Int. J. Oceanol. Limnol.*, **1**:237–245.

Fowler, T. K., and R. F. Post. 1966. Progress toward fusion power. *Sci. Amer.*, **215**:21–31.

Freiling, E. C., and N. E. Ballou. 1962. Nature of nuclear debris in sea-water. *Nature*, **195**:1283–1287.

Freke, A. M. 1967. A model for the approximate calculation of safe rates of discharge of radioactive wastes into marine environments.. *Health Phys.*, **13**:743–758.

Fukai, R. 1965. Chemical composition of shallow-water sediments in the Bay of Roquebrune. *Bull. Inst. Océanogr. Monaco*, **65**:1–15.

Goldberg, E. D. 1963. The oceans as a chemical system. In *The sea*, M. N. Hill (ed.), Interscience Publishers, New York, Vol. II, pp. 3–25.

Goldberg, E. D., M. Koide, R. A. Schmitt, and R. H. Smith. 1963. Rare-earth distributions in the marine environment. *J. Geophys. Res.*, **68**:4209–4217.

Gorbman, A., and M. S. James. 1963. An exploratory study of radiation damage in the thyroids of coral reef fishes from the Eniwetok Atoll. In *Radioecology*, V. Schultz and A. W. Klement (eds.), Reinhold, New York, pp. 385–399.

Greendale, A. E., and N. E. Ballou. 1954. Physical state of fission product elements following their vaporization in distilled water and seawater. U.S. Naval Radiological Defense Lab., Rept. 436.

Grosch, D. S. 1962. The survival of *Artemia* populations in radioactive seawater. *Biol. Bull.*, **123**:302–316.

Grosch, D. S. 1966. The reproductive capacity of *Artemia* subjected to successive contaminations with radiophosphorus. *Biol. Bull.*, **131**:261–271.

Gross, M. G., C. A. Barnes, and G. K. Riel. 1965. Radioactivity of the Columbia River effluent. *Science*, **149**:1088–1090.

Hagemann, F., J. Gray, Jr., L. Machta, and A. Turkevich. 1959. Stratospheric carbon-14, carbon dioxide, and tritium. *Science*, **130**:542–552.

Honstead, J. F. 1968. A survey of environmental dose evaluations. *Nucl. Safety*, **9**:383–393.

Honstead, J. F., and D. N. Brady. 1967. The uptake and retention of ^{32}P and ^{65}Zn from the consumption of Columbia River fish. *Health Phys.*, **13**:455–463.

Honstead, J. F., and P. W. Hildebrandt. 1967. Uptake and retention of zinc-65 from certain foods. *Health Phys.*, **13**:649–652.

Howells, H. 1966. Discharges of low-activity radioactive effluent from the Windscale Works into the Irish Sea. In *Disposal of radioactive wastes into seas, oceans and surface waters.* IAEA Proc. Ser., IAEA, Vienna, pp. 769–785.

International Commission on Radiological Protection. 1959. Recommendations of the International Commission on Radiological Protection (adopted September 9, 1958), Pergamon Press, New York. vi + 18 pp.

Isaacs, J. D. 1964. Discussion of R. T. P. Whipple. Considerations on the siting of outfalls for the sea disposal of radioactive effluent in tidal waters. *Adv. Water Pollut. Res.*, **3**:26–33.

Jefferies, D. F. 1968. Fission-product radionuclides in sediments from the North-East Irish Sea. *Helgoländer wiss. Meeresunters.*, **17**:280–290.

Jennings, C. D., and C. Osterberg. 1969. Sediment radioactivity in the Columbia River estuary. In *Symposium on radioecology*, D. J. Nelson and F. C. Evans (eds.), USAEC CONF-670503, Oak Ridge, Tenn., pp. 300–306.

Johnson, V., N. Cutshall, and C. Osterberg. 1967. Retention of ^{65}Zn by Columbia River sediment. *Water Resour. Res.*, **3**:99–102.

Junkins, R. L., E. C. Watson, I. C. Nelson, and R. C. Henle. 1960. Evaluation of radiological conditions in the vicinity of Hanford for 1959. HW-64371, UC-41, *Health and Safety* (TID-4500, 15th Ed.).

Klement, A. W., Jr. 1959. A review of potential radionuclides produced in weapons detonations. USAEC, Washington-1024. iii + 102 pp.

Kosourov, G. I. 1966. Atmospheric radioactivity over the Atlantic Ocean. In *Radioactive contamination of the sea*, V. I. Baranov and L. M. Khitrov (eds.), AEC-tr-6641, pp. 23–31.

Krey, P. W., M. T. Kleinman, and B. K. Krajewski. 1970. Sr^{90}, Zr^{95} and Pu^{238} stratospheric inventories 1967–1969, pp. I-39-69. In Health and Safety Laboratory Fallout Program Quarterly Summary Report July 1, 1970. USAEC HASL-227.

Kuenzler, E. J. 1969. Turnover and transport of cobalt by marine zooplankton. In *Symposium on radioecology*, D. J. Nelson and F. C. Evans (eds.), USAEC CONF-670503, Oak Ridge, Tenn., pp. 483–492.

Lewis, F. 1967. *One of our H-bombs is missing*. McGraw-Hill, New York. 270 pp.

Libby, W. F. 1956. Radioactive strontium fallout. *Proc. Natl. Acad. Sci. U.S.*, 42:365–390.

List, R. J., L. Machta, L. T. Alexander, J. S. Allen, M. W. Meyer, V. T. Valassis, and E. P. Hardy, Jr. 1965. Strontium-90 on the earth's surface. In *Radioactive fallout from nuclear weapons test*, A. W. Klement, Jr. (ed.), USAEC, Div. Tech. Inform., Oak Ridge, Tenn., pp. 359–368.

Little, J. B. 1966. Environmental hazards. Ionizing radiation. *New Engl. J. Med.*, 275:929–938.

Longley, H., and W. L. Templeton. 1965. Marine environmental monitoring in the vicinity of Windscale. In *Radiological monitoring of the environment*. Pergamon Press, New York, pp. 219–247.

Lowman, F. G. 1960. Marine biological investigations at the Eniwetok test site. In *Disposal of radioactive wastes*, vol. II, IAEA, Vienna, pp. 105–138.

Lowman, F. G. 1969. Radionuclides of interest in the specific activity approach. *Biosci.*, 19:993–999.

Lowman, F., T. R. Rice, and F. A. Richards. In press. Accumulation and redistribution of radionuclides by marine organisms. In *Radioactivity in the marine environments*, A. H. Seymour (ed.), NAS-NRC, Washington, D.C.

Lyerly, R. L., and W. Mitchell, III. 1967. Nuclear power plants, *Understanding the atom*. USAEC, Oak Ridge, Tenn.

Mamuro, T., K. Yoshikawa, T. Matsunami, A. Fujita, and T. Azuma. 1963. Fractionation phenomena in highly radioactive fall-out particles. *Nature*, 197:964–966.

Marshall, J. S. 1962. The effects of continuous gamma radiation on the intrinsic rate of natural increase of *Daphnia pulex*. *Ecology*, 43:598–607.

Marshall, J. S. 1966. Population dynamics of *Daphnia pulex* as modified by chronic radiation stress. *Ecology*, 47:561–571.

Martin, W. E. 1969. Bioenvironmental studies of the radiological-safety feasibility of nuclear excavation. *Biosci.* 19:135–137.

Matsuda, H., and K. Hayashi. 1959. Nuclear weapons and man. Foreign Lit. Press, Moscow. (translated into Russian from original Japanese version published 1956, Tokyo). Cited by Polikarpov (1966a), p. 6.

Mauchline, J. 1963. The biological and geographical distribution in the Irish Sea of radioactive effluents from Windscale Works 1959 to 1960. United Kingdom Atomic Energy Authority, Production Group, Windscale, Rep. AHSB(RP)-R-27. Cited by Polikarpov (1966a), pp. 7 and 8.

Mauchline, J., and W. L. Templeton. 1964. Artificial and natural radioisotopes in the marine environment. *Oceanogr. Mar. Biol. Annu. Rev.*, 2:229–279.

Menzel, R. G. 1960. Transport of strontium-90 in runoff. *Science*, **131**:499–500.

Miyake, Y. 1963. Artificial radioactivity in the sea. In *The sea*, M. N. Hill (ed.), Interscience Publishers, New York, Vol. II, pp. 78–87.

Miyake, Y., and K. Saruhashi. 1958. Distribution of man-made radioactivity in the North Pacific through summer 1955. *J. Mar. Res.*, **17**:383–389.

Miyake, Y., and K. Saruhashi. 1960. Vertical and horizontal mixing rates of radioactive material in the ocean. In *Disposal of radioactive wastes*, vol. II. IAEA, Vienna, pp. 167–173.

Miyake, Y., K. Saruhashi, Y. Katsuragi, and T. Kanazawa. 1961. Cesium-137 and strontium-90 in seawater. *J. Radiat. Res.*, **2**:25–28.

Miyake, Y., K. Saruhashi, Y. Katsuragi, and T. Kanazawa. 1962. Penetration of ^{90}Sr and ^{137}Cs in deep layers of the Pacific and vertical diffusion rate of deep water. *J. Radiat. Res.* **3**:141–147.

Moghissi, A. A., and C. R. Porter. 1968. Tritium in surface waters of the United States, 1966. *Radiol. Health Data Rept.*, **9**:337–339.

Müller, G. 1967. The HCl-soluble iron, manganese, and copper contents of recent Indian Ocean sediments off the eastern coast of Somalia. *Mineral. Deposita*, **2**:54–61.

National Academy of Sciences–National Research Council. 1959a. *Radioactive waste disposal into Atlantic and Gulf Coastal waters*. NAS-NRC Publ. 655. Washington, D.C., viii + 37 pp.

National Academy of Sciences–National Research Council. 1959b. *Considerations on the disposal of radioactive wastes from nuclear-powered ships into the marine environment*. NAS-NRC Publ. 658. Washington, D.C., xviii + 52 pp.

National Academy of Sciences–National Research Council. 1962. *Disposal of low-level radioactive waste into Pacific coastal waters*. NAS-NRC Publ. 985. Washington, D.C., xii + 87 pp.

National Committee on Radiation Protection. 1954. *Radioactive-waste disposal in the ocean*. Natl. Bur. Std. Handbook 58. Washington, D.C., iv + 31 pp.

National Committee on Radiation Protection. 1959. *Maximum permissible body burdens and maximum permissible concentrations of radionuclides in air and in water for occupational exposure*. Natl. Bur. Std. Handbook 69. Washington, D.C., viii + 95 pp.

Nelson, J. L., R. W. Perkins, and J. M. Nielson. 1964. *Progress in studies of radionuclides in Columbia River sediments*. A summary of Hanford achievements in this program under General Electric 1963–1964. USAEC Rept. HW-83614.

Nelson, J. L., R. W. Perkins, J. M. Nielsen, and W. L. Haushild. 1966. Reactions of radionuclides from the Hanford reactors with Columbia River sediments. In *Disposal of radioactive wastes into seas, oceans and surface waters*. IAEA Proc. Ser., IAEA, Vienna, pp. 139–161.

Ophel, I. L., and J. M. Judd. 1967. Skeletal distribution of strontium and calcium and strontium/calcium ratios in several species of fish. In *Strontium metabolism*, J. M. A. Lenihan, J. F. Loutit, and J. H. Martin (eds.), Academic Press, New York, pp. 103–109.

Osterberg, C. L. (ed.). 1965. Ecological studies of radioactivity in the Columbia River and adjacent Pacific Ocean. Progress Rep., FY 1964. AEC Contract AT(45-1)1750, Ref. 65–14.

Osterberg, C., A. G. Carey, and H. Curl. 1963. Acceleration of sinking rates of radionuclides in the ocean. *Nature*, **200**:1276–1277.

Osterberg, C. L., N. Cutshall, V. Johnson, J. Cronin, D. Jennings, and L. Frederick. 1966. Some non-biological aspects of Columbia River radioactivity. In *Disposal of radioactive wastes into seas, oceans and surface waters*. IAEA Proc. Ser., IAEA, Vienna, pp. 321–335.

Palmer, H. E., and T. M. Beasley. 1965. Iron-55 in humans and their foods *Science*, **149**:431–432.

Palmer, H. E., and T. M. Beasley. 1967. Iron-55 in man and the biosphere. *Health Phys.*, **13**:889–895.

Palmer, H. E., J. C. Langford, C. E. Jenkins, T. M. Beasley, and J. M. Aase. 1968. Levels of iron-55 in humans, animals, and food, 1964–1967. *Radiol. Health Data Rept.* (August 1968): 387–390.

Parker, F. L. 1965. United States and Soviet Union waste-disposal standards. *Nucl. Safety*, **6**:433–436.

Parker, F. L. 1967. Disposal of low-level radioactive wastes into the oceans. *Nucl. Safety*, **8**:376–382.

Parker, F. L. 1968. Radioactive wastes from fusion reactors. *Science*, **159**:83–84.

Parker, F. L., and D. J. Rose. 1968. Wastes from fusion reactors. *Science*, **159**:1376.

Pearcy, W. G., and C. L. Osterberg. 1964. Vertical distribution of radionuclides as measured in oceanic animals. *Nature*, **204**:440–441.

Perkins, R. W., J. L. Nelson, and W. L. Haushild. 1966. Behavior and transport of radionuclides in the Columbia River between Hanford and Vancouver, Washington. *Limnol. Oceanogr.*, **11**:235–248.

Polikarpov, G. G. 1966a. *Radioecology of aquatic organisms*. Reinhold, New York.

Polikarpov, G. G. 1966b. Regularities of uptake and accumulation of radionuclides in aquatic organisms. In *Radioecological concentration processes*, B. Åberg and F. P. Hungate (eds.), Pergamon Press, Oxford, pp. 819–825.

Polikarpov, G. G., and V. N. Ivanov. 1961. Injurious effect of strontium-90—yttrium-90 on early development of mullet, wrasse, horse mackerel, and anchovy. A translation of *Doklady Akad. Nauk SSSR*, **144**:491–494, issued by Consultant's Bureau, N.Y., 1962.

Preston, A. 1966a. Concentration of ^{65}Zn in the flesh of oysters related to the discharge of cooling pond effluent from the C.E.G.B. Nuclear Power Station at Bradwell-on-Sea, Essex. In *Radioecological concentration processes*, B. Åberg and F. P. Hungate (eds.), Pergamon Press, Oxford, pp. 995–1004.

Preston, A. 1966b. Site evaluations and the discharge of aqueous radioactive wastes from civil nuclear power stations in England and Wales. In *Disposal of radioactive wastes into seas, oceans and surface waters*. IAEA Proc. Ser., IAEA, Vienna, pp. 725–737.

Price, T. J. 1965. Accumulation of radionuclides and the effects of radiation on molluscs. In *Biological problems in water pollution*. Trans. 3rd Seminar. Publ. No. 999-WP-25, (Robert A. Taft Sanitary Engineering Center, Cincinnati, Ohio), pp. 202–210.

Revelle, R. R., and M. B. Schaffer. 1957. General considerations concerning the ocean as a receptacle for artificial radioactive materials. In *The effects of atomic radiation on oceanography and fisheries*. NAS-NRC, Publ. 551, Washington, D.C., pp. 1–25.

Rice, T. R., and V. M. Willis. 1959. Uptake, accumulation, and loss of radioactive cerium-144 by marine planktonic algae. *Limnol. Oceanogr.*, **4**:277–290.

Rocco, G. G., and W. S. Broecker. 1963. The vertical distribution of cesium-137 and strontium-90 in the oceans. *J. Geophys. Res.*, **68**:4501–4512.

Schelske, C. L., W. D. C. Smith, and J. Lewis. 1965. Radioactivity in the estuarine environment. *In* Annual report of the Bureau of Commercial Fisheries Radiobiological Laboratory, Beaufort, N.C., for the fiscal year ending June 30, 1964. *U.S. Fish Wildl. Serv.*, Circ. 217, pp. 8–13.

Schelske, C. L., W. D. C. Smith, and J. Lewis. 1966. Radio-assay. *In* Annual report of the Bureau of Commercial Fisheries Radiobiological Laboratory, Beaufort, N.C., for the fiscal year ending June 30, 1965. *U.S. Fish Wildl. Serv.*, Circ. 244, pp. 12–14.

Schelske, C. L., W. D. C. Smith, and J. Lewis. 1967. Radioactivity in the estuarine environment. *In* Annual report of the Bureau of Commercial Fisheries Radiobiological Laboratory, Beaufort, N.C., for the fiscal year ending June 30, 1966. *U.S. Fish Wildl. Serv.*, Circ. 270, pp. 9–12.

Schreiber, B. 1966. Radionuclides in marine plankton and in coastal sediments. In *Radioecological concentration processes*, B. Åberg and F. P. Hungate (eds.), Pergamon Press, Oxford, pp. 753–770.

Schutz, D. F., and K. K. Turekian. 1965. The investigation of the geographical and vertical distribution of several trace elements in sea water using neutron activation analysis. *Geochim. Cosmochim. Acta*, **29**:259–313.

Schweitzer, G. K. 1956. The radiocolloidal properties of the rare-earth elements, pp. 31–34. In G. C. Kyker and E. B. Anderson (eds.), *Rare earths in biochemical and medical research:* A conference sponsored by the Medical Division, Oak Ridge Institute of Nuclear Studies, October, 1955. ORINS-12.

Seymour, A. H. (ed.). *Radioactivity in the Marine Environment*, NAS-NRC, Washington, D.C. In press.

Sillén, L. G. 1961. The physical chemistry of sea water. In *Oceanography*, M. Sears (ed.), AAAS, Washington, D.C. Publ. 67, pp. 549–581.

Sillén, L. G. 1967. The ocean as a chemical system. *Science*, **156**:1189–1197.

Strutt, R. J. 1906. On the distribution of radium in the earth's crust. Part II.— Sedimentary rocks. *Proc. Roy. Soc. London*, **A78**:150–153.

Sugihara, T. T., and V. T. Bowen. 1962. Radioactive rare earths from fallout for study of particulate movement in the sea, pp. 57–65. In *Radioisotopes in the physical sciences and industry*. IAEA, Vienna.

Szulc, T. 1967. *The Bombs of Palomares*. Viking Press, New York.

Templeton, W. L. 1965. Ecological aspects of the disposal of radioactive wastes to the sea. In *Ecology and the industrial society*. G. T. Goodman, R. W. Edwards, and J. M. Lambert (eds.), John Wiley, New York, pp. 65–97.

Templeton, W. L., and V. M. Brown. 1963. Accumulation of calcium and strontium by brown trout from waters in the United Kingdom. *Nature*, **198**: 198–200.

Templeton, W. L., and V. M. Brown. 1964. The relationship between the concentrations of calcium, strontium and strontium-90 in wild brown trout, *Salmo trutta* L. and the concentrations of the stable elements in some waters of the United Kingdom, and the implications in radiological health studies. *Int. J. Air Water Pollut.*, **8**:49–75 (and errata **8**:609–610).

Ting, R. Y., and V. Roman deVega. 1969. The nature of the distribution of trace elements in longnose anchovy, Atlantic thread herring, and alga. In *Symposium on radioecology*, D. J. Nelson and F. C. Evans (eds.), USAEC CONF-670503, Oak Ridge, Tenn., pp. 527–534.

United Nations. 1962. Report of the U.N. Scientific Committee on the Effects of Atomic Radiation. General Assembly, Official Records, 17th Sess., Suppl. No. 16 (A/5216).

United Nations. 1966. Report of the U.N. Scientific Committee on the Effects of Atomic Radiation. General Assembly, Official Records, 21st Sess., Suppl. No. 14 (A/6314).

United Nations. 1969. Report of the U.N. Scientific Committee on the Effects of Atomic Radiation. General Assembly, Official Records, 24th Sess. Suppl. No. 13 (A/7613).

U.S. Atomic Energy Commission. 1968a. Forecast of growth of nuclear power, December 1967. Wash-1084. Reactors (TID-4500).

U.S. Atomic Energy Commission. 1968b. Nuclear reactors built, being built, or planned in the United States as of December 31, 1967. TID-8200 (17th Rev.).

U.S. Atomic Energy Commission. 1968c. Statement of Glenn T. Seaborg on the decision to allow former residents of Bikini to return home. News Release L-191, August 12, 1968.

U. S. Atomic Energy Commission. 1968d. Major activities in the atomic energy programs, January–December 1967, p. 50.

U.S. Atomic Energy Commission. 1968e. Fundamental nuclear energy research—1967. A supplemental report to the annual report to Congress for 1967 of the USAEC. pp. 292–293.

U.S. Atomic Energy Commission. 1970. Nuclear reactors built, being built, or planned in the United States as of June 30, 1970. TID-8200 (22nd Rev.).

Vogt, J. R. 1969. Radionuclide production for the nuclear excavation of an Isthmian Canal. Biosci., **19**:138–139.

Volchok, H. L. 1965. Balancing of the SR-90 budget, pp. 261–268. In Health and Safety Laboratory Fallout Program Quarterly Summary Report October 1, 1965. USAEG HASL 164.

Volchok, H. L. 1970. Worldwide deposition of Sr^{90} through 1969. In Health and Safety Laboratory Fallout Program Quarterly Summary Report July 1, 1970. USAEC HASL-227, pp. I-70–80.

Watson, D. G., J. J. Davis, and W. C. Hanson. 1963. Interspecies differences in accumulation of gamma emitters by marine organisms near the Columbia River mouth. *Limnol. Oceanogr.*, **8**:305–309.

Welander, A. D. 1954. Some effects of X-irradiation of different embryonic stages of the trout (*Salmo gairdnerii*). *Growth*, **18**:227–255.

Welander, A. D., L. R. Donaldson, R. F. Foster, K. Bonham, and A. H. Seymour. 1948. The effects of roentgen rays on the embryos and larvae of the chinook salmon. *Growth*, **12**:203–242.

Welander, A. D., L. R. Donaldson, R. F. Foster, K. Bonham, A. H. Seymour, and and F. G. Lowman. 1949. The effects of roentgen rays on adult rainbow trout. UWFL-17. University of Washington Laboratory of Radiation Biology, Seattle, Wash. 20 pp.

White, J. C., Jr., and J. W. Angelovic. 1966. Tolerances of several marine species to Co-60 irradiation. *Chesapeake Sci.*, **7**:36–39.

Williams, R. B. 1966. Annual phytoplanktonic production in a system of shallow temperate estuaries. In *Some contemporary studies in marine science*, Harold Barnes (ed.), George Allen and Unwin Ltd., London, pp. 699–716.

Williams, R. B., M. B. Murdoch, and L. K. Thomas. 1968. Standing crop and importance of zooplankton in a system of shallow estuaries. *Chesapeake Sci.*, **9**:42–51.

Wolfe, D. A. 1967. Seasonal variation of caesium-137 from fallout in a clam, *Rangia cuneata* Gray. *Nature*, **215**: 1270–1271.

Wolfe, D. A. 1970. Zinc enzymes in *Crassostrea virginica*. *J. Fish. Res. Bd. Can.*, **27**:59–69.

Wolfe, D. A., J. M. Lewis, and J. H. Brooks. 1969. Estuarine biogeochemistry of cesium 137. In Annual report of the Bureau of Commercial Fisheries Radiobiological Laboratory, Beaufort, N.C., for the fiscal year ending June 30, 1968. *U.S. Fish Wildl. Serv.*, Circ. 309, pp. 13–17.

Wolfe, D. A., and C. L. Schelske. 1969. Accumulation of fallout radioisotopes by bivalve molluscs from the lower Trent and Neuse Rivers. In *Symposium on radioecology*, D. J. Nelson and F. C. Evans (eds.), USAEC CONF-670503, Oak Ridge, Tenn., pp. 493–504.

Yamagata, N. 1961. The concentration of ^{90}Sr and ^{137}Cs in land waters in Japan. U.N. Document A/AC.82/G/L.690.

Yamagata, N. 1965. Evaluation of marine contamination with radioisotopes of long half-lives. *Health Phys.*, **11**:1005–1008.

Zaystev, Yu. P., and G. G. Polikarpov. 1964. The problems of radioecology of hyponeuston. In Radiochemical and ecological studies of the sea. U.S. Dept. Commerce, JPRS: 25,966. Washington, D.C., pp. 28–41.

13

Petroleum—the Problem

JAMES E. MOSS (*Retired*), *American Petroleum Institute, Washington, D.C.*

Over half a century has passed since the New York and Philadelphia pilots reported large patches of oil floating in these harbors. Some called it sewer grease but others said it came from ships. Oil on the water's surface occurred so infrequently that little attention was paid to it. Soon the wildlife people recognized a serious problem when ducks and gulls were killed along the Atlantic coast from oil pollution. The First World War had started and with it the Sea Pollution Era. Many ships were converted to oil burners in 1914 and practically all built thereafter burned oil. Oil to fuel ships had come to stay and oil to fuel the armed forces was beginning.

The world consumption of petroleum has increased twenty-five fold since 1914, and while great improvement has been made in reducing the amount of waste oil that reaches oceans, this effort has not kept pace with the fantastic increase in its use. Today, after three international conferences, a treaty, the attention of a Queen and our President, the worst sea pollution case in history, 800 learned papers, and a dozen bills in Congress, there is good reason to believe that some practical plan may bring it under control.

To deal effectively with this complex scientific, industrial problem with national and international implications in the space available, presents nearly insuperable difficulties for the author. Although he has spoken and written on oil pollution problems many times in the past half century, he must now add the momentous happenings of the last 3 yr: the wreck of the *Torrey Canyon*, the Santa Barbara accident, a report to the President on oil pollution, a study on oil spillage, and legislative proposals in the Congress. Nonetheless, an attempt will be made to reach those in both government and industry, who have authority to act, to correct an increasingly hazardous world pollution problem.

The following plan has been adopted in the preparation of this chapter. This history of the transportation of petroleum by sea is given together with its volume and character. The design, size, and operation of tank ships is described with the sources and volume of oily wastes. Waste reduction methods past and proposed are examined and observations on pollution control are presented. The text is supported by many references to four sources which, as supplements to this chapter, will give the reader a more detailed understanding of sea pollution by oil (Moss, 1963; Battelle, 1967; Zobell, 1962).

Sources of Petroleum Which Reach the Oceans

The oceans have always contained some petroleum. These natural sources are called seepages or submarine oil springs. There are two shown on U.S. Coast and Geodetic Survey charts about 2 miles off the California coast. There are at least seven in the Carribean. The U.S. Government made a survey of the submerged oil fields off the Texas coast as early as 1906. These natural seepages are not believed to make a significant contribution to general sea pollution by oil.

Vessels, accidents such as colliding, stranding, and foundering contribute to pollution. During the Second World War, 61 tank ships were lost off United States coasts alone. These vessels contained over 5 million barrels of oil. It is not believed that they have contributed substantially to sustained sea pollution.

In some parts of the world, mostly in the United States, oil wells are drilled on the continental slopes from platforms over the water. There are over 6000 such wells off the coasts of Louisiana and Texas alone. While these offshore drilling operations are potential sources of pollution they do not contribute substantially to general pollution except in those cases of an individual oil well "blow-out" or incidents such as at Santa Barbara, California, where leakage in the geological formation occurred. Offshore oil production operations are regulated for pollution control by the Department of the Interior and for safety by the U.S. Coast Guard under the Outer Continental Shelf Lands Act (43 USC, 1331–1343).

It is now generally agreed that the most significant petroleum source polluting the sea originates in ships. Tankers, when they discharge their oil cargoes, must ballast some of their tanks with seawater so that they will be stable and seaworthy. Dry cargo and passenger ships must ballast some of their fuel tanks for the same reason. Where these tanks are used alternately for oil and seawater ballast, some oil will normally be discharged into the sea with the ballast. Casualties to loaded tankers may cause a small part to the whole oil cargo to be released into the sea. In pilot waters or near a coast line, this may create heavy local pollution.

Tonnage and Character of Petroleum Moving by Sea

Of the total tonnage of all commodities moving by sea in world trade, petroleum represents over 55 per cent (API, 1967). At the present time in excess of 1 billion long tons of petroleum per year moves by sea, of which about 60 per cent is crude oil and 40 per cent, is products. At the major ports of the United States, ships transporting petroleum account for the following percentages of all other traffic (API, 1967):

Seattle	44
Portland	37
San Francisco	75
Los Angeles	76
San Diego	40
Galveston-Houston	65
New Orleans-Baton Rouge	42
Mobile	25
Savannah-Brunswick	42
Baltimore	25
Philadelphia	53
New York	69
Providence	91
Boston	84

About 35 per cent of the tonnage moving over our inland waterways consists of petroleum.

There are over 1000 kinds of crude oils, produced, stored, transported, and refined separately in world trade (Moss, 1963) and from these over 2000 separate and distinct petroleum products are made. In the last two decades a whole new petrochemical industry has developed that now produces several thousand derivatives of petroleum (Moss, 1963). Many of these products of the petroleum industry move by tank ships or barges. There are 1193 petrochemical plants in the free world, 567 of which are in the United States (API, 1967).

Tank Ships That Move Petroleum by Sea

Drake's oil well started to produce in Pennsylvania in August 1859, and in 1861 the first shipment of oil was made from Philadelphia to England in wooden barrels in a sailing ship. The barrels leaked so much that they created a serious fire hazard. They were also expensive both in cost and in labor for loading and unloading. It was obvious that if large tanks could be built into ships the petroleum could be carried in bulk and loaded and unloaded with pumps. This was first attempted in the *Ramsey* in 1863. It continued under various plans and patents for 23 yr. Many of these early bulk petroleum carriers burned or exploded. Leakage of the tanks filled the surrounding hold spaces with gas which required only a spark or flame to cause a disaster. Some dispaired of ever being able to transport petroleum in bulk by sea with safety.

In 1886 a new tank ship appeared at sea. She was the first ship designed exclusively for the carriage of liquid cargo in bulk. This vessel embodied all of the lessons learned in earlier practice. The hull was of iron. The engines and boilers were aft. The cargo space was separated from the rest of the ship by cofferdams. The cargo tanks were created by merely dividing the iron hull space by iron bulkheads. There were no spaces in the cargo area for gas to accumulate. The cargo piping ran through the bottom of the ship. The cargo pumps were in a pump room enclosed by iron bulkheads low in the ship. There were other features found by experience to contribute to safety. Many felt that this vessel would be a failure. The *S/S Gluckauf* was built to the order of Heinrich Reidman of Deutsche Amerikanische Petroleum Gasellschaft by Armstrong Whitworth Company, England. Her length was 300 ft, beam 37 ft and molded depth 23 ft. Her gross tonnage was 2307 and her deadweight tonnage (dwt) was 3000. Nearly a quarter century had elapsed between the first attempt to transport petroleum in bulk by sea and the appearance of the *Gluckauf* which embodied the experience of those years. While many additional safety features have been added to tanker design since the *Gluckauf*, nearly all tank ships after her time have included her basic safety features.

As might be expected, the increased demand for petroleum and products caused a similar increase in the number of ships that carried them. From 1900 to 1966 the number of tank ships of over 2000 gross tons in the world fleet increased from 109 to 3524. In this same period the deadweight tonnage of these ships increased from 530,725 to 102,908,800. In the period, 1900–1947, the United States tanker fleet increased from 3 to 951 and the weight from 10,350 to 14,035,000 dwt. Since 1947 the ships have decreased to 387 and the tonnage to 8,549,900 (API 1967). This decrease is explained later.

While the increase in the world tanker fleet during the past two-thirds of a century is spectacular, it is less so, perhaps, than the increase in the dimensions of some of the ships which have joined this fleet in the past decade. During the Second World War the United States Government built 525 tankers. These were the familiar T-2 tankers. Many are still in operation. The dimensions of these tankers are 16,765 dwt, 503 ft b.p., 14.5 knots, 68 ft beam, 30 ft draft. At the time they were built they were considered big tankers.

Writing only 5 yr ago, the author described the size of the tankers being built and in operation, and in 1963 stated (Moss, 1963):

> Since the Second World War the capacity of new tankers has greatly increased and ships of 50,000 d.w.t. and 60,000 d.w.t. are quite common. There are at the present time about 16 tankers having a d.w.t. within the range of 60,000 to 104,520.
> Recently in Japan the tanker Nissho Maru has been launched. This immense vessel is of 130,050 d.w.t. She has a gross tonnage only slightly less than the Queen Mary and the Queen Elizabeth, but a displacement loaded far exceeding these largest liners. Her overall length is 954'9", beam 141'1", and a draft loaded of 53'10". She will carry an average cargo of 900,000 barrels of oil.

Four years later (early 1967) there were over 50 tankers on order with capacities of over 100,000 dwt. Five years later (mid 1968) there were 6 tankers of 276,000 dwt nearing completion or completed, having the following dimensions: length overall 1135 ft, length b.p. 1082 ft 8 in., beam 174 ft $10\frac{1}{2}$ in.; depth 105 ft, propulsion 34,700 shaft horsepower (shp), twin screw. There were 6 tankers nearing completion of 312,000 dwt (nearly $2\frac{1}{2}$ times the capacity of the *Nissho Maru* of 5 yr ago). The first of these great ships, *Universe Ireland*, was placed in operation in 1968.

Three classification societies: American Bureau of Shipping, Lloyds, and Bureau Veritas studies have concluded, that tank ships of 500,000 dwt are feasible. Such ships would have dimensions of: length 1250–1300 ft, beam 218–225 ft, depth 106–112 ft, draft 80–86 ft, shp 55,000–60,000, speed 17–18 knots. One of these ships, if built, would have a capacity of about 3,750,000 barrels of cargo.

The largest tankers now under construction or on order could enter only very few world ports (none in the United States). In most cases they would probably load and discharge at so-called sea loading terminals. At such terminals they would be moored to buoys in deep water after which they would pick up several cargo hoses from the bottom of the sea. These hoses are connected by pipe lines to tanks on shore. Such terminals have been in successful use for many years where no harbors are available or where the water is too shallow to get a ship close to shore. Another device consists of a single mooring buoy around which the ship swings through a full circle. The cargo hose runs over the bow to a rotating manifold on the mooring buoy which is in turn connected to a hose running to the bottom, connected there to a pipe line running to the shore tanks.

The Operation of Tank Ships

If a new tank ship which had never contained an oil cargo were placed at an oil loading terminal and filled to its load line with oil and immediately if after loading the cargo was discharged back to the terminal tanks it would, on the average, be found to be 0.30 per cent short of oil. This shortage results from the oil required to wet the ship's tank surfaces, which for a tanker of 100,000 dwt might have an area in excess of 25 acres. In addition to this "wetting oil" there would be included a small puddle of oil around the suction line in each tank plus the oil remaining in the ship's pipe lines and pumps. But only a part of this difference of 0.30 per cent reaches the sea, as shown later. This will be understood better if a complete cycle in the operation of a tanker is described.

Before the tanker arrives at the loading terminal, the shore tanks containing the ship's cargo are gauged. These are very large cylindrical tanks for which there are tables showing the tank contents for each $\frac{1}{8}$ in. of liquid depth. These measurements are taken by a gauger who does not report to the marine department but to the shore organization. While taking these measurements the oil temperature is also taken and all volumes are corrected to "volume 60 F." When the tanker arrives at the terminal, it is boarded by the gauger, who with the Chief Mate of the ship, inspects all of the ship's cargo tanks for absence of liquid and cleanliness (other than oil recovered from the last cargo which is described later under "load on top" procedure).

When the ship's tanks are filled, the gauger and Chief Mate again inspect the tanks recording the gauges for each tank and its temperature. A copy of the gauger's ticket, signed by both parties, is given by the gauger to the Chief Mate before the ship proceeds to sea. Upon arrival at the port of discharge the local gauger gauges the shore tanks in preparation for receiving the arriving cargo. He then boards the ship where he and the Chief Mate take the "arriving gauges" of the ship's tanks and their temperatures as before. When the cargo has been pumped ashore, the gauger and Chief Mate make a final inspection of each tank. The ship's gauging ticket, showing that each tank has been pumped dry, is signed by both parties with a copy for the ship. After the ship has discharged its cargo, the shore tank gauges are taken together with the temperature. It is noted that in the preceding description the cargo has been measured four times together with the oil temperature. From these four measurements a closed diagram may be made in the event of abnormal loss which will show where this loss took place. This closed diagram will show the totals and differences for: (a) shore tanks and ship tanks at the loading port; (b) ship tanks at loading and at discharging port; (c) ship tanks and shore tanks at the discharging port; (d) shore tanks at loading port and shore tanks at discharging ports. When the loss for the voyage described is excessive in the judgment of the inspection department (which is independent of the marine department), an investigation is made and if the ship is found to be responsible the marine department or its owners (if an independent) may be billed for it. This cargo measurement procedure is followed by all domestic and foreign oil companies.

This measurement procedure is of great significance in connection with oil pollution. It establishes the fact that from early times in the petroleum industry precise measurement corrected for temperature was made with each movement and that abnormal losses are charged to the carrier. In other words there is a penalty, both inter and intracompany for loss of oil. Between the well and the customer, petroleum is measured at least 10 times. No other basic commodity approaches petroleum in quantity control.

After the cargo of oil has been discharged, the pumps and cargo lines are washed to a cargo tank to be used for settling out the water from the oil. Water is then drawn in from the harbor or river to ballast selected cargo tanks to provide stability and seaworthiness while the ship is on its ballast voyage. At sea other tanks are washed down and the oily water is discharged to the settling tank. Clean ocean water is pumped to the clean tanks. Thus by cleaning tanks at sea the waste oil is kept aboard and the ballast water in the clean tanks may be discharged at the dock. If the loading terminal has facilities for receiving oily ballast at the dock, the ballast may be discharged directly to the shore slop line. All of the oil loading terminals in the United States have waste oil facilities. In 1960 there were 111 such individual facilities with a total capacity of 900,000 barrels. (Moss, 1963.)

The washing of the ship's tanks is accomplished with washing machines using jets of water at a pressure of 160 psi and a temperature of 170–180 F. The washing machines consist of two opposed nozzles rotated by a built-in water turbine which rotate in vertical and horizontal planes. The nozzle assembly is attached to the end of a reinforced rubber hose which is lowered into the tank to be washed and clamped at selected levels. On some of the new tankers the tank cleaning apparatus is built in place.

It was stated earlier that it required about 0.30 per cent of the cargo to wet the surfaces of the tanks but that that quantity of oil was not lost even if the ballast water was all pumped to the sea without first having the oil settled out on the ship. The reasons for this are first, only about one third of the cargo weight is taken for ballast. This means that only one third of the cargo tanks would carry ballast so that the oil lost would be only 0.10 per cent. Second, about one half of this remaining oil would not go out with the ballast water because it would float on top and when the tank was finally drained about $\frac{1}{2}$ of the 0.10 per cent of the oil would remain in the ship, pipe lines, and pumps, and in a small puddle around the suctions in each tank.

The maximum quantity of oil reaching the sea each year as the result of ballasting and cleaning the tanks of the world fleet can be roughly approximated by knowing the number of times, on the average, tankers are ballasted and cleaned and the quantity of oil lost in each of these operations. This estimation appeared to be 4 to 5 hundred thousand tons/yr in 1963 (Moss, 1963). This would be the maximum loss if all ballast and tank washings were discharged into the sea without separation and recovery within the ship. This is a very minute part of the whole commerce, but significant in terms of the total discharged into the sea.

Since World War II great progress has been made in reducing the waste oil discharged into the sea by tankers. One of the greatest savings has resulted from the welding of ships. Each of the old riveted ships contained several hundred thousand rivets many of which, especially after storms, leaked oil to the sea. The corrosion of cargo tanks resulting from the carriage of salt water ballast is a very heavy source of expense. A 10-yr research program proved that cargo tanks could be coated to resist the combined effects of cargo, salt ballast water, and high pressure-temperature washing water. Before the use of coatings the steel tank surfaces were rough and pitted. The coated surfaces are smooth and they hold less oil and also reduce its rundown time. The advent of larger tankers has made it possible to arrange the tanks so that certain spaces can be dedicated exclusively to the carriage of ballast water. Thus when ballast is discharged, little or no liquid petroleum reaches the sea. While such tanks solve the ballast problem, they do not directly assist with the problem of cleaning waste oil from the tanks when the ship goes to a repair yard for overhaul or tank cleaning before a change of cargo. The so-called load on top plan has been practiced by some oil companies with their own ships for many years. The extension of this plan by oil companies to tankers they have under charter should substantially reduce petroleum which reaches the sea. It is, in addition, an excellent example of what may be accomplished by seeking to reduce oil pollution of the oceans on a world fleet basis. Finally it is true that the ratio of wetted tank surface to tank contents is smaller in the case of large tankers. Thus less oil per unit of cargo would adhere to the tank surfaces.

Waste Oil Problems Exterior to the Tanker

The sea pollution problem extends far beyond the operating problems of tankers; indeed those problems are probably easier to solve than those exterior to the ship. Only 25 per cent of the world tanker fleet is owned by oil companies. Nearly all the rest of this great fleet is owned by so-called independents. Independent tanker owners are only interested in transporting oil—they are oil carriers exclusively. They charter their ships to the oil companies. The independent tanker owners thus perform the important function of providing a flexible transportation service capable of adjusting to changing demands in the world market. The independent tanker owner seldom owns shore property. His ships are repaired, stored, and serviced in any world port where those services are required. Oil companies sometimes charter their ships to other oil companies. Thus one company may require a tanker to lift a cargo in the Middle East while another company may require a tanker to lift a cargo in the Far East. Neither company may have a ship of its own in the right position but by cross chartering the needs of one company may be provided for by the other.

Oil terminals where tankers load and discharge their cargoes are nearly all owned by oil companies. With few exceptions only the oil terminals associated with oil refineries have waste oil separators and treating facilities necessary for breaking oil-water emulsions, oils of all gravities, and petrochemical wastes.

Consider now an independent tanker chartered by an oil company to transport crude oil from the Middle East to New York. The tanker may load in Haifa or Tripoli, and discharge its cargo in New York, as planned, and return to the loading port in ballast for another cargo. Since the oily water ballast cannot be discharged into the sea at Haifa or Tripoli, some of the ship's cargo tanks must be cleaned at sea so that clean ballast water from clean tanks may be discharged into the sea at the loading port. But when the tanks are cleaned at sea to carry clean ballast water several hundred barrels of oil will be recovered. This oil cannot be discharged at sea by a tanker of over 20,000 dwt for this is prohibited by a 1962 amendment to the 1954 Oil Pollution Convention. The tanker must, therefore, bring this waste oil to the loading terminal at Haifa or Tripoli, but these loading terminals may have no facilities for reclaiming the waste oil and may decline it. The Ship Master must then decide what to do with the waste oil. To bring it back to the United States is difficult for several reasons.

First, he does not know whether the terminal in New York will receive it. Second, a customs duty would be charged on it. Third, since the ship could only be loaded to a certain depth under the International Load Line Treaty, the waste oil would shut out a similar part of the new cargo. Fourth, if the tanker was sailing under a new charter, the new charterer might not wish to load his cargo on top of waste oil recovered from a different crude oil on the previous voyage.

Returning to the initial cleaning of the ship at sea, there is another serious problem. Ships in international trade pay canal, lighthouse, and port dues based on the admeasurement tonnage for each ship determined by the nation of its flag. If cargo is capable of being carried in a compartment of the ship not expressly exempted under the rules, then a waste oil storage space or a space fitted with a separator would void the exemption for that space. Thus a ship owner who fitted his ship with an oil separator in a cofferdam or used the cofferdam or similar space for waste oil storage would place himself at a competitive disadvantage with his competitors who did not do so.

There is another waste oil problem associated with tankers which must clean all of their tanks because of a change in cargo or because they must go to a shipyard for periodic overhaul or repairs. This might be accomplished to the greatest advantage for the ship at the oil dock where the last cargo was discharged, but tank cleaning and gas freeing are hazardous undertakings, and, in addition, dock space is nearly always at a premium. The terminal owner is usually reluctant, therefore, to permit this operation to be performed at his docks unless the ship and terminal are owned by the same company and thus have the same safety standards. For this reason most tank cleaning for change of cargo or preparatory to entering a shipyard is done at sea. If going to a shipyard the waste oil will be discharged there. If the ship is cleaning its tanks before receiving a different kind of cargo, the owner may try to arrange with the loading terminal to take the waste oil or, failing this, arrange with a tank cleaning company to furnish a barge to receive the waste oil. Most important, better accommodations for waste oils must be provided by tanker companies, the oil industry, and the shipyards. An extension of the "load on top" practice for crude oils throughout the world fleet is ground for hope that better accommodations will result.

Present State of Sea Pollution Control

During the past 2 yr sea pollution has been the subject of more than one hundred learned papers, countless press, radio, and TV notices, government reports, legislative proposals, and Presidential notice. This is more attention than it has had during the previous 50 or more years. There is today a great pressure to end sea pollution by oil, to do *now* what three international conferences and thousands of sincere and able individuals and their governments have failed to do in 40 yr. If this author were asked (as he has been) to summarize his views on how sea pollution by oil may be brought under control, he would without hesitation suggest the following:

> Be silent until you understand the problem
> Develop a solution to all of its elements
> Establish a priority for each source of pollution
> Develop a plan and pursue it consistently.

The difficulties of controlling sea pollution have been continuously underestimated. This is because the problem is not understood. The planners have continuously proposed solutions which have fallen short of being effective. They then turn to outright prohibitions. These will not be effective either; the volume of the commerce is too great to expect perfection.

The correct approach is first to determine in considerable detail the conditions under which oil pollution is injurious. It will be shown that we do not know this. Second, attack the sources of pollution according to importance. Discharge of wastes facilities must be provided for handling ashore. Until now governments have encouraged the discharge of oily wastes at sea. The next most important source of oil pollution must then be considered and solutions found for it. In this case it may be that industry can produce the solution. There is no doubt that oil reaching the oceans can be very substantially reduced but there is also no doubt that we are dealing here with a physical problem having practical limitations.

The author, having participated in all efforts of consequence of government and industry prior to 1964 to regulate tank vessels for increased safety and decreased pollution, feels obliged to contribute his views on sea pollution based on experience going back to the First World War. In February 1968, "A Report on Pollution of the Nation's Waters by Oil and Other Hazardous Substances," (Sec. Depts. of the Interior & Transportation, 1967) made at the request of former President Johnson by the Secretaries of the Interior and Transportation was released. It was no doubt produced under considerable pressure and contains little new information. It contains an "Action Program" expressed in general terms and an important announcement that a literature search and review was underway to evaluate the present state of the art relative to the control of spills of oil.

In July 1967, the U.S. Coast Guard contracted with Battelle Northwestern (Battelle, 1967) to make a literature search on oil spills with comments. The report is dated November 20, 1967 and was released for public distribution on March 18, 1968. It is referred to hereafter as "Oil Spillage Study." This study, made in the incredible time of 4 months, is a major contribution to the literature of sea pollution by tanker casualty. It is well organized and documented by sections. Not only should it be maintained as recommended by its authors, but competent individuals should be invited to submit additional information and comments.

These two documents, the report and the study, are based mainly on information drawn from the oil and tanker industries, but it is not believed that either document was discussed with industry representatives prior to publication. Nothing whatever was mentioned about the responsibility for tanker design or the efforts of the tanker industry over the years to develop a safe and efficient vessel. Government standard for tanker design (Tanker Regulations) are complied with by all operating vessels. It was unfortunate indeed that the implications to an uninformed reader are that the industry is neglectful of its public obligations. While the government has the right to undertake any kind of study, the release of a study by the government to the public dealing exclusively with possible inadequacies of the equipment of an industry without reference to its compliance with present government requirements is most unfortunate. This criticism does not attach to those who made the Oil Spillage Study.

Under these circumstances the author feels compelled to supplement the government's Report and Study with additional information which may partly restore the damage already done.

1. Since 1936 the design, operation, and manning of tank vessels has been the exclusive responsibility of the United States government. The law as to responsibility is as follows:

". . . In order to secure effective provision against the hazards of life and property created by the vessels (tank vessels) to which this section applies the Commandant of the Coast Guard (this read Secretary of Commerce up to the Second World War) shall establish such additional rules and regulations as may be necessary with respect to the design and construction, alteration, or repair of such vessels including the superstructures, hulls, places for stowing and carrying such liquid cargo, fittings, equipment, appliances, propulsive machinery, auxiliary machinery and boilers thereof; and with respect to all materials used in such construction, alteration, or repair, and with respect to the handling and stowage of such liquid cargo; the manner of such handling and stowage, and the machinery and appliances used in such handling and stowage; and with respect to equipment and appliances for life saving and fire protection; and with respect to the operation of such vessels; and with respect to the requirements of the manning of such vessels and the duties and qualifications of the officers and crew thereof and with respect of all the foregoing . . . ."

46 USC 391a(RS4417a)

This law, drafted to be inclusive of every feature of tanker design and operation and to provide single responsibility for those carriers of dangerous cargo, is supported by the "Rules and Regulations for Tank Vessels" (C.G. 123) all of which have the force of law. That publication contained 133 pages. Every tank ship and tank barge in the United States and its possessions is inspected and certified under this law and its supporting detailed regulations. The regulations contain provisions relating to the avoidance of oil pollution and the U.S. Coast Guard Manual for the Safe Handling of Flammable and Combustible Liquids contains instructions for the same purpose. (C.G. 174)

2. At the beginning of the Second World War the United States government built to its own order 525 tankers following commercial tanker design. They were built under the current Tanker Regulations of the Government. These were the famous T2's, some of which are still in service.

3. Navy oilers designed by the Navy follow commercial practice with military features superimposed.

4. Tankers of the Military Sea Transportation Service are of commercial design built under the U.S. Tanker Regulations and inspected by the U.S. Coast Guard.

5. On July 25, 1925 the Standard Oil Co. (N.J.) tanker *Charles Pratt* arrived in Texas City, Texas, with 350 barrels of Regan County crude oil recovered enroute from New York as the result of cleaning tanks. The ship had a cargo capacity of 120,000 barrels. "Skimmers" had been fitted in the after cofferdam. A representative of the National Bureau of Standards made the voyage as an observer and reported on the oil recovery operation. Skimmers were fitted on two other ships and five additional ships were to be similarly equipped. The ship owner found that in using the cofferdam for anything but water ballast he had forfeited his right to deduct the cofferdam tonnage from the official tonnage of his ships. This official tonnage, fixed then by the U.S. Customs, but now by the U.S. Coast Guard, determines the Panama and Suez canal tolls and

the light and harbor dues of maritime nations. The owner appealed to the U.S. Collector of Customs but was told the law (R.S. 4153, 46 USC 77) provided for neither alternative nor interpretation. To have used this apparatus, to have even left it on the ship, would have placed this ship owner under substantial competitive disadvantage and for this reason the project had to be abandoned.

The International Conference on Oil Pollution held in Washington, D.C., in June 1926 (a year later) adopted eleven resolutions. Three of these resolutions read as follows:

"7. That no penalty or disability of any kind whatever in the matter of tonnage measurement or payment of dues be incurred by any vessel by reason only of the fitting of any device or apparatus for the separation of oil from water.
8. That dues based on tonnage shall not be charged in respect of any space rendered unavailable for cargo by the installation of any device or apparatus for separating oil from water.
9. That the term "device or apparatus for separating oil from water" as used in recommendation Nos. 7 and 8 shall include any tank or tanks or reasonable size used exclusively for receiving waste oil recovered from the device or apparatus, and also the piping and fittings necessary for its operation." (P.6-Proceedings)

Thus it is seen that to keep waste oil out of the sea, waste oil recovery apparatus must be installed. But to avoid penalty against the fitting of waste oil recovery apparatus and spaces for its storage, exceptions to admeasurement laws must be made. Further it is obvious that these exceptions must be uniform among maritime nations, otherwise the shipping of some nations may be penalized. Finally it will be observed that the First International Conference on Oil Pollution recognized the importance of this subject and made recommendations to provide for the installation of such apparatus without penalty to individual ships or nations. This was in June 1926 (Moss, 1963).

Forty-one years after the International Conference on Pollution of the Sea held in Washington, D.C., in 1926, at which resolutions on tonnage exemptions for waste oil separators and tanks were adopted (and nothing done about it), a resolution [Resolution A 115(V)] of October 25, 1967 was adopted by the Intergovernmental Maritime Consultative Organization (IMCO of the United Nations), reading in part as follows:

THE ASSEMBLY,
    Noting Article 16(i) of the IMCO Convention concerning the functions of the Assembly,
   FURTHER NOTING that the amendments to the International Convention for the Prevention of Pollution of the Seas by Oil, 1954, adopted 1962, came into force 18 May 1967 and 28 June 1967,

ALSO NOTING that those amendments result in a great increase in the number of areas and zones in which the discharge of oil and oily mixtures is prohibited,

DESIRING to encourage ship owners and operators to cooperate in the programme for the prevention of pollution of the seas by oil,

HAVING CONSIDERED the Recommendation of the Maritime Safety Committee (MSC XVI/4, paragraph 7 and Annex IV) on the treatment of spaces on board ships for the separation, clarification or purification and the carriage of slop oil,

DECIDES:
- (a) to adopt that Recommendation, the text of which is set out in the Annex to this Resolution, and
- (b) to invite governments concerned to include in their national tonnage measurement requirements provisions to give effect to the Recommendation.

ANNEX

RECOMMENDATION ON THE TREATMENT OF SPACES ON BOARD SHIPS FOR THE SEPARATION, CLARIFICATION OR PURIFICATION, AND THE CARRIAGE OF SLOP OIL.

1. In view of the coming into force of the amendments adopted by the London conference of 1962 to the International Convention for the Prevention of Pollution of the Seas by Oil, 1954, which greatly increased the number and areas of zones in which discharge of oil or oily mixtures is prohibited, provisions should be introduced into present national tonnage measurement requirements so that the following spaces, provided they are not available for any other purpose and are properly marked in accordance with the regulations of the pertinent national authority, may be deducted from gross tonnage in determining net tonnage:

Space taken up by machinery used exclusively to separate, clarify or purify a ships own slop oil mixture or tank cleaning residue, as defined in regulations of the pertinent national authority:

Space taken up by a tank or tanks used exclusively for the carriage of such slop oil mixture or residue.

2. The above provisions should be applicable to all ships whether existing or new . . . .

Assuming that the Washington Oil Pollution Conference of 1926 could have persuaded the United States and the other maritime nations to adopt its resolutions on admeasurement to exempt from the ship's official tonnage spaces used exclusively for the separation and storage of waste oil—as IMCO of the United Nations is now, 42 yr later, attempting to do—a rough calculation will show the maximum tonnage of petroleum which would have been saved from the sea and its gross value.

From 1926 to 1968 about 1.8×10^{11} barrels of crude oil were produced worldwide. This is equivalent to 2.4×10^{10} long tons of average gravity crude oil. Using the 1963 ratio, 34.0 per cent of this tonnage moved by sea, or 8.3×10^9 long tons. In 1925 the *S/S Charles Pratt* cleaned the whole ship after discharging a cargo of 1.2×10^5 barrels of Regan County crude and recovered 350 barrels or 0.3 per cent of the cargo. This resulted from cleaning the whole ship which is, of course, not representative of average losses for a year including deballasting. The loss for average conditions, including crude oil and products (both persistent and non-persistent oils), would be about 0.1 per cent of the cargo. Applying this factor to the tons moved by sea (1926–1968) there would have been a potential recovery of 8.5×10^6 long tons of crude and products. Assuming an average value of \$25.00/long ton for this recovered oil, its gross value would have been \$213 million dollars. It is believed that if maritime nations will act to remove the penalties resulting from the collection and discharge of waste oil ashore, very little will be discharged into the sea in the normal operation of tank ships.

The policy of government with reference to waste oil has been unrealistic. Inadvertently they have actually promoted the discharge of waste oil into the sea. Additional problems of this kind, which must also be cleared up, are considered next.

In 1930, an International Load Line Convention was adopted by a Load Line Conference. It was subscribed by the United States, January 1, 1933. This Convention limits the depth to which ships may be loaded by requiring marks to be cut and painted on their sides to indicate their legal draft. Thus a ship loaded to its legal draft will hold a definite tonnage of cargo. Earlier in this chapter it was shown that in most cases it would be more desirable to bring back waste oil recovered on the ballast voyage, to the discharge port and discharge it ashore where refinery equipment for processing the waste would normally be available. Under the Load Line Convention, however, the waste oil (although it would amount to less than 1 per cent of the cargo) cannot be brought back to the loading port without displacing an equal tonnage of good oil. Here then we find another penalty against the most logical procedure for disposing of the waste oil where it can be properly processed. The waste oil from cleaning the ship might increase the draft of the ship by not more than $1\frac{1}{2}$ in. This is only a very small increase in the total draft of the ship. In the past when tankers have been in short supply governments have waived the load line rules to permit loading to drafts greater by 4 to 8 in. and no adverse effects have been observed. Here then are other government requirements which, inadvertently promote the dumping of waste oil at sea and which should be carefully examined.

In most countries customs officials will wish to collect a duty on waste oil discharged from a vessel arriving from a foreign port. At the London

Oil Pollution Conference of 1962 attention was called to this customs problem. Resolution 9 of that Conference seeks to discourage this policy with respect to waste lubricating oil from ship's engines but fails to deal with the waste cargo oil (Moss, 1963). In all of these cases governments will wish to limit new provisions relating to waste oil to a fixed percentage of the basic cargo to avoid abuse.

That outright prohibitions against pollution are no substitute for a national or international policy which is practical and facilitates compliance is illustrated by the following U.S. Coast Guard study covering a 5-yr period (1958–1962) of oil pollution violations in the territorial waters of the United States. It was found that 47 per cent were caused by foreign flag vessels and of these nearly 50 per cent were of flags signatory to the London Convention of 1954. If this is the percentage of compliance in United States territorial waters extending only 3 miles off shore it is unlikely that there is any serious compliance in the zones created by the 1954 Convention and extended in 1962 having in mind that these zones extend from 50 to 100 miles off shore.

That tankers have no monopoly on pollution violations and that public vessels of the United States contribute more than their share is illustrated by the following reports quoted in the Report to the President on Oil Pollution. The U.S. Coast Guard reported that there were 371 oil pollution cases in 1966. Of these 279 vessels, 143 of which were of United States flag and 86 of foreign flag. Of the 279 vessels, 49 were tankers, 56 barges, 18 dry cargo ships, 91 naval vessels, and 15 of other categories.

The California State Department of Fish and Game reported 181 oil spills during 1966 in the Los Angeles and Long Beach areas alone, of which 59 were from merchant vessels, 19 from land based facilities, and 67 from naval vessels.

The Microbiology of the Oceans

Those who give thought to the oceans, to the vast quantity of wastes deposited in them by the rivers, and to the greater quantity of plants and animals that live and die in them, must conclude that without the power to reduce these wastes neither the oceans nor the land could now support life—for the oceans are nature's great septic tanks.

In 1926 a fundamental research project to study the origin and environment of source sediments of petroleum was organized in the American Petroleum Institute. For 5 yr this project was carried on under the directorship of Dr. Park Davis Trask and continued under him during his association with the United States Geological Survey from 1931 to

1948. Dr. Trask then became associated with the faculty of the University of California.

In 1942, a new project was undertaken by the American Petroleum Institute. It was Project 43, "The Transformation of Organic Material into Petroleum." Project 43 was divided into two parts: 43A under the directorship of Dr. Claude E. ZoBell at Scripps Institute of Oceanography (University of California), at LaJolla, California, was to investigate thoroughly the role of marine bacteria in the conversion of organic materials into petroleum. Project 43B, under the directorship of Dr. F. C. Whitmore of Pennsylvania State College, was concerned with the action of bacteria on hydrocarbons.

While Project 43A was directed at the reduction of organic matter to petroleum by bacteria in seawater and bottom sediments, it soon became evident that it was very important to know in detail the effect of microorganisms on petroleum or hydrocarbons. (ZoBell, 1946, 1946a.)

While none of this extensive research was directed at the problem of oil pollution of the sea, nevertheless, it furnished plausible answers to many sea pollution phenomena and suggested areas of inquiry which appeared to be rewarding if pursued.

Of all of the scientific and technological problems associated with sea pollution the most important is the capacity of the oceans for the biochemical degradation of wastes. Until this is better understood, pollution abatement planning cannot be scientifically undertaken, nor in the judgment of the author, will sea pollution by oil be brought under control. The reasons for this conclusion are simple and fundamental. The vast quantity of petroleum moving by sea is known. The casualty rates for ships of all categories are known. The rigors of the marine environment and the frailties of mankind are understood. The standing of the United States merchant marine (including tank ships) among the fleets of all nations with respect to casualties has seldom placed it lower than second. It is unlikely that this enviable record can be substantially improved. Since an accidental oil spill of any consequence hazards the lives of those involved, the negligence factor is low. While improvement is of course possible the question is whether the improvement attainable would ever be accepted as satisfactory according to present standards. These considerations suggest that every possible means for the reduction of oil reaching the oceans be reviewed and that the evidence for the commonly accepted reputation of oily wastes as destroyers of sea life should be carefully reexamined.

The biochemical degradation of petroleum by bacteria is now a well-established scientific fact. Since the oceans are known to contain vast numbers of microorganisms, what of the degradation of petroleum by bacteria in the oceans? The staff that prepared the Oil Spillage Study at Battelle Northwestern for the government raises this question in connection with the pollution caused by the loss of the tanker *Torrey Canyon*, which stranded on Pollard Rock near Lands End, England, and released 119,000 tons of Kuwait crude into the sea. Quoted below from the Oil Spillage Study is the question and their answer.

"4.4 BIOLOGICAL DEGRADATION OF CRUDE OIL AND OIL FRACTIONS IN THE OCEAN.

A significant question posed as the result of the Torrey Canyon incident is the following. If no measures were taken to remove oil spilled on the surface of the ocean, how long would the oil persist? In particular could the microbiological life in the sea metabolize the oil? ZoBell (1962) has spoken to these points at great length. His paper represents what must be regarded as the most complete compilation of information on the subject.

ZoBell concludes that: 'Virtually all kinds of oil are susceptible to microbial oxidation. The rate of such oxidation is influenced by the kinds and abundance of micro-organisms present, the availability of oxygen, temperature and the dispersion of the oil in water. Microbial oxidation is most rapid when the hydrocarbon molecule is in intimate contact with water and at temperatures ranging from 15 to 35 C; some oxidation occurs at temperatures as low as 0 C. An average of one-third of the hydrocarbon may be converted into bacterial cells, which provide food for many animals. The remaining two-thirds of the hydrocarbon is oxidized largely to CO_2 and H_2O. In the marine environment oil persists only when protected from bacterial action.

Based upon rates at which marine bacteria have been observed to oxidize various kinds of mineral oils under controlled laboratory conditions and upon information on the abundance of bacteria in the sea, it is estimated that oil might be oxidized in the sea at rates as high as 100 to 960 mg/m^3 day or 36 to 350 g/m^3 year.'"

While no one takes exception now to the degradation of petroleum by microorganisms of the ocean, some have expressed the opinion that Dr. ZoBell's rates may be high, or even if correct, too slow to be of substantial assistance.

In this connection attention is called to the following.

1. ZoBell's rates were based on "controlled laboratory conditions." He did not simulate sea conditions where petroleum in slicks is brought into contact with vast areas of sea surfaces.

2. Ocean winds and currents and a moving ship have the capacity to bring polluting petroleum into contact with unlimited volumes of seawater. In addition, the turbulence of the sea surface itself assists in pro-

ducing the condition of the most rapid oxidation described by ZoBell as follows: "Microbial oxidation is most rapid when the hydrocarbon molecule is in intimate contact with water." The tendency of oil to form a constantly thinning film on water has already started to promote this intimate oil and water mixture.

3. One slick from the Torrey Canyon had a surface area of more than one billion square meters.

4. The wake volume for a tanker moving at 18 knots for 24 hr is about 3.65×10^{10} ft^3 (Moss, 1963). If all of the waste oil resulting from the cleaning of a 45,000 dwt tanker (750 barrels) were discharged into the wake during one day, the oil concentration would be only 0.11 ppm compared with 100 ppm allowed under the London Convention of 1954.

5. Using ZoBell's rate of degradation data, the oil would be completely degraded in a sea temperature of 15 C in 2 days 4 hr and at the half rate in 4 days 8 hr.

6. In 1959, John Dennis conducted studies of polluted beaches on the east coast of Florida for the American Petroleum Institute. For 1 yr the size of oil particles cast upon the beaches was measured. Their average mass was 2 g. The average diameter of sea bacteria is from 0.2 to 2.0 μ. The diameter of the oil particles to provide, within practical limits, the molecular relationship for fast degradation cited above by ZoBell would be about 4 μ. The average diameter of the 1.5×10^4 particles of oil which for a year were cast upon the Florida beaches was 1.55×10^4 μ. Thus these particles of oil were about 4000 times too large in diameter for rapid degradation. These plastic particles of oil were once a fluid. The sea leached out the lighter ends of the oil and they became plastic lumps. Thus the longer the weathering the lower the degradation rate. It follows that for the highest degradation rate, oily wastes that must be discharged into the sea should be in the form of a finely divided oil-in-water emulsion. It also follows that since these wastes are not discharged into the sea in this form it cannot be said how fast the degradation rate would be under controlled conditions but it does indicate that this emulsion (mechanical) should be created at the earliest moment.

7. According to ZoBell (1962), "An average of one-third of the hydrocarbon may be converted into bacterial cells that provide food for many animals. The conversion of one-third of this waste petroleum into bacterial cells provide more food for protozoa and thus enriches the food chain of the oceans." It is ironic that some find it difficult to accept the facts of the microdegradation of petroleum by bacteria in the oceans, where one third of it becomes food for *fish* at the same time that some of the larger oil companies are producing protein from petroleum by bac-

teria for feeding *men!* The world nutrition problem is essentially one of finding sufficient sources of proteins. Several oil companies have been producing proteins from petroleum for several years. The world production of proteins is about 25 million tons/yr. It is estimated that by using 50 million tons of petroleum the world production of proteins could be doubled. This is only 2.5 per cent of world crude production. By doubling the protein production the grams per day per capita for underdeveloped countries representing nearly 75 per cent of the world's population would be nearly doubled. Turning back now to the section on the present state of sea pollution control, it was estimated that a potential loss of 8.5×10^6 tons of oil was discharged into the sea from 1926 to 1968 due to government discouragement of recovery techniques resulting from ship tonnage admeasurement requirements; with the processes now being used this quantity of waste oil would have produced 4.3×10^6 tons of protein. The present protein content per day per capita for underdeveloped countries of the world is about 10 g or 3650 g/capita/yr. Roughly, therefore, one ton of protein would furnish 300 persons/yr with their present protein diet. And thus had the wasted oil (1926–1968) been recovered and turned into 4.3×10^6 tons of protein, it would have doubled the protein diet of 1.3×10^8 people in the underdeveloped countries for 1 yr (Champagnat, 1965; Tanner, 1967).

As recommended by Batelle (1967) more accurate standards and measures of the effects of oil pollution on the flora and fauna of the sea and beaches should be established by research. Particular attention should be given to differences in pollution resulting from solid oil masses, such as oil slicks and finely divided oil in water emulsions when in motion. While there is a great volume of literature on the effects of oil pollution, little of it is usable for the making of pollution control policy.

The Future of Sea Pollution Control and a Warning

As already mentioned, the Liberian tanker *Torrey Canyon* United States owned, was stranded on Pollard Rock near Lands End, England, on March 18, 1967. She carried a cargo of 119,000 tons of Kuwait crude oil. About 50,000 tons of cargo leaked from the ship, ruptured from the stranding during the first 10 days. On March 26, in a gale, the tanker broke her back and another 50,000 tons of oil went into the sea. Subsequently she broke in three parts. On March 28 the salvage work was abandoned. On March 28, 29, and 30, the British Government ordered the wreck which still contained 20,000 tons of oil, bombed by military aircraft to ignite and burn the remaining oil. The oil came ashore for the most part along the coast of Cornwall in England and the Brittany coast of France (Gill et al., 1967; Committee of Scientists, 1967).

When it became evident that the *Torrey Canyon* could not be saved and that her entire cargo would soon reach the sea and pollute the coasts of southwest England and the harbors and possibly the French coast the government and the people became deeply concerned. The Prime Minister and others of high office visited the scene. A committee of scientists was appointed to advise on courses and plans. All branches of the armed forces were called upon for ships, equipment, and thousands of men. Civil branches of the government contributed their time and talents. Twenty-four vessels at sea and several thousand men on the beaches used $2\frac{1}{2}$ million gallons of detergent. On the French coast alone 600 naval and civil vessels and large beach parties worked to control the pollution. The total cost to the Crown for this effort is said to have been 8 to 12 million dollars.

The gallant men who labored to deny to the sea the cargo of the *Torrey Canyon* and having failed, labored again with the French to deny it to the beaches were heroes all. The technical historians will disagree on the total result of this grand effort but its value to those who face similar problems in the future is beyond estimate. This story, so well documented, is the greatest practical contribution to pollution control resulting from tanker casualties ever made. Over 65 papers and reports have been published on some phase of the loss of the *Torrey Canyon*. In the salvage of ships, success in the main is always contingent on the weather. In the salvage of tank ships success is contingent both on the weather and the grave hazards of dangerous cargo. A great virtue of the *Torrey Canyon* operation was that so many things were attempted. And where the problem is so complex and the experiences so limited, it will always be true that the value of finding out what does not work is only exceeded by finding out what does. The two most significant decisions made with respect to the *Torrey Canyon* were to bomb her and to use detergents.

What is to be done with out present store of information and experience accumulated over the years and now crowned by the *Torrey Canyon* casualty? Will it be relegated to the archives until the next great tanker strands? The first paragraph under "Conclusions and Recommendations" of the "Report of Scientists on the Scientific and Technological Aspects of the *Torrey Canyon* disaster could be prophetic:

> The Torrey Canyon disaster is, to all intents and purposes, a thing of the past. Having polluted the French coast at the beginning of April, the last large patches of oil from the tanker were disposed of by the French Navy in early June. The summer brought few complaints either from holiday makers or fishermen, or if

they were made, they were so muted as to have attracted no notice in the national press. There is hardly any memory of the public anxiety which prevailed during the period of crisis in the latter part of March and early part of April. (Committee of Scientists, 1967).

Will we, with the passage of time, gradually degrade the *Torrey Canyon* experience unconsciously from "disaster" to "episode" to "incident" as appears in the Oil Spillage Study? Or will we go to the other extreme and draft legislation to provide for a crash program which may ruin small barge owners, cross organize and confuse the responsibilities of the agencies, and ultimately prove to be unconstitutional?

Since World War II there has been an unprecedented increase in the use of petroleum, in the use of pleasure boats, in the use of big tankers and in the disuse of the Suez Canal and similar developments, but these changes are all part of the continuing problem. The immediate concern should be a program for the future. Many suggestions have been made and these should all receive careful consideration to avoid great waste of time and money. The petroleum industry has held the line on prices for many years. Most price increases in its products are the result of increased taxation. The closing of the Suez Canal forced the petroleum industry to carry oil from the Middle East around the south end of Africa, an increased distance of 5000 miles. This added expense could have been passed along to the customer but it has been covered, or nearly covered, by building much larger ships capable of moving oil as cheaply around the Cape of Good Hope as smaller tankers moved it through the Suez. There appears to be no other way for tankers to absorb further expense so that in the future most, if not all, must be paid by the consuming public. The per capita consumption of food in the United States is about 1 ton/yr. Over 200 million tons of food are required to feed the nation each year, but over 500 million tons of petroleum are required to fuel it. Unnecessary or capricious regulation of a commerce as vast as petroleum can become an expensive obstruction in short order but there is another more serious risk. Petroleum is regulated by the government as a dangerous commodity. It must be obvious therefore that changes of consequence in method and equipment must be examined by regulatory bodies with great care.

To the author the most significant trend in shipping since World War II is the tendency of some owners to waive proof by service at sea for theoretical efficiency. At the same time that this trend has become well established, the suggestions and recommendations for additional changes and innovations have reached an all time high. The excellent literature review contained in the Oil Spillage Study (Battelle, 1967) sheds much light on this trend in the area of tanker design, operation, and pollution from oil spillage. The authors of the Oil Spillage Study have made their recommendations (Battelle, 1967). In detail there are several hundred of these but in the area of ship design alone there are 30 recommendations for research and study. Some of the proposals have been tried and others might be unsafe or inappropriate for various reasons, but it is the opinion of the author that *all* recommendations should receive careful study by the United States Coast Guard and ship owners. Where full scale research is conducted at sea for a sufficient time the semi-empirical principle of ship design is of course maintained. The author has gone to some lengths to call attention to the unprecedented number of proposals for changes in tanker design and to warn against their adoption by avoidance of what has heretofore been considered "due process," namely, by proof at sea. He considers the preservation of this principle to take precedence over all of the plans, proposals, and recommendations that have been made relating to tank ships and for this reason he suggests to both government and industry to proceed with great caution.

It will be appropriate to sound a final warning for national defense. Without petroleum, military operations on land, sea, and in the air would come to a halt. Over 20 per cent of our present national petroleum requirement is imported, in time of war at least 25 per cent must be. This military petroleum must be supplied to our armed forces on a global plan. Two world wars have found us dangerously short of shipping. The shortage of tank ships was so great in World War II 525 new ones costing several billion dollars had to be constructed by the government. The war could not progress faster than these vessels became available. Shipbuilders were not available for such a project. Many had to be made. Millions of tons of steel were needed for merchant ships at the same time that steel was needed for naval ships, guns, and tanks. It cost nearly double the peacetime price to build these ships. When it came time to man them, the same shortage problem was faced with crews. When they finally got to sea nearly one-tenth of the ships suffered fractures in their hull structures. At one time this caused the war in the Pacific to be slowed down due to tanker shortage. This was from hasty design and construction which, with a bigger peacetime fleet, could have been avoided or at least cushioned.

Today, as we have seen, tank ships of the United States are on the decline. They receive no subsidy. They pay the highest wages in the world and their officers and men enjoy the best working conditions. In the last great war the United States provided for military purposes 2.7×10^9 barrels of gasoline of all kinds, 2.6×10^9 barrels of fuel (including kerosine), and 1.7×10^8 barrels of lubricating oil. Nearly all of this oil moved to the front in tankers. In tonnage petroleum represented well over half of all military stores of every kind shipped overseas.

Tankers are already regulated by more than 400,000 words in U.S. Statutes and 300,000 more words of regulations having the force of law and written under these statutes. No other vehicle of commerce is more completely regulated. Tankers are essential in *both* peace and war and yet are on the decline. The surviving tankers are kept alive by a statute which prohibits foreign-flag vessels from trading between United States ports. The moral of this statement is that additional burdens should not be imposed on tank vessels without assurance that the object sought is justified and that it will be achieved.

Observations

A National Policy for the Abatement of Pollution of the Coasts and Navigable Waters and Its Administration

1. The 1954 Oil Pollution Convention, to which the United States subscribes, contains a recommendation to all nations to appoint a National Committee on the Prevention of Pollution of the Seas by Oil to coordinate national effort and be the contact point, through the Department of State, for international cooperation. Thus the present National Committee On the Prevention of Pollution of the Seas by Oil has been approved by the Congress and has national and international recognition. It is an all-government committee with members representing all interested government agencies. The Commandant of the Coast Guard is Chairman. Since, by statute, the Commandant is also the government official that has complete authority over, and responsibility for, all tank vessels, the National Committee on the Prevention of Pollution of the Seas by Oil is, both logically and by statute, the proper custodian of pollution abatement planning and administration for the coastal and territorial waters of the United States for vessels and their cargoes. There are other persuasive reasons. The Coast Guard has jurisdiction over the ships and planes, small craft and their bases, and the communications system that would be used for search, rescue, and salvage. But perhaps of greatest importance is the relationship of the Coast Guard to the officers and seamen of all vessels in the United States Merchant Marine, tankers, dry cargo, and passenger ships. The officers of tank ships and passenger and dry cargo ships are sworn agents of the United States.

They are under oath to uphold all shipping statutes and regulations written under them. All have passed examinations for their licenses which are issued to them by the Coast Guard and may be revoked for cause. Among the unlicensed men who have responsibilities for the cargo, all are certificated as Tanker Men. These certificates are, after examination, issued by the Coast Guard and they may be revoked for cause. Both officers and men are responsible to the Coast Guard.

2. It is suggested that the National Committee on Prevention of Pollution of the Seas by Oil name a small committee of their members to prepare projects and agendas.

3. It is suggested that the Coast Guard designate an officer at Headquarters with whom representatives of tanker owners could cooperate in matters relating to ship salvage.

4. It is suggested that the petroleum industry prepare to cooperate with the Coast Guard by organizing to clean up oil spills originating with their own vessels by (a) appointing from their marine shore organizations on a rotating part-time basis by companies, experienced officers from the east, west, and Gulf coasts to cooperate with the designated Coast Guard officer; and (b) standardizing as much of salvage planning and procedure as feasible such as a standard form or book for casualty reporting by code. A communications book containing names, addresses, and office and home phone numbers of key operating personnel in the marine departments of tanker owners and at headquarters and in the districts of the Coast Guard; and (c) investigating and reporting on equipment, standards, and operations in aid of salvage work, such as dependability of communications. (Actual salvage work would be handled by the owner as at present.)

It is suggested that the petroleum industry (a) appoint advisory groups to the National Committee upon request; (b) have its appropriate committee meet from time to time with the designated Coast Guard Officer on salvage work; (c) hold joint meetings with Coast Guard, Naval Salvage Officer, and industry salvage group; and (d) invite speakers on salvage at tanker conferences.

Research and Oil Pollution of the Sea

Considering again the 760 papers and articles reviewed in the Oil Spillage Study, it was 8 yr after the first paper appeared on the effects of oil pollution until the first paper appeared on a subject relating to the control of sea pollution by oil. As late as 1961 there were twice as many papers and articles available on the effects and restoration after sea pollution as there were on subjects relating to its reduction and control. It was not until 1967 that those relating to "causes" exceeded by a small margin those relating to "effects."

At the time of the *Torrey Canyon* stranding, biological degradation of petroleum in the sea was an established fact and yet detergents, most of which were toxic biologically, were used in great quantities, causing many experts to report that most of the damage was done by the detergents.

Since the ultimate fate of petroleum in the sea is biological degradation it follows that all of its states between spillage and disappearance are transitory. Persistence, therefore, has no meaning unless it is identified with time. Time of degradation is in turn affected by bacterial count, oxygen, temperature, hydrocarbon dispersion, and distribution. Since other than by evaporation and solution with the sea the oil would persist indefinitely in the absence of bacteria, we cannot claim to know the mechanism of sea pollution by oil, for we do not know, except in the most general terms, the phenomena of its disappearance. In this age of science the absence of research at sea on a subject claimed to be so vital to the health and happiness of the people as sea pollution by oil is a reflection on those in government and industry who seek means for its control.

Therefore, the following suggestions are offered:

1. The petroleum industry should establish and finance a research project to determine the degradation rate of petroleum at sea and on the coasts under various conditions.

2. As a part of the above research project it is suggested that the petroleum industry study the best means for removing oil that is polluting beaches by returning it to the sea where its biological degradation will be continued. Representatives of the government should be invited to participate.

3. The petroleum industry should cooperate with the Coast Guard in establishing standards in aid of salvage operations.

4. It is also suggested that the Department of the Interior undertake a substantial research program to develop (*a*) pollution standards for static and flowing seawater under oil films and under oil-in-water emulsions; (*b*) the destruction of oil slicks without prejudice to the process of biological degradation; (*c*) the evaluation of nontoxic detergents; and, (*d*) the determination, in situ, of the effects of oil pollution from stranded tankers or similar sources of oil by the use of sink boxes containing living specimens of local flora and fauna to be compared with similar control specimens.

Major Tanker Casualties at Sea

So long as men go down to the sea in ships there will be casualties. No magic formula will reduce the rate. It will be reduced in the future, as in the past, by the same sustained effort which has maintained the standing of the United States Merchant Marine no lower than second over the years in safety in world shipping. Tankers are subject to all of the hazards of the sea and, in addition, to those of the dangerous cargo they transport. Unlike dry cargo ships, which spend several days in port on each voyage, tankers are underway 10 to 15 per cent more each year than general cargo vessels.

The morale among the officers and men of United States tankers has always been high. Their wages and working conditions, as already reported, are the highest. Many of the officers are Naval Reservists and among those of higher rank there are a substantial number who served on tankers throughout World War II. For the first year of United States participation in that war their casualty rate exceeded that of combat Marines fourfold. It will ever be true that the best insurance against casualty at sea prevails when crews are competent, ships seaworthy, owners enlightened, and government effective. The conduct of the affairs of the Merchant Marine by the United States Coast Guard should inspire the confidence of the government and the people as a result of the effective discharge of their responsibilities essential in peace and vital in war.

Since the loss of the *Torrey Canyon* it has been suggested that a contingency plan be established in the event of a similar accident in the United States. There is little similarity in marine salvage cases and while two cases may be similar to start with weather can intervene to make them substantially dissimilar. Advance planning should take the form of assisting "the dauntless three": the Ship Master, the Salvage Master, and the ranking Coast Guard Officer present. These men, working under the most difficult and dangerous conditions, may be depended upon to cooperate fully. The Master is in command of his ship until he abandons it or is superseded by due process. The Salvage Master uses his skill, crew, and equipment to free the ship so long as retained by the owners and Ship Master. The Coast Guard Officer will first concern himself with the saving of life and then counsel and assist with plans agreed upon by all three.

The author has been asked to state in general terms what plan he would follow in salvaging a large, loaded, stranded tanker that was leaking oil (or might later do so) in quantity to the sea with special reference to reducing as much as possible the resulting pollution of the seas by oil. Only with the hope that his own ideas may stimulate others to improve upon them does he comply.

Suggested Means for Reducing the Volume and Effects of Oil Pollution Resulting from a Leaking Loaded Tanker Stranded on the Coast

Before the Casualty

1. Operate a seaworthy fleet designed, built, and operated in strict conformity with government regulations and classification society rules.
2. Through industry associations, promote meetings between industry and Coast Guard representatives to provide cooperation in tanker salvage work.
3. Through industry associations supply tanker owners and the Coast Guard with information agreed to under (XDii).
4. Provide against failure of communications in emergency.
5. Produce and provide to all tank ships and owners and the Coast Guard a Tanker Casualty Book containing spaces for all information pertinent to salvage work. This book should be organized as a check off list which at the same time would code the information for fast dispatch by radio.
6. The saving of life and the care of the injured will, of course, be the first concern of the salvage party.

Attention is called to the dangers of salvage work on tank vessels. Only those of the salvage party or representatives of the owner or the Commandant of the Coast Guard should be allowed aboard during salvage operations. The Salvage Master on the *Torrey Canyon* was killed. A Lloyds surveyor was killed on another tanker. Many others have been killed or injured. The author knows of only one place more dangerous than a stranded tanker—a stranded tanker in a mine field. He knows first hand.

After the Casualty

The following plans may be considered when they do not conflict with the general salvage program. Their object is to reduce pollution to a minimum by taking advantage of the full microbiological capacity of the ocean for the degradation of oils. Some were used in World War II when tankers were under attack by enemy submarines.

1. Reduce the flow of oil to the sea by transferring cargo from leaking tanks to empty undamaged tanks where possible.

2. Break up oil leaking to the sea by streams of water to produce an oil-in-water emulsion in the sea.

3. If main engines can be turned over slowly ahead without danger this leaking oil will be washed through the wheel and further broken up. The rudder may be used to direct it offshore.

4. The main engines may be similarly used to cause a flow of leaking oil along the sides of the ship toward the bow or stern.

5. There are times when weather conditions permit the main engines of the rescue fleet to be similarly used to pull or push the forming oil slick further offshore. Most of these vessels as they arrive on the scene will anchor seaward. If, as they arrive, they take up positions seaward off the stranded tanker some 500 to 600 yd (depending on their lengths), and then anchor with a good scope of chain and go slow astern on their engines they can, by using their rudders, come around to a position holding against their anchors and headed seaward. Their engines going slow astern will pull the oil slick away from the tanker, churn it up with their wheels, and start it toward the open sea. A smart watch must be kept during such operations of course. The advantage of a plan of this kind is that the equipment will be at the scene nevertheless and can be used immediately if not required for more important assignments.

6. Crude and heavy oils when spread on the surface of the sea are practically impossible to ignite. Gasoline will ignite and burn on a calm sea if the film is fairly thick. After about an hour of exposure the film of gasoline will normally evaporate except that close to the leak which is being constantly replenished. Crude oil contains lighter fractions which usually ignite if the oil has just been exposed to the air but this is less likely if the oil film is thin. Usually a slick of crude oil on the surface of the sea which has weathered for 30 to 45 min is safe from ignition. This would normally be the case at any point of the slick a quarter mile or more from the leaking tanker. *This is the time to emulsify the oil slick.*

7. The oil slick may be emulsified, weather permitting, by calling on all of the high speed motor craft owners (and larger vessels also) in the community to volunteer their boats and services to run back and forth from the far end of the slick to within $\frac{1}{4}$ mile of the tanker. These vessels would be acting as "blenders" to produce an oil in water emulsion which will facilitate bacterial degradation of the oil at the most rapid rate running around the edges of the slick and ending at the center. With a large continuous spill this may take several days but for every day it is maintained shore pollution from the slick will be substantially reduced. This plan has several advantages. No special equipment is used. Motor boats are available at many points around the coast on short notice. The Coast

Guard is well known to motor boat owners. Motor boats would scare water fowl away from the slick at the same time that they were blending the oil. An oil-water emulsion would be far less harmful to the fowl. As the oil slick became mechanically emulsified, the bacteriological degradation rate would sharply increase and in a few days the slick should vanish unless it is constantly maintained by fresh oil from the tanker.

8. When a tanker is stranded the obvious thing to do is to "lighten ship." If the tanker is "in ballast" this is readily and safely accomplished by pumping ballast to the sea. What little oil goes over the side with the ballast would be of no consequence. On the other hand if the tanker stranded when loaded with oil, a far more serious problem is created. Many tankers stranded when in ballast have gotten themselves afloat with nothing more serious than some bottom damage. A stranding when loaded will nearly always require outside assistance unless cargo is jettisoned. Legislation is pending which would regulate the jettisoning of oil cargo. This should be avoided if there is evidence that fine weather will continue and a reasonable chance that a small tanker or oil barge can be brought alongside in reasonable time. It is of course assumed that before cargo is jettisoned as a last act of desperation, that all fresh water that can be spared has been run to the sea, that the ship's engines have been used at high water, and that available vessels have assisted on tow lines.

The most desperate situation of course is a loaded tanker stranded on a rocky coast with the weather and local conditions such that barges and small vessels cannot come alongside and holed in the engine and pump rooms and one or more cargo tanks. And yet with portable pumps, compressed air, and a brave and versatile salvage party, ships have literally been snatched from certain doom in this condition.

9. There is a phenomenon of ship movement which must always be offset when a tanker strands under full power. When a tanker moves ahead at normal speed she "squats" aft, that is to say the stern (if a big ship) will drag down or squat from 8 to 10 ft while the bow rises by the same amount. Where a ship strands on a rocky coast at full speed, it attempts to come to rest at normal trim (that is, without the squat). This usually results in the forward compartments of the ship becoming impaled on the rocks, resulting in a general setting up of the ship's bottom forward. This is a desperate situation requiring desperate remedies. The forward tanks must be blown with air if their bottoms are ruptured. This forces the oil and water in these tanks to be blown to the sea through the ruptures in the bottom shell. If some tanks forward remain undamaged

they must be pumped to the sea. It is thus in the death struggle of a tanker that oil reaches the ocean. If the vessel finally breaks in two, then a great quantity of the cargo may go into the sea at one time. Such was the predicament and the end of the *Torrey Canyon*.

10. When a loaded tanker is abandoned to the sea, or after stranding and breaking in two, all cargo spaces should be closed and secured. This is in keeping with the principle that if oil must be surrendered to the sea the rate should be as low as possible (unless it is being jettisoned) so that it will be biologically degraded before it creates much of a slick. A loaded tanker should never be destroyed unless it has become a hazard to navigation. The ends of a tanker broken in two are individually stronger than the original ship. They are therefore rugged oil containers which will release the oil to the sea slowly by the ship's vent system or ultimately by the rusting through of the thinnest plate.

11. Until more is known about the cleaning up of beaches polluted by oil it would appear that the most logical plan to follow would be to wash the oil back into the sea, with streams of seawater where its full bacterial degradation could continue. Extremely heavy oil deposits should be bulldozed into the sea. To dispose of the oil otherwise risks the pollution of other areas. Light deposits of liquid oil at or near low water should disappear in a few hours.

Attempts to determine whether wrecks of loaded tankers lost by enemy action in the war now contain oil is a dangerous waste of time and money. As soon as a loaded tanker sinks the oil starts to escape slowly as the result of the two way flow through the tank vent valves and lines. The oil, being lighter, exerts an upward pressure and as it flows up and out a similar quantity of seawater flows down and into the tank through the vent lines. The escape of the oil is very slow which is desirable and it is to some extent emulsified by flowing in a direction opposite to the seawater in the same pipe. It is for this reason that oil slicks from tankers which have been sunk for several days are seldom reported. In the case of very heavy oils (such as Bunker C fuel) the oil would hardly flow through the vent lines in cold water. In addition its gravity being nearly that of seawater does not create the hydrostatic pressure differential that would be the case with lighter oils.

The theory that large quantities of oil may be released from tanker wrecks during storms is unlikely. Practically all loaded tankers sunk by enemy action off the United States coasts went down in soundings of 40 fathoms or more. Surface disturbances, such as storms, are hardly noticeable at depths of 25 fathoms or more.

There has been much discussion as to whether the new large tankers will tend to decrease or increase the overall pollution of the oceans. It has been said that because they are larger fewer voyages will be required to move the same quantity of oil so that there should be fewer accidents and therefore less pollution. This is probably true, but it is not the prime reason why there should be less pollution with the big ships. Seventy-five per cent of all shipping accidents take place in pilot waters. The larger of these vessels will not enter pilot waters at all, or else they will move in pilot waters for very short distances. Their casualty rate for stranding and collision should thus be substantially less than that of smaller vessels on normal runs. When they do have accidents causing oil pollution, as of course some will, it will be more than likely that the pollution will take place in open water where it will have more time to degrade biologically. The future safety record of big tankers will be the resultant of two opposed considerations: (a) improvement resulting from fewer voyages for the whole fleet and more off shore navigation and (b) debasement of the record of smaller ships due to definite "big ship" characteristics such as deeper drafts or other things we may not even have anticipated.

Due to circumstances described at the beginning of this chapter it has been necessary to compress a half century of sea pollution history, 2 yr of national and international concern, the tone of 800 learned papers, and the behavior of petroleum in and out of ships into a very modest space. The author only hopes that he has left the reader with a better understanding of the reasons for the presence of petroleum in the oceans and the difficulties of the task of reducing it. That entirely too much petroleum reaches the oceans in a form unacceptable for rapid biological degradation, there can be no doubt. That it can be substantially reduced is certain provided there is a sound national program continuously and energetically pursued together with international cooperation.

Addendum

Since the completion of this chapter, two events believed to be worthy of special notice because they have some relation to sea pollution by oil have taken place. They will be briefly described.

January 28, 1969—Blow-Out of Well No. 5, Platform A, Santa Barbara Channel, California

While drilling this well the drill penetrated a pool of very high pressure gas and oil. The well was provided with all means for preventing blowouts. Because of the very high pressure, these devices failed. As a last resort the drill pipe was dropped back into the hole and blind rams were used to cap the well. Normally this would have stopped the flow of oil but now, due to the high gas and oil pressure, cracks opened in the ocean floor and oil and gas escaped to the sea.

The Santa Barbara Channel oil rights are owned by the United States Government. In February 1968 the government accepted bids from 20 oil companies totalling 603 million dollars for the drilling rights on 71 tracts totalling 363,181 acres of undersea lands along this channel. Well no. 5 was being drilled on one of these tracts.

During the past 45 yr about 8000 wells have been drilled in offshore areas of the continental United States. Of these, only 16 wells "blew out of control." Of the 16 blow-outs, 12 were gas wells which caused no problem. Of the remaining four oil wells only one caused significant pollution.

During the first 12 days following the blow-out of well no. 5, an average of about 500 barrels of oil a day flowed into the ocean. Although some of this oil was recovered, much of it formed a drifting oil slick that smeared boats, beaches, and harbors. By February 8, with the effort of hundreds of oil men, the well and surrounding areas were sealed with a 3500-ft column of cement. In a bottom structure such as this, however, a permanent sealing cannot be accomplished.

Following the blow-out of this well, fantastic stories of the damage appeared in the press and on television programs, yet 3 months after the accident, the beaches were clean and there was little evidence that there had been heavy oil pollution in the area. The effect of this pollution on the ecology of the Santa Barbara Channel was determined approximately by March 10 by W. North and his associates from the California Institute of Technology. The conclusion from this preliminary study was that there had been little apparent damage. North's final report was made in March 1970. The most recent report on this oil spill will be found in the Allan Hancock Foundation Publication (1970).

In the second section of this chapter (Sources of Petroleum Which Reach the Oceans), it is stated that there are two submarine oil springs off the California coast. These springs are shown on charts of the U.S. Coast and Geodetic Survey. One of these springs is in the Santa Barbara area. The Spanish explorers discovered these oil springs four centuries ago. One flows 20 barrels of oil each day.

What is the importance of these off shore oil wells? They are very important. The proved reserves of oil in the United States equal but a 10-yr supply of oil for the nation. Ten years ago 13,191 exploratory wells were drilled in the United States. Only 8879 were drilled in 1968. About 13 per cent of our oil and gas comes from wells on the sea bed off the coasts. The leasing of these offshore oil lands is an important source of revenue to the Federal and State governments. Since 1954, 3.5 billion dollars has been paid to the Federal government in sales and rentals. Another 950 million has been paid in oil and gas royalties. The states have collected about a billion dollars in oil revenues in the same period. The petroleum industry has invested 16.3 billions in offshore production of oil in two decades. It has grossed in that time but 8.8 billions. Thus the off shore operations of the petroleum industry show a deficit in two decades of 7.5 billions.

In summary, the United States has title to submerged lands. These include the submerged lands under the Santa Barbara Channel. The United States determines to lease or sell these Santa Barbara drilling rights. The owner knew there was oil and gas seepage in the area. An oil company purchased the right to a tract and proceeded to drill. The company was experienced. Its drill crew was experienced, having drilled many wells, some on this tract from this platform. The drilling equipment was modern in all respects; it complied with all regulations and was properly used. In the course of drilling, the well, blew out (or more properly, the geological structure through which the well passed blew out). Heavy pollution of beaches and boats took place for 2 or 3 weeks, the effects of which lasted for 2 or 3 months. The public reaction was to cease all drilling in the Santa Barbara Channel. The decision of the President's Special Committee was to continue all drilling, since it would reduce the pressures that caused the blow-out, and to adopt some additional safety precautions. This case illustrates the inherent complication of pollution cases and the consequent difficulty of making sound decisions and of establishing effective but practical policy. For example, nothing has been said of the technology of this accident, of safety problems, of the normal economics of the operation, of the methods and cost of clean-up, of the legal considerations, local, national, and international, and of the ecology of the area before and after the spill and the duration of its effects.

The lessons to be learned from *Torrey Canyon* and well no. 5 and many earlier oil spills on water that are not so well known are starting to form a pattern. The following observations are presented.

1. We know far less about oil spills than we should know. Let it not be forgotten that we bombed the *Torrey Canyon* to burn the remaining oil when it was known from war experience that crude oil under such conditions burns indifferently if at all. The bombing of the ship destroyed

its hull structure releasing several thousand additional tons of oil to the sea. We used thousands of gallons of detergents which were toxic to marine microorganisms, the only certain means of rendering oil harmless on the spot. We first ordered wells shut down on the Santa Barbara Channel to stop the spillage of oil and finally ordered them opened up for the same purpose! We also used toxic detergents and dispersants.

2. We have been inclined to overestimate the deleterious effects of oil spillage, and then in ignorance used cures more deleterious than the malady.

3. The single undisputed fact about oil spills is again that we know too little about them. A program of research should be undertaken promptly and pursued with energy. The further civilization advances, the more likely its problems will be settled by compromise. Sound compromise implies a balanced knowledge of the subject to be controlled. Unless these controls are balanced, they will range from the ineffective to the obstructive. With such an involved subject as oil pollution it is vital that we know what we are about when we establish controls, for ineffective controls help no one, while the cost of obstructive controls is paid for by the people.

4. Finally, if the oil and gas underground in the Santa Barbara Channel are at such a high pressure in a rock structure that can be fractured by the drilling of a well, then it might be better to reduce this pressure by drilling wells taking the risk of some pollution rather than to wait for a future earth tremor to release a very large quantity of oil and gas at one time.

September 14, 1969—S/S Manhattan, Icebreaker-Tanker, in Fog, Ice, and Darkness, Passed through Prince of Wales Strait and Entered Amundsen Gulf, Alaska, the First Commercial Ship in History to Traverse the Northwest Passage

Early in 1968 geologists and production men of Atlantic Richfield Oil Company and Humble Oil and Refining Company concluded that they had discovered what appeared to be the largest oil field in America and certainly the most inaccessible. It was located on the North Slope of Alaska, adjacent to Prudhoe Bay.

The first and most difficult problem posed by this great discovery was how to get the oil from Alaska to centers of consumption. No time could be lost. The proven petroleum reserves of the United States would only last for 10 yr and domestic production of crude oil was expected to peak in 1970. The United States is now importing 12.2 per cent of its domestic requirement of crude petroleum. Domestic consumption of petroleum, now at 14 million barrels/day, was expected to reach 20 million by 1980. Free world consumption now at 36 million barrels daily was expected to reach 63 million by 1980. It was possible that the Prudhoe Bay Field could restore the deficit in United States production to meet domestic needs. It was also possible that this new source of petroleum would be of great military importance by making it possible for us to assist European allies with their petroleum requirements in the event of war, provided, however, that a substantial part of this Alaskan oil could be delivered to the United States east coast. It was clear that this could be accomplished at lowest cost if tankers could be successfully operated through the Northwest Passage—a venture which no commercial vessel had ever accomplished. In the meantime, an immense pipeline from the new oil field to an ice free port on the Alaskan Panhandle—a distance of 800 miles—had been projected at a cost of nearly a billion dollars.

The Humble Oil Company and associates now made one of the boldest and most far reaching decisions in industrial history. They determined to modify the best available vessel to simulate the operation of a tank ship in the ice of the Northwest Passage! Earlier in this chapter the author, in commenting on the many suggestions which had recently been made for the modification of tank ships, pointed out that prudent owners, naval architects, and marine engineers had, from earliest times, rejected novel changes in ships unless they had been preceded by a period of "proof at sea." The adventurers in this case wisely decided to secure the information required on a scale of 12 in. to the foot. It would cost them 50 million dollars; every penny of which, in the judgment of the author, was well spent—not only in their interest, but in the interest of national defense and the development of Alaska and northern Canada.

Only a few months after the discovery of the Alaskan North Slope oil field, the Humble Company and associates decided to challenge the ice of the Northwest Passage. The *S/S Manhattan*, largest and most powerful ship in the U.S. Merchant Marine, was procured. This vessel was 940 ft long, 148 ft beam, 142,500 tons with engines of 43,000 shp delivered to twin screws giving a speed of 16 knots.

The strengthening of the hull, including bow designed for ice breaking, required 9000 tons of steel. Many other changes and additions were made including very elaborate navigational and radio equipment. The vessel was so large, alterations so extensive, and available time so short, that it was found to be desirable to cut the original ship into four pieces to hasten the completion of the work. Accordingly, Sun Shipbuilding Company, prime contractor, assigned one part of the ship to Newport News

Shipbuilding and another part to Alabama Dry Dock, keeping two parts itself. One of these parts, the original bow, was stored and Bath Iron Works was assigned the building of the new ice-breaking bow.

By mid-August 1969, the four parts of the *Manhattan* had been reassembled by Sun at its yard at Chester, Pennsylvania. Successful trials were held after which 8000 different items of stores and 184,000 barrels of fuel were taken aboard. And on August 21, the *S/S Manhattan* started for the great adventure—a voyage of 8800 miles through the Northwest Passage and back to New York. She was in Halifax on August 28. Many scientists and specialists were aboard, and the U.S. Coast Guard icebreaker, *Northwind*, and the Canadian breaker, *Sir John A. MacDonald*, joined up just before ice was encountered. The MIT ice-breaking bow of the *Manhattan* with entry angle of 18° increasing to 30° functioned well in ice 6 to 14 ft thick. On September 12, in attempting to take the most direct route through M'Clure Strait, the *Manhattan* was brought to a halt in an ice flow running parallel to Banks Island under a northwest wind of 30 mph. The *Manhattan* and Canadian icebreaker *MacDonald* then retraced their course and traversed Prince of Wales Strait and reached Prudhoe Bay on September 18, 1969, thus completing the first successful commercial voyage through the Northwest Passage. The return voyage was made without incident with arrival in New York on Wednesday, November 12. Due to the heavy construction of this tanker for this service and the near absence of other traffic, it is believed that pollution of the sea by oil will not be a problem in the navigation of the Northwest Passage.

The projectors of this historic voyage state that they will not know until mid-year 1970 whether estimates now in process will favor the development of this route from Prudhoe Bay to the east coast. If the economics of this venture support the development of this route, the adventurers have indicated that they would use especially designed ships of about 250,000 dead weight tons, that a round voyage New York/Prudhoe Bay would require about $2\frac{1}{2}$ months, and that a start would be made with five such ships costing about 59 million dollars each. One such ship would be required for each 50,000 barrels of oil produced per day.

The opening of the Northwest Passage is vital to the defense of Alaska and northern Canada and to their commercial development. The governments of both these areas should support by every reasonable means the pioneering effort of the oil companies for this in turn will serve to release the fabulous mineral wealth of these northern lands of promise. When the last computations are made and judgement is finally called upon, it is hoped that the *Manhattan* adventurers will recall that they have proven that safe transit of the Northwest Passage is possible and that in reducing these operations to practice costs should be reduced by 10 to 20 per cent.

May the industrial statesmen who conceived and carried out the *Manhattan* experiment be successful in establishing the first sustained service by ship through the Northwest Passage (Robertson and German, 1969; *Manhattan*, 1969; Anonymous, 1969a; Anonymous, 1969b).

References

Allan Hancock Foundation. 1970. Biological and Oceanographical survey of the Santa Barbara oil spill. University of Southern California. 2 vols.

American Petroleum Institute. 1967. *Petroleum facts and figures.* American Petroleum Institute, New York, N.Y.

Anonymous. 1969a. Ocean industry.

Anonymous. 1969b. Under sea technology.

Battelle Memorial Institute. 1967. *Oil Spillage study.* Department of Commerce, Clearing House, Washington, D.C.

Champagnat, Alfred. 1965. Proteins from petroleum. *Scientific Ame.*, October 1965. Paper presented before American Chemical Society, Atlantic City, N.J., September 1965.

Committee of Scientists. 1967. *The Torrey Canyon.* Her Majesty's Stationery Office, London, England.

Gill, C., F. Broker, and T. Soper. 1967. *The wreck of the Torrey Canyon.* David & Charles, New York, 128 pp.

Manhattan. 1969. News releases of Humble Oil & Refining Co.

Moss, James E. 1963. Character and control of sea pollution by oil. American Petroleum Institute, 1101 Seventeenth Street, N.W., Washington, D.C., 122 pp.

Robertson, C., and W. H. German. 1969. A discourse to prove a passage by the northwest to Cathila. Presented at the 14th Tanker Conference, American Petroleum Institute.

Secretaries, Department of the Interior and Transportation. 1967. Oil pollution. Superintendent of Documents, U.S. Government Printing Office, Washington, D.C.

Tanner, James C. 1967. Food from fuel. *Wall Street Journal*, June 6, 1967.

ZoBell, Claude E. 1946. Action of microorganisms on hydrocarbons. *Bacteriol. Rev.*, **10**:33–36.

ZoBell, Claude E. 1946a. *Marine bacteriology.* Chronica Botanica, Waltham, Mass., 240 pp.

ZoBell, Claude E. 1962. The occurrence, effects and fate of oil polluting the sea. *Intern. J. Air Water Pollution*, **7**:173–193.

14

Petroleum—Biological Effects in the Marine Environment

D. K. BUTTON, *Associate Professor, Institute of Marine Science, University of Alaska, College, Alaska*

Petroleum is a mixture of reduced carbon chains and rings having a variable amount of heterocyclic and unsaturated compounds. The straight chain-saturated hydrocarbons predominate. These highly reduced molecules are relatively stable chemically in the absence of oxidants such as oxygen and have, therefore, accumulated in nature, and in some cases collected in subterranean oil fields.

The principal hydrocarbon sources of the marine environment are accidental oil spills, fuel and lubricant escapement, primary phytoplankton production, and natural oil seeps. Accidental oil spills attract the greatest public attention because they are concentrated and visible for a time. The frequency of oil tankers lost at sea has been steadily increasing with increased oil transportation. Currently about 5×10^{13} g of oil are being transported by sea at any one time (Cooke, 1969). World oil tankers lost at sea in 1963 amounted to 6×10^{12} g (Secretaries of Interior and Transportation, 1968). In a 2-yr period the Federal Water Pollution Control Agency lists approximately 50 oil slick sightings in Alaska's Cook Inlet alone, 10 of which averaged an estimated 1000 barrels each.* A reasonable estimate for oil spill input is something like 0.1 per cent of the total oil transported (Blumer, 1969a; Moss, Chap. 13, this book).

Terrestrial oil spills contribute significantly to this quantity. Five incidents were reported within the continental United States during the

* Federal Water Pollution Control Agency, College, Alaska.

last year totaling 10^9 g (Smithsonian Institute, 1969), the Chicago River, Piscataqus River, Weymouth, Fore River, and the Allegheny River oil spills. Smaller additions regularly occur during the disposal of used lubricating oils.

Hydrocarbons identical with those found in petroleum are common constituents of plant hydrocarbons. Kelp has a typical plant and microorganism hydrocarbon content amounting to about 0.03 per cent (Davis, 1967). Based on the world ocean annual primary productivity of 5×10^{16} g/yr (Fairbridge, 1966), the recently produced photosynthetic hydrocarbons add some 10^{13} g/yr, or about the same amount as introduced by oil spills.

Oil seeps have been known throughout recorded history and were an early means of oil prospecting. The Trinidad tar pits are expected to yield 25 million tons of asphalt which on removal it is replaced from below (Davis, 1967). Oil slicks have been reported from seeps at Coal Oil Point in California, Cook Inlet, Alaska, and on the island of Tonga (Taylor, 1969).

It is thought that Santa Barbara has experienced larger oil slicks in geological time than in the 1969 drilling accident. Slicks have been repeatedly sighted on some of Alaska's artic north slope lakes. One of these covering about a half acre was sighted by a passing charter pilot on Oil Lake a few miles west of the Colville River delta east of Point Barrow in the fall of 1968. The following summer no oil could be found on the beach although the short cool summers retard microbial oxidation of plant material and a thick absorbent bed of peat-like vegetation accumulates on the Windward side and constantly turns in the steady breeze. This could obscure a small slick. The northwest or lee shore is cut through a sandy thermokarst rising some 15 ft above the otherwise flat tundra. Here three layers of coal-like material about 1 cm thick were apparent a few centimeters apart with an uppermost tarry layer of sand. There is no evidence however that such seeps and eruptions contribute significantly to the hydrocarbon content of marine waters.

Microbiological Hydrocarbon Oxidation

It is apparent that the oil input rates reported which are similar to the world's annual catch of fish are balanced with a removal mechanism operating at essentially the same rate. Otherwise, accumulation of oils at the air-sea-land interfaces would be more prevalent. However these removal processes, principally biodegradation, are not apparent to one viewing a high local concentration of oil such as the recent Santa Barbara oil spill of 10^9 g (Jones et al., 1969). This slick covered about 800 mile^2, persisted through a storm and lasted some 2 weeks. Oil spills in Cook Inlet on the other hand dissipate quite rapidly. The half-life of a Cook Inlet slick apparent to the aerial observer is about 1 day. Much of Cook Inlet is strongly mixed by 40-ft tides. Spilled crude is quickly bounded by a layer of emulsion, "chocolate mousse," with subsequent dissipation proceeding rapidly so that the primary removal mechanism is a function of mixing (Le Petit and Barthelemy, 1968). This has been substantiated in our laboratory with stirred and quiescent oil slicks in 5-gal bottles undergoing biodegradation with and without added silt. Even on nutritionally rich water a very thin oil slick will last months when incubated in an unstirred vessel with a mixed culture of hydrocarbon oxidizing microorganisms. Relatively heavy oil slicks will break up in a few days however when inoculated and gently stirred with a rotating magnet.

Microbiological petroleum oxidation kinetics are influenced by the number and molecular dissimilarity of the components, the insolubility of the higher boiling fraction, and the toxicity of the lower molecular weight fraction. Nevertheless biooxidation occurs in all so that continuous cultures can be grown on 40 mg/l phenol, and black-top highways and asphaltic pipe coatings are decomposed (Holberg, 1964). Crude oil contains many components. Mair (1962) lists 234 paraffins, cycloparaffins, and aromatics identified up through C_{33}. The 10 most abundant make up only 17 per cent of the total. Thus crude oil is a diverse substrate. Hydrocarbon oxidation mechanisms are however fairly general as compared with carbohydrate metabolism. Terminal methyl groups of normal hydrocarbons are enzymatically attacked in one of the two ways (Johnson, 1964): (a) by oxidation with molecular oxygen to form an alcohol through a hydroperoxide intermediate catalyzed by a mixed function oxidase, or (b) by a mechanism as yet obscure that will proceed anaerobically, presumably involving a coenzyme such as NAD^+. Subsequent oxidation occurs through conventional β-oxidation to acetate, a common biochemical pathway intermediate. Stepwise breakdown may proceed from either end of the chain. Many microorganisms possess the enzymes for ω-oxidation initially yielding dicarboxylic acids. Aromatic oxidation is characterized by the oxidation of benzene to catechol with the participation of molecular oxygen and possibly NAD^+ (Gibson et al., 1968). The vicinal hydroxyls formed are easily split to enter conventional oxidative pathways.

Normal hydrocarbons are metabolized in groups with preference for a molecular weight range but no odd–even carbon specificity. For example, organisms may use ^8C through ^{10}C, ^{12}C through ^{20}C, or ^8C through ^{20}C. The lower molecular weight limit is probably a function of toxicity stemming from disruption of the functional phospholipid components of the cell envelope. Very long hydrocarbon chains are perhaps more difficult to oxidize. Some of our mixed cultures growing on crude oil leave a residue of highly reduced nonaromatic hydrocarbons longer than ^{25}C.

Paraffinic hydrocarbon utilization patterns of specific isolates are frequently sufficiently different that they can be separated by the gas chromatogram of a culture medium containing a mixed substrate like kerosene. Specificity increases greatly as branching is introduced into the carbon chain. Some of our isolates prefer pristane (2, 6, 10, 14 tetramethylpentadecane). Others will use either branched or straight chain hydrocarbons. Several cultures form or enlarge a component in kerosene of the ^{11}C range. Detergents have been made more biodegradable by increasing the hydrocarbon chain length of their normal aliphatic portions to greater than ^{4}C. For a review see Johnson (1964).

The most significant portion of hydrocarbon oxidation occurs from the oil phase directly to the cell surface without aqueous solution. All of the oil oxidizing organisms available in this laboratory will emulsify oil, easily taking their own weight of crude oil into an oil-in-water dispersion. Organisms oxidizing kerosene are frequently not visible upon suitable magnification unless the nonaqueous phase is removed. In addition to metabolic activity such as CO_2 evolution, their presence is apparent by the distorted shape of the otherwise spherical oil emulsion droplets. Occasionally the organisms protrude from a droplet or just have one end attached as the other rotates freely. Low oil solubility has apparently directed the evolutionary development of mechanisms involving direct transport of the oil phase components into the organism. This allows for paraffin utilization in the range of molecular weights up through solid waxes at normal temperatures. Rate limitation for oil oxidation is probably not a function of diffusion transport of dissolved species. Johnson (1964) calculates the solubility of dodecane at 1.7×10^{-8} M to be below the range of effective utilization from solution. This is consistent with our apparent Michaelis constant for growth measurements of 4×10^{-13} M for vitamin B_1 (Button, 1970), 10^{-8} M for phosphate, and that of 4×10^{-7} M and $1 \times ^{-7}M$ for glucose and acetate measured by Wright and Hobbie (1965) as they are approximately related by their respective yield constants γ (Button, 1969).

$$K_s = \frac{1}{7 \times 10^7 \, \gamma}$$

The normal steady state level of hydrocarbons in seawater is very low. Blumer (1969b) has been able to resolve some of the chromatographically but quantitative measurements are not yet available. Most organics such as carbohydrates and amino acids are microbiologically oxidized with a great deal of specificity. This is imparted by stereospecific reactions involved in their transport and degradation. As pointed out earlier, hydrocarbon oxidation is a more general process and its transport may be just as general. We know that many oil oxidizing organisms have hydrophobic surfaces with a great affinity for oil so that the absorption of oil molecules colliding with the cell surface seems likely. Such indiscriminate oxidation would allow the concentration of individual species to function additively resulting in lower steady state hydrocarbon concentrations than total carbohydrates for example.

Oil spills emulsify fairly rapidly. Even the thick black portions of young slicks are water in oil emulsions. In laboratory culture these are visibly permeated by motile bacteria once a good starter culture develops. The organisms are capable of moving through the slick and breaking it up. Very gently stirred slicks break up in about 10 days at room temperature in the laboratory, whereas unstirred or poisoned controls are stable for extended periods. This emulsion is sufficiently dense to sink and mixes very easily. Vorshilova and Dianova (1950) described the fate of oil spillings in the Moscow River. They found part of the oil sinking to the bottom requiring concomitant denitrification or sulfate reduction for microbial oxidation. Photographs have been taken (Holme, 1969) of large clumps of oil that sank from the *Torrey Canyon* incident off the English Channel in 1967.

The fate of oil slicks is probably one of emulsification with permeation and multiplication of microorganisms. Mixing energy input appears to be the major factor determining how long the oil stays on the surface and to what size clumps it will break. Oil in protected calm areas may gradually impinge on the beach as some did in Santa Barbara. High concentrations with some wave action may lead to the sinking of some of the heavier fractions. Spills in highly turbulent areas such as upper Cook Inlet are probably completely dispersed into millimeter size particles in the order of 3 or 4 days. Biodegradation follows in all cases except where large clumps are buried in sediment and protected from oxidants. This is accomplished by a diverse flora, bacteria, and fungi, whose taxonomy is not extensively described. These are however rather small microorganisms of diverse morphology whose shorter dimension is normally less than 1 μ. Growth rates run about 0.35/hr on nutrient broth and 0.08 on mineral salts with kerosene at 25 C. Growth rates change with temperature according to the Arrhenius equation with an activation energy of about 15,000 cal/mole giving a generation time of about 14 hr at 5 C. Thus a single microorganism capable of oxidizing oil components of an oil droplet would require about 1 week to do so after initial contact. The bulk of hydrocarbon oxidizers found in Cook Inlet do not appear to have psychrophilic characteristics by any of the usual criteria. A noteworthy characteristic is that they grow quite nicely in the most nearly organic-free synthetic seawater media we can prepare.

An interesting problem is posed by the potential shipping of oil in the arctic. Biodegradation is the ultimate fate of oil spills. This proceeds as a function of dispersal so that nutrients can move to the site of oxidation, and of temperature. The ice pack imposes a barrier to wind mixing and tidal currents are nil. It is thus possible that oil spills will remain as a continuous phase on the water and in the ice, remaining for extended periods preserved by low temperatures and subject to gradual release. What effect such a sequence would have on arctic fauna and albedo has not been predicted.

Toxicity

Most of the low molecular weight hydrocarbons are hazardous to organisms in high concentration. Normal heptane for example is regarded as moderately hazardous to inhalation and is narcotic in high concentrations (Sax, 1963). This is not surprising in view of the high lipid content of membranes functional in active transport and nerve impulse transmission. Any freely diffusible hydrophobic species would be expected to alter such functions. In fact cyclopropane and ethene are effective surgical anesthetics.

Freshwater bass subjected to water mixed with crude oil lose their equilibrium. Normal behavior can be restored by stripping low molecular weight hydrocarbons back out of their environment with a stream of air bubbles (Wiebe, 1952). Concentrations of a number of solvents such as phenol and benzene in the range of 10 to 100 mg/l cause mortality of representative freshwater fishes after a 96-hr exposure (Pickering and Henderson, 1966). Amoebae exposed to water saturated with lower hydrocarbons show microscopically visible membrane damage and death. However higher molecular weight normal hydrocarbons induce no appreciable effect after microinjection (Goldacre, 1968). Many microorganisms capable of growth on normal hydrocarbons will not grow on the lower molecular weight range, hexane and heptane, for example, unless the concentration of these solvents is reduced to below saturation (Johnson, 1964). Reservoirs as unlikely as aircraft fuel tanks are however subject to fouling by microorganisms growing on condensed water vapor so that biocidal agents are common jet fuel additives.

Seawater has the very real elusive property of toxicity toward microorganisms. Although identification of this factor has not been successful, a very likely explanation is the cumulative effects of heavy metal inhibition on a relatively small biomass in a medium devoid of sufficient chelating capacity. Unbuffered phosphate limited continuous yeast cultures are inhibited by 10^{-7} M copper for example (Button, 1970). Nickel is one of the more toxic of the heavy metals (Fuhrmann and Rothstein, 1968) and crude oil can contain on the order of ppm nickel. This is a cumulative-type inhibitor that can build in an area frequented by oil pollution similar to the rise of lead in San Francisco Bay due to automobile exhaust (Tatsumoto and Patterson, 1963).

Taste tainting has been a recurrent problem to the shellfish industry (Simpson, 1968). Oil droplets are ingested by these filter feeders dissolving in the animal's lipids causing an off flavor. Normally the taste can be removed by holding the shellfish for a few months in oil-free water. Rainbow trout with an oily flavor have been caught downstream from a Canadian oil refinery (Krishnaswami and Kupchanko, 1969). It was found that the refinery effluent diluted 1:100 or to one quarter the minimum dilution detectable by odor would produce an oily taste in fish after a 24-hr exposure.

Conclusions

Oil spills are one of the many ways man changes the chemistry and physics of the earth. In the ocean they are emulsified by mixing energy or, less frequently, impinge on shore while still rafted on the surface. Biooxidation proceeds most effectively in the oil phase of the emulsified fraction.

Although many metabolites probably contribute to the cyclic nature of the ocean blooms, the overall controlling factor in euphotic zone biomass is combined nitrogen. Thus a well-dispersed oil spill probably has little effect on the productivity of a region. Oil chemistry is sufficiently similar to animal and plant lipids that the normal degradative processes are probably indiscriminate and steady state dissolved hydrocarbon levels not changed significantly by spills.

Spills in calm near shore water have been the most troublesome in fouling shore lines and killing birds. Frequent spills contribute to oxygen depletion and the aesthetic quality of surface and subsurface waters. Cumulative effects include the build up of relatively toxic heavy metals such as nickel. Potential arctic spills present a whole new spectrum of problems, that must be vigorously attacked in view of the potential of this region for large-scale oil production and transportation.

References

Blumer, M. 1969a. Oil pollution of the ocean. Massachusetts Institute of Technology, Woods Hole Oceanographic Institution Symposium. In press.

Blumer, M. 1969b. Personal communication.

Button, D. K. 1970. Some factors influencing kinetic constants for microbial growth in dilute solution. In *Proceedings of the symposium on organic matter in natural waters*, D. W. Hood (ed.). University of Alaska, Institute of Marine Science, Occasional Publ. No. **1**:537–547.

Button, D. K. 1969. Thiamine limited steady state growth of the yeast. *Cryptococcus albidus. J. Gen. Microbiol.*, **58**:15–21.

Cooke, R. F. 1969. Oil transportation by sea. Massachusetts Institute of Technology, Woods Hole Oceanographic Institution Symposium. In press.

Davis, J. B. 1967. *Petroleum microbiology*. Elsevier Publishing Company, New York, 604 pp.

Fairbridge, R. W. (ed.). 1966. *The encyclopedia of oceanography*. Reinhold Publishing Corp., New York, 1021 pp.

Fuhrmann, G. F., and A. Rothstein. 1968. The mechanism of the partial inhibition of fermentation in yeast by nickel ions. *Biochim. Biophys. Acta*, **163**: 331–338.

Gibson, D. T., J. R. Koch, and R. E. Kallio. 1968. Oxidative degradation of aromatic hydrocarbons by microorganisms. I. Enzymatic formation of catechol free benzene. *Biochem.*, **7**:2653–2662.

Goldacre, R. J. 1968. Effects of detergents and oils on the cell membrane. In *The biological effects of oil pollution on littoral communities*, Field Study Council (eds.), E. W. Classey, Ltd., Hampton, England, Vol. 2, 198 pp.

Holberg, A. J. 1964. *Bituminous materials: Asphalts, tars and pitches*. Interscience, New York. Vol. I, 432 pp.

Holme, N. A. 1969. Effects of Torrey Canyon pollution of marine life. Massachusetts Institute of Technology, Woods Hole Oceanographic Institution Symposium. In press.

Johnson, M. J. 1964. Utilization of hydrocarbons by microorganisms. *Chem. Ind. (London)*, **36**:1532–1537.

Jones, L. G., C. T. Mitchell, E. K. Anderson, and W. J. North. 1969. Just how serious was the Santa Barbara oil spill? *Ocean Ind.*, **4**:53–56.

Krishnaswami, S. K., and E. E. Kupchanko. 1969. Relationship between odor of petroleum refinery waste water and occurrence of "oily" taste-flavor in trout, *Salmo gairdnerii. J. Water Pollution Fed.*, **41**:189–196.

Le Petit, J., and M. H. Barthelemy. 1968. Les hydrocarbures en mer: Le probleme de l'epuration des zones littorales par les microorganismes. *Ann. Inst. Pasteur*, **114**:149–158.

Mair, B. J. 1962. Hydrocarbons isolated from petroleum. *Oil Gas J.*, **62**:130–134.

Pickering, Q. H., and C. Henderson. 1966. Acute toxicity of some important petrochemicals to fish. *J. Water Pollution Control Federation*, **38**:1419–1429.

Sax, N. I. 1963. *Dangerous properties of industrial materials*, 14th Ed., Reinhold Book Corp., New York. 1343 pp.

Secretaries of Interior and Transportation. 1968. A report on pollution of the nation's waters by oil and other hazardous substances. U.S. Government Printing Office, Washington, D.C., 31 pp.

Simpson, A. C. 1968. Oil, emulsifiers, and commercial shellfish. In *The biological effects of oil pollution on littoral communities*, Field Study Council (eds.), Vol. 2, E. W. Classey, Ltd., Hampton, England, pp. 91–98.

Smithsonian Institution Center for Short Lived Phenomena. 1969. Event Notification Reports No. 211, 474, 555, 581, 583.

Tatsumoto, M., and C. C. Patterson. 1963. In *Earth science and meteorities*, J. Geiss and E. D. Goldberg (eds.), John Wiley, New York. 312 pp.

Taylor, D. M. 1969. World report, gas and oil. *Ocean Ind.*, **4**:26–30.

Vorshilova, A. A., and E. V. Dianova. 1950. Bacterial degradation of oil and its migration in natural waters. *Mikrobiologiya*, **19**:203–210.

Wiebe, A. H. 1952. The effect of crude oil on fresh water fish. *Amer. Fish. Soc.*, **6**:324–331.

Wright, R. T., and J. E. Hobbie. 1965. The uptake of organic solutes in lake water. *Limnol. Oceanog.*, **10**:22–28.

15

Petroleum—Stable Isotope Ratio Variations

PATRICK L. PARKER, *Associate Professor of Chemistry, The University of Texas, Marine Science Institute, Port Aransas, Texas*

Stable isotope ratio variations for several of the light elements have provided insight into many natural geochemical processes. These variations can also serve as indicators of man's activity in cases where the natural stable isotope ratios have been shifted by significant addition or removal of material with different and characteristic stable isotope ratios. This effect has so far been demonstrated only for stable carbon isotopes. However, the prospects of detecting very subtle changes in geochemical reservoirs of major biologically active elements by this technique appear promising.

In this chapter a brief review of the theoretical foundations of modern isotope geochemistry is given. These foundations are the principles which describe chemical isotope effects. Next, the operation of isotope effects in natural systems to produce stable isotope ratio variations in geochemical reservoirs is illustrated. This discussion is based on the isotope chemistry of carbon, but the principles apply to other light elements. Finally, specific cases in which stable carbon isotope ratio variations have served as indicators of man's activity are described. The prospects of making use of stable isotope ratio data for elements other than carbon as similar indicators are discussed.

Background

The major elements which show nonradiogenic isotope ratio variations in nature along with their stable isotope abundances are shown in Table

Table 15.1. Per cent Isotopic Abundances for Biologically Active Elements

Carbon		Oxygen	
12	98.893	16	99.759
13	1.017	17	0.0374
		18	0.2039

Nitrogen		Sulfur	
14	99.634	32	95.0
15	0.366	33	0.760
		34	4.22
		36	0.014

From Friedlander et al. (1964).

15.1. If there were no isotope effects in physical and chemical processes, all natural materials would have these exact isotopic abundances. Isotope effects do occur so the exact isotope ratio of a given sample for these elements will depend on the chemical and physical history of the sample.

Although physical processes, especially evaporation, do cause isotope fractionation, many natural variations result from chemical isotope effects. Two types of chemical isotope effects operate in natural systems; the equilibrium isotope effect and the kinetic isotope effect. The chemical basis of the equilibrium isotope effect is well understood (Urey, 1947). Isotope equilibrium constants may be calculated provided the necessary spectral data are available (Bigeleisen and Mayer, 1947; Bigeleisen, 1958). In exchange reactions which have an equilibrium isotope effect the heavy isotope concentrates in one of the chemical species because the chemical properties of each isotopic species are slightly different. As an example, consider the exchange between carbon dioxide and biocarbonate at equilibrium.

$$^{13}CO_2(g) + H^{12}CO_3^-(aq) = {}^{12}CO_2(g) + H^{13}CO_3^-(aq) \tag{1}$$

The equilibrium constant for this reaction is

$$K = \frac{[H^{13}CO_3/H^{12}CO_3]}{[^{13}CO_2/^{12}CO_2]}$$

where the concentrations are the same as abundance ratios. If there were no isotope effect K would be a small whole number for all exchange reactions. The references already given form the basis for the calculation of this type of equilibrium constant. In practice it is usually more useful to measure equilibrium constants than to calculate them. The isotope equilibrium constant for (1) has been measured by several workers (Hoering, 1960; Vogel, 1961; Deuser and Degens, 1967; Wendt, 1968). The experimental quantity is the fractionation factor, α. The equilibrium constant is readily derived from α.

For (1),

$$\alpha = \frac{(^{13}C/^{12}C)HCO_3^-}{(^{13}C/^{12}C)CO_2} \tag{2}$$

and

$\alpha = K$, but in general
$\alpha = K^{1/n}$

where n is the number of equivalent exchangeable atoms under consideration. Then K, and hence α, are dependent on temperature but not on the conditions of the reaction or its mechanism.

Data in the field of isotope geochemistry are expressed in terms of δ (del), the per mil difference between the isotope ratio of the sample and a standard material. While this terminology may seem awkward at first, its merits soon become obvious. Del is the quantity that most isotope ratio mass spectrometers yield. Del is all that is required to relate calculated equilibrium constants and experimentally determined fractionation factors. Del—^{13}C is defined as

$$\delta^{13}C = \frac{(^{13}C/^{12}C)\ \text{sample} - (^{13}C/^{12}C)\ \text{std}}{(^{13}C/^{12}C)\ \text{std}} \times 1000 \tag{3}$$

Similar definitions can be written for isotope pairs of other elements. Isotope reference standards are available from the National Bureau of Standards. NBS20 is the carbon standard used in this chapter. Based on this definition stable isotope ratio variations will be expressed as small positive and negative numbers. A negative δ means the sample is enriched in the light isotope relative to the standard while a positive δ means the sample is likewise depleted in the light isotope. Modern isotope ratio mass spectrometers will measure δ to within ± 0.1 to 0.5 on a routine basis.

As a result of the equilibrium isotope effect in (1), atmospheric CO_2 is about 7/mil enriched in ^{12}C relative to the bicarbonate of seawater. This is illustrated in Fig. 15.1 where atmospheric $CO_2(g)$ is -7 and seawater

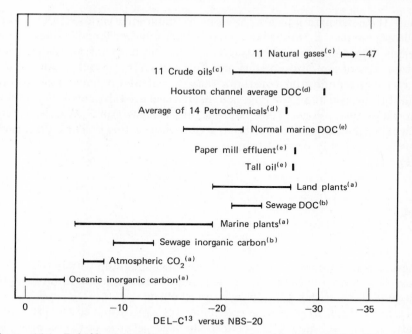

Figure 15.1 Del-^{13}C of biological carbon reservoirs. (*a*) Craig (1953); (*b*) Reimers (1968); (*c*) Silverman (1967); (*d*) Calder and Parker (1968); (*e*) Calder (1969).

bicarbonate is near zero. The small amount of $CO_2(g)$ dissolved in seawater, and the large amount in freshwater, will remain at -7/mil so long as it is in isotopic equilibrium with atmospheric CO_2. These two reservoirs constitute the two major biologically active reservoirs of inorganic carbon. It is seen later that the activities of man have changed the $^{13}C/^{12}C$ ratio, that is, $\delta^{13}C$, of both these reservoirs in some environments.

The kinetic isotope effect is the second type of chemical process which brings about variations in δ values in natural systems. Kinetic isotope effects, being mechanism-dependent, are not as readily calculated as equilibrium isotope effects. Organic chemists have studied reaction mechanisms by comparing measured and calculated isotope effects (Bigeleisen and Wolfsberg, 1958).

Kinetic isotope effects are the result of competitive isotope reactions of the general type

$$A_1 + B \xrightarrow{k_1} P_1$$
$$A_2 + B \xrightarrow{k_2} P_2$$
(4)

where the subscripts 1 and 2 refer to the light and heavy isotopic species, A and B are the reactants and P is the product. Due to the kinetic isotope effect the molecule with the light isotope generally reacts significantly faster than the heavy one. In this case the isotope effect is given by the ratio of rate constants, k_1/k_2. If the reservoir of A is large enough to not become depleted, the fractionation factor, α, is

$$\alpha = \frac{\left(\frac{\text{heavy}}{\text{light}}\right)_A}{\left(\frac{\text{heavy}}{\text{light}}\right)_P} \tag{5}$$

If the product is taken as the sample and the reactant as the standard in (3), then

$$\alpha \cong 1 + \frac{\text{sample}}{1000}$$

Kinetic isotope effects in biological cycles are responsible for most of the isotope ratio variations observed for the elements carbon, nitrogen, and sulfur, and for the $\delta^{18}O$ of atmospheric O_2. A very excellent review of stable isotope chemistry and geochemistry as well as a discussion of experimental techniques are given in a recent publication by McMullen and Thode (1963).

The overall results of the equilibrium and kinetic isotope effects in the various chemical reactions of carbon, which constitute the carbon cycle in nature, are shown in Fig. 15.1. Natural carbon has been separated into a series of reservoirs each with a more or less characteristic isotope ratio. This figure does not include all known data. It is intended to help us focus on the problem of how to recognize man's activity.

Stable Isotope Ratios as Indicators

Del—^{13}C of atmospheric CO_2 has been shown to be almost constant with regard to time and location (Keeling, 1961). This is not the case for aquatic inorganic carbon (IOC). IOC is the total CO_2 that can be released by an excess of acid. Del—^{13}C of IOC is most constant for open marine waters and most variable for estuaries and bays. In open ocean waters it is close to zero. Since most IOC in freshwater at pH 7 is present as $CO_2(aq)$ it should have the same $\delta^{13}C$ as atmospheric CO_2, -7. The values for estuaries then will vary depending on the mixture of fresh and marine waters. Freshwater IOC—$\delta^{13}C$ values as negative as -8 and -9 have been reported (Sackett and Moore, 1966). They are more negative than -7 due to the contribution of CO_2 derived from the oxidation of organic matter. If normal marine waters with IOC—$\delta^{13}C$ values near zero or even normal freshwaters with IOC—^{13}C values of -7 to -9 receive large amounts of IOC derived from organic carbon as a result of man's activity, the IOC—$\delta^{13}C$ of the normal system will be shifted. This is especially so for the marine system.

Reimers (1968) studied the $\delta^{13}C$ values of various chemical factions of the effluents of sewage treatment plans. These results are summarized in Fig. 15.1. The effluent IOC—$\delta^{13}C$ range was -9.3 to -13.9. The average value was -11.2. The marine bay, Corpus Christi Bay, which received these effluents had an IOC—$\delta^{13}C$ range from $+0.3$ to -4.7/mil with -2.0 being the average over a 5-month period. The open Gulf of Mexico has IOC—$\delta^{13}C$ values of ± 0.5. The negative values of the Bay are the result of the sewage effluent and the input of normal river IOC. The bay was always near pH of 8.0 so that biocarbonate would be the chief molecular form of inorganic carbon. It is clear that even in this shallow bay the IOC is not in isotopic equilibrium with the atmospheric CO_2. Exchange is not as fast as input. At equilibrium, (1) would generate IOC with a $\delta^{13}C$ close to zero. In cases where large amounts of sewage effluent are being put directly into marine waters, as in offshore California, the IOC—$\delta^{13}C$ may well serve as an indicator of man's activity. If the effluent is introduced well below the air-sea boundary, so that atmospheric exchange is eliminated, then the IOC—$\delta^{13}C$ may serve as a tracer of the water mass receiving the effluent.

Petroleum and natural gas have the most negative $\delta^{13}C$ values of any of the major geochemical reservoirs of carbon. Calder and Parker (1968) demonstrated that petrochemicals derived from petroleum and natural gas retain this negative $\delta^{13}C$. They found that fourteen petrochemicals taken from a plant had an average $\delta^{13}C$ of -27.2 while the dissolved organic carbon (DOC) in the effluent from the same plant ranged between -25.7 and -39.3. The same relationship holds for refinery effluents because crude oil is almost as negative as petrochemical carbon. Most of the 128 crude oils reported by Eckelman et al. (1962) had $\delta^{13}C$ values between -26 and -29 with a good many in the -30 to -33 range.

Normal marine organic carbon has the most positive $\delta^{13}C$ value of the organic reservoirs. Therefore the aquatic system consisting of a marine bay receiving effluents from oil refineries and petrochemical plants is an ideal case to demonstrate the use of stable isotope ratio variations as indicators of man's activity. Calder and Parker found that the organic carbon in the Houston Ship Channel shows the predicted relationships.

The $\delta^{13}C$ of the organic carbon for the Houston Ship Channel is shown in Table 15.2. The average DOC—$\delta^{13}C$ was -30.5. This contrasts with a

Table 15.2. Del-^{13}C and Concentration of Dissolved and Particulate Organic Carbon in the Houston Ship Channel

Station	DOC (mg C/l)	DOC (δ^{13}C)	POC (mg C/l)	POC (δ^{13}C)
1	5.8	−26	20	−19.8
2	5.6	−31.2	4.2	−21.3
3	3.9	−31.5	3.2	−24.2
4	4.0	−29.3	2.6	−23.1
5	9.0	−30.1	19	−26.3
6	26	−26.9	12	−24.4
7	3.1	—	4.0	−23.3
8	11	−48.8	8.8	−27.4
9	3.8	−29	2.4	−24.7
10	8.4	−27.1	3.8	−25.9
11	6.0	−27.5	2.9	−24
12	4.1	−28.2	3.6	−25.2
13	2.7	−28.2	4.0	−26
14	19	−32	16	−25.8
15	2.9	—	2.2	−23.0
16	1.4	−24.8	2.7	−23.3
17	2.1	−24.9	4.4	−21.1

From Calder and Parker (1968).

value of −20 for DOC—δ^{13}C of normal marine water found by these authors, and −22 found for Pacific Ocean water (Williams, 1968). When normal and pollutant carbon are mixed, as in the Ship Channel, the δ^{13}C of the mixture is given by the expression

$$\delta_m = \frac{C_n \delta_n + C_p \delta_p}{C_n + C_p} \tag{6}$$

where C = concentration of carbon mg/l, $\delta = \delta^{13}$C, and n refers to normal carbon, p refers to pollutant carbon, and m refers to the mixture. Still following Calder and Parker, the ratio of the amount of pollutant carbon to normal carbon in the mixture is given by

$$\frac{C_p}{C_n} = \frac{\delta_n - \delta_m}{\delta_m - \delta_p}$$

Trial values put into (6) indicated that between 1 out of 2 and 6 out of 7 of the carbon atoms in the Houston Ship Channel is derived from petrochemical pollution.

Much more research needs to be done on this approach. It would be very important to establish base line data for DOC—δ^{13}C for a number of our major rivers, estuaries, and near-shore marine waters. As man continues to use the marine environment it may well be that the DOC—δ^{13}C and IOC—δ^{13}C values will gradually shift, if indeed they have not already done so.

Reimers (1968) studied DOC—δ^{13}C of sewage plant effluents (Fig. 15.1). However it appears that this ratio is not characteristic enough to be useful as a tracer. The potential use of IOC—δ^{13}C of sewage effluents has already been pointed out. We have done a preliminary study of paper mill effluent DOC—δ^{13}C values which suggest that this may be a useful parameter for these systems (Table 15.3). It was found that the tall oil fraction of the plant studied had a δ^{13}C of -27. DOC—δ^{13}C of the plant effluent was -27.4. This plant did not recover its tall oil. A marine bay receiving substantial quantities of this type of effluent would have its DOC—δ^{13}C shifted. The same effect might be observed on a river by comparing data taken upstream and downstream from a papermill.

Friedman and Irsa (1967) have shown that air rich in automobile exhaust can show an 8/mil enrichment in ^{12}C for carbon dioxide. Again this shift toward a more negative δ^{13}C is due to the introduction of carbon dioxide derived from petrochemical products.

The δ^{13}C method gives no information about the chemical structure of organic pollution or its toxic properties. Its greatest usefulness is that it gives insight into the overall carbon flux of very large systems from

Table 15.3. Del ^{13}C Values from a Paper Mill

Fraction	Hardwood		Pine
Chips	-24.5		-24.8
Black liquor	-25.2		-24.9
Bleached cellulose	-23.4		-23.0
Lime mud		-23.4	
Tall oil		-27.0	
Effluent		-27.4	

From Calder (1969).

relatively few and simple measurements. It is a subtle indicator of man's impact on nature.

Other Elements

This chapter is intended to describe the state of our knowledge regarding the use of stable isotope ratios to detect man's influence on natural waters. It would not be correct to leave the impression that a great deal is known about this subject. It appears to be a useful approach. In the hope that other investigators will find it useful, the remainder of this chapter is devoted to providing background data and suggestions which hopefully will lead to further research on the use of stable isotope ratio variations in detecting short-term changes in geochemical cycles especially as they result from man's activities.

Sulfur

The stable isotope geochemistry of sulfur has received a great deal of study. This has resulted in one of the more profound uses of stable isotope data. I refer to the isotope data which indicate that sulfur dome deposits, such as occur along the Texas and Louisiana coasts, are the result of the biological reduction of sulfate (McMullen and Thode, 1963). Figure 15.2 summarizes the pertinent data on sulfur isotopes. Normal seawater sulfate is positive with respect to reduced sulfur. Following the line of reasoning that was used in the carbon isotope work, it appears

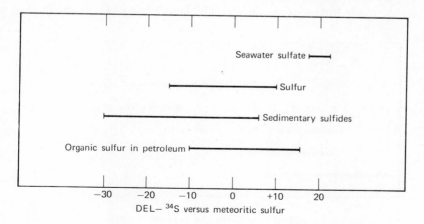

Figure 15.2 Del-^{34}S of biological sulfur reservoirs. From McMullen and Thode, 1963.

possible that normal marine waters receiving effluents rich in sulfur derived from elemental sulfur, sulfides, or organic sulfur would have their sulfate shifted in a negative direction. An example of this might be a petroleum refinery which oxidizes organic sulfur to sulfate and releases it into marine waters. More specific studies might include the $\delta^{34}S$ of sulfate upstream and downstream from a papermill which uses the Kraft process. It is possible that restricted marine estuaries could have the $\delta^{34}S$ of their sulfate shifted due to man's heavy use of reduced sulfur.

Oxygen

The stable isotope geochemistry of oxygen has been studied more than that of any other element. Oxygen isotope data have been very useful in solving geological and meteorological problems (Epstein, 1959; O'Neil and Clayton, 1964; Craig and Gordon, 1965). Figure 15.3 shows the oxygen isotope ratio for a few of the major biological reservoirs. Unfortunately the isotopic relationships in the biological oxygen cycle are not very well understood. The most different and characteristic reservoir is atmospheric O_2. The positive $\delta^{18}O$ of O_2 is the steady state result of kinetic isotope effects in photosynthesis and respiration (Dole, 1956). At the present time $+23$ is the balance point. It is known that $\delta^{18}O$ of O_2 can be shifted in the case where a small O_2 reservoir is partially used up due to biological respiration. The $\delta^{18}O$ of the O_2 dissolved in seawater in Fig. 15.3 is such a case.

It may be that $\delta^{18}O$ of atmospheric O_2 can be shifted due to the changes man has made in the oxygen cycle. The heavy use of O_2 in the oxidation

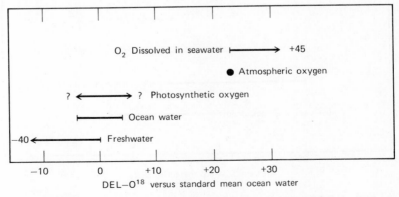

Figure 15.3 Del-^{18}O of biological oxygen reservoirs. From Boato, 1961.

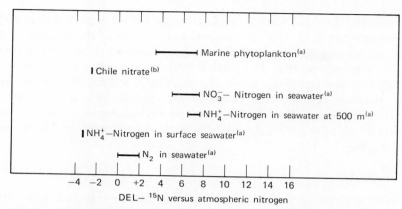

Figure 15.4 Del-¹⁴N of biological nitrogen reservoirs. (*a*) Miyake and Wada (1967); (*b*) Hoering (1955).

of sulfur and of fossil fuels might shift $\delta^{18}O$ of O_2 in restricted locations. The $\delta^{18}O$ of O_2 in our major estuaries, polluted and normal, remain to be investigated.

Nitrogen

The isotope chemistry of the biological nitrogen cycle is beginning to emerge (Fig. 15.4). All data indicate that $\delta^{15}N$ of atmospheric N_2 is the same at all locations. Atmospheric nitrogen is taken as the standard, that is, $\delta^{15}N$ equal to zero. Biological nitrogen has plus $\delta^{15}N$ values ranging from 0 to 20. Hoering and Ford (1960) have shown that there is essentially no nitrogen isotope effect in the fixation of N_2 by *Azotobacter*. If this is generally true for fixation then the fractionation observed in organic nitrogen must arise through some other reaction. A large kinetic isotope effect has been reported in the microbiological reduction of NO_3^- and NO_2^- (Wellman et al., 1968). The evolved N_2 becomes negative while the unreacted reservoir nitrogen, the NO_3^-, becomes positive. If this reservoir nitrogen (NO_3^-) is then used by other organisms to make organic nitrogen, then indeed the organic nitrogen would be plus. Miyake and Wada (1967) reported that NO_3^- and some NH_4^+ in the seawater samples studied by them had a $\delta^{15}N$ ranging between +3 and +10. A −4 value was reported for a single surface NH_4^+ sample.

One would very much like to know $\delta^{15}N$ of the vast amount of nitrogen fertilizer being produced by industrial processes from N_2. Since isotope effects decrease with increasing temperature, this fertilizer probably has the same $\delta^{15}N$ as N_2, that is, zero. Hoering (1955) reported that Chile nitrate had a $\delta^{15}N$ of -2.6. A detailed study of $\delta^{15}N$ of the N_2, NO_3^-, and NH_4^+ of aquatic systems receiving sewage effluents or agricultural runoff might demonstrate that the isotope ratio could serve as a tracer for nitrogen. Industrial NO_3^- from N_2 may have a $\delta^{15}N$ near zero while NO_3^- that is part of a normal biological system may be 5–10/mil positive. It is almost certain that N_2 dissolved in seawater near a large sewage outfall would have a $\delta^{15}N$ different from the $+1$ reported by Miyake and Wada (1967).

Conclusion

Man's activities have changed the stable isotope ratios of some of the major carbon reservoirs in aquatic systems. These changes are not, in themselves, bad. This perhaps makes them very useful objects of study, since no emotion need be involved. They do point to the fact that we may be changing the environment, especially around the sea shores, in some profound ways. It is possible that isotope ratio shifts in carbon, and other elements, may be used as an index or indicator of man's activity.

References

Bigeleisen, J. 1958. The significance of the product and sum rules to isotopic fractionation process. In *Proc. int. symp. on isotope separation*, J. Kistemaker, J. Bigeleisen and A. O. C. Nier (eds.), Interscience Publishers, Inc., New York.

Bigeleisen, J., and M. Mayer. 1947. Calculation of equilibrium constants for isotopic exchange reactions. *J. Chem. Phys.*, **15**:261.

Bigeleisen, J., and M. Wolfsberg. 1958. Theoretical and experimental aspects of isotope effects in chemical kinetics. In *Advances in chemical physics*, I. Prigogine (ed.), Interscience Publishers, Inc., New York.

Boato, G. 1961. Isotope fractionation processes in nature. Comitato Nazionale per l'Energia Nucleare, Summer Course on Nuclear Geology, Laboratoria di Geologia Nucleare, Pisa.

Calder, J. A. 1969. Carbon isotope effects in biochemical and geochemical systems. Ph.D. Dissertation, University of Texas.

Calder, J. A., and P. L. Parker. 1968. Stable carbon isotope ratios as indices of petrochemical pollution of aquatic systems. *Environ. Sci. Tech.*, **2**:535.

Craig, H. 1953. The geochemistry of the stable carbon isotopes. *Geochim. Cosmochim. Acta*, **3**:53.

Craig, H., and L. I. Gordon. 1965. Isotopic oceanography: deuterium and oxygen-18 variations in the ocean and the marine atmosphere. In *Marine geochemistry*. D. R. Schink and J. T. Corless (eds.) University of Rhode Island Graduate School of Oceanography, Occasional Publ. No. 3, Kingston, R.I.

Deuser, W. G., and E. T. Degens. 1967. Carbon isotope fractionation in the system CO_2 (gas)—CO_2 (aqueous)—HCO_3^- (aqueous). *Nature*, **215**:1033.

Dole, M. 1956. The oxygen isotope cycle in nature. Nuclear processes in geologic settings, NAS–NRC, Publ. 400.

Eckelmann, W. R., W. S. Broecker, D. W. Whitlock, and J. R. Allsup. 1962. Implications of carbon isotopic composition of total organic carbon of some recent sediments and ancient oils. *Bull. Amer. Assoc. Petrol. Geologists*, **46**:699.

Epstein, S. 1959. The variations of the O^{18}/O^{16} ratio in nature and some geologic implications. In *Researches in geochemistry*, P. H. Abelson (ed.), John Wiley, New York, Vol. I.

Friedlander, G., J. W. Kennedy, and J. M. Miller. 1964. *Nuclear and radiochemistry*. 2nd Ed. John Wiley & Sons, New York.

Friedman, L., and A. P. Irsa. 1967. Variations in the isotopic composition of carbon in urban atmospheric carbon dioxide. *Science*, **158**:263.

Hoering, T. C. 1955. Variations of nitrogen-15 abundance in naturally occurring substances. *Science*, **122**:1233.

Hoering, T. C. 1960. The biogeochemistry of the stable isotopes of carbon. *Carnegie Inst. Wash.*, **59**:158.

Hoering, T. C. and H. T. Ford. 1960. The isotope effect in the fixation of nitrogen by *Azotobacter*. *J. Amer. Chem. Soc.*, **82**:376.

Keeling, C. 1961. A mechanism for cyclic enrichment of carbon-12 by terrestrial plants. *Geochim. Cosmochim. Acta*, **24**:299.

McMullen, C. C., and H. G. Thode. 1963. Isotope abundance measurements and application to chemistry. In *Mass spectrometry*, C. A. McDowell (ed.) McGraw-Hill Book Co., New York.

Miyake, Y., and E. Wada. 1967. The abundance ratio of N^{15}/N^{14} in marine environments. *Records of Oceanographic Works in Japan*, **9**:37.

O'Neil, J. R., and R. N. Clayton. 1964. Oxygen isotope geothermometry. In *Isotopic and cosmic chemistry*. H. Craig, S. L. Miller, and G. J. Wasserburg (eds.) North-Holland Publishing Co., Amsterdam.

Reimers, R. S. 1968. A stable carbon isotopic study of a marine bay and domestic waste treatment plant. M.S. Thesis, University of Texas.

Sackett, W. M. and W. S. Moore. 1966. Isotope variations of dissolved inorganic carbon. *Chem. Geol.*, **1**:323.

Silverman, S. R. 1967. Carbon isotopic evidence for role of lipids in petroleum formation. *J. Amer. Oil Chemists' Soc.*, **44**:691.

Urey, H. C. 1947. The thermodynamic properties of isotopic substances. *J. Chem. Soc.*, **41**:562.

Vogel, J. C. 1961. Isotope separation factors of carbon in the equilibrium system $CO_2-HCO_3^--CO_3^-$. Comitato Nazionale per l'Energia Nucleare, Summer course on Nuclear Geology, Laboratorio di Geologia Nucleare, Pisa.

Wellman, R. P., F. D. Cook, and H. R. Krouse. 1968. Nitrogen-15: microbiological alteration of abundance. *Science*, **161**:269.

Wendt, I. 1968. Fractionation of carbon isotopes and its temperature dependence in the system CO_2-gas–CO_2 in solution and HCO_3^-–CO_2 in solution. *Earth Planetary Sci. Lett.*, **4**:64.

Williams, P. M. 1968. Stable carbon isotopes in the dissolved organic matter of the sea. *Nature*, **219**:152.

16

Municipal Wastes

PROFESSOR ERNST FØYN, *Institutt for Marin Biologi,* AVD.C, *Kjemisk Oseanografi, Oslo, Norway*

This chapter deals with the disposal of municipal wastes in the sea and discusses the effects of such wastes in the marine environment.

In practice, however, human and industrial activities are closely connected. A great amount of industrial activity will always go on in a community and is, indeed, often the basis on which the community is founded. It is, therefore, impossible to make a perfect separation between municipal and industrial waste. In fact, when dealing with waste to be released, it is often preferable to mix the waste from households and factories, and dispose of it together with the municipal sewage in the same system.

When studying the effects of municipal wastes disposed in the sea, it is necessary to search localities where the household waste is the dominating component. The main constituent of municipal wastes is the water. The physiologist has calculated the amount of water necessary to maintain the water balance in the human body to be 2 or 3 l/day. The "water workers" will know, however, that the amount of water necessary to satisfy the requirements for a person in a modern community is at least 100 times this amount with water for industrial use not included.

Water is needed for bathing and washing as well as for kitchen and sanitary equipment and for various other uses, but above all for the one main task of washing and transporting away unwanted waste from human activities. Enormous amounts of water are needed for these purposes. Two questions, therefore, arise: how are sufficient amounts of clean fresh water procured, and how should we release the used water with a waste content from different sources in a safe way? The cheapest way of disposal certainly is to release the waste water untreated in the nearest water way, river, or lake, or in the open sea. In some places this can still be safely done.

However, the amount of waste is increasing day by day. We must, therefore, accept that due care is to be taken by such waste disposal to protect not only the neighbors, but also the neighboring communities and the neighboring countries. It seems more and more necessary to protect the environmental conditions in the world from pollution which may follow the disposal of waste in natural waters.

To be able to do so it is necessary to achieve a knowledge of the amount and the composition of the waste water which is to be disposed and to consider the different effects of this waste on the environment after its disposal.

Physical and Chemical Properties

The amount of waste water released from different communities may vary within wide limits even in the same country, as shown in Table 16.1, giving the effluent flow on a per capita basis at 10 different sampling places in the city of Oslo. Influence of industrial systems may explain the great differences observed.

The value—240 l/(individual day)—obtained at sampling Station 1 was for a new complete sewage system which carried the waste from

Table 16.1. Volume of Per Capita Effluent at Ten Different Locations Oslo, Norway

Sampling Place	l/(Person Day)
1	240
2	516
3	2700
4	216
5	1185
6	367
7	234
8	625
9	386
10	290

average families whose household had modern sanitary equipment and is probably a good average value for municipal waste volumes.

The effluent flow is closely connected with the daily life and the habits of the inhabitants. Time of vacation and holidays as well as day and night activity are also markedly reflected by the variations in the waste water flow, as shown in Fig. 16.1. This figure gives the effluence in volume per day at a sampling place which receives water from a population of about 50,000 people in Oslo, Norway. The flow rate of about 80 l/sec during night time is shown to increase rapidly during the first hours of the day to about 130 l/sec to reach a high level during the day and again to decrease as night approaches.

The specific gravity of domestic waste water does not usually vary very much from that of clean, fresh water. In most places the temperature during the summer is lower, and during the winter higher than the temperature in the receiving water body.

Municipal sewage is a turbid liquid with a content of particulated, finely dispersed, colloidal, and dissolved material from three main sources: kitchens, washrooms, and water closets.

Waste water from the kitchen is a mixture of carbohydrate, fat, and proteins together with small amounts of other wastes which vary in com-

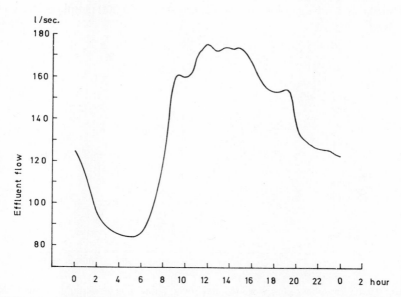

Figure 16.1 Variation of sewage flow over 1 day from a population of 50,000 peoples.

position from house to house depending on different circumstances. In some places, garbage grinders are, for instance, commonly used, whereas in other places these are forbidden due to the low capacity of sewage systems.

Waste from wash rooms consists of small amounts of sand, dust, traces of oil, fat, textile fibres, and so on. There is, however, reason to believe that washing agents used in the washing process are the most dominating wastes in the effluents from wash rooms: different kinds of soaps, chemicals as sodium hypoclorite, perborate, and especially the detergents which now are used all over the world on an increasing scale.

The gravest waste material in a modern community, however, is derived from the sanitary equipment of the homes where different excretion products from the human body are released into the sewer system. A human body in equilibrium accumulates neither organic nor inorganic matter. The constituents of the food not used in the energy metabolism are, therefore, released in the same amount as they are taken in. Phosphorus, for instance, in the human body is released exactly in the same proportion as it is present in the food, and will, therefore, always remain a constituent of this type of waste water. As discussed later, this will have special consequences when released into the sea.

The daily variability in the amount of total phosphorus in the inlet and outlet of a sewage purification plant serving 50,000 people is shown in Fig. 16.2.

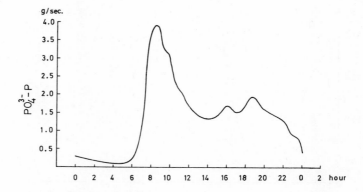

Figure 16.2 Phosphate phosphorous inflow over 1 day to a sewage purification plant receiving waste from 50,000 peoples.

Table 16.2. Average Wastewater Quality During Specific Periods of Three Different Test Homes in the United States

Characteristics of test homes	Home I		Home II		Home III	
Total people	5		5		5	
Children	2		3		1	
Home laundries	1		1		1	
Dishwashers	1		1		1	
Baths	5		3		1	
Analysis	Without Disposer	With Disposer	Without Disposer	With Disposer	Without Disposer	With Disposer
Total solids[a]	866	997	788	859	1249	1536
Suspended solids[a]	363	478	293	360	473	602
Total volatile solids[a]	468	571	414	485	659	942
COD[a]	705	959	540	640	882	1133
BOD$_5$[a]	542	518	284	356	479	598
Detergent[a]	5.3	23.2	5.2	5.7	6.9	7.4
Total nitrogen[a]	69	63	61	62	121	189
NH$_3$-nitrogen[a]	53	47	48	41	92	154
Total PO$_4$[a]	47	40	70	51	65	79
Ortho PO$_4$[a]	31	36	40	32	40	52
Grease[a]	95	134	33	41	66	92
pH	8.0	7.6	8.0	8.2	8.3	8.4
Flow[b]	388	374	331	258	123	148

[a] Given in mg/l.
[b] Given in gal./day.

Different sorts of human parasites and bacteria are also released with the fecal products and are eventually transported into the sea. The fate and effect of all these components are factors which govern the pollution process that takes place when waste is discharged into the marine environment.

The composition of household waste will certainly vary from house to house, but the main constituents are the same, at least in houses with modern sanitary equipment. This may be seen from Table 16.2 which shows the composition of waste water from three different houses in the

United States. The values are calculated from analyses published by Watson et al. (1967).

Purification of Waste Containing Water

For many years intensive work has gone on to find suitable purification methods for municipal waste and treatment processes for waste water, mostly in order to protect the inland water resources. In communities where direct release of this waste to the open sea is possible, there may, in fact, be a question as to what extent purification is necessary, what is possible and even what is the best from an economic point of view.

A completely satisfactory purification can only be achieved by processes which are able to do the following:

1. Remove sludge and suspended particles.
2. Oxidize polluted organic material.
3. Sterilize the effluent.
4. Remove plant fertilizing compounds.

Removal of more than 90 per cent of organic material and bacteria is possible in the activated sludge process. The important removal of plant fertilizing elements such as phosphorus and nitrogen compounds is poor in most of the existing treatment plants. Methods, however, have been worked out that are effective in removing the nutrients of phosphorus and nitrogen from the effluent streams. Wuhrmann (1957) and Rohlich (1961) have, for instance, introduced chemical steps in the activated sludge process in which the nitrogen and phosphorus compounds from the waste water can be removed to a high degree. The electrolytic sewage purification process (Føyn 1956, 1964) removes sludge, organic material, bacteria, and plant fertilizing compounds by a combined chemical and mechanical treatment technique.

The final stage in the sewage purification process is the disposal of the large quantities of sludge which in a biological treatment plant may be about 3 kg/(capita)(day). The sludge is a mixture of inorganic and organic material often with a water content of about 97 per cent. Inland treatment plants have to dry and store their sludge on land. When possible, however, a disposal of the material into the sea seems to be preferred. In some places the sludge is released through pipelines at a safe distance from the shore as at the Hyperian plant in Los Angeles. In other places the material is pumped on board tankers and transported to the ocean where it is released at special dumping sites.

PURIFICATION OF WASTE CONTAINING WATER 451

When solid sludge is disposed at sea, several factors affect the behavior of the sludge waste after disposal. Currents and turbulent diffusion as well as stratification in the receiving water influence the speed by which the sludge is distributed. The solid material, however, sinks and may in some places cover the bottom including the fauna and flora like a thick carpet. After disposal, the waste appears to settle rather fast according to investigations carried out on the New York Bight waste disposal area and in Delaware Bay (Buelow et al., 1968). A heavy accumulation of waste was found on restricted bottom areas. In some places sludge-thickness of at least 5 ft were observed by core samples in the New York Bight disposal site.

Beyer (1955) investigating the dumping of sewage sludge from primary and secondary treatment plants in Oslo studied the fate of the disposed material by photographing the sludge cloud from a helicopter (Fig. 16.3). The sinking speed of the particles was determined by equipment on board our research vessels. Wind and currents extended the cloud of waste over the sea. After about 50 min, nearly 90,000 m² were covered with sludge. Subsurface currents and variations in the seawater density influenced the downward speed of the sludge. Accumulation of the sinking particles was found in the pycnocline, but this was also the level in the sea where horizontal spreading was most pronounced.

Figure 16.3 Sludge "cloud" settling at a dumping place in the Oslofjord. Photo by Beyer (1955).

Many, perhaps the majority of the communities in the world, are situated near inland rivers and lakes. Even if the greater part of the waste water from these localities is subject to purification processes of some kind or other, it is obvious that a great amount of the waste introduced into freshwater must follow the stream and, therefore, finally will be mixed into the ocean. Natural purification processes are certainly active on the way downstream, but today many rivers are heavily loaded with waste from human activities. This must be taken into consideration when evaluating the effects of municipal waste on the ocean.

Behavior and Effect of Waste Disposal

The topographical and hydrographical conditions in the recipient as well as the way by which the waste is released in the water are responsible for the primary distribution of the waste in the receiving water masses. Turbulent movements and currents govern the further transport and interchange with greater water masses in the ocean.

The physical, chemical, and biological conditions both in the seawater and in the waste containing water must, therefore, be considered when studying the fate and effects of municipal waste after being disposed. Seawater is a solution of different salts of which the major constituents are always present in such high amounts that they are excluded from being limiting factors in any of those reactions, chemical or biological, which may take place in the sea. On the other hand, the buffer capacity and the ionic strength of seawater are so high that certain reactions are bound to take place. Precipitation, coagulation, and flocculation processes are generally observed when waste water, with its contents of particulate and soluted organic material, is mixed with seawater. Fatty acids, in the form of soap for instance, react with calcium and magnesium ions and form insoluble products, colloidal solutes of different kinds denature when negative charged particles in the waste are neutralized by the positive charged ions of seawater. Such reactions may form precipitates in the receiving water which is of importance when evaluating the effects of municipal waste on the marine environment.

The first effect of disposal of sewage and sludge into seawater is mainly of a physical and chemical nature. Depending on the specific gravity of the waste products in relation to the density of the receiving seawater, the material is dispersed into the water masses, settled to the bottom or floated to the surface. A decrease of the light penetration into the water is inevitable. Waste disposal into the sea may have different unwanted consequences. There are many harmful effects that may follow when wastes from municipalities are released into the sea. Some of these are decrease of photosynthesis and plant production caused by reduced light, pollution of fish and fishing gear by sludge material, destruction of bottom fauna and flora by settled sludge, reduction of amenities by floating material washed on shore, and destruction of valuable recreation sites. These effects are, however, of more or less local nature and concentrated around outlets and dumping places. More serious are the primary and secondary consequences of the chemical and biological destruction of the large amounts of organic material which are in reality the main constituents of municipal waste.

The primary effect of the biodegradation is a reduction of the available oxygen in the water. In restricted water masses a full depletion of all the dissolved oxygen may occur. The biological oxidation processes do, however, continue even below this stage, and the oxidation goes on at the expense of the oxygen, first gained from the nitrate, and later from sulphate molecules, which are thereby reduced. Poisonous hydrogen sulphide in such water masses may then appear and prevent all higher life in the waters. The reduced compounds discharged in this way are in reality also an oxygen debt which has to be paid before accumulation of oxygen in the water can take place.

The biological consequences of oxygen depletion are well known. Some organisms may live or survive in water masses with very poor oxygen conditions; certain fishes will, however, thrive only in water with an oxygen content of 5 mg/l or more. This may influence the probability of fishing in marine waters polluted by municipal waste. Figure 16.4

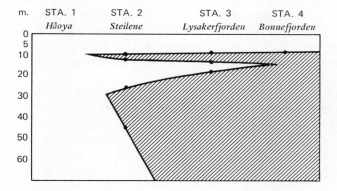

Figure 16.4 Distribution of oxygen in the inner Oslofjord. Depth with oxygen content less than 2.7 ml-O_2/l, hatched.

illustrates the distribution of oxygen north of Drøbak in August 1957. Isolines for the oxygen value critical to cod, 2.7 ml O_2/l are drawn from Station 1, about 25 km below Oslo Harbor, to Station 4, about 10 km inside the harbor.

A second consequence of the biodegradation is the phosphorus and nitrogen compounds remaining after degradation of the organic material of the waste. These compounds are plant fertilizers which, when released in too high concentrations, may induce an excessive growth of phytoplankton in the surface layers of the water. The turbidity caused by the plants may increase to such a degree that practically no light penetrates and the photosynthetic activity in the water masses below is reduced to zero.

The turnover rate in phytoplankton is, however, rapid. The plant cells die, and their compounds of phosphorus and nitrogen are again set free and used when the carbonaceous matter is sinking through the deeper layers and is oxidized. *The amounts of organic material produced in this way may often be many times higher than that of the material which is released with the waste water directly, even if this is released untreated.* It must be realized that the pollution caused by these processes which may affect great water masses is much more serious than the primary pollution which is of local character only.

The harmful effects of phosphorus and nitrogen compounds in freshwater have been recognized for a long time and are described as a eutrophication process. Observations of eutrophic conditions in saltwater are also reported, among others by Braarud (1945) from Oslofjord and Edmondson (1961) from Lake Washington. In a recent publication Korringa (1968) describes such conditions and their effect at a Dutch oyster district. At some distance from the harbor the water had a greenish color; large amounts of flagellates predominated. Oyster larvae carried to this section failed to settle. They developed symptoms of ill health, lost their attractive natural pigmentation, developed a hyaline vesica, and finally disintegrated.

Plankton blooms leading to toxic conditions in the water are well known from the phenomenon which is called the "red tide." Appearance of red tide is associated with overfertilization and is generally caused by upwelling of nutrient-rich water masses from the depth, but may as well be caused by nutritional rich effluent from land. Sometimes these algae may directly poison and kill fish and even hurt man when the algae are blown inshore with the foam from the sea. Filter feeders, which are not poisoned themselves by toxic algae, might accumulate the toxin in their flesh, and when eaten cause grave paralytic poisoning of man.

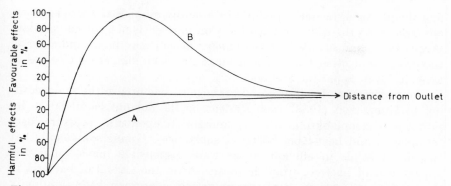

Figure 16.5 Harmful and favorable effects of waste as a function of distance from the outlet.

The favorable effect caused by nutritional support of increased primary production in the open sea is, however, obvious, and Pentelow (1961) has made observations which indicate that although sewage discharge has short-term enhancement of productivity, it is, however, questionable whether the use of these plant fertilizers on land could not give better economic results. This points to the necessity of removal of plant fertilizers from the sewage in treatment processes.

It is obvious that the effect of any disposal of waste into the sea depends on the dilution of the waste in the recipient and the adjacent water masses. Decreasing effects with the increasing distances from the outlet are, therefore, generally observed. Sometimes, however, the effects occurring some distance from a waste disposal site are of a more complex character. This could be the case with wastes rich in nutrients. Depending on the concentration and the hydrographic and biological conditions of the sea, such elements achieve superfertilization and thereby induce excessive and harmful plant growth in the water. On the other hand, these elements are natural fertilizers, which when present in the right concentrations, establish favorable and even optimum growth conditions. Depending on the distance from the outlet, such elements, therefore, cause either harmful or favorable effects, contrary to poisonous material, for instance which causes harmful effects only. In Fig. 16.5 an attempt is made to describe these various effects graphically.

Bacteria and Viruses

The amount of bacteria released in human feces are estimated to be of the magnitude of about 10^7 organisms/g. They are mainly colibacteria and the degree of bacterial pollution caused by municipal waste is generally related to the coliform bacteria. Coli in itself is not considered dangerous to human health, but the *number* of coli may be an indicator of the presence of other bacteria and viruses causing dangerous diseases as typhoid fever, hepatitis, and polio.

The persistence of coliform and pathogenic bacteria in the sea according to Gunnerson (1958) is dependent on the amount of predation by protozoa or zooplankton, salinity, sunlight, temperature, pressure, bacteriophage, and heat-label bacterial substances. Gunnerson found that suspended solids in effluent water were responsible for a rapid disappearance of bacteria from the water. After the arrival of bacteria in the surface zone, however, the reduction continued at a gradually decreasing rate.

The marine pollution problems in the southern California area were also discussed by ZoBell (1960), who had collected during the last two decades several thousand samples for bacterial analyses of seawater, bottom deposits, and plankton hauls from the Pacific Ocean. Less than 1 per cent of the samples of open ocean material gave positive presumptive tests. *Escherichia coli* was not present in any of them. Numerous samples from fish, sea lions, and shellfish as well as seawater and bottom deposits collected near the shore indicated, however, that colibacteria did occur in the marine environment.

A much more rapid disappearance of bacteria was observed in the open sea than was found during laboratory experiments with seawater. The survival of coliform bacteria in seawater was found to be influenced by a multiplicity of interrelated factors: sedimentation, dilution, flocculation, aggregation, adsorption, predation, antibiotics, antagonism, phage, sunlight, and other mechanisms. Further it was evident to some degree that the bacterial population of the surface film of water was considerably larger per unit volume than in the underlying water.

Rheinheimer (1968) reports similar results from his research of the bacteria in the river Elbe (Germany) and found that high amounts of freshwater bacteria are released with the water of the river. When reaching the saline water masses, the population of bacteria changes fairly rapidly to halophile bacteria. This may also be seen in connection with the presence of bactericide compounds produced in seawater.

As mentioned above, coli bacteria, which are not dangerous themselves, may indicate the presence of other bacteria, viruses, streptococci, staphylococci, and yeast, which cause diseases of various kinds. It is obvious that such germs may possibly be concentrated by filter feeders. The fact that gastric diseases are transferred to man when raw shellfish have been consumed, is common knowledge. Even viruses may be transferred in this way.

Thus Liu et al. (1966), who have studied the fate of poliovirus in Northern Quahaugs, found that several species of shellfish were capable of accumulating significant amounts of virus very fast in the digestive diverticula and hemolymph. The first infectious outbreaks of hepatitis, associated with the consumption of raw oysters were, according to the above authors, reported in Sweden by Roos (1956). Four outbreaks which involved approximately 900 cases of illness in the United States were also traced back to consumption of raw quahaugs and hard clams (Mason and McLean, 1962).

Parasites

Parasitological problems should probably also be considered in connection with disposal of municipal waste in the sea. It is a well-known fact that a large percentage of zooparasites of importance to human health gains entrance into the human body by contaminated food. Certainly the possibility also exists that marine animals may be infected by human parasites released with the waste. By feeding on infected animals, a transmittance to man may again take place.

Such cases are well known from river and lake parasites, but have been less investigated concerning saltwater forms. Cheng (1965) discusses parasitological problems associated with food protection, and he points to the fact that both parasitic protozoa, trematodes, cestodes and nematodes are present in seawater. Various cases where marine animals serve as intermediate hosts for parasites which also attack man were described. The nematode Anisakis, for instance, which may be derived from poorly cooked herrings, causes tumorous-like growths in the stomach and intestines of man. In Holland several persons were recently reported to have become seriously ill and many of them died from the attack of this parasite. The part played by parasitism in the municipal waste disposal problem seems to need further investigation.

As mentioned earlier, it is impossible to make a perfect separation between municipal and industrial waste as a great deal of industrial activity always goes on in a community. Some wastes are poisonous as is the case with wastes containing heavy metals. Wastes from breweries, slaughterhouses, and other such factories are more like household wastes. Large amounts of water are often necessary for industrial purposes. The production of 1 l of beer needs, for instance, 24 l of water, and 1 l of milk needs about 5 l of water.

Table 16.3. Person Equivalents of Various Types of Waste

Industry	Unit of Processed Material	Person Equivalent
Dairy without cheese production	1000 l milk	30–80
Dairy with cheese-production	1000 l milk	100–250
Brewery	1000 l beer	300–2000
Laundry	1 ton laundry	700–2300
Sugar factory	1 ton beet sugar	120–400
Slaughterhouse	1 ox = $2\frac{1}{2}$ pigs	70–200

From Sierp (1959).

The effect of the effluent is often measured in units called "person equivalents" and is based on the 5-day biological oxygen demand, BOD_5 analysis. Table 16.3 is extracted from Sierp (1959) and gives the values of various types of industrial waste measured as person equivalents. From these data it is apparent that when discussing the effect of waste from a municipality, it seems necessary to pay some attention to the part various types of industrial factories play when releasing their waste water into the main sewage system.

References

Beyer, F. 1955. The spreading of sinking particles in the sea. *Sewage Ind. Wastes*, **27**:1073–1078.

Braarud, T. 1945. A phytoplankton survey of the polluted waters of inner Oslo Fjord. *Hvalråd. Skr.*, **28**:142 pp.

Buelow, R. W., B. H. Pringle, and J. L. Verber. 1968. Preliminary investigation of sewage sludge dumping off Delaware Bay. U.S. Dept. Health, Education, and Welfare. January 1968. 21 pp.

Cheng, T. C. 1965. Parasitological problems associated with food protection. *J. Environ. Health*, **28**:208–214.

Edmondson, W. T. 1961. Changes in Lake Washington following an increase in the nutrient income. *Verhandl. Intern. Verein. theoret. Angew. Limnol.*, **14**:167–175.

Føyn, E. 1956. Elektrolytisk kloakkrensing. *Tek. Ukebl.*, **19**:433–437.

Føyn, E. 1964. Removal of sewage nutrients by electrolytic treatment. *Verhandl. Intern. Verein. Theoret. Angew. Limnol.*, **15**:569–579.

Gunnerson, C. G. 1958. Sewage disposal in Santa Monica Bay, California. *J. Sanit. Eng. Div. Amer. Soc. Civil. Engrs.* **84**:No. SA 1, Proc. paper 1534.

Korringa, P. 1968. Biological consequences of marine pollution with special reference to the North Sea fisheries. *Helgoländer wiss. Meeresunters.*, **17**: 126–140.

Liu, O. C., H. R. Seraichekas, and B. L. Murphy. 1966. Fate of poliovirus in northern Quahaugs. *Proc. Soc. Exp. Biol. Med.*, **121**:601–607.

Mason, J. O., and W. R. McLean. 1962. Infectious hepatitis traced to the consumption of raw oysters. *Amer. J. Hyg.*, **75**:90–111.

Pentelow, F. T. K. 1961. Fish, sewage and sea outfalls. *Instit. Publ. Health Engrs. J.*, **60**:232. (*Wat. Pollut. Abstr.*, **35**:1005, 1962)

Rheinheimer, G. 1968. Die Bedeutung des Elbe-Ästuars für die Abwasserbelastung der südlichen Nordsee in bakteriologischer Sicht. *Helgoländer wiss. Meeresunters.*, **17**:445–454.

Rohlich, G. A. 1961. Chemical methods for the removal of nitrogen and phosphorus from sewage plant effluents. In *Algae and Metropolitan Wastes. Transactions of the Seminar on Algae and Metropolitan Wastes, Cincinnati, Ohio, 1960.* U.S. Dept. of Health, Education, and Welfare, pp. 130–135.

Roos, B. 1956. Från Göteborgs hälsovårdsnämnd, hepatit epidemi, spridd genom ostron. *Svenska Läkartidn.*, **53**:989–1003.

Sierp, F. 1959. *Die gewerblichen und industriellen Abwässer*, Springer Verlag, Berlin.

Watson, K. S., R. P. Farrell, and J. S. Anderson. 1967. The contribution from the individual home to the sewer system. *J. Water Pollution Control Federation*, **39**:2039–2054.

Wuhrmann, K. 1957. Die dritte Reinigungsstufe: Wege und bisherige Erfolge in der Eliminierung eutrophierender Stoffe. *Schweiz. Z. Hydrol.*, **19**:409–427.

ZoBell, C. E. 1960. Marine pollution problems in the Southern California area. In *Biological problems in water pollution. Transactions of the Second Seminar on Biological problems in water pollution, Cincinnati, Ohio, 1959.* U.S. Dept. of Health, Education, and Welfare, pp. 177–183.

17

Heavy-Metal Contamination

MARGARET MERLINI, *Biology Division—Euratom Joint Research Centre, Ispra, Italy*

It is difficult to assess the entirety of the burden of heavy metals the oceans must bear from man's indiscriminate use and disposal of them in seas, inland waters, and the air. Although no comprehensive survey has been made on the nature and all of the possible sources of heavy-metal contamination, one can gleam information from studies of different types of pollution, some of which are discussed in this book. One can consider the introduction of such elements into the oceans to be indirect, that is, from contaminated rivers, streams, and runoff; or they can be direct, that is, when metallic pollutants are pumped into estuaries or the sea from the shore, discharged into the water by vessels, or added to the water from aerial fallout.

Metals and Their Sources

Perhaps the universal offenders are industrial wastes and pesticides because of worldwide industrial growth and technological assistance among nations. As McKee (1967) pointed out, "the number and diversity of chemical pollutants added to marine waters are legion," and wherever industrial wastes are discharged, the heavy metals, especially copper, zinc, chromium, cadmium, nickel, and lead, are found. It is significant that heavy metals are part of wastes from most of the 10 major United States industrial groupings as listed by the Department of the Interior in a report "The Cost of Clean Water" (Anonymous, 1968). Here, the elements most often encountered are zinc, copper, chromium, and iron.

In January 1966 a Symposium on the Mercury Problem was held in Stockholm where it was pointed out that the dispersal of mercury in the environment by man stems from its wide agricultural use as a germicide or fungicide; its general use as a slimicide, fungicide, and paint mildew preventer; and its use in electrical, chemical, and pharmaceutical industries. In addition, mercury can be set free during the heating or burning of ores, soils, and fuels (Walker, 1967; Ashworth, 1967; Johnels et al., 1967; Westöö, 1967).

Most of the metals are found in industrial wastes all over the world, as in the case of the Shinkashi River in Japan on the shores of which there are 2000 factories (Nishihara, 1967). From an investigation in Europe of the discharge of sewage into the sea, one is given an impressive picture of the number of kilometers of polluted coast in each of the countries examined caused by wastes from industries, domestic sewage, ships, and other factors (Koch, 1960; Biannuci and De Stefani, 1968). For many of these countries, expansion of industrial areas has meant that waste problems have been solved by pumping pollutants over long distances into the sea or coastal rivers near estuaries. Recently Korringa (1968) noted that there is an unmistakable trend toward the location of new industries at the border of the sea.

While pesticides with chlorinated hydrocarbons are primarily under fire for their detrimental side effects, the ones with metal components are usually cited less often. Of the many and varied kinds of pesticides, the fungicides are more likely to have metals such as copper (as sulfate or naphthenate), zinc (as sulfate or oxide), or mercury (as methyl-, ethyl-, alkoxyethyl-, aryl-mercury compounds, or mercuric chloride). The molluscicides are likely to have copper sulfate, and insecticides often contain lead arsenate or arsenic (Grimanis, 1968).

In 1957 it was reported that in the United States alone, 35 million lb of undiluted arsenical salts and 45 million lb of copper sulfate were used with other types of insecticides and fungicides (Cottam, 1960). More alarming were the figures given by Cottam indicating that 5 per cent of the United States had been treated with chemical pesticides mixed with dusts, oils, water, emulsifiers, and other carriers with a total volume between 2 and 3 billion lb. A fourfold expansion was expected over the ensuing 10 to 15 yr. From the figures given in *The Pesticide Review* in 1967, this expectation was not fulfilled, but the quantity in use has remained at the same level. If in the United States the utilization of certain fungicides and insecticides has not increased, the production and quantities exported all over the world continue to augment (*The Pesticide Review*, 1967).

Other possible sources of heavy metal pollution of the marine environment are mine residues, known to have poisoned streams and rivers

(Brinkhurst, 1965; Anonymous, 1967), and crude or heavy fuel oil pollution from ships, pipelines, offshore wells, refining industries, and so on. Vanadium, manganese, cobalt, and nickel occur in crude oils as nonvolatile metal porphyrins (Johannesson, 1955; Guinn and Bellanca, 1968). It is possible that these metals are available to certain flagellated protozoans and bacteria known to require preformed porphyrins (Fruton and Simmonds, 1958), and that emulsified oil droplets can reach benthonic organisms as well (Clendenning and North, 1960). Of worldwide concern is domestic pollution, and one cannot exclude the possibility that heavy metals are involved in this universal contaminant.

These, then, may be considered the major sources of heavy metals which either directly or indirectly reach the ocean: zinc, copper, cadmium, chromium, lead, nickel, iron, vanadium, and mercury. Up to now, the real impact of industrialization and technological advancement with regard to pollution from wastes has been borne by streams, rivers, lakes, and estuaries. It is from the disquieting state of these bodies of water that concern for the well-being of the oceans should arise.

Pollution of Land Waters

The innumerable examples of heavy-metal pollution of lakes, rivers, and streams in the literature are testimony to the gravity of the situation. A classic example of what can happen to a body of water in the industrial era when water masses seem to present the best and cheapest way to get rid of by-products, irrespective of their toxicity to the aquatic flora and fauna, is Lago d'Orta in northern Italy. Since the initial pollution of the lake took place 40 yr ago, it has been possible to follow the evolution of the situation.

In 1929 an artificial silk factory was built on the southern end of Lago d'Orta without taking into account the fact that the emissary was at the northern extremity. As a result, the discharged copper and ammonia had to traverse the entire lake before reaching the outflowing river. Nor did the relatively large mass of water (the lake volume is 1308 million m^3) provide the dilution one might expect since complete water renewal is of the order of 11 to 12 yr. Within 2 yr all of the plankton, benthos, and fishes were dead, despite a warning published by Monti (1930) soon after the initial contamination. The events that followed are documented in a series of scientific publications (Baldi, 1949; Moretti, 1954; Corbella et al., 1958; Tonolli and Vollenweider, 1960; Tonolli, 1961; Vollenweider, 1963).

In 1958 Corbella et al., after a series of core analyses, pointed out that the disastrous effects of the pollution were still evident in the complete lack of benthonic fauna within the top 6 cm of sediment. The once abundant species of diatoms *Cyclotella* and *Fragiliaria* had been immediately and drastically affected, and later the *Synedra* species. *Achnantes nodosa*, on the other hand, survived their less tolerant competitors.

Vollenweider (1963) reviewed the situation and explained some of the consequences of the pollution despite the fact that over the years copper had been eliminated from the wastes discharged into the lake. Once the element was reduced or removed, littoral growth near stream inlets appeared and there was a fairly rapid recolonization of the epilimnic algal populations, primarily blue-green and green algae. The result was eutrophication of the lake with pH fluctuations and oxygen supersaturation during thermal stratification. The zooplankton populations are still poor, composed primarily of rotifers and *Cyclops strenuus*. The ionic composition of the water has changed from an oligotrophic bicarbonate to an artificial sulphate-nitrate type with cation deficiencies. Vollenweider concluded by stating that the acidification phenomenon (related to ammonia oxidation and the increasing oxygen consumption) "are the determining vectors governing the further limnological development and give rise to serious apprehension as to whether there will be an efficient recuperation of Lake Orta."

The above example is not unique, neither for Europe nor elsewhere. In the United States, a recent article reviewed the complex situation involving the Great Lakes (Anonymous, 1967a). Lake Erie, once one of the most fertile and productive of all of the Great Lakes, has changed so drastically chemically and physically that the more precious species of fish have virtually disappeared and the future of the remaining species is seriously threatened. Obviously, the unfavorable situation is not due entirely to heavy-metal contamination but industrial wastes containing such elements are known to be a contributing factor. Indeed, it was pointed out after an analysis of the situation, that little is known about the accumulation of heavy metals, pesticides, and detergents in the biota and its environment. It should be added that Lake Michigan and Lake Ontario have similar problems of industrial and domestic pollution and the possibility of eutrophication is of great concern.

Many of the consequences of heavy-metal pollution observed for the land waters have found their counterparts in the marine environment.

Pollution of the Marine Environment

A dramatic case of pollution of the marine environment occurred in March 1965 in Holland, between Scheveningen and IJmuiden where about 100,000 fishes were found dead on the beaches and others were found swimming about with uncoordinated movements (Roskam, 1965). Copper sulfate had been deposited on the shores and an analysis of the water gave a value of copper of 500 µg/l whereas the normal content for the North Sea is about 3 µg/l. As the polluted water streamed by, the eulittoral mussel population was killed and some of the species of plankton were affected such as *S. costata* and *A. japonica*. On the other hand, there was an increase in the number of oligotrichous ciliates. The intestinal tracts of the dead fishes had high concentrations of copper and the author suggests that the element had been accumulated in one of the links of the food chain. Korringa (1968) later commented that this case clearly demonstrates how erroneous it is to make a decision to discharge a pollutant into the sea on the basis of calculations of the eventual concentration of the pollutant following dispersion and dilution. Despite the fact that the Atlantic Ocean annually displaces about 23,000 km^3 of North Sea water, the dispersion of copper sulfate in the narrow strip of coastal water barely diluted the chemical before it reached the Waddenzee.

The important mussel farms in the Dutch episode fortunately escaped contamination and destruction; but there are many instances of heavy-metal accumulation in edible molluscs due to contamination of their environment. "Green oysters" are readily recognized by their color and taste as having high amounts of copper. In Japan, a factory and exhausted mine contaminated the Nobeoka Bay at one end, and the soft tissues of "green oysters" (*Ostrea gigas, O. circunspecta,* and *O. spinosa*) gave values for copper from 320 to 687 mg/kg wet weight of soft tissues. Oysters from the uncontaminated area of the same bay had 40 to 99 mg Cu/kg wet weight of soft tissues (Ikuta, 1967). In another polluted area, Akamizu Inlet, the zinc content reached a value of 575 mg/kg wet weight of soft tissues for *Ostrea*, whereas the highest value for oysters from an unpolluted area was 191 mg/kg (Ikuta, 1968).

Industrial wastes in the Bay of Kohsiung in Taiwan have not only produced "green oysters" due to copper, but also the elimination of plankton and benthonic organisms, with subsequent decrease in coastal fishing (Chang, 1965).

Fish and shellfish are part of the diet of the inhabitants of Minamata Village in Japan. Continuous discharges of mercury into Minamata Bay resulted in an excessive accumulation of the element in fish and particularly in the mollusc *Hormomya mutabilis*. Unlike copper, mercury does not discolor or impart an unpleasant taste to the animal's soft tissues, and as a consequence, there was no overt warning for the population which began to show symptoms of a malady that became known as

"Minamata disease" due to the difficulties encountered in diagnosing it. Irukayama et al. (1961) reported on studies which ascertained the origin of the causative agent. From their investigations it became apparent that the extracts from the food items collected in Minamata Bay contained organic mercury, which was responsible for the disease. Autopsies of the poisoned victims revealed elevated amounts of mercury in the liver, also in the organic form.

Mercury contamination was not limited to Minamata Bay. In 1965 there was another outbreak of Minamata disease in individuals who ate fish from the Agano River in Japan. From investigations that followed it was established that the fish were contaminated with methyl mercury which was discharged by an acetaldehyde factory (Sumino, 1968). In the Scandinavian countries where elevated quantities are utilized in the manufacture of seed disinfectants, but primarily for the preservation of pulp (Ulfvarson, 1967), the problem became acute when mercurial poisoning was discovered in the wildlife. In 1964 high levels of mercury were detected in birds and freshwater fishes found dead (Lundholm, 1967). Birds which feed on contaminated fish have great quantities of mercury in their feathers; the sea eagle was one of the birds found to have large amounts of the element. Salt water fishes have not been extensively investigated, but of those examined, mackerel, herring, and cod did not have the high levels of mercury found for the freshwater species. Of the three, the cod had the greatest amount, 150 to 190 ng/g. According to Eriksson (1967) there is about 30 ng/1 of mercury in ocean water mostly as the $HgCl_4^{-2}$ complex.

Direct pollution of the marine environment by heavy metals from the atmosphere was recently demonstrated by Dudey et al. (1968). Aerosol samples taken over the Atlantic Ocean in 1966 were found to contain "such toxic elements as zinc, cadmium, mercury, and selenium in amounts which are respectable percentages of their concentrations in the presumably contaminated air of several coastal cities." There is an increasing awareness that modern industry is adding heavy metals to the atmosphere. Copper, iron, and chromium were among the six suspended metals in air samples from two locations, Cincinnati and Fairfax (Lee et al., 1968) with concentrations of the metals 3 to 5 times higher over the more industrial Cincinnati than suburban Fairfax. Airborne particulate matter collected over 22 different locations throughout metropolitan Chicago also revealed such metals as manganese, aluminum, and vanadium along with bromine, sodium, and chlorine (Brar et al., 1968). Pollution of the urban atmosphere in England by hexavalent chromium was signalized by Bowen (1968) as coming from the painted yellow lines denoting parking restrictions. Pesticides have been detected in the atmosphere and rain (Tarrant and Tatton, 1968), and are known to be carried by the trade winds (Risebrough et al., 1968; Goldberg and Risebrough, Chaps. 3 and 10, respectively, in this book). Hence, the presence of heavy metals in marine aerosols is not surprising, and is likely to increase.

Some Considerations on Heavy Metals in the Marine Hydrosphere and Biosphere

The vastness of the ocean is often used as an argument for its use in the disposal of man's waste products. *It is because of its vastness that care must be taken not to damage the biota since small and often large derangements are not as perceptible as in restricted environments such as lakes and streams.* Examples have been given for estuaries, bays, and coastal waters where the animal or plant populations were decimated or completely eliminated due to metal pollution. It is more difficult to find changes in biotic structure in the open sea since we know so little about the natural construction of populations, their interrelationships, and the numerous factors which constitute the total environment. The information we do have about many of these parameters is piecemeal even though they have been dealt with in studies stimulated by problems associated with radioactive pollutants. Since all the elements found in marine organisms come from the surrounding medium, it is obvious that the first point to be considered is the presence of heavy metals in the environment in which they live.

The Hydrosphere

Heavy metals are insidious due to the fact that they are ever present and practically indestructible in an aqueous environment. Even if precipitated or adsorbed, the process can be reversible and the metals return to the ionic form. If not, the precipitate or particles settle on the bottom to be ingested by benthic organisms. Dilution and dispersion help minimize the danger to some of the biota, but filter-feeding organisms are notorious for their ability to concentrate many polyvalent ions—in particular, copper, zinc, iron, and manganese—from very dilute solutions. Therefore, the quantities, distribution, and physical and chemical behavior of metallic trace elements in seawater is of great importance.

It is not within the scope of this chapter to discuss in detail the chemistry of the oceans, but it can be considered, nevertheless, a starting point for an understanding of the availability of the metals to marine organisms, their eventual accumulation, concentration, and cycling from water to organisms.

Fukai and Huynh-Ngoc (1968) reviewed what is known about the chemical behavior of chromium, manganese, iron, cobalt, zinc, and cerium in seawater. They aptly point out that our knowledge of trace elements is so limited that a general distribution pattern for most of them cannot be made. In addition, the actual chemical states of the elements are rarely taken into consideration when data are reported, although some authors have already used filtering techniques with success (Goldberg et al., 1952; Fox et al., 1953; Strickland and Parsons, 1960; Lowman et al., 1966). More recently, Hood (1967) determined the vertical distribution of copper, manganese, and zinc (from 0 to about 3000 m) in a column of water from Sigsbee Deep, Gulf of Mexico. He reported great variation both in kinds and amounts of the elements at different depths, with copper and zinc associated with the particulate fraction, and manganese in solution. Much of the zinc, substantial amounts of copper and small amounts of manganese were nondialyzable, that is, firmly bound to organic complexes. Such information is extremely important since quantitative data are lacking on trace elements in organic particulates and on the amount of trace elements associated with dissolved organic substances. Some progress is being made in that direction, however (Koyama, 1962; Duursma, 1965). There is also indirect evidence of an organic-bound fraction for elements such as manganese and zinc (Rona et al., 1962), iron (Laevastu and Thompson, 1958), mercury (Hosohara et al., 1961; Hanya et al., 1963), and copper (Slowey et al., 1967).

From the studies undertaken by Fukai and Huynh-Ngoc (1968) on the precipitation of trace amounts of radioelements, the following facts emerged:

1. Radiochromium in the trivalent state may precipitate in the presence of iron hydroxide; in the hexavalent state it remains in solution and its stability is dependent upon the amount of organic substances in the seawater.

2. There is a negligible amount of radioactive ferric iron in soluble form, although results on nonradioactive iron indicate that some of it exists in the soluble state. In any case, there is evidence that particulate iron can be utilized by phytoplankton (Harvey, 1938; Goldberg, 1952).

3. Radiomanganese is in solution in the divalent state initially and as such can be accumulated and metabolized by marine organisms.

4. Radiozinc is probably initially adsorbed onto surfaces, but an appreciable amount remains in solution and is thus available to organisms.

A logical corollary to the problem of trace metal elements in seawater is that dealing with the influx of these metals from freshwaters. What happens when soluble elements in river water enter the sea is of great importance since so much of the indirect contamination of the oceans has its origin in the land waters.

Lowman et al. (1966) reported on such an environmental situation on the west coast of Puerto Rico. They determined the amounts of particulate and soluble trace elements in the three major rivers under investigation and found that the metallic elements such as iron, manganese, copper, cobalt, zinc, and nickel were in amounts several times greater than that found in the seawater. After a series of experiments they showed that iron, manganese, and scandium underwent rapid precipitation upon entering seawater. Cobalt, zinc, nickel, and copper precipitated in a linear manner with time. This, in turn, affected the accumulation of elements by epibenthic species because the availability of the elements was dependent upon the geographical location and the physical and chemical forms of the oligoelements before and after mixing of the two types of water. Other factors, such as the dissolved and suspended materials in the river outflows also helped to influence distribution patterns of the trace elements in the sediments where the organisms were living. In addition, the feeding habits of the animals determined the accumulation of one element rather than another.

The Biosphere

One of the striking features of the marine biosphere, recognized for many years, is its higher concentration of heavy metals than the hydrosphere (Noddack and Noddack, 1939). Many of the early data were obtained because of the ease with which some of the metals could be determined, qualitatively or quantitatively, in particular species with high concentrations of one or more elements (Vinogradov, 1953). From the standpoint of pollution of the marine environment with heavy metals, it is highly desirable to know with certitude the amounts and distribution of these metals in the biota. This information would contribute to an understanding of the cycling of metals from water to organisms, their vertical and horizontal displacement in water, and it would help to judge the extent of abnormal accumulation in the biota. Fortunately, the use of more refined techniques for the determination of trace elements in biological materials has produced, and continues to produce, a large mass of accurate data. Recently Goldberg (1965) collected in a single volume information on trace element concentrations in marine organisms from published work following the period reported by Vinogradov (1953). Continued work in this field produced general surveys of elements for many marine species (Lowman et al., 1966; Bowen et al., 1967; Robertson et al., 1968; Stevenson et al., 1964), surveys of particular metals in marine organisms, such as for cobalt (Fukai, 1968), chromium (Fukai and Broquet, 1965), and methyl mercury (Sumino, 1968), and studies for a particular phylum or group of animals: heavy metals in marine molluscs (Brooks and Rumsby, 1965), metals in skeletons of sea urchins (Stevenson and Lugo Ufret, 1966), and vanadium and other metals in ascidians (Papadopoulou et al., 1967; Carlisle, 1968). These are just a few of the numerous publications, and they do not take into account results reported for stable metals in connection with investigations of their radioactive isotopes (see, for example, *Radioecological Concentration Processes*, Pergamon Press, Oxford and New York, 1966).

From the accurate data obtained, enrichment factors or concentration factors, which are defined,

$$CF = \frac{\mu g \text{ element/g organism or part, wet weight}}{\mu g \text{ element/g seawater}}$$

have been reported by many authors (Goldberg, 1965). These factors put into evidence the degree to which marine organisms concentrate elements

Table 17.1. Enrichment Factors for the Trace Element Composition of Shellfish Compared with the Marine Environment

Element	Enrichment Factors		
	Scallop	Oyster	Mussel
Ag	2,300	18,700	330
Cd	2,260,000	318,000	100,000
Cr	200,000	60,000	320,000
Cu	3,000	13,700	3,000
Fe	291,500	68,200	196,000
Mn	55,500	4,000	13,500
Mo	90	30	60
Ni	12,000	4,000	14,000
Pb	5,300	3,300	4,000
V	4,500	1,500	2,500
Zn	28,000	110,300	9,100

From Brooks and Rumsby (1965).

from the water, the differences among species, and the predilection of individual parts of the organism for particular elements. The table published by Brooks and Rumsby (1965) is given below in its entirety as an example of heavy-metal concentration in marine molluscs over that in the water.

Stevenson et al. (1964) reported concentration factors for nickel, manganese, chromium, and iron in different species of marine organisms. A resumé of their tables II and III is presented to illustrate the ranges found, the differences among species, and concentrations in various parts of the animals (Table 17.2).

Despite the information available, however, the mechanisms whereby heavy metals enter and become part of the biota are still being elucidated. The ocean as an environment for any one species is not just the aqueous medium with its content of heavy metals, minerals, luminosity, salinity,

Table 17.2.

Species	Nickel Conc. Factor ($\times 10^3$)	Manganese Conc. Factor ($\times 10^3$)	Chromium Conc. Factor ($\times 10^3$)	Iron Conc. Factor ($\times 10^3$)
Echinoderms (urchins)				
Tripneustes esculenta				
Testis	43	5	400	5
Ovary	3	2	—	—
Echinometra lucunter	74	7	860	9
Echinoderms (brittle stars)				
Ophiothrix suensoni	32	8	480	17
Gorgonians				
Eunicia sp.				
sp. 1	29	4	340	5
sp. 2	26	5	380	7
sp. 3	27	4	420	7
Pseudapterygorgia sp.	34	5	320	8
Molluscs				
Tectarius muricatus				
Animal	24	10	1,400	5
Shell	67	1	870	4
Marine algae				
12 species (range)	2 to 40	2 to 30	12 to 260	10 to 80

temperature, and nutrients, but includes other species of animals or plants as food, symbionts, or predators. From these components the long food chain, characteristic of life in the sea, begins and ends for this environment. Phytoplankton are the microscopic plants which initiate the food chain, deriving energy from sunlight and the necessary nutrients from the water. These, in turn, are eaten by zooplankton, minute animals which are then devoured by other larger animals, on up through the food web. Thus each species, after the plants, can accumulate heavy metals from the surrounding medium and from its food source.

Notwithstanding the complexities and interrelationships, different aspects of the problem have been tackled, resulting in a mélange of facts dealing with the mode of entry and the ways the elements are either fixed or metabolized by the organism. However, to explain each aspect in detail would mean treating each species separately. In general, the modes of entry of metals into animals can be listed as follows:

1. As soluble ions, or ions bound to organic molecules.
2. With food items in which certain metals were preconcentrated before ingestion by the animal (Bowen and Sutton, 1951).
3. With seston, the surfaces of which have adsorbed metals, (Armstrong and Atkins, 1950). Particulate matter, provided the particle size is acceptable to the animal, can enter with water and/or food.
4. With symbionts. Other organisms, having a positive or negative (parasitic) relationship with the organism can be carriers of metals.

Uptake and accumulation can be the following:

1. An active biological process when certain metal ions are absorbed against a concentration gradient due to physiological need.
2. Absorption due to an attraction between the ions and substances produced by the organism, such as the uptake of polyvalent ions by mucus in molluscs (Korringa, 1952) and in tunicates (Goldberg, 1957; Carlisle, 1968).
3. A passive process with physical adsorption onto exposed surfaces.
4. A matter of uncontrolled absorption with an inefficient mechanism for excretion of the metals, and detoxification of the excess (discussed for nonmarine animals by Bowen, 1950; Cavalloro and Merlini, 1967).

For any one species, the above mechanisms will be conditioned by food habits (which establish the animal at a certain trophic level), physiology (conditioned by seasonal cycles), age, and sex, plus the many environmental factors already touched upon (availability of metals, temperature, oxygen, salinity, light, pH, etc.), and the physical forces: currents, tides, and changes due to climatic conditions.

Heavy metals such as chromium, manganese, iron, copper, zinc, cobalt, molybdenum, and others are considered essential elements, and in many instances, the quantities necessary for the growth and well-being of organisms are known to be small. Thus it becomes clear that certain pathways will depend upon the degree of homeostasis achieved by the animal, while others will be governed by factors extraneous to physiological control. It has been noted, in fact, that vertebrates with more integrated systems accumulate less trace elements than invertebrates. For lower animals, it is difficult to individualize the self-regulating mechanisms since the systems which have an integrative action have not been delineated.

Despite this fact, the more recent investigations of the role of trace elements in marine organisms have tried to give some insight into the problem of heavy-metal accumulation. One approach has been to consider what is known about the stability of the complexes that divalent metal ions form with organic ligands; that is, that the stability of the complex increases with increasing basicity of the metal ion $Pd > Cu > Ni > Pb > Co > Zn > Fe > Cd > Mn$ (Schubert, 1954). Some researchers have found that the rank order of metal enrichment in tissues and organs of organisms follows this trend (Goldberg, 1957). Bowen and Sutton (1951) working with marine sponges found $Cu > Ni > Co$; and for brown algae the order was $Ni > Zn$ (Black and Mitchell, 1952). Although Brooks and Rumsby (1965) failed to confirm this trend for the scallop, mussel, and oyster, they pointed out that ingested particulate matter and adhered sediment, difficult to remove from the gills, could mask the direct coordination bonds of silver, cadmium, chromium, copper, iron, manganese, molybdenum, nickel, and zinc with the suitable organic ligands. The investigation of elements in marine fishes appears to confirm the tendency heavy metals have to form organic complexes (Goldberg, 1962). Metals such as manganese, copper, nickel, cadmium, and zinc concentrate in the internal organs where these complexes can be formed. Only iron was associated with the blood and circulatory systems. Also, metal concentrations were found to be higher in the dark muscle than in the light muscle.

Toxicity and Tolerance of Heavy Metals

It is unfortunate that marine organisms have not been the subject of more recent research along this line. Most toxicity studies are done with freshwater fishes (Doudoroff and Katz, 1953; Lloyd, 1965), with an occasional mention of euryhaline or saltwater fishes. Two old reports dealing with the effects of metals on marine species are those of Richet (1881) and Thomas (1915). Richet gave limits of tolerance of metals by the animal as grams of metal per liter of water based on tests that lasted 48 hr. According to his data, the metals were tolerated in this order: Cd > Zn > Fe > Cu, with mercury the least-tolerated element. Thomas (1915) studied metallic salts and concluded that mercuric chloride, cadmium nitrate, and cupric sulfate or chloride were quite toxic in seawater, whereas zinc sulfate and manganese chloride were less toxic.

Despite the dearth of information on marine fishes, some of the qualitative results reviewed by Doudoroff and Katz (1953) and the more quantitative data of Lloyd (1965) are of general significance.

1. There is an increase in survival time of the animal, with no change in the lethal threshold concentration, when there is a decrease in the water temperature.

2. Dissolved gases and calcium in the water appear to modify the quantity of heavy metal necessary to kill the animal. From experimental evidence, the protective action of calcium is internal or cellular, rather than at the epithelial surfaces of the gills as once thought.

3. Low oxygen content increases the toxicity of heavy metals due to the physiological response by the animal of pumping water more rapidly through its gills. As Lloyd pointed out, this is important since any factor in the environment which produces an increased respiratory flow will also increase the rate at which the metal arrives at the gill surface and is eventually taken in by the fish. This observation seems to confirm the hypothesis that there is a lethal threshold concentration in fish for metals in solution (toxicity due to food poisoning is not considered here). If the blood removes the metal ions at a rate equal to or greater than the rate in which they enter the animal there is no buildup of the element, and the fish will survive.

When Lloyd (1965) expressed the concentrations of the heavy metals as the multiple of their lethal threshold concentration, the log concentration-log survival time curves for copper and zinc salts for rainbow trout were the same, irrespective of the hardness of the water, suggesting a similar toxic action for these two metals. Lloyd (1965) elaborated the data of Schweiger (1957) in the same fashion and it appears that, at least for the rainbow trout, the toxic action of cadmium, nickel, and cobalt is the same, whereas that of mercury is more toxic and that of manganese is less.

4. Synergism of cupric and zinc sulfates was reported by Bandt (1946) and Doudoroff (1952) and confirmed by Lloyd (1965) in experiments in which the survival times for fish were considerably shorter in water containing both elements. Thus, mixtures of heavy metals may be more toxic than would be expected if each metal were considered singly.

Bioassay studies for the determination of environmental contaminants have brought to light some interesting facts about heavy metals. For instance, Okubo and Okubo (1962) demonstrated a very high degree of sensitivity in larvae of sea urchins (*Hemicentrotus pulcherrimus* and *Anthocidaris crassispina*) and those of the molluscs (*Crassostrea gigas* and *Mytilus edulis*) to mercuric chloride (as well as to sodium cyanide and picric acid). When comparative toxicity tests with brine shrimp *Artemia salina* and *Balanus* were made, it was found that the embryos of the sea urchins and the bivalves were 10 times as sensitive as *Balanus* nauplii and 100 times as sensitive as *Artemia*.

Another investigation aimed at setting up a bioassay technique reported by Crandall and Goodnight (1963) deals with the histopathological effects of sublethal concentrations of lead nitrate, zinc sulfate, and sodium pentachlorophenate in *Lebistes reticulatus*. They found a lack of mesenteric fat, reduction of renal peritubular lymphoid tissue, apparent dilation of renal tubules with degeneration of tubular epithelial cells, retardation and aberrant gonadal development, liver damage, and blood cell destruction.

Hasselrot (1965) undertook the study of the effects of heavy metals on caged salmon fry (*Salmo salar*) exposed to water from mine drainage. In this fashion he had a perfect continuous-flow system with all the "normal" environmental fluctuations. Aside from mortality data, he found that 14 days of exposure produced a depression of hematocrit values in the fry, an indication of a pathological state.

An example of the effects of sublethal concentrations of zinc and copper was recorded by Sprague (1965) in the field. In 1960 it was noted that 22 per cent instead of the usual 1.5 to 3.0 per cent of the adult salmon moving upstream to spawn in the Atlantic from the Miramichi River returned downstream through the counting fence. The cause was traced to the presence of zinc and copper in a tributary which came from a mine. When the pollution was reduced the percentage of salmon which returned downstream also diminished.

Extended studies by Sprague et al. (1965) of the polluted area brought forth the fact that the young salmon population was depressed due to the limited number of eggs produced each year and the poor survival rates of underyearlings and parr. Interesting was the finding that in 1961 there was an increase (2.5 times) in the number of caddisflies in the polluted areas of the Miramichi as compared with the control area. The scarcity of fish that eat the caddisflies was proposed as an explanation.

Swedish researchers have studied many aspects of the uptake and distribution of mercury compounds in organisms. Boëtius (1960) investigated the lethal action of mercuric chloride and phenylmercuric acetate on several species of fishes, taking into account different parameters. He was able to demonstrate that the toxic ions were picked up by the fishes from solutions of the two mercury compounds and that they were accumulated to the point where they were lethal to the animals. The poisonous action of the compounds is temperature-dependent, with survival time reduced with increasing temperature. The euryhaline species *Gasterosteus aculeatus* and *Salmo gairdnerii* were more sensitive to mercuric chloride than the freshwater fishes studied; and *Salmo* was particularly sensitive to solutions of phenylmercuric acetate. In a study of the toxicity of these compounds in seawater, it was learned that the survival curves for *Gasterosteus* followed those found for fishes in freshwater solutions. Phenylmercuric acetate, however, had a somewhat stronger effect than mercuric chloride in seawater due, perhaps, to the chemical changes mercuric chloride undergoes (dissociation of this weak electrolyte accelerated by dilution).

Recently, Johnels et al. (1967) were able to demonstrate that the high values for mercury found in the pike were due to industrial discharges. Also, there was evidence that plankton, insect larvae, crayfish, and macrophytes can accumulate mercury either directly or indirectly from water.

Using whole-body autoradiography, Bäckström (1967) was able to localize the mercury in different tissues and organs of the pike. High concentrations of the element were found in the kidneys, liver, and spleen after administration of mercury chloride by means other than immersion in contaminated water. The gills and skin showed intense accumulation when the fish were kept in contaminated water. Organic compounds given to fish produced a greater concentration of mercury in the skeletal muscles than the chloride form, and also very high concentrations in the liver and kidney.

The researches of Sumino (1968) corroborate and extend the work of the previous authors. Using the goldfish as the test animal he found the following:

1. Goldfish are able to pick up and concentrate methyl mercury from water containing minimal amounts of the compound (0.0003 ppm) over a period of 10 days.

2. There is uptake of alkyl mercury by the fish from water with peak accumulation after 12 days; and 50 per cent was lost after 10 to 11 days in uncontaminated water (calculated from his graph). In these fish, methyl and ethyl mercury were detected, whereas no methyl mercury was found in goldfish kept in various inorganic mercury solutions.

Sumino (1968) also found traces of methyl mercury in various species of marine fishes, seaweed, and a mollusc collected from the Indian Ocean, Pacific Ocean, the Bering Sea, and the African Coast. He found the results difficult to explain since there was no apparent source of contamination in the vicinity of the collection sites. It is less surprising if one considers that the trade winds carry pesticides and that mercury was found in marine aerosols, as reported above.

In California the giant kelp, *Macrocystis pyrifera*, forms beds from which abalones, sport fishes, and the surface canopy of this plant are harvested (Clendenning and North, 1960). Large kelp beds along the coast near Los Angeles and San Diego have been greatly reduced due not only to natural causes, but also to noxious elements in the water from industrial and domestic sewage. In an effort to individualize these elements, experimental studies were carried out on the inactivation of photosynthesis in kelp fronds by the inorganic ions commonly found in the wastes. It was found that 50 per cent inactivation was obtained after a 4-day exposure to the following metal concentrations: mercury, 0.05 ppm; copper, 0.1 ppm; nickel, 2.0 ppm; chromium (hexavalent), 5.0 ppm; and zinc, 10 ppm.

Interesting were the results of studies done to explain observations in the field where benthic animals were found dead 10 wk following gross contamination of a bay with diesel oil from a tanker which had run aground. It was learned that oil is easily emulsified in seawater by surf action, and deeply immersed animals received the oil droplets following emulsification. In addition, kelp fronds react to oil emulsions after 2 days or longer, the time necessary to penetrate the plant. It is not yet clear how the oil manages to get through the thick, hydrated cell walls, but it does, and damages the seaweed as a result.

Beyond the Sea

It would be unreasonable to leave the problem of heavy-metal contamination of marine organisms without considering some of its far-reaching effects. There is no need to emphasize the increasing dependence of man upon the sea for alimentation, and therefore, the need for uncontaminated food items for his diet. This has been amply illustrated.

The perfection of methods for the conservation and transport of these foods means an ever-increasing distribution among inland populations. In addition, the possible use of processed meals made of marine organisms as a source of protein for underfed populations renders more urgent the considerations of heavy-metal contamination of the marine environment.

Before the advent of nuclear energy, interest in trace elements in the human body was confined to those elements which were possible to detect with conventional techniques and which were considered essential. Radioactive nuclides, whether detected in animals or plants due to fallout or used as a research tool, made possible the investigation of many elements previously disregarded. In addition, the appearance of sensitive techniques provided the necessary impetus for the investigation of stable trace elements in problems of human health and disease, from the tissue to the cellular level.

From the standpoint of health and safety, required in an era when hazards due to the manifold peaceful uses of nuclear energy are evident, studies have been undertaken to learn as much as possible about the concentration and distribution of stable elements in the normal human body (Tipton, 1960; Harrison et al., 1968). The data accumulated need guidelines for their biological interpretation since the occurrence and wide ranges of some elements are often difficult to explain.

Schroeder (1960) gives an informative picture of trace metals in man. He has argued that if an element is not present in plants it will not be essential for man. Hence those elements present in man which do not come from plants can be considered contaminants. After a perusal of the Periodic Table of Elements and due consideration of what is known about the possible metabolic role of each element, Schroeder puts forth the following ideas about metals: manganese, cobalt, copper, zinc, and molybdenum are trace metals present in plants and animals and are considered essential for growth. Rubidium, boron, barium, strontium, vanadium, chromium, nickel, and aluminium are ubiquitous with suspected metabolic functions. The metals not usually present in plants (unless from fertilizers) but found in human tissues are cadmium, mercury, lead, gold, silver, tin, bismuth, and titanium which are most likely environmental contaminants since they are not usually detected in infants but are apparently acquired with age.

For many metals there is an established role in enzyme systems and the investigation of these metalloenzymes continues at a great rate. Evidence is being accumulated on the relationship between abnormal trace element levels in the human body and certain pathological conditions (the interested reader is referred to the numerous publications of Dr.

Henry A. Schroeder in the *Journal of Chronic Diseases;* those of Dr. P. O. Wester, Aktiebolaget Atomenergi Reports, Sweden, Stockholm; and Linekin et al., 1968; Strain et al., 1968). More apropos is the research on the distribution of trace elements through organs and muscles in terrestrial and aquatic (marine) organisms and in man (Rancitelli et al., 1968).

Due consideration should also be given to the plight of migratory birds which occupy, at one time or another, almost every body of water. The unfortunate loss of some species due to the high level of mercury in fish should serve as a warning concerning other metals which can be equally lethal. Thus ecological niches can be disrupted on land, just as in the sea, due to metal pollution of a food chain which extends outside the realm of the ocean.

Summary

The need for a closer look at heavy-metal pollution of the marine environment has been illustrated. Only some of the multiple factors involved have been touched upon, and these points can be summarized.

1. Heavy metals are by-products of modern industrial and technological advancements. They find their way into the sea by direct disposal (including aerial fallout) and by indirect pathways.

2. The solution to problems of metal waste disposal might be expected on the basis of the vastness of the sea. Unfortunately, the waters of the oceans are not thoroughly mixed, nor uniform in temperature or movement, and as a result, the lack of uniform dilution causes local concentrations of metals with subsequent changes in the environment.

3. Not all of the ocean's production is available to man. Those areas from which man harvests food are, all too often, the hardest hit when pollutants are indiscriminately added to the water. Although some fisheries extend to great depths, the near-shore areas and continental shelf provide large populations of edible molluscs, seaweeds, fishes, and crustacea. Examples are given of the detrimental effects of heavy metals on these organisms.

4. Metals are nondestructible, and even when discharged into the ocean in small quantities can be accumulated to an alarming and lethal level by certain species. The uptake, retention, toxicity, and tolerance of metals by organisms are governed by many physiological and nonphysiological factors.

5. Once heavy metals have been absorbed or adsorbed by certain organisms their distribution in the sea can be greater vertically and horizontally than if they were carried by the surface currents.

6. The derangement of ecological niches in the sea go unnoticed at first. Nonetheless, it is well to bear in mind that each step in the food chain, from phytoplankton to the consumer (often it is man), must be protected in order to provide that continuity necessary to safeguard both this valuable resource and man himself.

7. The detrimental side effects of metals on man are becoming increasingly evident. Land pollution creates sea pollution and should be rectified, for the passage of metals from seawater to marine organisms to man often can be brief.

8. More research on the effects of heavy metals in marine organisms is expected for an understanding of the cycling of metals through the marine environment and for the adoption of necessary preventative measures.

If metals helped the course of history and the progress of man, their use (with misuse) is now being governed by the law of diminishing returns. Ketchum (1967), in a talk on man's resources and the marine environment, aptly stated: "Potentially progress carries within it the threat of destruction of health and welfare of man himself." It should not be necessary to make this potential a reality.

References

Anonymous. 1967a. Lake Erie—dying but not dead. *Environ. Sci. Tech.*, **1**: 212–218.

Anonymous. 1967b. Pollution—not for cities alone. *Environ. Sci. Tech.*, **1**:686.

Anonymous. 1968. The cost of clean water. *Environ. Sci. Technol.*, **2**:257–266.

Armstrong, F. A. J., and W. R. G. Atkins. 1950. The suspended matter of seawater. *J. Marine Biol. Assoc. U.K.*, **29**:139–143.

Ashworth, De B. 1967. Use of organo-mercurials in crop protection in the United Kingdom. *Oikos*, Suppl. **9**:19–20.

Bäckström, J. 1967. Distribution of mercury compounds in fish and birds. *Oikos*, Suppl. **9**:30–31.

Baldi, E. 1949. Il Lago d'Orta, suo declino biologico e condizioni attuali. *Mem. Ist. Ital. Idrobiol.*, **5**:147–188.

Bandt, H. J. 1946. Ueber verstarkte Schadwirkungen auf Fische, insbesondere uber erhohte Giftwirkung durch Kombination von Abwassergiften. *Beitr. Wasser-, Abwasser-u. Fischereichemie*, No. **1**:15–23.

Bianucci, G., and G. DeStefani. 1968. *Il trattamento delle acque per uso industriale*. Ulrico Hoepli, Milano, 355 pp.

Black, W. A. P., and R. L. Mitchell. 1952. Trace elements in the common brown algae and seawater. *J. Marine Biol. Assoc. U.K.*, **30**:575–584.

Boëtius, J. 1960. Lethal action of mercuric chloride and phenylmercuric acetate on fishes. *Medd. Danmarks Fiskeri-og Havundersøgelser*, **3**:93–115.

Bowen, H. J. M. 1968. Urban pollution. *Nature*, **218**:106.

Bowen, V. T. 1950. Manganese metabolism of social Vespidae. *J. Exp. Zool.*, **115**:175–205.

Bowen, V. T., H. Curl, Jr., and G. D. Nicholls. 1967. Elemental composition of marine plankton species. Report NYO-2174-60, 5 pp. plus tables.

Bowen, V. T., and D. Sutton. 1951. Mineral constituents of marine sponges. *J. Marine Res.*, **10**:153–169.

Brar, S. S., D. M. Nelson, E. L. Kanabrocki, E. E. Moore, C. D. Burnham, and D. M. Hattori. 1968. Thermal neutron activation analysis of airborne particulate matter in Chicago metropolitan area, pp. 560–568. In *Preprints of contributed condensations, the 1968 international conference on modern trends in activation analysis*. National Bureau of Standards, Washington, D.C.

Brinkhurst, R. O. 1965. Observations on the recovery of a British River from gross organic pollution. *Hydrobiologia*, **XXV**:9–51.

Brooks, R. R., and M. G. Rumsby. 1965. The biogeochemistry of trace element uptake by some New Zealand bivalves. *Limnol. Oceanogr.*, **10**:521–527.

Carlisle, D. B. 1968. Vanadium and other metals in ascidians. *Proc. Roy. Soc.*, *B* **171**:31–42.

Cavalloro, R., and Margaret Merlini. 1967. Stable manganese and fallout radiomanganese in animals from irrigated ecosystems of the Po Valley. *Ecology*, **48**:924–928.

Chang, P. S. 1965. Effects of pollution on oysters and fish in Taiwan, pp. 368–370. In *Third seminar biological problems in water pollution*, C. M. Tarzwell (ed.), U.S. Dept. Health, Education, Welfare, Public Health Service. Cincinnati, Ohio.

Clendenning, K. A., and W. J. North. 1960. Effects of wastes on the giant kelp, *Macrocystis pyrifera*. In *Proceedings of the first international conference on waste disposal in the marine environment*, E. A. Pearson (ed.), Pergamon Press, Oxford, London, New York, Paris, pp. 82–91.

Corbella, C., V. Tonolli, and L. Tonolli. 1958. I sedimenti del Lago d'Orta, testimoni di una diastrosa polluzione cupro-ammoniacale. *Mem. Ist. Ital. Idrobiol.*, **10**:9–50.

Cottam, Clarence. 1960. A conservationist's view on the new insecticides. In *Transactions of the 1959 seminar, biological problems in water pollution*, C. M. Tarzwell (ed.), U.S. Dept. Health, Education, and Welfare Tech. Report 1060–3, 285 pp.

Crandall, C. A., and C. J. Goodnight. 1963. The effects of sublethal concentrations of several toxicants to the common guppy, *Lebistes reticulatus*. *Trans. Amer. Microscop. Soc.*, **62**: 59–73.

Doudoroff, P. 1952. Some recent developments in the study of toxic industrial wastes. In *Proceedings of the fourth Pacific Northwest industrial waste conference*, State College of Washington, Pullman, Washington, pp. 21–25.

Doudoroff, P., and M. Katz. 1953. Critical review of the literature on the toxicity of industrial wastes and their components to fish. II. The metals, as salts. *Sewage Ind. Wastes*, **25**:802–839.

Dudey, N. D., E. E. Ross, and V. E. Noshkin. 1968. Application of activation analysis and Ge (Li) detection techniques for the determination of stable elements in marine aerosols. In *Preprints of contributed condensations, the 1968 international conference modern trends in activation analysis*, National Bureau of Standards, Washington, D.C., pp. 569–577.

Duursma, E. K. 1965. The dissolved organic constituents of seawater. In *Chemical oceanography*. J. P. Riley and G. Skirrow (eds.), Academic Press, London and New York, pp. 433–475.

Eriksson, E. 1967. Mercury in nature. *Oikos*, Suppl. **9**:13.

Fox, D. L., C. H. Oppenheimer, and J. S. Kettredge. 1953. Microfiltration in oceanographic research. II. Retention of colloidal micelles by adsorptive filters and by filter-feeding invertebrates, proportions of dispersed organic to dispersed inorganic matter and to organic solutes. *J. Marine Res.*, **12**:233.

Fruton, J. S., and S. Simmonds. 1958. *General biochemistry*, 2nd ed., Wiley, New York and London, 1077 pp.

Fukai, R. 1968. Distribution of cobalt in marine organisms. *Radioactivity in the sea*, No. 23, I.A.E.A., Vienna, 19 pp.

Fukai, R., and D. Broquet. 1965. Distribution of chromium in marine organisms. *Bull. Inst. Océanogr., Monaco*, **65**:1–19.

Fukai, R., and L. Huynh-Ngoc. 1968. Studies on the chemical behaviour of radionuclides in sea water. I. General considerations, and study of precipitation of trace amounts of chromium, manganese, iron, cobalt, zinc, and cerium. *Radioactivity in the sea*, No. 22, I.A.E.A., Vienna, 27 pp.

Goldberg, E. D. 1952. Iron assimilation by marine diatoms. *Biol. Bull.*, **102**: 243–248.

Goldberg, E. D. 1957. Ch. 12. Biogeochemistry of trace metals. In *Treatise on marine ecology and paleoecology*, J. W. Hedgpeth (ed.), Vol. 1. Geol. Soc. Amer. Mem. **67**, pp. 345–357.

Goldberg, E. D. 1962. Elemental composition of some pelagic fishes. *Limnol. Oceanogr.* Suppl. to **7**:lxii–lxxv.

Goldberg, E. D. 1965. *Review of trace element concentrations in marine organisms*, Puerto Rico Nuclear Center, Puerto Rico, 535 pp.

Goldberg, E. D., M. Baker, and D. L. Fox. 1952. Microfiltration in oceanographic research. I. Marine sampling with molecular filter. *J. Marine Res.*, **11**:194.

Grimanis, A. P. 1968. Simultaneous determination of arsenic and copper in wine and biological materials by neutron activation analysis. In *Preprints of contributed condensations, the 1968 international conference modern trends in activation analysis*, National Bureau of Standards, Washington, D.C., pp. 996–1003.

Guinn, V. P., and S. C. Bellanca. 1968. Neutron activation analysis identification of the source of oil pollution of the waterways. In *Preprints of contributed condensations, the 1968 international conference modern trends in activation analysis*. National Bureau of Standards, Washington, D.C., pp. 614–619.

Hanya, T., R. Ishiwatari, and H. Ichikuni. 1963. The mechanism of removal of mercury from seawater to bottom muds in Minamata Bay. *J. Oceanog. Soc. Japan*, **19**:20–26.

Harrison, W. W., M. G. Netsky, and M. D. Brown. 1968. Trace elements in human brain: copper, zinc, iron, and magnesium. *Clin. Chim. Acta*, **21**:55–60.

Harvey, H. W. 1938. The supply of iron to diatoms. *J. Marine Biol. Assoc. U.K.*, **22**:205.

Hasselrot, T. B. 1965. A study of remaining water pollution from a metal mine with caged fish as indicators. *Vattenhygien*, **1**:11–16.

Hood, D. W. 1967. Chemistry of the oceans: some trace metal-organic associations and chemical parameter differences in top one meter of surface. *Environ. Sci. Tech.*, **1**:303–305.

Hosohara, K., H. Kozuma, K. Kawasaki, and T. Tsuruta. 1961. Studies on the total amount of mercury in sea waters. *J. Chem. Soc. Japan, Pure Chem. Sect.* **82**:1479 (in Japanese, quoted by Fukai and Huynh-Ngoc, 1968).

Ikuta, K. 1967. Studies on accumulation of heavy metals in aquatic organisms. I. On the copper contents in oysters. *Bull. Japan. Soc. Sci. Fish.*, **33**:405–409.

Ikuta, K. 1968. Studies on accumulation of heavy metals in aquatic organisms. II. On accumulation of copper and zinc in oysters. *Bull. Japan. Soc. Sci. Fish.*, **34**:112–116.

Irukayama, K., T. Kondo, F. Kai, and M. Fujiki. 1961. Studies on the origin of the causative agent of Minamata disease. I. Organic mercury compound in the fish and shellfish from Minamata Bay. *Kumamoto Med. J.*, **14**:157–169.

Johannesson, J. K. 1955. The identification of fuel oils polluting coastal waters. *Analyst*, **80**:840–841.

Johnels, A. G., T. Westermark, W. Berg, P. I. Persson, and B. Sjöstrand. 1967. Pike (*Esox lucius*, L.) and some other aquatic organisms in Sweden as indicators of mercury contamination in the environment. *Oikos*, **18**:323–333.

Ketchum, B. H. 1967. Man's resources in the marine environment. In *Pollution and marine ecology*, T. A. Olson and F. J. Burgess (eds.), Interscience, New York, pp. 1–11.

Koch, P. 1960. Discharge of wastes into the sea in European coastal areas. In *Proceedings of the first international conference on waste disposal in the marine environment*, E. A. Pearson (ed.), Pergamon Press, Oxford, London, New York, Paris, pp. 122–163.

Korringa, P. 1952. Recent advances in oyster biology. *Quart. Rev. Biol.*, **27**: 266–308.

Korringa, P. 1968. Biological consequences of marine pollution with special reference to the North Sea fisheries. *Helgolaender Wiss. Meeresuntersuch.*, **17**:126–140.

Koyama, T. 1962. Organic compounds in seawater. *J. Oceanogr. Soc. Japan*, 20th Anniv. vol.

Laevastu, T., and T. G. Thompson. 1958. Soluble iron in coastal waters. *J. Marine Res.*, **16**:192.

Lee, R. E., Jr., R. K. Patterson, and J. Wagman. 1968. Particle-size distribution of metal components in urban air. *Environ. Sci. Technol.*, **2**:288–290.

Linekin, D. M., J. F. Balcius, R. D. Cooper, and G. L. Brownell. 1968. Multi-element analysis of pathological tissue. In *Preprints of contributed condensations, the 1968 international conference modern trends in activation analysis*. National Bureau of Standards, Washington, D.C., pp. 813–817.

Lloyd, R. 1965. Factors that affect the tolerance of fish to heavy metal poisoning. In *Third seminar biological problems in water pollution*, C. M. Tarzwell (ed.). U.S. Dept. Health, Education, Welfare, Public Health Service, Cincinnati, Ohio, pp. 181–186.

Lowman, F. G., D. K. Phelps, R. McClin, V. Roman DeVega, I. Oliver De Padovani, and R. J. Garcia. 1966. Interactions of the environmental and biological factors on the distribution of trace elements in the marine environment. In *Disposal of radioactive wastes into seas, oceans, and surface waters*. I.A.E.A., Vienna, pp. 249–266.

Lundholm, B. 1967. Background information: a survey of the situation in Sweden. *Oikos*, Suppl., **9**:45–49.

McKee, J. E. 1967. Parameters of marine pollution—an overall evaluation. In *Pollution and marine ecology*, T. A. Olson and F. J. Burgess (eds.), Interscience, New York, pp. 259–266.

Monti, R. 1930. La graduale estinzione della vita nel Lago d'Orta. *Rend. R. Ist. Lomb. Sci. Lett.*, **63**:75–94.

Moretti, G. P. 1954. Il limnobio neritico dei tricotteri a testimonianza dell'attuale situazione biologica del Lago d'Orta. *Boll. Soc. Eustachiana*, **47**:59–123.

Nishihara, S. 1967. A river can be made to help itself. *Environ. Sci. and Tech.*, **1**:710–716.

Noddack, I., and W. Noddack. 1939. Die Haufigkeiten der Schwermetalle in Meeres Tieren. *Arkiv. Zool.*, **32A**:1–35.

Okubo, K., and T. Okubo. 1962. Study on the bioassay method for the evaluation of water pollution. II. Use of the fertilized eggs of sea urchins and bivalves. *Bull. Tokai Regional Fisheries Res. Lab.*, Tokyo, Japan, No. 32.

Papadopoulou, C. P., C. T. Cazianis, and A. P. Grimanis. 1967. Neutron activation analysis of vanadium, copper, zinc, bromine and iodine in *Pyura microcosmus*. In *Nuclear activation techniques in the life sciences*. I.A.E.A., Vienna, pp. 365–377.

Rancitelli, L. A., J. A. Cooper, and R. W. Perkins. 1968. The multielement analysis of biological materials by neutron activation and direct instrumental techniques. In *Preprints of contributed condensations, the 1968 international conference modern trends in activation analysis*, National Bureau of Standards, Washington, D.C., pp. 802–812.

Richet, C. 1881. De la toxicité comparée des différents métaux. *Compt. Rend. Acad. Sci.*, 93:649–651.

Risebrough, R. W., R. J. Huggett, J. J. Griffin, E. D. Goldberg. 1968. Pesticides: transatlantic movements in the northeast trades. *Science*, 159:1233–1236.

Robertson, D. E., L. A. Rancitelli, and R. W. Perkins. 1968. Multielement analysis of seawater, marine organisms and sediments by neutron activation without chemical separations. Batelle Memorial Institute report BNWL-SA-1776 REV, Pacific Northwest Laboratory Richland, Washington, 18 pp. plus tables and graphs.

Rona, E., D. W. Hood, L. Muse, and B. Buglio. 1962. Activation analysis of manganese and zinc in seawater. *Limnol. Oceanogr.*, 7:201–206.

Roskam, R. Th. 1965. A case of copper pollution along the Dutch shore. In *C.M. Council Meeting*. Int. Council for the Exploration of the Sea, Sect. C. Near Northern Seas Committee, Amsterdam, Holland.

Schroeder, H. A. 1960. Possible relationships between trace metals and chronic diseases. In *metal binding in medicine*, M. J. Seven and L. A. Johnson (eds.), Lippincott, Philadelphia and Montreal, pp. 59–67.

Schubert, J. 1954. Interactions of metals with small molecules in relation to metal-protein complexes. In *Chemical specificity in biological interactions*, F. R. N. Gurd (ed.), Academic Press, New York, pp. 114–163.

Schweiger, G. 1957. Die toxikologische Einwirking von Schwermetallsalzen auf Fische und Fischnahrtiere. *Arch. Fischereiwissenschaft*, 8:54–78.

Slowey, J. F., L. M. Jeffrey, and D. W. Hood. 1967. Evidence for organic complexed copper in seawater. *Nature*, 214:377–378.

Sprague, J. B. 1965. Effects of sublethal concentrations of zinc and copper on migration of Atlantic salmon. In *Biological problems in water pollution*, C. M. Tarzwell (ed.), U.S. Dept. of Health, Education and Welfare. Div. of Water Supply and Pollution Control. Cincinnati, Ohio, pp. 332–333.

Sprague, J. B., P. F. Elson, and R. L. Saunders. 1965. Sublethal copper-zinc pollution in a salmon river—a field and laboratory study. In *Advances in water pollution research*, Vol. 1, Pergamon Press, Oxford, 373 pp.

Stevenson, R. A., and S. Lugo Ufret. 1966. Iron, manganese, and nickel in skeletons and food of the sea urchins *Tripneustes esculentus* and *Echinometra lucunter*. *Limnol. Oceanog.*, **11**:11–17.

Stevenson, R. A., S. Lugo Ufret, and A. T. Diecidue. 1964. Trace element analyses of some marine organisms. In *5th Inter-American Symposium on the Peaceful Application of Nuclear Energy*.

Strain, W. H., C. G. Rob, W. J. Pories, R. C. Childers, M. F. Thompson, Jr. J. A. Hennessen, and F. M. Graber. 1968. Element imbalances of artherosclerotic aortas. In *Preprints of contributed condensations, the 1968 international conference modern trends in activation analysis*, National Bureau of Standards, Washington, D.C., pp. 819–827.

Strickland, J. D. H., and T. R. Parsons. 1960. A manual of seawater analysis—with special reference to the more common micronutrients and to particulate organic material. *Fish. Res. Board Canada*, Bull. No. 125.

Sumino, K. 1968. Analysis of organic mercury compounds by gas chromatography. Part II. Determination of organic mercury compounds in various samples. *Kobe J. Med. Sci.*, **14**:131–148.

Tarrant, K. R., and J. O'G. Tatton. 1968. Organochlorine pesticides in rainwater in the British Isles. Nature, **219**:725–726.

The Pesticide Review. 1967. U.S. Dept. of Agriculture, Agricultural Stabilization and Conservation Service, Washington, D.C.

Thomas, A. 1915. Effects of certain metallic salts upon fishes. *Trans. Amer. Fisheries Soc.*, **44**:120–124.

Tipton, I. H. 1960. The distribution of trace metals in the human body. In *Metal binding in medicine*. M. J. Seven and L. A. Johnson (eds.), Lippincott, Philadelphia and Montreal, pp. 27–42.

Tonolli, L. 1961. La polluzione cuprica del Lago d'Orta: comportamento di alcune popolazioni di Diatomee. *Proc. S.I.L.*, **14**:900–904.

Tonolli, L., and R. A. Vollenweider. 1960. Le vicende del Lago d'Orta: inquinemento da scarichi cuproammoniacali. In *Convegno sulle acque di scarico industriali*, Milano, pp. 99–109.

Ulfvarson, U. 1967. Mercury consumption. *Oikos*, Suppl. **9**:22.

Vinogradov, A. P. 1953. *The elementary composition of marine organisms*. Sears Found. Marine Res., Mem. 2. Yale Univ., New Haven, Conn. 647 pp.

Vollenweider, R. A. 1963. Studi sulla situazione attuale del regime chimico e biologico del Lago d'Orta. *Mem. Ist. Ital. Idrobiol.*, **16**:21–125.

Walker, K. C. 1967. Legislative restrictions by federal action in the United States. *Oikos*, Suppl. **9**:21–22.

Westöö, G. 1967. Kvicksilver i fisk. *Var Foda*, **1**:1–7.

Part IV

Man's Alteration of Coastal Environment

18

Coastal Engineering Practices

CHARLES BRETSCHNEIDER, *Professor of Ocean Engineering, University of Hawaii, Honolulu, Hawaii*

> There rolls the deep where grew the trees,
> O earth, what changes hast thou seen!
> There where the long street roars, hath been
> The stillness of the central sea.
>
> Alfred Lord Tennyson in "The Princess" (XX 111)

Nature through geologic time has continuously changed the topography of the earth. Man over the last few thousand years has added a new force, the construction engineer, who is now playing a major role.

In this chapter only the problems of beach erosion and coastal engineering are considered. Their adverse effects are a type alteration of the environment that may lead to pollution. Much is said about coastal and offshore contamination of water in other chapters of this book. This chapter is concerned with the causes of beach erosion and discusses such problems relevant to man's impingement on the sea. Specifically, it can be concluded that a lot of "bungling" has occurred, both at the level of administration and scientific and engineering endeavor. However, it is now obvious that various Federal government agencies with funds (from the taxpayer) for development, reclamation, conservation, and preservation, have counteracted each other's activities. It is a fact of life that relates to the selfishness of man, a relation which is carried outside to the professional world to those dedicated to the solution of these same problems. These alerting facts strikingly point out why it is necessary that we take a good hard look at *The Impingement of Man on the Oceans*.

Actually, man's impingement on the sea begins at a far distance inland from the coastline. In this light one would also include the many inland waterways and the lakes of the world. What is happening, for example, on the Great Lakes of the United States today may happen on the great oceans of the world tomorrow—unless we stop our unwise destruction of our coastal and ocean environment. The Great Lakes might represent a technical hydraulic model for many aspects of the oceans.

If the water level in Lake Michigan, for example, is not controlled, then one should expect periods of major erosion of the beaches, dunes, and cliffs, and much property damage similar to that which took place some years ago at St. Joseph, Michigan. At that time the lake level reached a record high; a rising lake level is occurring at the date of this writing.

Certain man-made structures along Southern California have caused erosion downstream to such an extent that the beaches, dunes, and cliffs started to disappear into the sea. Private homes were damaged or lost to the sea (Johnson, 1951).

Even in the Arctic regions of Alaska, man is leaving his mark. Hume and Schalk (1964) have shown that almost 40,000 yd^3 of beach borrowed at Barrow, Alaska, for use on the airport runway, resulted in a shoreline retreat of 10.2 ft. This actually created a threat to the lives and homes of the native population.

The recovery of vast quantities of oil on the Alaska north slope should be controlled by proper management to make certain that the ecology of the system is not disturbed to the disadvantage of future man. It has taken thousands of years for the permafrost to become established. Nevertheless, as in so many similar cited cases, the north slope can be destroyed in less than a decade unless proper precautions are taken.

What the ecologists, marine biologists, fisheries experts, and interested laymen often fail to realize is that engineers can, through cooperative efforts and consorted guidance, help control and rectify the situation. First, let us accept the fact that there must be progress or else the world will become dormant. Progress for the benefit of mankind will naturally have a strong influence on ecology. Enlightenment, compromise, and cooperation with engineers who are given the responsibility of designing and building structures to protect mankind from the hazards and forces of nature and to provide for the rational development of the earth's resources to accommodate the increasing needs of mankind is essential.

Is it important, for example, to protect people from hurricane disaster by building a hurricane barrier even if it means a loss in fish catch? Is it more important to build a harbor by use of explosives (either high explosives or nuclear explosives) for the benefit of all people than to protect the marine ecology of a relatively small area? The areas affected are so often very minute compared with the total untapped marine resource areas of the country and the world. One might look at the Project Chariot study near Cape Thompson region, Alaska (U.S. Atomic Energy Commission Report, 1966). This study may have had some bearing on the decision not to build the harbor near Cape Thompson, Alaska, by use of nuclear explosives. However, one might ask, would such a harbor really be necessary even if it had to be built by conventional means?

Natural Sources of Beach Sands

The natural source of beach sands and materials is the land that is being eroded by natural means, such as caused by rainfall runoff and wind drift acting on the particles of sand. The rainfall runoff and the snowmelt runoff during floods causes a large transport of sand and deposition of sediments on the beaches by many rivers, brooks, and streams. There are no other original natural sources for beach sands, except for coral sand in the tropical and semitropical areas of the ocean. Man's transport of sand from one area to another is not a natural phenomenon. Making sand by mechanically grinding rock in a factory does not constitute a natural source of sand. This is an artificial source and, by present methods, produces only a very small contribution to the sand budget.

During floods or peak discharge periods most of the sands and sediments from the continents or islands to the beaches are transported. In some cases, ordinary coastal run-off will yield deposits of sand to the beaches. Wind action also has certain dynamic effects depending upon the directions and the strength of the wind relative to the coastline. Man's transporting from one area of the world by boat, train, or barge shipment does not increase the world balance of sand. As a matter of fact, taking sand from one seemingly overabundant area to be deposited on an undernourished beach will cause a reduction in the available world balance of sand if such sand is later eroded and lost to the bottom of the sea.

A continuing supply of sand results from the weathering processes of rocks. However, the net available sand may not be increasing if as much sand is irrevocably lost to the bottom of the sea as is provided by natural or artificial means. Johnson (1959) gives a comprehensive report on the supply and loss of sand to the coast. The reader should give great attention to this publication.

The Disrupting of Sand Transport to the Coastline

Trees and other shrubs planted to prevent land erosion might have a deleterious effect on beach erosion. If such action prevents the products of land erosions from reaching the sea, this action will disrupt and adversely affect the normal equilibrium balance of nature.

Man-made structures built for reclamation, flood control, hydroelectric power plants, or other purposes will tend to disrupt the normal equilibrium balance of nature's forces. To enhance certain aspects of man's environment some sacrifices will have to be made. If proper precautions are not taken, reclamation and/or conservation in one area can cause destruction in other areas.

Watts (1963) discusses the sediment discharge to the coast as related to shore processes, and Norris (1964) discusses dams and beach-sand supply in Southern California. Both papers relate to man's impingement on the sea; both papers discuss man's prevention of sand from reaching the beach.

Norris (1964) reports numerous studies which have shown that most beach sand is delivered by streams. Beaches are in equilibrium with their surroundings and require periodic nourishment to balance their natural losses of sand. Construction of dams, flood control channels, and settling basins tend to intercept sand and prevent it from reaching the beaches.

Norris (1964) lists numerous beach-sand supply barriers in Southern California. Many of the important sand-supplying streams in Southern California are affected by dams. Other man-made alterations of drainage basins include construction of concrete-lined channels which encourage rapid runoff of water but usually carry little sand. These channels often begin at the spillway of a flood-control dam or other structure that traps sediment. The concrete-lined channels prevent erosion of the stream channel and, in this manner, may seal off sand sources and prevent the sand from reaching the beach under natural conditions. Methods should be included in any flood control or multiple purpose project to permit the trapped sand to be flushed into the sea. This could be done periodically using peak discharges instead of average discharges of water from the reservoir into the channels.

There is a seasonal change in sediment transport and one should not construe that losses during one season are related to man's involvement. For example, Moberly (1963) reports that certain Hawaiian beaches erode during spring and summer tradewind activity and then are again built up during the winter Kona weather and when the tradewinds become more northerly. In Southern California, the beaches tend to become flatter and thinner during the winter months, often resulting in the exposure of rocks along the inshore areas. This occurs when the storm waves arrive from the west and northwest. During the summer months the beach sands move inshore, covering the exposed rocks and producing a steeper profile. This occurs when the long period waves or swells arrive on the coast from the Southern Hemisphere.

Beach Borrow Near Barrow, Alaska

Hume and Schalk (1964, 1967) discuss the effects of beach borrow in the Arctic. The longshore transportation of sediments in Arctic regions is slight compared with temperate regions, although shoreline processes of the Arctic are fundamentally the same. The difference between the processes of the two climates is chiefly one of magnitude. There is greater erosion and longshore transport in the temperate regions because of greater wave activity than in the Arctic and because of the effect of shore-fast ice and frozen ground of the Arctic. The effect of tidal range (6 in. in the Arctic) is less than in temperate regions.

The effects of the Arctic climate have been to slow erosion, protect the coast and reduce net longshore transport. The ocean at Barrow is frozen for 9 months a year and on this basis alone it will take about four times as many years for natural longshore transport to replenish beach borrow in the Arctic than in the temperate zones.

Additional information on Arctic beaches is given by Carsola (1952), Ray (1885), and Rex (1964) among others.

Responsibility

One might state that the land conservationists and reclamationists, the Parks Service of the U.S. Department of the Interior, the Forest Service of the Department of Agriculture, the U.S. Army Corps of Engineers, the Internal Revenue Service, and others are all responsible for not solving problems peculiar to man's impingement on the sea.

The Corps of Engineers tries to prevent coastal and beach erosion. The soil conservationists' programs attempt to prevent the natural erosion of land. If lands were allowed to erode as the result of flood waters coursing the rivers and streams leading to the sea, there would be much less unnatural beach erosion. It is unfortunate indeed that the people are having their tax monies spent by different federal factions—one prevents land erosion on the beaches and the other captures the sand that does not arrive at the beaches.

Without beaches to protect the headlands from the dynamic action of tides, waves, and currents the headlands would also soon disappear and the main continents would then be subjected to the direct action of the sea. It is important that the normal flow of sands from the land to the sea not be disrupted; a balance is necessary.

It is important that adequate sand comes to the beach, but it is more important this same sand does not leave the continental shelf. If excessive sand and other sediments reach the bottom of the sea there could be a rise in sea level, eventually resulting in higher tide and wave elevations which in turn would accelerate coastal erosion.

On the other hand, if such sands were retained at the coast on the beach, or even on the continental shelf, there would be additional protection offered against coastal and beach erosion. The water depth over the continental shelf would be slightly decreased so that the wave action reaching the beaches would be reduced. The wave activity at the coastline and the beaches depends upon the depth and width of the continental shelf, as indicated by Bretschneider (1954) and Bretschneider and Thompson (1955).

This shallow water would be conducive to less wave activity, although storm surges would be higher. The net result would be less damage during severe storms such as hurricanes in the Gulf of Mexico. For example, the wide and shallow continental shelf off the West Coast of Florida has helped to offset the more severe hurricane damage which the East Coast of Texas, with its narrow, deep shelf, has suffered.

The U.S. Bureau of Reclamation, the U.S. Army Engineers, the Forestry Service, the Park Service, the Highway Department, and many other Federal and State agencies as well as local governments, industries, and politicians are all responsible since they are all taking part in programs which lead to preservation and rehabilitation of the land and by so doing lead to destruction of the beaches and coastlines.

Specifically, people themselves and not their representative should be blamed. Likewise, politicians should not pass the buck to people, industry, and government agencies since, in general, these groups adhere to laws and guidelines established by the Federal, State, and local governments.

Protection Against Beach Erosion

Sand transport to the beach can be controlled by proper design and construction of sluice gate structures and/or dredging the sediment deposits in the reservoirs. Such should be provided for all flood control and reclamation projects that otherwise would have adverse effects on beach erosion.

The controlled discharge of water, instead of peak discharge, will not effectively carry sediments. Instead, controlled peak discharges should be made at reasonable intervals of time to provide for the transport of the impounded sands and sediments to the sea. Peak discharges should be controlled to remain well below flood stage of the low-lying homes and property of those living in potential flood-zone areas.

The Disruption of Beach Equilibrium

The disruption of beach equilibrium can take place when a man-made structure is built on, off, or along the coast—for example, seawalls, breakwaters, jetties, or groins. The sediments which reach the coastline (with or without dammed reservoirs) will be affected by the action of waves and currents. The action of waves and currents will cause a sorting of the particle sizes and the conveyance and distribution of the sediments along the coastline. This is a natural and normal process until man builds structures along and/or seaward of the coastline. Such structures will tend to serve as a means for trapping sand upstream of the longshore current along the coastline. Erosion will take place immediately downstream of the structure. The region further downstream will be replenished by this beach material and will tend to remain in natural equilibrium. To avoid downstream erosion, it is necessary to have a bypassing means to transport the sands from the upstream side to the downstream side of the structure. This can be accomplished by dredging or by use of a hydraulic sand bypassing plant. Otherwise, a resupply of the trapped sand should be hauled to and distributed on the beach.

What About Vegetation to Prevent Land Erosion?

Stone (1965) gives an excellent paper on preserving vegetation in parks and wilderness. His paper is very relevant to the problem at hand, for it shows how poorly we have understood and managed natural erosion problems in the past. Although Stone (1965) does not discuss beach erosion, there the problem is even more formidable. Stone (1965) points out that "this fragile natural resource (land) is endangered by the lack of trained specialists and the lack of clear objectives." A classic example is the big cone spruce stands in the Chaparral of Southern California maintained by natural firebreaks provided by intermittent low brush shallow soils. Today these natural firebreaks are overgrown with brush grass and other fire hazards and no longer offer protection for the big cone trees against fire. It is true that natural fires are a necessary part of forest-life ecology; however, in this particular case an unnatural fire hazard situation was promoted. The vegetation-preservation manager should think long before vegetation is planted that would prevent natural erosion of the land to the beach. This type of prevention of normally deposited sands going to the beach is a problem similar to the destruction of forests caused by man's improper management.

It may be necessary and desirable to grow vegetation which would prevent erosion in some locations but this should occur only where the sands were not originally destined for the beaches. It is important that conservationists and preservationists conduct experiments where the results do not cause damage elsewhere. A good place to start such a program is along the coastal dunes of the United States, remembering that the coastal dunes are our first line of defense against great sea storms.

The Dichotomy of Resource Management

Allen (1938) reported in *An Introduction to American Forestry*, the 1872 Yellowstone National Park Establishment Act, that the Federal government's efforts to preserve natural vegetation goes back only 100 yr. Within the following decade, state governments also participated.

In 1905 the Forest Service was organized within the Department of Agriculture to manage the forest reserves, secure favorable watershed conditions, and furnish a continuous supply of timber. It also recognized the importance of recreation. In 1916 the Park Service was organized within the Department of the Interior. Vegetation preservation constituted a minor part of the Department of Agriculture Forest Service, but a major part of the Department of the Interior Park Service.

The problem in both services was that of obtaining and/or training competent personnel and scientists in their respective fields. The Forest Service was only 11 yr ahead of the Park Service. The latter moved more slowly than the Forest Service and failed to benefit from the other's experience. Furthermore, the Park Service had a very difficult task, for they were neglected almost from the beginning and played a subordinate role to the Forest Service. Other government agencies, including the Forest Service, and private industries wanted to open the park lands for various dams and reservoirs, hydroelectric power, irrigation, flood control, mining, hunting, logging, and other commercial interests, all of which can have effects on beach erosion.

The National Academy of Sciences—National Research Council (1963) Advisory Committee looked into the research program of the National Park Service and found that the Park Service was both vastly understaffed and underbudgeted. The Park Service was only able to apply research in a piecemeal fashion and therefore failed to ensure the implementation of management. Trained specialists, including administrators, were required.

Current Status of the Problem

The U.S. Department of Agriculture (1963) held a Federal Inter-Agency Sedimentation Conference in which Symposium No. 3—Sedimentation in Estuaries, Harbors and Coastal Areas—is particularly germane to the present subject.

The supply of sediments to the shore from the streamflows is very important for the stability of the shoreline. When this supply is altered, littoral forces will accordingly try to adjust the shoreline. The development of works on upland drainage basins for irrigation, flood control, hydroelectric power, and so on, will reduce the sediment supply to the coast and the resulting effect to the shoreline and beach will be erosion. Unless an appropriate plan of shore protection is developed and constructed, valuable coastal property may be lost. Watts (1963) discusses this problem for Ventura and Northern Orange County of California.

On July 3, 1930 the Beach Erosion Board of the U.S. Army Corps of Engineers was established. The scope of the Board's work was at first restricted largely to the preparation and review of cooperative reports. In 1945 legislation was enacted authorizing the Board to undertake a program of general investigation and research and to publish technical reports relating to the problem of beach erosion and its control. On November 7, 1963, Public Law 172, enacted by the 88th Congress, approved the establishment of the Coastal Engineering Research Center as successor to the Beach Erosion Board.

During the 33 yr of the existence of the Beach Erosion Board, numerous studies, including theoretical, analytical, laboratory, and field studies have been made. Publications since 1940 to 1963 included 135 technical memoranda, 5 miscellaneous papers, 4 technical reports and 17 volumes of the Bulletin, not including the numerous papers presented in professional engineering and scientific journals. Between 1963 and 1968 the newly formed Coastal Engineering Research Center published 24 technical memoranda, 10 miscellaneous papers, and 2 bulletins.

The first conference on coastal engineering was held in October 1950 in Long Beach, California, and the proceedings were edited by Johnson (1951). There had been a total of 11 such conferences by 1968, and the proceedings numbered more than 10,000 pages. The earlier proceedings were published by the Council on Wave Research of the Engineering Foundation and the more recent ones by the American Society of Civil Engineers. Thus there has been a large fund of knowledge gained on beach erosion and coastal engineering during the last 20 yr.

Ecology and Beach Erosion

There are certain factors involving sand and sediment transport that can be devastating to ecology. These include the shifting of beach sand and sediments and the corresponding change of exposure of biota to the energy of the waves, the changing of water level, tides, and currents. Some of these causes may be natural, but others are man-made. The geological record reveals that natural changes have occurred throughout our history on earth, and further changes can be expected.

For example, a very recent natural phenomenon was a result of the great Alaskan earthquake of 1964. The shoreline of Northern Cook Inlet slumped into the bay, land and beach sands north of the epicenter sank as much as 6 ft and south up to 38 ft at McCloud Harbor on Monteque Island in Prince William Sound. Homer Spit dropped, and now at high tide the end of the Spit is unreachable since most of the leading edge is totally submerged.

In these cases the normal beach biota were either diminished or totally eliminated. The damage is still under repair by the indegenous organisms of the area and it will be many years, if ever, before the new beaches will be repopulated, spawning of salmon grounds developed, sea plants revegetate the changed regions, and the beaches again reach a status of equilibrium.

Man can cause similar and sometimes more dramatic effects even than a mighty earthquake. A beach or shallow area may be destroyed for specific organisms by changing the kind and supply of sediments. The sediments reach a beach by way of current transport and are sorted by wave energy. A coarse sandy beach that has its wave energy reduced by structures soon becomes muddy. Organisms requiring sand in support of their ecosystem will give way to those compatible with fine sediments. If, for example, clams characterized the resources of the original beach and only worms populated the muddy result, then the clam resource has been destroyed. Unlike much of the earthquake damage, recovery is unlikely because the energy of the system has been diverted and no force is at work to repair the change that has been wrought.

Further discussion on these points seems unnecessary, and is treated by others in this book, but hopefully the point is made that man and nature influence the normal biota inhabiting the shorelines and beaches of the world. These truly sensitive areas are drastically altered by what may appear to be only minor changes in their local environments.

Importance of Reference Material

A review of the reference material cited at the end of this chapter is very important for the interested reader. The references reveal numerous stories and photographs of case histories involving man's interaction with nature. There are too many such case history documents to reproduce and include in this chapter. However, a limited amount of discussion has been given here so that the reader may be encouraged to go to the original material and seek greater details for himself. One reason for not going into great detail of previously published material is based upon the context of the excellent paper on "Information Pollution" by Court (1969). It is not because I am lazy that I do not reproduce the material; but if the reader does not seek the original manuscripts, then only he can be accused of laziness or else otherwise noninterested in the subject.

Bibliography

Allen, S. W. 1938. *An introduction to american forestry*, 2nd ed., McGraw-Hill, New York.

Bretschneider, C. L. 1954. Generation of wind waves in shallow water, U.S. Army Corps of Engineers, Beach Erosion Board, Tech. Memo, **51**:24 pp.

Bretschneider, C. L., and W. C. Thompson. 1955. Dissipation of wave energy on Continental Shelf, Gulf of Mexico. Tech. Report Texas A & M Research Foundation, College Station, Texas, 25 pp.

Carsola, A. J. 1952. Marine geology of the Arctic Ocean and adjacent seas off Alaska and Northwestern Canada, Ph.D. thesis (unpublished), Univ. of Calif., Los Angeles.

Hume, J. D., and M. Schalk. 1964. The effects of beach borrow in the Arctic *Shore and Beach*. April 1964. 5 pp.

Hume, J. D., and M. Schalk. 1967. Shoreline processes near Barrow, Alaska; a comparison of the normal and the catastrophic. *Arctic*, **20**:86–103.

Johnson, D. W. 1951. *Coastal engineering, Proceedings of first conference*, Council on Wave Research, The Engineering Foundation.

Johnson, J. W. 1959. The supply and loss of sand to the coast. ASCE WW3 **2177**: 227–251.

Moberly Jr., Ralph. 1963. Coastal geology of Hawaii, Appendix I, Hawaii Shoreline, HIG Report No. **41**:216 pp.

National Academy of Science—National Research Council. 1963. A report by the Advisory Committee to the National Park Service on Research, Washington, D.C.

Norris, R. M. 1964. Dams and beach-sand supply in Southern California. In *Papers in marine geology*, R. L. Miller (ed.) MacMillan, New York, Chap. 9, pp. 154–171.

Ray, P. H. 1885. Report of the international polar expedition to Point Barrow, Alaska, 1881–1883, 48th Congress, 2nd Session, House of Rep. Ex. Doc. No. **44**:685 pp.

Rex, R. W. 1964. Arctic beaches, Barrow, Alaska. In *Papers in marine geology*, R. L. Miller (ed.), MacMillan, New York, Chap. 17, pp. 384–400.

Schalk, M. 1963. Study of near-shore bottom profiles east and southwest of Point Barrow, Alaska. Final Report to Arctic Institute of North America, Tech. Report, Smith College, Northampton, Mass.

Stone, E. C. 1965. Preserving vegetation in parks and wilderness. *Science*, **150**:1261–1267.

U.S. Atomic Energy Commission, Div. of Tech. Info. 1966. *Environment of Cape Thompson Region, Alaska*. N. J. Wilimovsky and J. N. Wolfe (eds.).

Watts, G. W. 1963. *Sediment discharge to the coast as related to shore processes.* U.S. Dept. of Agri., Proceedings of the Federal Inter-Agency Sedimentation Conference, U.S. Government Printing Office, Miscl. Publ. **970**:738–747.

Selected References

Bruun, P., and F. Gerritsen. 1960. *Stability of coastal inlets*, Vol. 1, University of Florida, Coastal Engineering, Gainesville, Florida.

Bruun, P. 1966. Tidal inlets and littoral drift, stability of coastal inlets, Vol. 2, H. Skipnes Offsettrykkeri, Trondheim, Norway.

Burns, R. E. 1963. *Importance of marine influences in estuarine sedimentation.* U.S. Dept. of Agri. Proceedings of the Federal Inter-Agency Sedimentation Conference. U.S. Government Printing Office, Miscl. Publ. **970**:593–598.

Court, A. 1969. Control of information pollution, the President's page, guest editorial, *Trans. Amer. Geophys. Union*, **50**:375.

Handin, J. W. 1951. *The source, transportation, and deposition of beach sediment in Southern California.* U.S. Army Corps of Engineers, Beach Erosion Board, Tech. Memo **22**:113 pp.

Ippen, A. T. 1969. *Estuary and coastline hydrodynamics*, McGraw-Hill, New York, 744 pp.

SELECTED REFERENCES

Johnson, D. W. 1919. *Shore processes and shoreline development*, Habner, New York, 584 pp.

Johnson, J. W. 1963. *Sand movement on coastal dunes.* U.S. Dept. of Agri. Proceedings of the Federal Inter-Agency Sedimentation Conference. U.S. Government Printing Office, Miscl. Publ. No. **970**:747–755.

Shepard, F. P. 1948. *Submarine geology*, Harper, New York, 348 pp.

Trask, P. D. 1955. *Movement of sand around Southern California promontories.* U.S. Army Corps of Engineers, Beach Erosion Board, Tech. Memo. **76**:60 pp.

Trask, P. D. 1932. *Origin and environment of source sediments of petroleum.* Amer. Petrol. Inst., Gulf Pub. Co., 323 pp.

U.S. Army Corps of Engineers. 1968. Coastal Engineering Research Center. Shore Protection Planning and Design. Tech. Rept. T.R.-4.

U.S. Congress House Doc. 761. 1948. Santa Barbara, California, Beach Erosion Control Study, 80th Congress, 2nd Sess., 35 pp.

U.S. Congress House Doc. 29. 1953. Appendix I, Coast of California, Carpinteria to Point Mugu. Beach Erosion Control Study, 83rd Congress, 1st Sess., 92 pp.

U.S. Department of Agriculture. 1963. Proceedings of the Federal Inter-Agency Sedimentation Conference, Agri. Res. Ser., U.S. Government Printing Office, Miscl. Publ. **970**:993 pp.

Wiegel, R. L. 1964. *Oceanographical engineering.* Prentice-Hall, Englewood Cliffs, N.J., 532 pp.

19

Forest Management and Agricultural Practices

DALE W. COLE, *Associate Professor of Forest Soils, College of Forest Resources, University of Washington, Seattle, Washington*

Management practices imposed on forest and agricultural lands obviously influence coastal waters since aquatic properties are partially dependent on the terrestrial environment. Manipulating this terrestrial environment to produce crops will in all probability change this delicate, complex relationship. Current agricultural and forestry practices require the harvesting, fertilizing, burning, scarification, and addition of biocides to control parts of the terrestrial biota. Such manipulations may, at times, be part of a well-conceived management program. But they are also practiced out of ignorance, by tradition, or simply for the sake of "convenience." However well-conceived these alterations not only potentially but typically change the chemical, physical, and biological properties of river, estuary, and coastal water systems.

Countless studies have been made to demonstrate the effects of basic agricultural and forestry practices on water bodies of all sizes, from farm ponds to major river estuary and lake systems (e.g., Ohle, 1955; Thomas, 1957; Sylvester, 1961; Hedgpeth, 1966). The deleterious effects of these treatments are diluted to some extent by the purity of water draining from untreated forested lands. As long as these lands remain a reserve unmolested and unmanipulated they will aid in stabilizing the quality and quantity of water reaching the ocean. In today's concept of forest management, encompassing the economic importance of forest land, it is unrealistic to consider forest areas in this sense. The future principles and practices of forest management will not differ substantially from those accepted today on agricultural lands. Another threat to the aquatic system posed by management of forest lands results from the fact that such lands are often associated with steeper, broken terrain and higher rates of precipitation. Disrupt this type of ecological unit and the consequence is a far greater risk factor than is found on agricultural lands.

The relationship between the terrestrial and oceanic environment must be considered before it becomes possible to sort out the significance of man's alterations of the first upon the second.

Relationship between Management of the Terrestrial Environment and Conditions of the Coastal Waterways

The relationship of terrestrial environments to aquatic systems is neither simple nor unidirectional. It is best understood by considering the many pathways of transfer between the two systems and the processes responsible for them. The actual magnitude of transfer is not fully known in natural systems or predictable in manipulated ones. This dearth of knowledge is particularly apparent in forestry, where observational studies have been the rule.

Only recently have comprehensive programs been established to examine the chain of events linking processes within a terrestrial ecosystem to specific properties of drainage waters. Probably the most significant current effort to develop an integrated research program is that organized under the auspices of the International Biological Program (IBP). Because so little is known, it is not surprising that the aquatic environment has suffered from practices imposed for the sake of growing both agricultural and forest crops.

The following discussion pays particular attention to transfer within the liquid phase of the terrestrial system, for it is here that the major transport of water, nutrients, organic residues, and particulate matter occurs. Where the information is available, these transport mechanisms are defined within the framework of the total ecosystem and the specific mechanisms involved are discussed. Many of these mechanisms and the rates involved are unknown in natural systems so it becomes highly speculative to suggest how these systems will behave when manipulated.

Discharge of Water

Because of its obvious economic value, the rate and quantity of discharge water from drainage basins have been extensively studied throughout the world. Historically such studies were initiated in Switzerland in the Emmenthal Mountains in 1900 where runoff from pasture and forested watershed was compared (Engler, 1919). Interest in this country was stimulated by Zon (1912). A controlled watershed approach was first initiated at Wagon Wheel Gap in the Rocky Mountains of Colorado in 1911. (Flow from two similar catchment basins is compared during a calibration period. This is followed by the treatment of one basin, maintaining the second basin as a control.) Since these very early attempts at evaluating water yield, there have been numerous additional studies under a variety of environmental and land management conditions. Basically, discharge characteristics of water from a drainage system are dependent on properties of incoming precipitation, physical configuration of the drainage basin, depth and permeability of the soil and parent material, and type and structure of the vegetative cover. These studies have been summarized and tabulated by Hibbert (1967).

Forest lands play a disproportionate role in the overall picture of providing drainage water to river systems. Although they comprise only $\frac{1}{3}$ of the total land area of this country (6.4 $\times$ 10^8 acres), forest lands yield about $\frac{2}{3}$ of the total flow. Forest lands receive an average of 42 in. of annual precipitation—more than twice the amount falling on other land areas. Of this, approximately 17 in. is runoff. This is almost 7 times the water yield from lands used in other ways (Storey and Reigner, 1966). It is no surprise that forest lands have long been regarded as a major source of water supply. Nor should it be surprising that a disproportionate amount of research on water yields and water quality has focused on forested watersheds.

Basically, the discharge of water from terrestrial ecosystems can be studied within the framework of the unit watershed as mentioned above. While this approach provides excellent data on the behavior of an entire drainage system, it also integrates many of these functions in the system, often making it difficult, if not impossible, to delineate specific processes within it. To describe specific functions within a system, additional techniques are available. The direct evaluation of the water budget can be obtained by using lysimeter systems or by measuring changes in the soil moisture percentages with instruments such as the neutron probe. The water budget can also be determined indirectly through energy-balance studies. All of these techniques have been successfully used under agricultural conditions. Because of the size and the relative heterogeneity of a forest, the techniques have not been as successfully applied to forest conditions.

Even the earliest studies, such as those at Wagon Wheel Gap (Bates and Henry, 1928), made it apparent that vegetative composition played

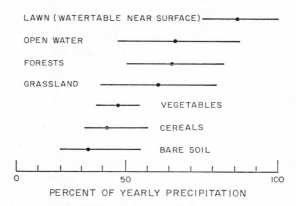

Figure 19.1 Annual vaporization among various cover types as percentage of precipitation. From Baumgartner, 1967.

a significant role in regulating the quantity and rate of water discharge. Recently this differential use of water by vegetative types has been conclusively demonstrated over a wide range of climatic conditions. Baumgartner (1967) has summarized much of this hydrological research. By comparing the consumptive use for the various vegetative cover types, it became obvious that a far greater quantity of water was re-vaporized from a forest stand than from cultivated or fallow conditions (Fig. 19.1). It has also been demonstrated that changing the forest species composition will influence water yield. Swank and Miner (1968) found that converting a hardwood-covered forest to white pine reduced the annual water yield by 3.7 in. (a reduction of about 10%).

By compiling hydrological data from experimental watersheds in the United States, Hibbert (1967) was able to evaluate management practices that significantly affect water yield. He concluded at that time that:

1. Reduction of forest cover increases water yield.
2. Establishment of forest cover on sparsely vegetated land decreases water yield.
3. Response to treatment is highly variable and for the most part unpredictable.

Recent studies at Coweeta (North Carolina), Fernow (West Virginia), H. J. Andrews (Oregon), and in chaparral sites in Arizona have all indicated that the influence of treatment is now forming a logical pattern. For example, at both the Coweeta and Fernow experimental watersheds the relationship between stream flow during the first year following cutting and the percentage of the forest removed by the cutting was nearly linear. A systematic difference was also found between the north and south aspects, apparently due to differences in evapotranspiration (Hibbert, 1967) (Fig. 19.2).

Perhaps the most conclusive evidence substantiating the contention that removing the forest will increase water yield was derived from Coweeta Watershed 13 (Hibbert, 1967) (Fig. 19.3). This watershed has received two identical treatments spaced 24 yr apart. The forest cover was systematically removed in 1940 after a calibration period of 4 yr. A very significant increase in water yield (approximately 14 in.) was found the first year after cutting. After 20 yr this increase in water yield had diminished to less than half the initial response. Presumably this resulted from reestablishment of forest cover. A second clearcutting of the forest in 1964 reestablished the water yield pattern found 24 yr earlier (after the first logging operation). Similar results were found by Cole (1966) for second-growth Douglas-fir using lysimetric techniques.

Unfortunately such increases in water yield can at times result in an increase in peak discharge rates. Consequently, removing the forest presents both a flooding and an erosional problem in downstream areas. In the Tennessee Valley Authority's Pine Tree Branch Watershed, peak discharge rates after a reforestation program on a poor agricultural soil

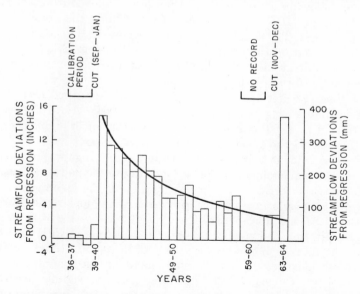

Figure 19.2 First-year streamflow increases from Coweeta and Fernow treatments versus reduction in forest cover, showing influence of aspect and the linear relation between area cut and amount of yield increase. From Hibbert, 1967.

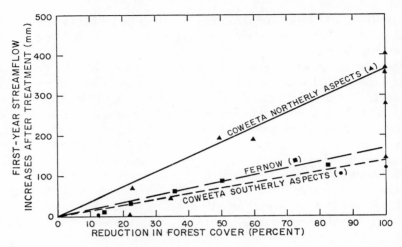

Figure 19.3 Deviations from regression of annual streamflow for Coweeta Watershed 13 on annual streamflow for control watershed during calibration (1936–1939) and treatment (1939–1964) periods. From Hibbert, 1967.

were reduced by about 90 per cent following summer storms and by 25 per cent following winter storms (Tennessee Valley Authority, 1962).

Peak discharge rates will be increased by burning the organic debris on the soil surface. The effect of fire on water flow depends on the intensity of the fire and on the properties of the soil system. Normally a fire will decrease infiltration rates as well as the storage capacity of the soil. In southern California, peak flows from a burned area were approximately 4 times greater than from a comparable unburned area during the first year following burning (Rowe et al, 1964). As in the case of clearcutting, the reestablishment of vegetation on the burned area rapidly diminished the effect of fire on discharge rates.

It should be evident that a real opportunity to change water yield will come from the management of forest land, due principally to these two reasons. First, the forest uses a higher fraction of the incoming precipitation through evapotranspirative processes than does agricultural or pasture land. Second, forest lands are typically associated with areas of higher precipitation and steeper topography. Associated with the opportunity is the danger of producing water at the wrong times or at the wrong rates. Changes in runoff rates have obvious implications to estuary ecology. Copeland (1968) pointed out the significance of decreased river flow on the biotic component of estuaries, including oyster, shrimp, and crab production.

Nutrient Losses in Ionic Form

The reader is aware of the striking differences in the nutritional content of river waters. Although such observations have been made throughout the world, it is difficult to determine a cause-effect relationship from them for drainage basins are incredibly complex. Drainage basins that are uniform in composition and relatively impervious in underlying geological structure are essential to an understanding of the relationship between nutrient release from the terrestrial environment and the chemistry of drainage waters.

Unfortunately, few hydrological studies are being made on watersheds meeting these criteria. The Hubbard Brook Watershed (New Hampshire) is one of the best known examples. Conditions causing the transfer of elements from the terrestrial environment to the drainage system under both natural and manipulated states have been evaluated (Bormann and Likens, 1967; Bormann et al., 1968). In general, ionic losses from the basin have been low despite very high rates of ionic input from atmospheric additions. During the year 1965–1966, 18 kg/ha of nitrate and 42 kg/ha of sulfate were deposited during precipitation (Fisher et al., 1968). High ionic losses (specifically nitrate) were found only after the removal of the vegetation (Bormann et al., 1968).

Although such studies clearly demonstrate the terrestrial release rates of ions into the aquatic system, they do not explain the causative processes. The influence of man on the nutrient content of river systems cannot be evaluated from release per se; the specific factors within the terrestrial environment causing this release must also be explained. As a case example, I will describe the pathways and transfer rates of a series of nutrient elements in a second-growth Douglas-fir ecosystem on the Cedar River Watershed (located in the Puget Sound Basin and serving as a principal water source for the City of Seattle).

Within this ecosystem the distribution, accumulation, loss, and internal cycling patterns for nitrogen, calcium, and potassium have been established (Cole et al., 1968). The low-nutrient content of the Cedar River cannot be directly attributed to a general lack of these elements in the ecosystem. As evidenced in Table 19.1, a substantial amount of all of these elements is found in either the ionic or organic form in the terrestrial phase of the system. A high fraction is actually located in the soil system, where transport to the river would most readily be accomplished. However, as Table 19.2 shows, the ecosystem remains virtually a closed cycle and only a minimal loss occurs beyond the rooting zone of this forest.

Table 19.1. Distribution of Nitrogen, Phosphorus, Potassium, and Calcium within the Major Components of the Ecosystem

Ecosystem Component	N		P		K		Ca	
	kg/ha	Per Cent of Total	kg/ha	Per Cent of Total	kg/ha	Per Cent of Total	kg/ha	Per Cent of Total
Forest	320	9.7	66	1.7	220	44.6	333	27.3
Subordinate vegetation	6	0.2	1	0.1	7	1.4	9	0.7
Forest floor	175	5.3	26	0.6	32	6.5	137	11.2
Soil	2809[a]	84.8	3878[a]	97.6	234[b]	47.5	741[b]	60.8
Total	3310		3971		493		1220	

From Cole et al. (1968).
[a] Total.
[b] Exchangeable with pH 7 ammonium acetate.

Table 19.2. Annual Transfer of Nitrogen, Phosphorus, Potassium, and Calcium (kg/ha) between Components of the Ecosystem

	N	P	K	Ca
Input (precipitation)	1.1	T[a]	0.8	2.8
Uptake by forest	38.8	7.23	29.4	24.4
Total return to forest floor	16.4	0.60	15.8	18.5
Leached from forest floor	4.8	0.95	10.5	17.4
Leached beyond rooting zone	0.6	0.02	1.0	4.5

From Cole et al. (1968).
[a] Trace.

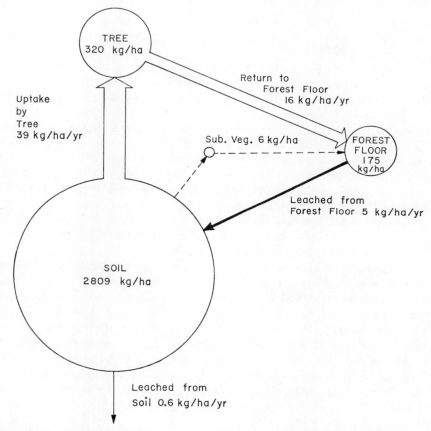

Figure 19.4 Distribution and cycling of nitrogen in a second-growth Douglas-fir ecosystem. From Cole et al., 1968.

By combining the distribution data of Table 19.1 and the transfer data of Table 19.2, cycling diagrams for the various elements have been constructed (e.g., Fig. 19.4).

The retention of these elements in the terrestrial phase is understandable when the specific processes regulating ion mobility in the ecosystem are considered. Before nitrogen can pass from a terrestrial phase it must first be mineralized to an ionic form through a biological process. Once in ionic form, it is still not free to actively leach through the soil. It can be retained biologically by reentering the nitrogen cycle and physically by the negative potential of the colloidal fraction of the soil. It is difficult for nitrogen to avoid these basic retention mechanisms. Both the microorganisms and the forest itself (Fig. 19.4) rapidly reenter nitrogen into the biological cycle.

Physical retention through the cation exchange mechanism of the soil is also significant. Again, the relationship is not simple. Nitrogen will be retained on the exchange capacity of the soil in the ammonium (NH_4^+) form. If, however, the ammonium is converted to nitrate (NO_3^-) through the biological processes of nitrification, the exchange capacity of the soil will be largely ineffective in stopping any leaching losses. However, nitrates are not naturally abundant in soils of the temperate forest region. For ammonium (or any other cation) to leach from the soil it has to be accompanied by an equivalent number of mobile negative charges. There are many sources of these. Generally, they are a consequence of some specific process such as respiration or nitrification. Limited amounts can also enter the system with incoming precipitation and through the solubility of the soil minerals.

The leaching of cations in the glaciated soils of the Cedar River Watershed is closely governed by the bicarbonate level of the soil solution (McColl and Cole, 1968). The presence of bicarbonate ion in a forest soil can be explained as follows. Carbon dioxide accumulates in the soil primarily through the processes of respiration and through the decomposition of organic matter. Approximately 1 per cent of the dissolved carbon dioxide from these processes hydrolizes and forms carbonic acid. This acid in turn ionizes into either bicarbonate or carbonate anions, the specific form being a function of pH at a given partial pressure of carbon dioxide at a given temperature. Where the bicarbonate ion comprises the major portion of the total anions in the soil solutions, which is typically the case, the associated hydrogen ion can replace readily available cations (such as ammonium, potassium, calcium, and magnesium) from the soil-exchange complex. These cations can then be transported along with the mobile bicarbonate ions through the soil into the stream or river system. A critical relationship therefore exists between soil respiration, formation of carbonic acid, ionization of carbonic acid into its ionic forms, and the leaching of cations in the soil system. Any process that adds anions to the system will react the same way. For each anion equivalent added to the system, potentially one equivalent of cation can be lost in the leaching process and thereby added to the aquatic system.

Certain land management practices can significantly influence the anion source. Ion additions with fertilizer salts will increase the anion level and consequently the cation discharge from the terrestrial phase.

In all probability this is one of the principal reasons for the relatively high leaching rate of nutrient elements through agricultural soils. Numerous studies have been made describing these losses as they are influenced by soil, crop, and fertilizer conditions (Allison, 1955, 1966; Barrows and Kilmer, 1963).

Under forest conditions there is no indication that parallel losses are occurring (Heilman, 1961). In fact, additions of 200 kg/ha of nitrogen in both the urea and ammonium sulfate form failed to significantly accelerate the leaching of nitrogen or any other element from the soil profile (Cole and Gessel, 1965). However, this could change as interest in forest fertilization grows, bringing with it repeated applications over large land areas. This could eventually affect nutrient levels in drainage waters.

During 1970, from 75,000 to 80,000 acres of forest land were fertilized in the State of Washington. A far more extensive forest fertilization program than this is currently part of the silvicultural programs in many countries in Europe (Tamm, 1968).

Vegetative cover also influences a number of critical mechanisms controlling cation transport in the soil. These mechanisms include cycling of minerals between the vegetative and soil components of the ecosystem, respirational rate within the soil, and processes of nitrification. Consequently, removing the vegetation can be expected to change the patterns of nutrient loss from the terrestrial phase. The extent of this increased loss has been studied under a variety of agricultural conditions (Barrows and Kilmer, 1963). Such losses usually are erosional and will be discussed later. Under forest conditions little increase in total loss was found by Gessel and Cole (1965) on the Cedar River Watershed. Since removing the trees increased total drainage, an actual decrease in ion concentration developed.

On the Hubbard Brook Forest, clearcutting the forest crop resulted in an apparent increase in the rate of nitrification (Bormann et al., 1968). The addition of nitrate ions to the soil system through such a process could play a significant role in accelerating cation losses from the soil profile. An increase in the rate of nitrification following harvesting has also been found by other investigators under both agricultural and forest conditions. One can only speculate why nitrification should be affected by crop removal. It could simply be that a source of nitrate utilization has been removed or the processes of nitrification have been enhanced.

In all likelihood, other management practices such as scarification of the soil surface and burning of surface organic residues—now commonly accepted in both forestry and agriculture—will further influence nutrient losses in the soil system. Specific studies are, however, far more limited.

Normally these treatments affect the stability of the soil surface and, therefore, surface erosional processes. These treatments are considered later in this chapter.

The addition of biocides will also affect ion transport. First, they will obviously change the cycling patterns by removing part of the biota. Second, addition of these substances may influence the activity of organisms in the soil surface and, therefore, the production of CO_2. It is not immediately obvious what the collective difference in changing these two basic mechanisms affecting ion transport in the soil will be. Direct observational data are lacking.

It is apparent that under certain conditions the management of agricultural and forest lands will influence the nutrient content of river systems (Sylvester, 1961). If this increase is appreciable it could cause eutrophic conditions in the drainage systems. In many lake systems, such as Lake Zurich, the Madison (Wisconsin) lakes, and Lake Washington (Seattle), the principal source of nutrient input is from sewage treatment plants rather than from agricultural practices (Edmondson, 1968). Only in some very specialized cases where drainage from a heavily fertilized field is passing directly into a small lake will such additions have a significant effect (Ohle, 1955).

Erosion and Sedimentation

The process of erosion is probably the single most significant reaction in the terrestrial system that directly affects the coastal environment. Associated with it is not only the physical displacement and redeposition of particulate matter, but also the principal transport mechanism for nutrient elements and persistent pesticides. The significance of erosion was recognized long ago on agricultural lands. Experiment stations, especially in the southeastern United States, were established to study this problem, but it was not until after World War II that parallel studies were initiated on forested lands (Dyrness, 1965).

The following four factors causing terrestrial surface erosion have been discussed by Smith and Wischmeier (1962):

1. Rainfall characteristics and their relationship to erosion—this is detachment in overland transport of soil particulate matter as influenced by rainfall intensity.

2. Soil erodibility—includes infiltration and percolation rates of soil together with those soil properties affecting the resistance of the soil to detachment and transport.

3. Topography—includes per cent slope and slope length.

4. Plant and litter cover—parameters that intercept rain and therefore directly affect the detachment and transport rates of soil.

All of these are extremely variable; thus, erosional conditions are also extremely variable. In addition, since many of these parameters are directly influenced by management practices, man can and does regulate (by design or by default) the amount of erosion that takes place. Smith and Wischmeier (1962) concluded that the greatest deterrent to erosion was plant cover.

Characteristics of parent material (Wallis and Willen, 1963) and soil (Anderson, 1954) significantly influence erosional processes, but the presence of a vegetative cover (Dyrness, 1965) is the critical factor. Consequently, management practices involving harvesting of crops, tilling or scarifying soil, and burning will all influence the extent of surface erosion.

This is a particular problem in forestry simply because forest lands are associated with higher and more intense rainfalls and steeper topographies. Thus removing vegetation or burning the surface organic layer on these sites carries a greater risk. All too often this risk has been ignored because the depletion in soil mass and loss of nutrient elements are regarded by some as quite small compared to the total capital. However, accumulation of this material in the drainage systems does have major implications for the coastal water systems.

The extent of erosional losses of both particulate matter and nutrient elements in agriculture was summarized by Barrows and Kilmer (1963). As might be expected, the greatest losses occurred under fallow conditions, and the extent of the losses was dependent on the specific soil and type of cropping.

Some of the highest losses of organic residues and nutrients have had little to do with the soil directly. Runoff from cattle feedlots is extremely high in organic residues and nitrogen (Miner et al., 1966). Hard surfacing of feedlots only adds to the pollution problem.

Under forest conditions, soils tend to be remarkably stable. This is probably due to the greater protection provided the soil surface. The kinetic energy of the incoming rain is fully dissipated by the time the drops have filtered through the crowns of the trees, the subordinate vegetative cover, and finally the organic layer directly overlying the soil. But many forest management practices result in total or partial destruction of this protective cover. One of the first studies to determine the effects of logging on surface erosion was conducted at Coweeta, North Carolina. With no restrictions placed on logging and roading techniques, sedimentation in the drainage water increased about fortyfold (Lieberman and

Hoover, 1948). Such studies have subsequently been made at many localities and with similar results (Rothacher et al., 1967; Reinhart and Eschner, 1962).

Generally the amount of erosion from the terrestrial landscape is strongly related to the removal of the forest tree and not simply to its cutting (Dils, 1957). It is the *removal* operation that exposes the soil surface and increases erosional potential.

Fires are often followed by large erosional losses. In areas where rainfalls are very intense, such as in the southwest, extremely high erosional rates are found after an area is burned (Rowe, 1941; Rich et al., 1961). Basically two factors affecting erosion potential are changed by burning the soil surface. The protective cover is removed, and there is a decrease in the infiltration rate of water into the soil. This decrease is caused in part by the development of hydrophobic properties (Osborn et al., 1964), the decrease in macropores (Tarrant, 1956), and destruction of the water-stable aggregates (Dyrness and Youngberg, 1957).

It is obvious that manipulating the terrestrial landscape will change the sediment and nutrient composition of drainage waters. But it is not so obvious how this will in turn affect the physical and biological conditions of the coastal water systems. The salmon fisheries industry has long been concerned with increased sedimentation from logging operations upstream from spawning beds. It has been established that fine sediments can be detrimental to embryos and alevins of salmon in the spawning grounds (Chapman, 1962; Cooper, 1965) but this does not necessarily result in a decreased density of spawners and fry (Sheridan and McNeil, 1968; Salo, 1966). In an addendum, Salo explained his results as follows (p. 61):

> After due deliberation of the effects of logging in Alaska and elsewhere, I am not sure that I can set minimum standards for the quality of water that is used for production of fish any more than I can set a minimum for the standards of living for humans in Africa, Asia, or anywhere else. It is becoming increasingly evident that nature has adapted these creatures to a situation that is endowed in the basic essentials to a minimum degree. She has compensated for this by a high fecundity rate and a resilience that is at times fascinating to behold.

While Salo is by no means arguing for increased sedimentation as a way of increasing fry survival, Stone and Vasey (1968) have argued that sedimentation was a necessary ecological process for the development and survival of the redwood forest.

Deleterious effects of increased sedimentation are well documented and have been discussed at great length. While much of the concern centers around siltation of navigable waterways, there is also concern for the biota of lakes (Edmondson, 1968) and estuary systems (Butler, 1966).

Retention and Transport of Biocides

The use of pesticides is summarized in Table 19.3 for forest, urban, agricultural, and other land areas (Tarrant, 1966). Although these chemicals are rarely applied directly to water systems, they represent an extreme pollution hazard to the aquatic habitat (Tarzwell, 1965; Butler, 1966; Cottam and Higgins, 1946). Terrestrial biocide applications have contaminated aquatic environments both by accidents during application and through transport processes between the terrestrial and aquatic systems after application.

Persistent pesticides such as DDT retain their toxic effect long after application. Since DDT will in part volatize after application, concentrations of this pesticide are found not only in streams draining treated areas but also in streams from drainage basins that have never received any pesticide application (Cope, 1961; Butler and Springer, 1963). It is common to find saltwater fish, shrimp, and oysters that contain some form of synthetic organic pesticide residue (cited in Westlake and Gunther, 1966).

Other organisms are also accumulating these residues with perilous effects. Wurster and Wingate (1968) have shown that certain carniverous and scavanging birds the world over are suffering a rapid reduction in reproductive capacity that is clearly related to the level of pesticides in their body fat. This applies even to oceanic birds such as the Bermuda petrel that have no contact with man or with sprayed areas. Woodwell (1968) has argued that if this indiscriminate terrestrial application of pesticides continues, we are in jeopardy of losing the biotic components of many of our water bodies, including the ocean. This is substantiated in part by Wurster (1968) who found DDT residues in the food chains of oceanic fish, and perhaps even in the phytoplankton. The U.S. Department of Interior first appropriated funds in 1958 specifically for pesticide research. Such attention to pesticides in water systems has resulted in more rigorous control of their use. However, it is still necessary to evaluate the use patterns of the hard pesticides and determine the possible mechanisms of transport in the terrestrial systems.

Table 19.3 indicates that 2.25×10^8 lb of insecticides were applied in the contiguous United States in 1962. Of this, 73 per cent was applied to agricultural lands, while only 0.8 per cent was applied to forest lands. In forest lands where applications are typically sporadic and light (from

Table 19.3. Total Acreage of Contiguous United States Area Receiving Insecticide Application in an Average Year, and Amount of Insecticides Used, by Land-use Type

Land-Use Type	Total United States Area in Land-Use Type (lb × 10⁶)	United States Area in Land-Use Type (% of Total)	Area Receiving Insecticides in an Average Year[a] (lb × 10⁶)	(% of Total)	Insecticides Used in United States[b] (lb × 10⁶)	(% of Total)
Forest	640	33.0	1.8	0.3	1.7	0.8
Urban and other built-up areas	53	2.7	15.0	28.3	55.8	24.8
Agriculture	457	23.6	68.6	15.0	164.8	73.2
All others	787	40.7	4.1	0.5	2.7	1.2
Totals	1935	100.0	89.5	—	225.0	100.0

[a] In an average year 4.6 per cent of total United States acres receive insecticides.
[b] From Tarrant, 1966.

1 to 5 lb/acre) there is some indication that appreciable contamination of adjacent water systems does not occur. The principal hazard occurs during application. By avoiding stream channels during DDT application for elm stanworm in Georgia, Frey (1962) avoided immediate detectable damage to the aquatic biota. When streams were not avoided, there was a serious depletion in aquatic insects. Similar results using DDT for elm stanworm were found by Grzenda et al. (1964). When they applied DDT to an entire drainage basin, residues up to 0.346 ppb immediately appeared in the drainage water. This contaminant rapidly decreased to less than 0.005 ppb after 2 months. When only 46 per cent of the watershed was sprayed during the subsequent year, no DDT residues were detected in drainage water or in the suspended sediment samples.

Large-scale operational pesticide programs have been carried out in Oregon (Perkins and Dolph, 1966) and Washington (Hemlock Looper Study Commission, 1964) without any reported damage to the water resource or to the aquatic environment. Numerous studies with various herbicides on forested land indicate the same response pattern (Krammes and Willets, 1964; Norris et al., 1965; Reigner et al., 1964; Reinhart, 1965). Contamination of drainage water occurred primarily when the herbicides were applied directly to the water system.

This behavior of biocides is predictable considering their reaction pattern in the soil system. Organic compounds of this type are readily adsorbed on the organic residues within and on top of the soil profile. Transfer of the persistent pesticides is then predicated on the mineralization of organic soil material or the physical transport of a dispersed organic colloid. Woodwell and Martin (1964) found that nearly all the DDT they applied to the soil surface was retained within a few inches of the surface. Similarly Riekerk and Gessel (1968), using a tension lysimeter system, found that less than $\frac{1}{3}$ of 1 per cent of the applied DDT leached beyond the 4-in. depth although the application was made on a very coarse glaciated soil. This low transfer occurred even at the high application rate of 50 lb/acre. As indicated, the usual application for the control of forest insects is from 1 to 5 lb/acre.

These losses of DDT from the terrestrial environment would certainly have been significantly higher if surface erosion had followed application. Consequently what might appear to be relatively safe techniques for insect control at the time of application could be extremely hazardous to aquatic biota if surface erosion eventually occurred—even as much as 10 yr later. The losses cited above would be significantly higher if the application followed any one of the many management practices used in cultivating terrestrial crops. Logging, scarification, road building, and agricultural cultivation all increase erosion potential. Fertilizers such as urea will peptize the organic soil colloids and thereby allow vertical transport through the soil system. In addition substantial proportions of pesticides might be volatized into the atmosphere after application and redeposited in subsequent precipitation. For example Tarrant et al. (1969) found that only 32 per cent of the DDT applied to a forest stand by aerial spraying ever reached the soil surface. There is evidence that DDT residues are now contained in nearly all natural ecological niches. For instance, accumulations of DDT have been found in the Blue Glacier of Mt. Olympus (Olympic National Park) in western Washington (cited in Westlake and Gunther, 1966). Since all the precipitation falling on this glacier comes from westerly weather patterns, DDT has circumnavigated the entire world, crossing in the process the barrier posed by the Pacific Ocean.

The significance of pesticide application to the terrestrial environment as related to long-term effects on the aquatic biota is not easy to determine with certainty. Most of the literature has addressed itself to contamination caused during application or immediately thereafter. Obviously, the persistent, hard pesticides could create a global contamination problem; Woodwell (1968) maintains such a problem already exists. Recent reports of heavy metal (Cd, Hg) concentrations in coastal fish seem to substantiate Woodwell's concern. Although the source of such concentrations is primarily thought to be industrial processes, these or similar substances are also used in control agents on agricultural and forest nursery lands. Investigations to establish the sources and transfer mechanisms of such contaminations have barely begun (Jernelöv, 1970), yet they may prove vital if we expect to continue to utilize the food resources of coastal water systems.

Future Implications

Intensive use of the terrestrial environment for the production of food and forest resources is both essential and inevitable. If current trends are any index of future directions, three principal changes will ensue.

1. A greater proportion of forest land will be placed under rigorous management regimes. These could involve a hazard to the aquatic habitats of river and estuary systems through the addition of pollutants, changes in nutrient regimes, or changes in rates and physical characteristics of runoff. However, the future of forestry is also bringing with it an increasing emphasis on the management of forest lands for a variety of products, where such additions or changes would be neither desirable nor

lawful. For example, many of the forest basins are either owned or managed by municipal water districts supplying the residential and industrial water needs of a community. This type of forest management has as its objective many criteria of water quality parallel to those of the aquatic biologist, in that both are equally concerned with organic, inorganic, and thermal changes, sediment load, turbidity, nutrient composition, and rate of runoff. Thus at times the forest manager inadvertently serves the needs of those directly responsible for the quality of the coastal water environment. Such nonrelated and unintentional cooperation is likely to increase, perhaps even become deliberate, in the future as increased emphasis is placed on the water quality of river systems used for residential and industrial purposes.

2. The intensity of agricultural land use will continue to change. During the past 10 yr, the number of acres in farm use has diminished. Associated with this decrease is an increase in farming technology, including the liberal use of fertilizers and biocides. For example, nitrogen fertilizer applications have increased by 200 per cent (cited in Engelstad, 1968) and biocide use has nearly doubled (cited in Westlake and Gunther, 1966) during this same decade. The focusing and intensifying of management activities on more limited areas will increase the potentiality of some of the by-products escaping into the drainage waters.

3. An ever-increasing percentage of the total land area will be diverted from agricultural or forest production and rezoned for other uses. Many of the fertile bottom lands of the Puget Sound Basin in the State of Washington have been rezoned from agricultural use and are now part of a large urban, industrial, and residential development. In King County (Seattle), this change in use has decreased farm acreage 70 per cent in the past 10 yr.

Management of agricultural and forest lands is not necessarily incompatible with the management of the aquatic environment of our coastal waters. However bad the terrestrial land manager must at times appear to the aquatic biologist, he might take on the stature of a saint when compared to the planners who "manage" our urban sprawl. The agricultural and forest manager is, at least, interested in the production of living systems and therefore has some common bond to his aquatic counterpart.

Because of the dependent relationship existing between the terrestrial environment and the coastal water systems, changes in use patterns will have serious, long-term effects. I do not believe that the long-term problem of our aquatic environment will originate with either agricultural or forest practices. Although such management practices will undoubtedly bring about changes in thermal properties, nutrient composition, and sediment loads of river systems, these changes will not be as great as those originating in residential and industrial areas nor will the problems of correcting agricultural and forestry practices be so painful. The future hazards stem from urbanization. The terrestrial manager might have a limited awareness of the problems in the coastal water systems, but he is not in the business of deliberately degrading the aquatic environment.

Residential and industrial developers have not shown that they carry with them this commonality of interests. The concentrated discharge of pollutants by industrial plants and municipalities will never be matched by the farmer or forest manager. The significance of this problem is discussed in other chapters of this book.

Conclusions

If we expect to understand the influence of land management practices on coastal, estuary, and marine waters, we must first understand the behavior of such ecosystems in their natural state. Only then can we differentiate the effects of management practices. The study of this relationship is so complex that it is only in recent years that we have had the sophisticated methods and background information to adequately address ourselves to the processes involved. Workers in aquatic ecology, and to a lesser extent in terrestrial ecology, have originated these studies or formulated the ideas leading to them. With the development of such programs as the International Biological Program (IBP), inclusive studies are now being constructed that involve the disciplines related to the collective problem.

Because man is rapidly changing the intensity of management of forest and agricultural lands, it is critical that inclusive studies are undertaken. For example, forest management now includes many techniques that were once exclusive to agriculture. Besides the systematic harvesting of forest crops on a rigorous rotational schedule, forest land management also includes land scarification for the establishment of the new forest crop, biocide additions for weed and insect control, and the addition of fertilizers for the improvement of productivity. Such practices will undoubtedly change the biological, chemical, and physical features of the coastal water systems. Such changes are not exclusively detrimental, for the nutrient capital and the physical features of the coastal water systems cannot survive without the erosion of the terrestrial landscape. Unfortunately, the current state of our knowledge limits us to little more than speculation concerning the consequences of further intensifying and expanding terrestrial management.

The change in land use appears to have a more predictable pattern: a strong and seemingly continuing tendency to expand urban areas at the expense of forest and agricultural lands. While certain agricultural and forestry practices are hazardous to the coastal environment, at their worst they fall far short of those hazards derived from urbanization.

References

Allison, F. E. 1955. The enigma of soil nitrogen balance sheets. *Advan. Agron.*, **7**:213–247.

Allison, F. E. 1966. The fate of nitrogen applied to soils. *Advan. Agron.*, **18**: 219–254.

Anderson, H. W. 1954. Suspended sediment discharge as related to streamflow, topography, soil and land use. *Trans. Amer. Geophys. Union*, **35**:268–281.

Barrows, H. L., and V. J. Kilmer. 1963. Plant nutrient losses from soil by water erosion. *Advan. Agron.*, **15**:203–315.

Bates, C. G., and A. J. Henry. 1928. Forest and streamflow experiments at Wagon Wheel Gap, Colorado. *Monthly Weather Rev.* Suppl. No. 20.

Baumgartner, A. 1967. Energetic bases for differential vaporization from forest and agricultural lands. In *Int. symp. on forest hydrology* W. E. Sooper and W. L. Lull (eds.), Pergamon, New York, pp. 381–388.

Bormann, F. H., and G. E. Likens. 1967. Nutrient cycling. *Science*, **155**:424–429.

Bormann, F. H., G. E. Likens, D. W. Fisher, and R. S. Pierce. 1968. Nutrient loss accelerated by clear-cutting of a forest ecosystem. *Science*, **159**:882–884.

Butler, P. A. 1966. The problems of pesticides in estuaries. In *A symposium on estuarine fisheries*, Amer. Fish. Soc. Spec. Publ. **3**:110–115.

Butler, P. A., and P. F. Springer. 1963. Pesticides—A new factor in coastal environments. In *Trans. 28th North Amer. wildlife natur. res. conf.*, J. B. Trefethen (ed.), pp. 378–390.

Chapman, D. W. 1962. Effects of logging upon fish resources of the West Coast. *J. Forest.*, **60**:533–537.

Cole, D. W. 1966. The forest soil-retention and flow of water. *Proc. Soc. Amer. Forest.*, **1966**:150–154.

Cole, D. W., and S. P. Gessel. 1965. Movement of elements through a forest soil as influenced by tree removal and fertilizer additions. In *Forest-soil relationships in North America* C. T. Youngberg (ed.), Oregon State University Press, Corvallis, Ore., pp. 95–104.

Cole, D. W., S. P. Gessel, and S. F. Dice. 1968. Distribution and cycling of nitrogen, phosphorus, potassium, and calcium in a second growth Douglas-fir ecosystem, pp. 197–232. In *Proc. symp. on primary productivity and mineral* cycling in natural ecosystems, AAAS Annual Meeting, New York, December 1967, Univ. of Maine Press, Orono, Maine.

Cooper, A. C. 1965. The effect of transported stream sediments on the survival of sockeye and pink salmon eggs and alevin. *Int. Pac. Salmon Fish. Comm. Bull.* **180**:71 pp.

Cope, O. B. 1961. Effects of DDT spraying for spruce budworm on fish in the Yellowstone River system. *Trans. Amer. Fish. Soc.*, **90**:239–251.

Copeland, B. J. 1968. Effects of decreased river flow on estuarine ecology. *J. Water Pollution Control Federation*, **38**:1831–1839.

Cottam, C., and E. Higgins. 1946. DDT: its effect on fish and wildlife. *U.S. Fish Wildlife Serv.* Circ. 11.

Dils, R. E. 1957. *A guide to the Coweeta Hydrologic Laboratory.* U.S. Forest Serv., Southeast. Forest Exp. Sta., Asheville, N.C.

Dyrness, C. T. 1965. Soil surface condition following tractor and high-lead logging in the Oregon Cascades. *J. Forest.*, **63**:272–275.

Dyrness, C. T., and C. T. Youngberg. 1957. The effect of logging and slash-burning on soil structure. *Soil Sci. Soc. Amer. Proc.*, **21**:444–447.

Edmondson, W. T. 1968. Water-quality management and lake eutrophication: the Lake Washington case. In *Water resources management and public policy* T. H. Campbell and R. O. Sylvester (eds.), University of Washington Press, Seattle, Wash., pp. 139–178.

Engelstad, O. P. 1968. Recent trends in fertilizer development. In *Forest fertilization theory and practice*, Tennessee Valley Authority, Natl. Fert. Develop. Center, Muscle Shoals, Ala., pp. 248–254.

Engler, A. 1919. Untersuchunger uber den Einfluss des Waldes auf den Stand der Gewasser. *Mitt. Schweiz. Zentr. Forstl. Versuchsw.* 12.

Fisher, D. W., A. W. Gambell, G. E. Likens, and F. H. Bormann. 1968. Atmospheric contribution to water quality of streams in the Hubbard Brook Experimental Forest, New Hampshire. *Water Resour. Res.*, **4**:1115–1126.

Frey, P. J. 1962. Effects of DDT spray on stream bottom organisms in two mountain streams in Georgia. *Biol. Abstr.*, **37**:492.

Gessel, S. P., and D. W. Cole. 1965. Influence of removal of forest cover on movement of water and associated elements through soil. *J. Amer. Water Works Assoc.*, **57**:1301–1310.

Grzenda, A. R., H. P. Nicholson, J. I. Teasley, and J. H. Patric. 1964. DDT residues in mountain stream water as influenced by treatment practices. *Econ. Entomol.*, **57**:615–618.

Hedgpeth, J. W. 1966. Aspects of the estuarine ecosystem. *A symposium on estuarine fisheries.* Amer. Fisheries Soc. Spec. Publ. **3**:3–11.

Heilman, P. E. 1961. Effects of nitrogen fertilization on the growth and nitrogen nutrition of low-site, Douglas-fir stands. Ph.D. dissertation, Univ. of Wash., Seattle.

Hemlock Looper Study Comm. 1964. Willapa hemlock looper infestation control project, Washington State Dept. Natur. Resour., Olympia, Wash.

Hibbert, A. R. 1967. Forest treatment effects on water yield, pp. 527–543. In *International symposium on forest hydrology* W. E. Sooper and W. L. Lull (eds.), Pergamon, New York.

Jernelöv, A. 1970. Release of methyl mercury from sediments with layers containing inorganic mercury at different depths. *Limnol. Oceanogr.*, **15**:958–960.

Krammes, J. S., and D. B. Willets. 1964. Effect of 2,4-D and 2,4,5-T on water quality after a spraying treatment. *U.S. Forest Serv. Res. Note* PSW-52.

Lieberman, J. A., and M. D. Hoover. 1948. Protecting quality of streamflow by better logging. *Southern Lumberman*, Dec. 15, 236–240.

McColl, J. G., and D. W. Cole. 1968. A mechanism of cation transport in a forest soil. *Northwest Sci.*, **42**:134–140.

Miner, J. R., R. I. Lipper, L. R. Fina, and J. W. Funk. 1966. Cattle feedlot runoff—its nature and variation. *J. Water Pollution Control Federation*, **38**:1582–1591.

Norris, L. A., M. Newton, and J. Zavitkowski. 1965. Streamwater contamination by herbicides resulting from brush-control operations on forest lands. (Absr.) *Western Weed Contr. Conf. Res. Comm. Res. Program Rep.*, pp. 35–37.

Ohle, W. 1955. Die Ursachen der rasanter Seeneutrophierung. *Int. Ver. Theoret. Angew. Limnol. Verh.*, **12**:377–382.

Osborn, F. J., R. E. Pelishek, J. S. Krammes, and J. Letey. 1964. Soil wettability as a factor in erodibility. *Soil Sci. Soc. Amer. Proc.*, **28**:294–295.

Perkins, R. F., and R. E. Dolph. 1966. Report of the 1965 Burns Douglas-fir tussock moth control project. U.S. Forest Serv., Pacific Northwest Forest Range Exp. Sta., Portland, Ore.

Reigner, I. C., W. E. Stopper, and R. R. Johnson. 1964. Control of riparian vegetation with phenoxy herbicides and the effect on streamflow quality. *Northeast. Weed Contr. Conf. Proc.*, **18**:563–570.

Reinhart, K. G. 1965. Herbicidal treatment of watersheds to increase water yield. *Northeast. Weed Contr. Conf. Proc.*, **19**:546–551.

Reinhart, K. G., and A. R. Eschner. 1962. Effect on streamflow of four different forest practices in the Allegheny Mountains. *J. Geophys. Res.*, **67**:2433–2445.

Rich, L. R., H. G. Reynolds, and J. A. West. 1961. The Workman Creek experimental watershed. *U.S. Forest Serv. Rocky Mountain Forest Exp. Sta. Paper* **65**.

Riekerk, H., and S. P. Gessel. 1968. The movement of DDT in forest soil solutions. *Soil Sci. Soc. Amer. Proc.*, **32**:595–596.

Rothacher, J., C. T. Dyrness and R. Fredriksen. 1967. Hydrologic and related characteristics of three small watersheds in the Oregon Cascades. U.S. Forest Serv. Pacific Northwest Forest Range Exp. Sta., Ogden, Utah, 54 pp.

Rowe, P. B. 1941. Some factors of the hydrology of the Sierra Nevada foothills. *Trans. Amer. Geophys. Union*, **22**:90–100.

Rowe, P. B., C. M. Countryman, and H. C. Storey. 1964. Hydrologic analysis to determine effects of fire on peak discharge and erosion rates. *Handbook Appl. Hydrol.*, **22**:31–44.

Salo, E. O. 1966. Study of the effects of logging on pink salmon in Alaska. *Proc. Soc. Amer. Forest.*, **1966**:59–62.

Sheridan, W. L., and W. J. McNeil. 1968. Some effects of logging on two salmon streams in Alaska. *J. Forest.*, **66**:128–133.

Smith, D. C., and W. H. Wischmeier. 1962. Rainfall erosion. *Advan. Agron.*, **14**:109–148.

Stone, E. C., and B. Vasey. 1968. Alluvial-flat redwoods: impact of flood control. *Science*, **160**:835–837.

Storey, H. C., and I. C. Reigner. 1966. Forests—national supplies and demands for water. *Proc. Soc. Amer. Forest.*, **1966**:143–146.

Swank, W. T., and N. H. Miner. 1968. Conversion of hardwood-covered watersheds to White Pine reduces water yield. *Water Resour. Res.*, **4**:947–954.

Sylvester, R. O. 1961. Nutrient content of drainage water from forested urban, and agricultural areas. In *Algae and metropolitan wastes*. Tech. Rept. W61-3. Robert A. Taft Sanitary Engineering Center, Cincinnati, Ohio, pp. 80–87.

Tamm, C. O. 1968. The evolution of forest fertilization in European silviculture. In *Forest fertilization theory and practice*, Tennessee Valley Authority, Natl. Fert. Develop. Center, Muscle Shoals, Ala., pp. 242–247.

Tarrant, R. F. 1956. Effects of slash burning on some soils of the Douglas-fir region. *Soil Sci. Soc. Amer. Proc.*, **20**:408–411.

Tarrant, R. F. 1966. Pesticides in forest water—symptom of a growing problem. *Proc. Soc. Amer. Forest.*, **1966**:159–163.

Tarrant, R. F., D. G. Moore, and W. B. Bollen. 1969. DDT residues in forest and soil after aerial spraying. *Agron. Abstr.* 1969.

Tarzwell, C. M. 1965. The toxicity of synthetic pesticides to aquatic organisms and suggestions for meeting the problem. In *A symposium of the British Ecological Society* G. T. Goodman and R. W. Edwards (eds.), Blackwell Sci. Publ., Oxford.

Tennessee Valley Authority. 1962. *Reforestation and erosion control influence upon the hydrology of pine tree branch watershed*, 1941–1960. Div. Water Contr. Plann. Hydraul. Data Branch, Knoxville, Tenn. 98 pp.

Thomas, E. A. 1957. Der Zurichsee, sein Wasser und sein Boden. *Jahrb. Zurichsee*, **17**:173–208.

Wallis, J. R., and D. W. Willen. 1963. Variation in dispersion ratio, surface aggregation ratio, and texture of some California surface soils as related to soil forming factors. *Int. Assoc. Sci. Hydrol.*, **8**:48–58.

Westlake, W. E. and F. A. Gunther. 1966. Occurrence and mode of introduction of pesticides in the environment. In *Organic pesticides in the environment*, Amer. Chem. Soc., Advan. Chem. Ser., pp. 110–121.

Woodwell, G. M. 1968. Radioactivity and fallout: the model pollution. *Garden J.*, **18**:100–104.

Woodwell, G. M., and F. T. Martin. 1964. Persistence of DDT in soils of heavily sprayed forest stands. *Science*, **145**:481–483.

Wurster, C. F. 1968. DDT reduces photosynthesis by marine phytoplankton. *Science*, **159**:1474–1475.

Wurster, C. F., and D. B. Wingate. 1968. DDT residues and declining reproduction in the Bermuda Petrel. *Science*, **159**:979–981.

Zon, R. 1912. Forest and water in the light of scientific investigation. Appendix V, Final Rep. Natl. Waterways Comm. Senate Document 469, 62d Congr. 2nd Session. (Reprinted 1927, U.S. Govt. Printing Office, Washington, D.C., 106 pp.)

20

Resource Exploitation—Living

TIMOTHY JOYNER, *National Marine Fisheries Service, Biological Laboratory, 2725 Montlake Blvd. East, Seattle, Washington*

The Ecological Impact of Hunting and Farming

Surely one of the significant turns in the path of the evolution of life on our planet must have occurred when one of our primate ancestors picked up a stone and bashed some unsuspecting small animal into a pulpy mass as soft and easy to eat as the fruit upon which he had formerly been content to feed. From that brutal act there followed a chain of events which led to the evolution of a unique superpredator, man. The havoc he has since wrought on other life forms with which he shares living space on this planet is astounding.

In nature, the persistence of complex chemical, biological, and ecological systems is dependent on the stability of the structures by which the components are bound together. In a wilderness, when there are no significant changes in the forces which bear upon it, the plants and animals will eventually become arranged into what is known as a climax community, a balanced ecological structure. The living components of such a system—the breeding populations—resist changes with respect to the other components in proportion to their numbers much as the inertia of a component of a simple mechanical system is proportional to its mass. The greater this inertial property of a breeding population in an ecosystem the greater will be the level at which it may be regularly cropped. This quantity is commonly known as the "sustained yield."

Among animals, the success of food-gathering activity can be defined in terms of the amounts of energy (*a*) derived from the food, (*b*) spent in the gathering of it, and (*c*) required to drive the basic life processes. A surplus of (*a*) over the sum of (*b*) and (*c*) yields a net energy asset available to extend the range and magnitude of an animal's interactions with its surroundings.

The development of human society has followed rather closely the evolution of food-gathering behavior. From the random search for edible parts of wild plants by which man's primitive forbears probably lived in equilibrium with their surroundings, human food-gathering has evolved into increasingly selective patterns which markedly alter the environments upon which they are imposed. Two basic categories have emerged: hunting and farming. The former implies a search for natural stocks of foodstuffs, and the latter some form of husbandry by which control is exerted over the distribution of a selected stock.

Hunting

The relationship between the hunter and the hunted is a delicate one. The predatory activity of one must be balanced by the breeding and growth of the other if the community of which both are a part is to remain in ecological equilibrium. When this balance is upset by excessive predatory activity on the part of the hunter, the result is overkill and the sustained yield is exceeded. It is unfortunate that throughout human history the appetite of the hunter for killing has grown at least as rapidly as his ability to kill. The human hunter apparently has not been as subject to the physiological and behavioral restraints which tend to inhibit other predatory animals from indulging in overkill.

Primitive Pleistocene hunters following game across Asia apparently first penetrated the North American continent some 10,500 to 12,000 yr ago. Within this same time period, some 70 per cent of the native North American mammals exceeding 100 lb in adult body weight became extinct. There is still considerable controversy over the causes of this massive extinction, but with considerable logic, Paul S. Martin (1967) has presented a case accusing predatory human invaders. Nonselective hunting techniques such as the use of fire to drive large herds into cul-de-sacs or over cliffs must have resulted in the slaughter of numbers of animals out of all proportion to the needs of the hunters. The restriction of the Pleistocene extinctions primarily to large animals suggests that desire for social prestige stimulated the appetites of the hunters as much as physiological hunger. More recently, the disappearance of 27 post-

Pleistocene species of moa and the appearance of Eastern Polynesians in New Zealand about 1000 yr ago further suggests yet another instance of overkill on the part of human hunters.

Apologists for our kind may point to the inadequacies of prehistoric records for conclusive proof of man's destruction of whole species of his neighbor creatures, but it is difficult to ignore the evidence of history. By the time of the invasion of North America by Europeans, the descendents of the prehistoric hunters who had earlier crossed over from Asia had settled into relatively stable niches in the ecology of the ranges they occupied. Hunting and fishing had developed into highly skilled crafts in which selectivity was enhanced by sophisticated tools. Recognition of the interaction of species in maintaining the balance of nature was evident in almost every facet of tribal behavior, and few, if any, species of game or fish were threatened with extinction by overkill.

With the coming of the Europeans, however, new factors were added to unbalance the equation. To the hunter's side were added a new generation of tools which increased his killing radius and mobility: firearms, beasts of burden, wheeled vehicles, wind-powered boats, navigation, and the deadliest of all—a new kind of hunger—economic, stimulated by social rather than physiological appetites. In this sense, it may not have been a new kind of hunger at all, but rather a throwback to the same sort of craving for social prestige which drove paleolithic hunters to their ruthless pursuit of the largest game. Whatever its origin, within a few hundred years the commercial hunger of the Europeans had reduced a number of the remaining native North American species nearly to the point of extinction. Commercial demand for hides had decimated the Great Plains bison and the Florida crocodile; for oil, whales; and for fur, the fur seal and the sea otter.

By the end of the nineteenth century Europeans in North America were beginning to rediscover the lesson learned 9000 yr earlier by the Indians of the Eastern forests—not to kill the goose that lays golden eggs. Peter Farb (1968) calls the Eastern Archaic Period, as revealed in the Modoc Rock Shelter on the Mississippi River southeast of St. Louis, "an Arcadian time in the history of man in North America, during which he utilized his resources to the fullest, yet still lived in harmony with his environment"—cropping only surplus resources as they became seasonally available. For over 300 yr the Europeans and their descendents had plundered the resources of the continent with no more regard for the ecological consequences than the stone-age hunters of the Pleistocene. By the beginning of the twentieth century, however, a few of the more thoughtful men of European stock in North America were beginning to recognize the need to preserve some of the diversity of life which had characterized the disappearing wilderness. A system of national forests, parks, and wildlife refuges was established. Guided by men like Gifford Pinchot with the enthusiastic support of President Theodore Roosevelt, the system was vigorously expanded and the principle of conservation became the foundation of national policy with respect to public lands. Thanks to the protection of the system, the once abundant Great Plains bison, which had been very nearly exterminated by mounted hunters with guns within the short space of 100 yr, has been saved from extinction, and protection is now being afforded other endangered species. In Africa another step forward to intelligent resource management is beginning to take place. This is the development of game ranching in which a diversity of selected herbivores, each with different feeding habits, is kept along with the cattle to maximize the production of meat from a given range.

Conservation in the twentieth century has become a matter of government policy in most of the world's nations. That the concept is nothing new has already been noted in the previous discussion of the 9000-yr-old Eastern Archaic culture on the Mississippi River. Lionel Walford (1958) pointed out that laws regulating fisheries were on the books in Sumeria 4000 yr ago, in China 2000 yr ago, in England during the fourteenth century, and in Holland in the seventeenth century. Civilized men grasp the concept readily enough when their food supply becomes threatened, but when a new, previously unexploited resource is discovered, this concept has been just as readily forgotten, displaced by the insatiable economic hunger generated by commerce. So it was when the Europeans arrived in North America, and so it has been as ocean-going ships and modern gear have exposed the great sea fisheries to the threat of overkill. In the case of the sea fisheries, however, the danger of overkill has been diminished by development of conservation policies by governments alerted to the dangers of resource depletion by the past plunder of the land. There are today intergovernmental commissions and councils for studying problems related to the harvest of the fisheries in each of the major fishing regions of the world. These are as follows: Indo-Pacific Fisheries Council (16 nations); Inter-American Tropical Tuna Commission (3 nations); International Commission for Northwest Atlantic Fisheries (10 nations); International Council for the Exploration of the Sea (13 nations); International North Pacific Fisheries Commission (3 nations); International Pacific Halibut Commission (2 nations); International Pacific Salmon Fisheries Commission (2 nations); International Whaling Commission (17 nations); Atlantic States Marine Fisheries Commission (15 states); Gulf States Marine Fisheries Commission (5 states);

and Pacific States Marine Fisheries Commission (3 states). This proliferation of intergovernmental compacts is largely due to the fact that fishing on the high seas is still in the hunting stage, much of it beyond the sovereign jurisdiction of any single government. The purpose is to achieve for each fishery a sustained-yield harvest shared by the participants. At the hunting level of food gathering this is the best that can be expected.

Farming

At the farming level, the husbandry of selected food organisms can produce yields far beyond those sustained under natural conditions. To bring this about requires reduction of conditions unfavorable and enhancement of conditions favorable to the survival, growth, and reproduction of the desired stocks. That farming has been generally a successful development in the evolution of human behavior is attested by the growth of population and culture which has followed the transition from hunting to agrarian societies throughout human history. Whenever the efficiency of the farmer released the energies of a large proportion of the population of which he was a part from daily exhaustion exacted by the exigencies of food gathering, the redirection of these energies into diverse channels almost inevitably produced a flowering of culture.

The record, however, has not all been good. Just as the greed of unsophisticated hunters unbalanced ecosystems to the extent that at the worst entire species became extinct, and at the least the productivity of hunting and fishing grounds were reduced below the level of minimum subsistence for the hunters, so farmers, by their unsophisticated excesses, have often squandered the productivity of their farmlands and ranges. It is now thought that removal of plant cover by overgrazing in areas of low rainfall has accelerated the formation of desert wastes in vast areas of North Africa, the Near East, and Western India.

The repeated planting and harvesting of a single crop will soon exhaust the soil of its nutrients and destroy its structure. Examples can be found the world over, but perhaps nowhere to such a profound degree as in Mexico. Long before recorded history, Indians of the highlands of Central America or Peru began cultivating an edible wild grass. Lesley Byrd Simpson (1963) calls the discovery of this edible grass "one of the most important achievements of mankind anywhere." Through many thousands of years of patient selection, this wild grass from Central America has evolved into the fantastically productive, giant corn stalks of Iowa. In Mexico, maize came to dominate man. Abundant yields of the grain generated larger populations, which once established, had to be supported, and the Indians were enslaved by their own invention. Maize made Mexico, essentially, into a one-crop country, for it is the one grain which flourished both on the high Plateau and in the lowlands. As late as 1930, Mexican production of maize exceeded by four times the three other major food crops together. Until recently, the use of fertilizers was not widespread in Mexico, and slash and burn techniques were employed since prehistoric times to clear new milpas (plots) as the old ones became exhausted. The bare slopes of abandoned fields, battered by summer rains, are eroded away and the soil washed into the sea. Professor Simpson writes, "It is a melancholy thing to see once-cultivated and once-prosperous countryside now thrusting out its fleshless bones in unheeded protest against the vandalism of man. The rich province of Mixteca Alta in western Oaxaca, was, four hundred years ago, one of the fairest spots in New Spain, renowned for its high culture and industrious population. It is today an almost unrelieved stretch of badlands . . . . The pitiful stands of maize growing in pocket-handkerchief milpas on the tops of mountains, or in the cracks and crevasses of their rocky slopes, and the undisguised poverty of the communities depending on them, are fierce reminders that maize is a savage taskmistress. To break her iron rule will require inventiveness, patience, and fortitude as great as those of the ancients who first harnessed her to the service of mankind."

Another example of the destruction of soil in Mexico by intensive monoculture is that brought about by the cultivation of the maguey cactus (*Agave americana*) from which the highlanders make an intoxicating beverage, pulque. It requires vast acreages, with the plants set out in widely spaced rows. The spaces between the rows are kept weeded, and the soil is eroded by wind and rain. The spaces become gullied and the land made useless for anything but growing more maguey.

From these examples it can be seen that single-crop husbandry will soon exhaust the soil and destroy its structure, making it more susceptible to erosion; and by loss of diversity it increases the likelihood of blight unchecked by the balance of a natural ecosystem.

Effects of Maritime Activities on the Coastal Environment

Because of its inaccessability to man relative to dry land, and the three-dimensional, interconnected fluid nature of the marine environment which readily affords means of interchange among the great diversity of its living populations, the effects of man's activities in the sea and at its margins have developed more slowly and are less easy to observe than on land. The collection of food organisms from beaches exposed at low tide is probably one of the oldest forms of food gathering employed by man since his development as a distinct species. The effects of even this most primitive kind of exploitation of marine resources are evident from the following quotation from Smith's *The Sea Fishes of Southern Africa* (1953):

> Only 20 years ago computation showed that near East London the natives removed over 1,000 tons of molluscs annually from one stretch of rocks only about 800 yards in length, while from that same area today the annual yield is not more than 20 tons. The teeming intertidal life of 30 years ago has given place to bare rock, and little of anything is to be found there. It is noteworthy that in these parts, shore fishes of angling significance have declined notably in numbers in recent years.

As marine food gathering progressed from the beaches to sheltered bays and sounds and eventually out to sea, the effects became increasingly difficult to observe. The technology necessary to support the gradual encroachment by man into the oceanic environment of the sea fisheries became increasingly sophisticated, and although it has only been in this century that man has seriously attempted to evaluate the effects of fishing on offshore fish stocks, it is interesting to delve back into history for clues as to how the evolution of man's ability to fish offshore waters has indirectly affected the coastal environment.

Development of Seafaring Communities

In the classical world of the Mediterranean, seafaring tradition grew in response to the needs of commerce and the conduct of warfare between the elements of the highly developed, agriculturally supported city states of Greece, Rome, and Carthage. The ships were either cargo or war galleys. Manpower was cheap, and efficiency in its use for driving ships was not a primary consideration in ship design.

On the shores of the Baltic and North Seas at that time, however, life was considerably harder. Compressed close to the beaches by impenetrable forests on one side and by stormy northern seas on the other, the inhabitants practiced only limited agriculture (mainly grazing a few cattle in the small pastures available) and had to depend heavily on hunting the game in the forests and the fish offshore. The latter necessity gave rise to the development of considerable skill in the design and operation of ships capable of weathering the stormy waters of the region. Since the small populations supported by this harsh environment put a premium on the value of manpower, ships were designed for greater efficiency in its application to propulsion and for better use of the power of the wind. And so, in time, there developed around the shores of the Baltic and North Seas, a society, rude in many of its aspects, but sophisticated in the skills of fishing and sailing in rough water.

On the island of Great Britain at the western margin of the North Sea Romans with their galleys had succeeded in transplanting Mediterranean civilization to the lush and fertile lowlands. During four centuries of Roman dominion, cities, towns, roads, and agriculture developed rapidly. Even after the Roman legions left as the power of the Empire diminished under the assaults of the German barbarians on Imperial Rome itself, the Romanized Celts of Britain kept the Roman traditions and way of life. They had never been a seafaring people; neither had the Romans after whom they now patterned their existence. They were, therefore, quite unprepared for the amphibious assaults of the Angles, Saxons, and Jutes in the 5th century.

These were the rude fishing folk of the Baltic region who had discovered that their seaworthy vessels and seafaring skills gave them marked advantages as pirates and raiders, advantages they pursued with vigor in ravaging the civilization the Romans had left behind in Celtic Britain.

Whether the conversion of the Baltic tribes from a fishing-hunting economy to one based on piracy was due to greed resulting from new awareness, made possible by their sea travels, of the niceties produced by civilized communities, or whether to failure of food supplies at home is debatable. Otto Petterson (1905) related biological phenomena to variations in hydrographic conditions in the Atlantic is reported by Peter Freuchen (1957) to have held that movement of herring stocks in and out of the Baltic may have been due to long-term fluctuations in tidal ranges. Failure of traditional fisheries at home could have driven the Vikings to plunder more fortunate lands using the seafaring skills they had developed as fishermen in better times. The motivation, however, is immaterial. The important point is that during five to six centuries of repeated forays and settlement in Great Britain, first by Angles, Saxons, and Jutes, and later by Vikings, seafaring skills were infused—along with a new language and system of law—into the fabric of the culture which evolved from British to English. The origin of the latter terms is interesting. It is derived from the Old Saxon *Englisc*, the language of the Angli, a people from Angla-land, now modern Schleswig in north Germany, where there still is today a district called Angel, the Old Saxon term for fish hook, from which is derived the modern English angle, to fish.

The settlement of the seafaring Germanic tribes along the east coast of Great Britain brought with it the development of towns in which fishing, and the building, servicing, and sailing of ships became primary activities. From these bases a vigorous maritime commerce grew, and spread across the island with the Saxon conquest to the mouth of the Avon on the Atlantic side, where developed the port of Bristol. In later centuries, Bristol sent explorers and colonists across the Atlantic to North America and fishermen to the Grand Banks of Newfoundland. Maritime commerce and fisheries flourished as the English nation evolved.

Maritime Commerce

Across the North Sea and along the southern shore of the Baltic a league of German towns centered on Lubeck began to dominate the economic life of the region. The trade upon which the Hanseatic League—as it was called—was based, centered on three staples: furs from Russia, cloth from Flanders, and salt herring from Norway and Sweden. The wealth of the Baltic herring fisheries in the fourteenth and fifteenth centuries conveniently coincides with Petterson's calculation of the maximum of the 1800-yr tidal range cycle which, according to his hypothesis, should have brought North Sea herring again into the Baltic along with the increased intrusion of saltwater.

Military Sea Power

The next century, the sixteenth, saw the rise of British sea power under the Tudor monarchs. Henry VII had subsidized the voyages of the Cabots which laid the foundation for the extension of British commercial interests to the continent of North America. Henry VIII, with considerable foresight, saw the promise of national strength in the seafaring skills of his subjects and began construction of a navy. His daughter, Elizabeth I, put it to use to destroy the predominance of Spanish sea power in the Atlantic and to extend British influence into the Pacific.

Exploration and Colonization

A direct outgrowth of the voyage of John Cabot to the vicinity of Newfoundland in 1497 was the development, early in the sixteenth century, of the great North Atlantic cod fishery. Cabot, on his return to Bristol, had reported enthusiastically on the abundance of codfish in the waters he had explored in the vicinity of Newfoundland. Corte-Real, a Portuguese who landed in Newfoundland in 1501, confirmed Cabot's report on the prodigious quantities of cod in these waters. By 1506 taxes were being collected by the Portuguese government on fish brought back from Newfoundland by its fishermen who apparently were the first to exploit the newly discovered fishing grounds (de Loture, 1949). Hard on their heels came the French and English, who along with the Portuguese had been fishing for cod in Icelandic waters a half-century earlier. Expansion of British and French settlement in North America was stimulated by the enormous economic value of the nearby cod fishery, and fishing ports were among the earliest commercial centers to .develop. Louisburg on Cape Breton, St. Johns in Newfoundland, and Gloucester in Massachusetts were founded by cod fishermen, and the cod fishery played a significant role in the development of the great port of Boston.

Feedback to the Environment

Using the North Atlantic region as a model, it can be seen that fishing has had a profound influence on the political and economic development of human societies which have had to look to the sea for sustenance. The skills of seamanship, developed among a rude fisher folk, contributed to the growth of a great commercial empire, the crossing of an ocean, and the peopling of a continent. It is appropriate now to consider the feedback. If the maritime environment spawned the human behavior patterns which led to the growth of mercantile societies and the subsequent explosion of North Sea peoples into the Atlantic and across to the shores of North America, what then has been the effect of all this activity on the environment which produced it?

Modification of the Shoreline and Seabed

Without interference from man a delicate equilibrium tends to maintain the condition of the shoreline and the near-shore seabed in coastal areas. The supply of sediment brought to a given stretch of coastline by wind, longshore transport, and rivers is balanced by removal processes generated by the energy and direction of the waves. The development of harbors and shipping facilities can easily upset this balance. Dams on the rivers, jetties, and groins, the dredging of bay-mouth channels, and the construction of offshore facilities all contribute to the disruption of coasts and beaches (Byrne, 1969; Chapter 18, this book).

Disruption of the physical configuration of the shoreline invariably upsets the ecological balance of its populations of living organisms. Spawning, nursery, feeding, and breeding grounds are destroyed or shifted. One has only to compare the barrenness of the filled-in marshes of the Jersey shore of the Port of New York to the teeming life of the North Carolina marshes to sense the impact that two centuries of harbor development have had on the Hudson estuary.

Pollution

The discharge of wastes resulting from human activity can also severely modify the coastal environment. Direct discharge of sewage, industrial, and mining wastes; the indirect contributions of drainage from land altered by urban and industrial development, agriculture, and forest use; accidental spills of oil and other chemicals; waste heat from the cooling systems of power and manufacturing plants; the fallout of chemical and radioactive wastes; and the absorption from the atmosphere of gaseous residues from the burning of fossil fuels and chemical processing all contribute to the pollution of the coastal environment. The fluidity and dissolving power of water brings about a more rapid and uniform distribution of polluting materials than occurs either on land or in the air. The relatively delicate and porous structures of marine organisms compared with land creatures or birds renders them the most susceptible to damage from pollution of their environment.

Some unfortunate examples of the destruction of valuable living resources by urban and industrial development can be found in the oyster fisheries of the United States. Figure 20.1 shows an evaluation of United States' oyster production prepared by Westley (1967) from data by Engle (1966). The decline of the East Coast fishery has been primarily due to pollution from industrial and urban developments on Chesapeake Bay, although poor management and disease have also taken their toll. On the West Coast, the production of native (Olympia) oysters declined to insignificance, largely due to pollution. The increase in West Coast production, beginning in 1930, resulted from the introduction of a different species, the Pacific oyster, from Japan. Present production consists almost entirely of this species.

Overfishing

It has only been since the nineteenth century that man's technical prowess at sea enabled him to develop overkill capabilities with respect to high-seas fisheries. As mentioned previously in this chapter, it is fortunate that at about the same time an awareness developed of the wisdom of applying the concept of sustained yield to the exploitation of fishery resources. Herring, plaice, sole, cod, halibut, salmon, tuna, whales, and fur-bearing marine mammals have all benefited to some extent from recognition of this concept.

Halibut. The history of the halibut fishery of the Pacific Northwest (Thompson and Freeman, 1930) provides the clearest example of measurable depletion of an ocean species brought about by overfishing.

Halibut was one of the mainstays in the diet of the coastal Indians of the Pacific Northwest who fished for it with a high degree of skill, using hooks and lines of remarkably sophisticated construction. Catches of 100 large fish per day per two-man canoe were not unusual, but the total Indian catch, though adequate to satisfy their needs, did not significantly affect the abundance of the stocks.

The birth of the present great commercial halibut fishery coincided with the coming, just before the end of the nineteenth century, of men and boats from the older halibut fisheries of the East and the completion of the Northern Pacific (1888) and the Canadian Pacific (1892) railroads over which the early catches were shipped. The fishery first exploited the shallow, protected banks close inshore accessible to small sailing schooners. By 1910 auxiliary power and steam propulsion on fishing vessels had made it possible to shift to the deeper, offshore banks as intensive exploitation brought about rapid depletion of inshore stocks. In 1915 the Report of the British Columbia Department of Fisheries presented the results of an investigation which showed that formerly productive banks were losing 75 per cent of their fish each decade.

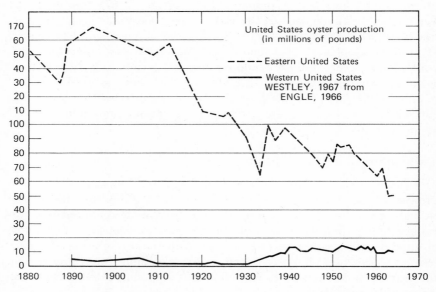

Figure 20.1 United States oyster production. From data of Engle, 1966 as prepared by Westley, 1967.

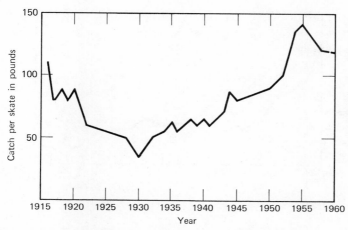

Figure 20.2 Decline and recovery of halibut catch per unit of effort showing effect of regulations begun in 1932 under the provisions of the 1930 treaty. From Royce, 1965.

Thompson, et al. (1931) evaluated the decline in catch from all banks along the entire Pacific Coast from 1913 to 1928. Figure 20.2 shows the average catch per standard skate (unit of gear). The 1913 catch refers only to the long exploited southern banks. The increased catches of the next 2 yr reflect the discovery of new banks in the Gulf of Alaska. Thereafter, in spite of continued expansion to the westward, the fishery was not able to prevent declining returns per unit of gear fished. Consequently, in 1924 a treaty between the governments of the United States and Canada created the International Fisheries Commission to undertake the conservation of the Pacific halibut. In 1930 a second treaty enabled the Commission to impose regulations on the fishery which set limits to fishing areas and seasons and on the total weight of fish caught. The success of the regulations, which went into effect in 1932, is reflected in the subsequent increase of catch per unit of effort.

Haddock and Plaice. Records of British trawlers working the North Sea and Icelandic fishing grounds after World War I also provide clear evidence of the effect of fishing on exploited fish populations. The characteristic decline of catch per unit of effort as stocks of North Sea haddock and Icelandic plaice were subjected to increasing fishing pressure by British trawlers can be seen in Figs. 20.3 and 20.4. These examples show that the effects of intensive fishing on high seas stocks of bottom fish. It has been questioned, however, whether high seas stocks can really

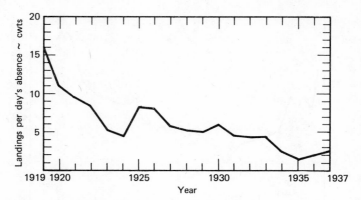

Figure 20.3 Haddock landings per day's absence of British North Sea trawlers. From Russell, 1939.

be threatened with extinction by overfishing. Crutchfield and Zellner (1962) have pointed out, for example, that declining catches per unit of gear with increased fishing effort are not necessarily the result of fewer recruits from declining numbers of progeny as the fishery crops off the adult spawners. The fecundity of many high seas species is such that relatively few adults can produce enough progeny to maintain the population. The reduced weight of the catch from heavily fished populations of such species can be due to the increasingly smaller size of fish available to the gear.

Biological or Economic Extinction. In considering the problem of extinction, however, it is necessary to define whether biological or eco-

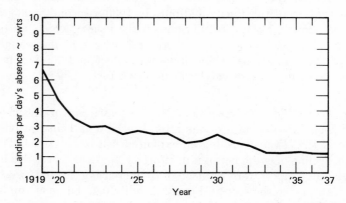

Figure 20.4 Plaice landings per day's absence of British trawlers, Icelandic grounds.

nomic extinction is involved. There should be little doubt that the former could be (and in some cases nearly has been) brought about as readily by an overenthusiastic fishery for marine mammals as by overkill of their terrestrial relatives. In the case of the fishery for halibut, a slow-growing species for which there is a considerable length of time between hatching of the eggs and attainment of size sufficient for recruitment to the fishery, it appears highly likely that the danger of economic extinction was averted only by the imposition of regulatory measures. Populations of highly prolific species with short life cycles can be endangered by fluctuating conditions of the environment and by pressures from species competing for the same habitat and food supply. The decline of the Pacific sardine fishery off the coast of California may have been such a case.

California Sardine. In a study of the population biology of this species (*Sardinops caerula*), Murphy (1966) summarized the history of the fishery. Figure 20.5 shows the yearly catch along the Pacific Coast, which rose from 28,000 tons in 1916 to nearly 800,000 tons in 1936. After a brief stable period ending in 1945, the landings diminished until they reached the present level which is on the order of 20,000 tons. Murphy's analysis estimated the 1932 population at 4.0 million tons with an equilibrium value of 2.4 million tons, and showed it to be improbable that the population would have declined in the absence of fishing. Application of fishing rates to the population estimate lowered reproduction to an extent that decline was inevitable. The spawning population of the pre-1949 population was estimated at 1 million tons with a maximum sus-

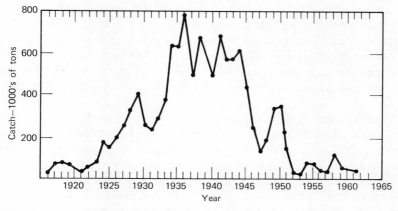

Figure 20.5 Seasonal catch of sardines along the Pacific coast. From Murphy, 1966.

tained yield calculated at 471,000 tons. The catch during this period averaged 570,000 tons, approximately 25 per cent higher than the maximum sustained yield. Successive, environmentally induced reproductive failures of the dominant race in 1949 and 1950, coupled with this high fishing intensity, could have brought about collapse of the population. The ecological void created by the decline of the sardine was quickly filled by the northern anchovy which occupied a similar ecological niche. Murphy felt that the rise of the anchovy population so surely altered the parameters of the sardine population that the sardine fishery is not likely to recover unless man or nature intervenes to reduce the anchovy population.

Need for Regulation. With the present state-of-the-art it is unlikely that offshore fishing threatens any existing species with biological extinction. If advances in fishing technology develop at rates comparable to the expected encroachment of man into the seas over the continental shelves, the likelihood that this happy situation will persist is sure to diminish. Severe depletion to a level at which fishing is no longer profitable (economic extinction) is a recognized possibility with existing technology and has been approached for several stocks. It is conceivable that where the survival of a given stock of fish in the face of natural pressures is already marginal, that overfishing could reduce the population to the point where it would no longer be capable of coping with environmental pressures and it would be threatened with biological extinction. In the absence of the self-regulating economic features of privately owned resources, some form of regulation to prevent overfishing of public offshore resources is desirable. The choice of objectives of such management, however, has been the subject of lively debate. Crutchfield and Zellner (1962) have treated the arguments in some detail in arriving at the conclusion that the basic objective should be economic "to see to it that the fisheries maximize the net economic yield—the difference between the aggregate money value of output and the aggregate money cost of input needed to produce it (excluding of course, money returns based on monopolistic restriction of output)."

I cannot argue against efficiency in the use of a resource being in the public interest, and an excellent criterion upon which to base management programs. There is, in the development of their argument, nevertheless, an issue which I do contest emphatically: that only the immediate needs of this society at this point in history need be considered in our approach to conservation. They wrote, "Unless the end products of a fishery are valued at more than enough to cover production costs, no commercial operation would arise, *nor would it be a matter of concern if these species ceased to exist* (italics mine)." The essence of the wonder that stimulates the minds of men is in the infinite diversity that surrounds them. The essence of the continuity of life lies in its adaptability to changing conditions in a dynamic universe. In the variety of species which maintain diversity in the planetary gene pool lies the genetic reserve to protect life itself from extinction by a changing environment. It would be shortsighted to consider only the needs of the present, but since we cannot fathom those of the distant future the least we can do is to preserve the materials from which they may be supplied.

Design of measures to conserve offshore stocks of fish subjected to fishing pressure should be guided by the concept of *eumetric fishing* developed by Beverton and Holt (1957). By their definition, a plot of mesh size (and consequently the age and size at which the fish first become accessible to the fishing gear) adjusted to produce the greatest yield for each level of fishing effort produces a eumetric fishing curve (Fig. 20.6).

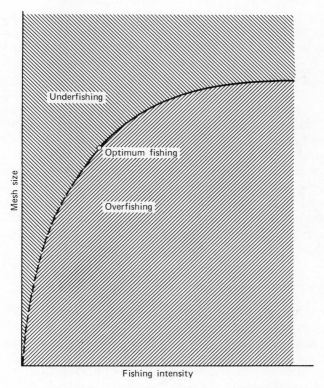

Figure 20.6 Eumetric fishing. From Beverton and Holt, 1957.

Eumetric fishing will generate the greatest yield for a given cost and will approach as a limit the yield that would be achieved if an entire year-class were taken when its total weight reached a maximum. Since the fishing effort required to reach this condition approaches infinity, it is obvious that the *optimum* catch must lie along the curve at some point below the maximum where the difference between returns and costs is greatest.

Problem for the Future

The rapidly increasing rate of man's encroachment into the coastal environment presents him with a problem of a kind he has faced many times before throughout the history of his expansion across the face of the globe—*how to occupy and use a range without destroying its ability to sustain its occupants.* In addressing himself to this problem on land, his failures, with a few exceptions, have outweighed his successes. Earlier in this chapter we noted how hunting and agricultural excesses destroyed the productivity of vast ranges into which man had intruded. It was also pointed out that application of ecological principles to hunting and fishing can lead to sustained-yield management of wildlife and fishery resources, and in husbandry, to well-managed, continuously productive farms, forests, and ranges.

With the rapid development of undersea technology now making imminent the massive intrusion by man into the coastal environment, it is critically important that management of the resources of the vast new underwater ranges about to be exploited be planned with full regard for the ecological consequences. For the foreseeable future, the logistics of human undersea activity decree that its intensity will vary inversely with depth and distance from shore, excluding of course the immediate vicinity of the surf. Demand for the resources of the coastal environment is increasing among a variety of users so that the needs of each must be weighed carefully against the needs of the others and against the ecological impact of the sum as well as each of the parts of such multiple use.

Jurisdictional Zones

The ocean areas of the world traditionally have been divided into three zones; (*a*) internal waters, (*b*) the territorial sea, and (*c*) the high seas. Due to increasing pressures on marine resources, international law now recognizes two additional jurisdictional zones between the territorial sea and the high seas: the contiguous zone and the continental shelf (Ely, 1968). Occasionally it will be convenient to lump the elements of the coastal region into two categories: (*a*) an exclusive fisheries zone comprising internal waters, the territorial sea and the contiguous zone, and (*b*) the continental shelf on which coastal nations may exercise sovereign rights in the exploration and exploitation of the resources of the seabed beyond the territorial sea.

Resource Use Compatibility

Within these two categories it is reasonable to expect vigorous competition for resources among users with such interests as aquaculture, commercial fishing, recreation, waste disposal, shipping, petroleum production, marine mining, and electric power generation.

Internal Waters

The protected waters of estuaries, bays, and sounds offer immediate potential for development of aquaculture. Existing aquacultural operations are now conducted in such waters, but Ryther and Bardach (1968), after a world-wide survey of aquaculture, concluded that "the practise of aquaculture may not only be greatly expanded, particularly in those parts of the world most in need of its products, but also that its yields may be appreciably increased through the use of modern science and technology." Spectacular increases in production of marine animal protein are possible through application of already proved aquacultural techniques to available acreage either not or not fully utilized.

Potential of Mussel Culture. An example of a vastly underutilized resource which, if developed along the lines of cultural techniques already demonstrably effective, should be able to provide a substantial reduction of the world protein deficit, can be found in mussels. These hardy bivalves are distributed along the coasts of all the temperate and tropic seas of the world. They are successfully cultured in the Low Countries, France, Italy, Spain, and the Philippines. The yield per acre in northwestern Spain, where suspended culture methods take full advantage of the cool, deep, nutrient-rich waters of protected bays, is prodigious—660 tons per acre of rafts per year. In the Philippines a promising start has been made on the off-bottom culture of green mussels in the tropical waters of Manila Bay where they are proving hardier and easier to cultivate than oysters.

At the Technological Laboratory of the National Marine Fisheries Service, samples of a dry, powdered protein concentrate have been prepared from Puget Sound mussels. Assay shows 70 per cent protein, 15 per cent glycogen and 12 per cent mineral ash. Low levels of fat and

fluoride simplify the processing which does not require shucking of
the shell. After extraction of water and lipids, ground shell fragments
may be separated from the protein by air-classification. That mussel
culture should be considered a promising source of protein concentrate
is evident from the fact that at the Spanish rate of production, 116.5
acres in rafts suspending mussel strings 33 ft long, would produce 76,900
tons of whole mussels annually—sufficient for the yearly manufacture
of 5,000 tons of protein concentrate.

Territorial Sea and Contiguous Zone

Beyond the internal waters of the bays, sounds, and estuaries, yet
still within the exclusive fishing zone including the territorial sea and
the contiguous zone, lies another large, relatively shallow area. Among
the activities presently competing for space in this zone are commercial
fishing, shipping, oil and mineral extraction, and various kinds of saltwater recreation. Although aquaculture is not now on the list, the development of techniques for underwater aquaculture, protected from wave
and storm action (i.e., sea-bottom ranching and submerged, off-bottom,
suspended systems for shellfish culture) would make available sufficient
additional acreage for truly prodigious aquacultural production. Underwater aquaculture beyond the internal waters would be more compatible
with some of the other potential uses of the same area. It should not
interfere significantly with surface or shallow fishing, shipping, or recreational use. On the other hand, bottom fishing, mooring of surface
vessels, and movement of undersea craft would be limited wherever it
might be carried on. Within internal waters, the full array of water and
shoreside land uses are potential sources of legal conflict in their
interactions.

Rationale for Aquaculture

In addition to legal considerations, development of a rationale for
aquacultural use of coastal waters must take into account the chief ecological lesson learned from experience with the husbandry of land resources—that loss of diversity through overemphasis on monoculture
leads to (a) inefficient use of available range, (b) increases the suscepti-
bility of the crop to destruction from unwanted effects of ecological
imbalance, and (c) tends to disrupt nutrient cycling, causing substrate
exhaustion. The first difficulty can be overcome by multiple zone culture
in each plot, range, or enclosure. This involves the simultaneous culture
of compatible, noncompeting organisms within the system, each taking

advantage of living requirements ignored by the other occupants. An example would be the simultaneous occupancy of a culture enclosure by fish and filter-feeding bivalve molluscs feeding on different size ranges of food organisms in the water above the bottom, edible benthic crustaceans feeding on bottom detritus, and abalone or limpets feeding on the algae on the walls of the enclosure. The second difficulty can be overcome by reserving "wild" areas or areas of low-intensity cultivation for the purpose of maintaining healthy, viable populations of native organisms ecologically adjusted to survival without cultivation. This would provide ensurance against the destruction of life forms of no immediate commercial value, but with aesthetic or possible future values as yet undetermined. The third difficulty is avoided in agriculture by means of crop diversification and use of fertilizers, and the same approach will serve for aquaculture. A plan for use of heated effluents from thermal power plants for temperature control and of sewage treatment effluents for fertilizer in multicomponent aquacultural systems has been proposed by Gaucher (1968).

Fishing

The waters of the coastal zone from the shoreline to the high seas, for the foreseeable future, will always be legitimate grounds for fishing. Competition from other uses will diminish progressively seaward. It is not likely that aquacultural practices will extend beyond the contiguous zone, since the exclusive harvesting rights which encourage development of this type of activity would be difficult to ensure beyond the limits of national jurisdiction.

Improvements in the efficiency of fishing methods are likely to stem from new ship and gear designs and the application of underwater acoustics, light, electrical fields, and chemicals to the roundup and herding of fish populations.

When the proficiency of the fisherman exceeds the ability of the fished stock to replace losses from fishing, careful attention must be given to developing the restraints necessary to ensure optimum sustained yield from the fishery. One approach, employed in the American salmon fishery, consists of enforcing the necessary restraint by gear limitations restricting the efficiency of an unlimited number of fishermen permitted to enter the fishery. This has little merit. A proper rationale should encourage maximum fishing efficiency to benefit the producers with increased profits and the consumers with lower prices. Fishing with maximum efficiency, however, will demand the discipline of limited entry to make effective catch limitation possible. The moral and legal problems raised by this necessity, however, are complex and will require exercise of the most exquisite judgment on the part of the legislators and jurists who will ultimately be responsible for resolving them.

It is hoped that in the development of scientific, legal, and moral rationales for the exploitation of the living resources of coastal waters, the right of future generations to select from a diversity of organisms to fill their as yet unimagined needs will be considered as seriously as the urgent, yet fleeting demands of the present.

References

Byrne, John V. 1969. *N.Y. Acad. Sci.*, **9**:36–37.

Beverton, R. J. H., and S. J. Holt. 1957. *On the dynamics of exploited fish populations.* Her Majesty's Stationery Office, London, 533 pp.

Crutchfield, J., and A. Zellner. 1962. Economic aspects of the halibut fishery. *Fish. Ind. Res.*, **1**:173 pp.

de Loture, R. 1949. History of the great fishery of Newfoundland. *U.S. Fish Wildlife Serv. Spec. Sci. Rept., Fisheries* No. **213**. 147 pp.

Ely, N. 1968. Legal problems in undersea mineral development. Prepared for Coastal States Conference on Ocean Mining Law, Portland, Ore. December 11–13, 1968. 28 pp.

Engle, J. B. 1966. The molluscan shellfish industry. Current status and trends. *Proc. Natl. Shellfish Assoc.*, **56**:13–21.

Farb, P. 1968. *Man's rise to civilization as shown by the indians of North America from primeval times to the coming of the industrial state.* Dutton, New York, 332 pp.

Freuchen, P. 1957. *Peter Freuchen's book of the seven seas.* Simon and Schuster, New York, 512 pp.

Gaucher, T. A. 1968. *Potential for aquiculture.* In *Study of means to revitalize the Connecticut fishing industry.* General Dynamics Co. Electric Boat Div. Rept. U413-68-016, Contract RSA-66-8, pp. 4-1–4-22.

Martin, P. S. 1967. Pleistocene overkill. *Nat. Hist.* **LXXVI**:32–38.

Murphy, G. I. 1966. Population biology of the Pacific sardine (*Sardinops caerula*). *Proc. Cal. Acad. Sci.*, 4th ser. **XXXIV**:1.

Petterson, O. 1905. On the occurrence in the Atlantic Current of variations, periodical and otherwise, and their bearing on meteorological and biological phenomena. *I.C.E.S. Rapp. Proc. Verb.*, III, A:1.

Royce, W. F. 1965. Conservation and practices of fishery resources. In *Problems in resource management* J. A. Crutchfield (ed.), University of Washington Press, Seattle, Wash.

Ryther, J. H., and J. E. Bardach. 1968. *The status and potential of aquaculture.* American Institute of Biological Science, Washington, D.C. vol. I, 261 pp.; vol. II, 225 pp.

Russell, E. S. 1939. An elementary treatment of the overfishing problem. *Cons. Perm. Int. Expl. Mer. Rapp. Prol. Verb.,* **1939**:5–14.

Simpson, L. B. 1963. *Many Mexicos.* Univ. Calif. Press, Berkeley and Los Angeles. 349 pp.

Smith, J. L. B. 1953. *The sea fishes of southern Africa.* Central News Agency Ltd., South Africa.

Thompson, W. F., and N. L. Freeman. 1930. History of the Pacific halibut fishery. *Intern. Fish. Comm. Rept.* No. **5**. 61 pp.

Thompson, W. F., H. A. Dunlop, and F. H. Bell. 1931. Biological statistics of the Pacific Halibut Fishery (1) changes in yield of a standardized unit of gear. *Intern. Fish. Comm. Rept.* No. **6**. 108 pp.

Walford, Lionel A. 1958. *Living resources of the sea.* Ronald Press, New York. 321 pp.

Westley, R. E. 1967. The oyster producing potential of Puget Sound. Presented at 21st Ann. Conv. Pacific Oyster Growers Assoc., Aug. 24, 1967, Olympia, Wash.

21

Marine Mining and the Environment

JOHN W. PADAN, *Director, Marine Minerals Technology Center, National Oceanic and Atmospheric Administration, U.S. Department of Commerce, Tiburon, California*

The pressures of an ever-expanding world population coupled with an omnipresent desire for higher standards of living have created tremendous demands for increased supplies of raw materials. As living standards have improved, the demand for resources has accelerated faster than population and will continue to do so for some time to come. By 1980 consumption of nonmetals in the United States alone is expected to be double the 1965 level. Domestic needs for metals and petroleum products will probably increase by 50 per cent during this same interval. Yet there is no shortage of mineral resources in sight.

Man's engineering abilities have thus far enabled him to discover and utilize larger and larger mineral deposits of lower and lower quality. This traditional capability, plus an ability to obtain from other countries the commodities not immediately available at reasonable cost, the imaginative use of substitute materials, and the recycling of selected material for reuse, will continue to provide raw materials to meet man's needs.

Technology cannot be taken for granted: therefore, federal (U.S. Senate, 1968) and industrial planners are looking far into the future to identify weaknesses in the supply situation. After potential problems and contingencies are defined, research can be planned which hopefully will obviate all but minor fluctuations in the supply-demand relationship. This is a critical goal because raw materials such as mineral resources are the very foundation of our sophisticated society. Any unforseen long-term shortage of a key commodity could have a far-reaching and damaging impact on our economy.

Technology has helped us to supply our growing mineral requirements from deposits that are steadily diminishing in grade. But where technology has given, it has also frequently taken away. Too seldom in the past have we had the wisdom to look beyond our immediate needs for materials and appreciate our ultimate need for a quality environment. Too often, as a consequence, our technology has been inadequate or improperly applied. In our quest for essential minerals and fuels we have polluted our air and water, and created threats to the safety and health of our people. The growing public awareness of these problems is placing severe restraints on the industries that supply us with vital raw materials. And those restraints pose a major engineering challenge in the development of marine mineral resources.

As we look to sea for new sources of minerals, we remain constantly aware of the many other human needs that the marine environment can fulfill and of the necessity for preserving its capacity to do so.

The value of minerals recovered from oceans bordering the United States in recent years is shown in Table 21.1. Oil, gas, and sulfur comprise by far the major commodities produced. While this trend will continue for years to come, man must begin to show concern for the impact of mining solid minerals from the sea, the sea floor, and beneath the sea floor, because experience with the hydrocarbons suggests that problems will occur.

Recent marine mining activities embrace the globe (Table 21.2) but primarily have involved exploration rather than production. The focus of this exploration is upon placer deposits of heavy minerals such as gold and tin. Such occurrences of minerals were created by beach and stream action when sea levels were lower; they are now found in unconsolidated sediments of the continental shelf.

At present offshore placer deposits actually are being mined in only a few areas (Fig. 21.1) and this situation will not change drastically in the next few years. Nevertheless, it is this near shore, shallow-water environment that offers the most promise for expanded exploitation. In such areas the problems of working at sea are least severe as evidenced by the production of construction materials from this zone (Table 21.1). The Interior Department's Bureau of Mines recognized the opportunity in 1963 for expanded activity in this environment and began to define the problems to be overcome in tapping offshore placer deposits of heavy metals (Howard and Padan, 1966).

Table 21.1. Value of Mineral Production from Oceans Bordering the United States, 1960–1969 (Millions of Dollars)

Type of Marine Mineral Deposit	Year									
	1960	1961	1962	1963	1964	1965	1966	1967	1968	1969
Unconsolidated deposits (sand, gravel, zircon, feldspar, cement, rock, and limestone[a]	46.8	46.2	44.3	42.5	43.6	51.4	51.6	55.9	52.8	56.0
Seawater (magnesium metal and compounds, salt, and bromine)	69.0	73.0	89.1	84.6	94.5	102.6	117.0	113.8	146.1	151.7
Production through wells penetrating seafloor (oil, gas, and sulphur)	423.6	496.6	620.7	730.8	820.3	933.3	1177.7	1450.9	1980.0	2327.3

[a] Some production from beaches is included.

Table 21.2. Recent Marine Mining Activities

Location	Activity[a]	Water Depth (ft)[b]	Interest
Africa			
Red Sea	Exploration	6,000±	Sulfide muds
Territory of Southwest Africa	Dredging	100−	Diamonds
Republic of South Africa	Exploration	[c]	Phosphate
Asia			
Borneo	Exploration	600−	Tin
Indian Ocean	Exploration	[c]	Chrome, iron
Indonesia	Dredging	150−	Tin
Japan	Dredging	30−	Iron sands
Malaysia	Exploration	600−	Tin
Papua and New Guinea	Exploration	600−	Iron sands
Philippines	Exploration	600−	Iron, gold, titanium
Thailand	Dredging	150−	Tin
Europe			
Iceland	Dredging	150±	Shell sands
Great Britain	Dredging	100−	Tin, sand
North America			
Bahamas	Dredging	[c]	Aragonite
Canada (British Columbia)	Exploration	12,000±	Manganese nodules
Caribbean Sea	Exploration	200−	[c]
Mexico	Exploration	600−	Phosphate sands
Pacific Ocean	Exploration	12,000+	Manganese nodules
United States			
Alaska	Exploration	200−	Gold
Blake Plateau	Exploration	600–2,400	Manganese, phosphate
California	Dredging	30±	Shells
California	Exploration	600±	Phosphate
North Carolina	Exploration	600−	Phosphate sands
Oceania			
Australia	Exploration	600−	Phosphate, heavy metals, gold
Fiji	Exploration	600−	Gold
New Zealand	Exploration	600−	Heavy metals, gold, phosphate
Solomon Islands	Exploration	600−	Tin
Tasmania	Exploration	600−	Heavy metals

[a] Dredging operations generally include exploration activity, does not include mines originating on land and drifted out under the sea floor.
[b] Less than is represented by −; more than is represented by +; approximately is represented by ±.
[c] Unknown.

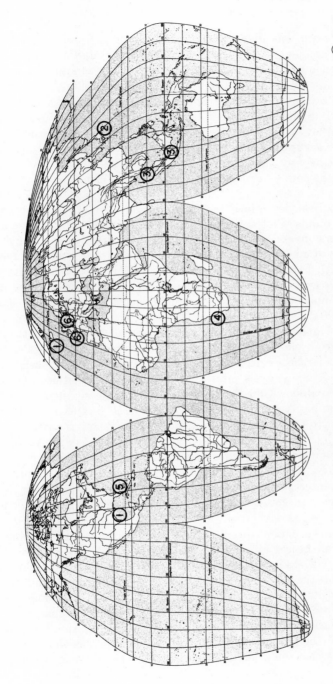

Figure 21.1 Marine mining in these and other areas consists of mining unconsolidated sediments (exclusive of sand and gravel). ① Shells; ② iron sands; ③ tin; ④ diamonds; ⑤ aragonite; ⑥ sand and gravel.

The program was given added stimulus in 1966 with the initiation of the Department of the Interior's Heavy Metals Program. With the intent of encouraging the development of offshore mining, the Bureau of Mines in 1967 sampled undersea placer deposits, located by the Geological Survey, in Norton Sound near Nome, Alaska (Jenkins and Lense, 1968). Samples from each of these test holes were found on analysis to contain low concentrations of gold, but because of the problems inherent in accurate sampling of offshore placers (Howard and Padan, 1966) and the fact that the costs of mining offshore in the Arctic are unknown, the commercial significance is not yet clear (National Council, 1968).

Seawater itself is "mined" for sodium, magnesium, calcium, and bromine compounds. This activity is expected to increase as more and more saline water conversion plants come on stream. Other interesting developments suggest an eventual increased use of seawater for mineral recovery. For example, British Efforts at the Atomic Energy Research Establishment in Harwell indicate that selective resins may someday be utilized in tidal channels or lagoons to concentrate specific ions found in seawater (Spence, 1964).

Within a generation ore will be mined from deep beneath the sea floor, far from shore. Already coal is recovered from beneath the sea floor in 57 nearshore areas, and iron ore in two others (Cruickshank et al., 1968). The U.S. Navy is interested in manned, one-atmosphere installations excavated within the bedrock beneath the sea floor. While only in the planning stage, the Navy's Project *Rock Site* (Austin, 1966) will give a boost to the technology needed to mine in similar environments.

Eventually man will learn to harvest the blanket deposits of precipitates lying under 2 to 3 miles of water. Such deposits, known to exist for a century, have recently sparked speculation in the popular press. Industry is interested because of the apparent magnitude of the deposits, and is actively exploring the feasibility of mining, but so far no commercial operations have resulted.

Although we cannot predict the rate at which marine mining will grow, we can be certain that as technology continues to advance man ultimately will equip himself with the means to exploit all the mineral resources of the sea: sea-floor occurrences of construction material, placer deposits of heavy metals, seawater, sub-sea-floor lode deposits, and deep-water blanket deposits of phosphorite and manganese.

While all this will come to pass in time, the true ocean mining industry will probably be born on the continental shelves in the sublittoral zone— the area between the intertidal zone and the edge of the continental shelf.

This is of concern to scientists and engineers alike because this area comprises the bulk of the zone of light penetration—the zone teeming, where conditions are favorable, with marine life.

At the moment science cannot anticipate what effect marine mining will have on the environment. It is certain, nevertheless, that there is a tremendous potential effect due simply to the magnitude of the operation. For example, a single dredge of the size used for gold dredging on land excavates several million cubic yards (My^3) of sediment per year. A conventional dredge would cover only about 10 acres to excavate each 1 My^3 of sediment. But a dredge designed to recover, for example, a blanket of nodules one nodule in thickness would have to cover over 6 square miles to mine at the same rate of production.

Marine mining may effect a series of changes in the physical environment. The locally deeper sea floor may cause alterations in wave refraction that can erode beaches. Fine particles from the mine, if they do not muddy the swimming water, may settle into new beaches in unwanted places or into silt beds hazardous to navigation.

Possible effects on marine life are far more complex. Excavation will destroy bottom-dwelling organisms and soil-stabilizing vegetation. Excavation explosives will damage air bladders and vascular systems, particularly in very small fish and larvae. Some nutrients and chemicals could have a beneficial effect (Joyner, 1968), depending on the chemistry of the water mass, but heavy metals, released and not completely recovered by the mining system, could be concentrated in certain fish in lethal amounts.

Turbidity will retard photosynthesis and thereby reduce the basic productivity of the area. Fish, whose gills become clogged with the suspended particles, may die. A change in heat radiation can be expected, also changing the ecological character of the area. During settling, the suspended particles can scavenge organic waste and, upon deposition, drastically reduce the oxygen available to sea-floor organisms, smothering fish food, shell fish, and spawning grounds.

After a sea-floor mine has been worked out, the environment will reach equilibrium. But the deposition of fine-grained material over a large area down current can constitute an unstable mass, with the resultant problem of frequent resuspension, and can frequently eliminate biologic succession.

Marine mining has offered a splendid opportunity to "do it right." In all probability the industry will develop and also any given operation will indeed change the physical, chemical, and biological characteristics of the area. The social cost, or ultimate cost to society, of the operation is what now must be considered. The goal must be to understand the probable relationship between marine mining and the environment. From this understanding can flow recommendations for legislation and reclamation regulations to ensure the compatible developments of all marine resources. While such regulations will tend to increase the costs of mining offshore, at least they will be costs that can be estimated in advance of capital investment. This situation is more sound fiscally than commencing an operation, killing a school of fish, and then being shut down by public response.

Fortunately time is on man's side. There is no apparent critical shortage of minerals in sight whereby one can postulate a hurried turn to the sea. Rather, the activities offshore will be stepped up gradually as economics permit. But facts are needed. Some are available but are buried in a variety of references, whereas others must be developed by means of research.

The research needs are twofold: techniques for predicting what may go wrong under specific sets of circumstances, and operational criteria to prevent mistakes. The end result of the research will be the creation of well-founded regulations for offshore mining. The following paragraphs briefly outline the nature of this aspect of the Marine Minerals Technology Center's program in marine minerals research.

As a first step, the nature of environmental disturbances caused by offshore industries similar to mining has been examined. This will be followed by state-of-art analyses of offshore sediment transport and deposition, as well as the establishment of dynamic ecological baselines.

The development of prediction techniques will result in the ability to model mathematically the changes to be expected in all three elements of a marine mine (ore body, water mass, and marine life) during the simulated life of the mine. Already a start has been made in the analysis of variables to be used in the model. Top priority has been placed on that part of the problem that appears to offer the greatest enigma: the effect of mining on marine life. While the research will stress the avoidance of damage to the environment, through prediction techniques, it will assess opportunities to enhance local conditions (such as churning up nutrients into the water mass and utilization of waste material).

Operational criteria too can be defined only through research. Prior to mining, an operator must be able to identify the salient characteristics of the ore body, its water overburden, and the marine life. He must be able to establish—or, more likely, furnish data for someone else to establish—a dynamic baseline against which the changes due to mining can be measured. Indicator parameters (Joyner, 1968) must be identified, their allowable variations determined, and standards established. All these limiting factors must be considered as design criteria in the development of the marine mining system. Finally, a prediction model can enable the operator—or a government representative—to analyze the planned mining scheme vis-a-vis the local environment.

Research must be performed to enable the operator to plan the procedure of mining, such as the production rate, excavation pattern, and method of waste disposal. Instrumentation must be available for the monitoring of indicator parameters to ensure that standards are being met.

After the ore body has been worked out, regulatory agencies must have the knowledge to establish the nature and extent of reclamation that should be performed. Bottom stabilization by vegetation may be required as well as the deposition of sand and/or gravel to enhance biologic succession.

The National Oceanic and Atmospheric Administration believes that learning how to predict the probable environmental effects of marine mining is a logical goal for federal research.

The Marine Minerals Technology Center's program is aimed at the development of recommended procedures, which may result in regulations, to be followed during offshore mining. Possibly, by this means, man cannot only avoid the errors he has made on land but also ultimately enjoy the compatible development of all of the sea's resources.

References

Austin, Carl F. 1966. Manned undersea structures—the rock site concept. Naval Ordnance Test Station Tech. Paper No. 4162, 28 pp.

Clarke, T. A., A. O. Flecksig, and R. W. Grigg. 1967. Ecological studies during project Sealab II. *Science*, **157**:1381–1389.

Cruickshank, M. J., C. M. Romanowitz, and M. P. Overall. 1968. Offshore mining—present and future. *Eng. Mining J.*, **169**:84–91.

Howard, T. E., and J. W. Padan. 1966. Problems in evaluating marine mineral resources. *Mining Engr.*, **18**:57–61.

Jenkins, R. L., and A. H. Lense. 1968. Sampling minerals of the ocean floor. Presentation of AIChE Meeting, Tampa, Fla. 12 pp.

Joyner, Timothy. 1968. Memorandum to A. P. Nelson, Research Director, U.S. Bureau of Mines Marine Minerals Technology Center, Tiburon, Calif.

National Council on Marine Resources and Engineering Development. 1968. *Marine Science Affairs—A Year of Plans and Progress.* U.S. Government Printing Office, Washington, D.C., 228 pp.

Spence, R. 1964. Extraction of uranium of sea water. *Nature*, **1964**:1110–1115.

United States Senate. 1968. Hearing before the subcommittee on minerals, materials, and fuels of the committee on interior and insular affairs, United States Senate, 19th Cong., 2nd Sess., on mineral shortages. 83 pp.

Part V

Models for Studying Future Alterations of the Ocean by Input of Nonindigenous Substances

22

Radioactive Models

YASUO MIYAKE, *Department of Chemistry, Tokyo Kyoiku University, Tokyo, Japan*

The Sources of the Artificial, Radioactive Substances in the Oceans

Impingement of man-made radioactive substances on the oceans was first recognized after the larger nuclear tests carried out in 1954 at the equatorial North Pacific. The development both in military and peaceful uses of the atomic energy in the last decade has increased further the risk of radioactive contamination of the oceans.

There are three main sources of contamination of the oceans by the artificial radioactive substances: nuclear plants on land, nuclear powered ships, and nuclear weapon tests.

Radioactive wastes from nuclear plants on land are disposed into the ocean in two ways: (*a*) the packaged disposal to the deep sea floor and (*b*) the liquid effluents on coasts. In (*a*) wastes are packaged with steel containers filled with concrete cement and are disposed into the sea, usually deeper than 2000 m. Until recently some 10^4 Ci of the radioactive materials were put into the oceans annually by this method. The nuclidic composition of packaged wastes varies widely, but main parts are neutron-induced nuclides such as ^{60}Co which are produced in the nuclear power plants in which ion-exchange resins are used for decontamination of the primary coolant.

The lifetime of the packages at the ocean bottom is estimated to be 10 to 20 yr. Therefore, there is a gradual leakage of radioactive materials from packages. Methods of solidification or vitrification of the radioactive wastes are under investigation in order to keep the radioactivity at the sea bottom for a longer time.

Among the direct effluents of radioactive wastes into the oceans, the largest is from the reprocessing plants for spent resin. Now, from several hundreds to a few tens of thousands of curies in a year of low-level liquid wastes are disposed through a pipeline of 1 to 5 km length out of each plant. The liquid wastes consist mainly of fission products of varying compositions (^{90}Sr, ^{144}Ce, ^{103}Ru-^{103}Rh, ^{106}Ru-^{106}Rh, etc.) with considerable quantities of tritium and small amounts of plutonium and uranium.

From nuclear plants other than reprocessing, small amounts of fission products, neutron-activated and other radioactive isotopes, are disposed into rivers and coastal waters. The total activity ranges from a few tens of millicuries to a few of curies annually from each plant.

The main waste from nuclear-powered ships is the spent ion-exchange resin which has been used for decontamination of the primary coolant water. The resin contains neutron-activated nuclides, such as ^{60}Co, ^{55}Fe, ^{59}Fe, ^{51}Cr, and ^{182}Ta. Exchange resins containing 10 to 100 Ci of ^{60}Co and other radioisotopes are disposed from ships to the sea surface at least 12 mi from shore at intervals of 6 to 9 months.

In the case of nuclear submarines for military uses, decontaminated primary coolants of a few tons are expelled due to the thermal expansion of water at the start of the engine. By U.S. Navy regulation, the primary coolant which is allowed to vent contains less than 3 μCi/ml. The radioactivity consists for the most part of neutron-induced nuclides.

Radioactive debris from nuclear weapon tests enter the oceans via many routes. Most of the fission materials are dissolved in seawater by underwater explosions. Ground-test explosions on atolls or small islands sometimes bring about serious contamination of the ocean. Castle Test, carried out in the spring of 1954, is one of the most well-known examples.

Radioactive fallout is classified into three categories: (a) immediate fallout near testing sites; (b) tropospheric fallout; and (c) stratospheric fallout. The residence time of the radioactive debris in the troposphere is about 0.1 yr, while that of the stratosphere is 1 to 2 yr. In the middle latitudes in the northern hemisphere, the tropospheric fallout is transported mainly by the westerly wind around the earth in 2 to 3 wk. The speed of transport is in accordance with the wind velocity at the altitude near 300 to 500 mb. The occurrence of the spring maximum and the maximum settling in the middle latitude belt between 30° and 50° are interesting problems in stratospheric fallout.

In either tropospheric or stratospheric fallout the dry settling rate is small. Most of the radioactive fallout is associated with meteoric precipitations. According to our observations 90 per cent of the fallout is due to wet precipitation. In this connection, the amount of oceanic fallout is suspected to be larger than continental fallout owing to scavenging by sea sprays. Radioactive debris from nuclear testings consists mainly of fission products. But, in addition, there are various kinds of neutron-induced nuclides including neptunium (^{239}Np), uranium isotope (^{237}U), plutonium (^{240}Pu, ^{239}Pu, ^{238}Pu), and other transuranium elements. Radiocarbon, tritium, and beryllium isotope (^{7}Be) are also produced by nuclear detonation. A more detailed treatment of sources of artificial radioactivity in the oceans is given in Chapter 12 of this book.

The Behavior of the Artificial, Radioactive Substances in the Oceans

Most of radioactive materials from different sources are put into the oceans at the surface, the coasts, or the sea floor. In other words, they enter the oceans at the boundaries between the air, land, bottom, and water. Therefore, the behavior or fate of radioactive substances in the oceans is largely controlled by these boundary conditions.

Radioactive effluent on the coast is transported by tidal and coastal currents and the eddy diffusion on the horizontal and vertical directions which result in the dilution of the radioactive materials. Some of the substances (^{60}Co, ^{65}Zn, ^{59}Fe, etc.) are taken up by the sea organisms—first by plants and then by animals, with wide ranges of the concentration factors (Table 22.1). Some radioactive nuclides enter directly into fish bodies through skin and gills. Others (^{106}Ru-^{106}Rh, ^{103}Ru-^{103}Rh, ^{59}Fe, ^{55}Fe, ^{141}Ce, ^{144}Ce, ^{137}Cs, etc.) are adsorbed on particulated matter and precipitate down to the shallow bottom. The radioactive materials in deposits are then taken up by bottom fauna, some of which are prey to fishes. It is often observed that in some biological materials the specific activity of a certain nuclide exceeds that in seawater owing to the deposition of highly radioactive particulate matter. Sometimes the shore sand is contaminated by water, sea spray, or creeping contaminated sands. Some radioactive materials adhere to fishing nets and other gear which exposes fishermen to irradiation.

The extent of surface area along the coast which suffers radioactive contamination depends on the rate of disposal, concentrations of radio-nuclides, and physical conditions of coastal waters. Usually the con-

Table 22.1. Concentration Factors (CF) of Some Chemical Elements in Marine Organisms

Element	Organism[a]	CF	Element	Organism[a]	CF
Cr	w	60–120,000	Sr	w	0.05–1,600
	i	2–9,000		i	0.1–60,000
	f	2,000		f	0.03–20
Mn	w	88–42,000	Zr	w	170–2,960
	i	120–550,000		i	0.6–600
	f	70–126,000		f	0.008–247
Fe	w	100–45,000	Ru	w	30–1,210
	i	13–78,000		i	1–3,200
	f	0.05–3,000		f	0.01–20
Co	w	45–3,700	Cs	w	1.3–240
	i	5.4–20,000		i	0.1–72
	f	0.5–560		f	2–244
Zn	w	80–50,000	Ce	w	100–4,500
	i	0.05–40,000		i	2–1,000
	f	1.4–15,500		f	0.26–611
Y	w	160–900	I	w	140–140,000
	i	12–250		i	0.4–20,000
	f	0.5–10,000		f	1.3–15

[a] The w denotes seaweed; i, invertebrates; f, fish.

taminated area ranges from a few kilometers to several hundred kilometers along and off shores. The coastal areas are generally productive areas and great amounts of seafood are produced there. Therefore, a careful monitoring for contamination of fishes and seaweed is important to protect man from radiation. Up to the present, there are many studies on the effects of radioactivity on coastal organisms. Figure 22.1 illustrates the gradual increase of ^{90}Sr in clam shells since 1955 (Hiyama, 1960). Radioactive substances are transported to other places by migratory fish as well as by marine products which are transported to distant markets.

The radioactive substances which enters the surface layer of the open oceans are dissolved or suspended in water and transported horizontally by the ocean currents, and then the horizontal turbulent diffusion, like a river in the ocean, gradually diffuses into wider areas. Vertically, they are quickly mixed well within the mixed layer, but owing to the steep

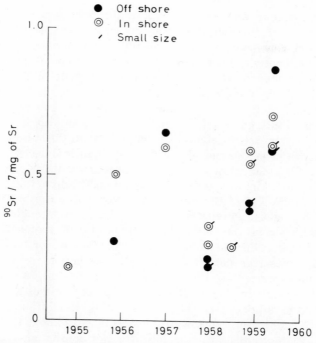

Figure 22.1 The increase of ^{90}Sr in clam shells (unit: pCi). From Hiyama, 1960.

gradient in water density, the rate of penetration into the deep layer is slow.

Patterns of Radioactive Contamination of the Oceans

Horizontal

The large-scale transport of radioactive materials by horizontal water movement in the oceans was first observed by Miyake et al. (1955) after observing an area of sea near Bikini-Eniwetok atolls aboard the survey boat *Shunkotsu-maru* in 1954.

The U.S. Atomic Energy Commission conducted a series of nuclear-explosion tests at Bikini-Eniwetok atolls from March 1, 1954 to the middle of May 1954. These operations were called the "Castle Test." A small Japanese fishing boat *Daigo Fukuryu-maru* (*Lucky Dragon V*) was working about 110 km east of the atolls on March 1, 1954. Due to the fallout of highly radioactive dust, mainly consisting of CaO and CaCO$_3$, fishermen and crews were heavily exposed to radiation. They returned to Japan on March 10. Fish which were raised by the *Lucky Dragon*, and many other fishing boats were also heavily contaminated. In order to survey the radioactivity near the atolls, the R. V. *Shunkotsu-maru*, which belonged to the Japanese Fisheries Agency, was sent to the equatorial North Pacific in the middle of May 1954.

The artificial radioactive substances in seawater were coprecipitated with barium sulfate and ferric hydroxide and the total β activity in the precipitates was measured aboard the ship. Though the survey was done about 1 month after the end of Castle Test, high activity was detected in seawater. The maximum activity in seawater was 9.1×10^4 dpm/l at 450 km west of Bikini and activity higher than 10^3 dpm/l distributed as far as 2000 km WNW of Bikini along the North Equatorial Current. In plankton, the highest activity was 8×10^5 dpm/g wet weight.

Figure 22.2 shows the horizontal distribution of radioactivity near Bikini-Eniwetok atolls. As shown in Fig. 22.2, the major portion of the radioactivity was transported in the WNW direction from the atolls, while a counter current in the WSW direction was also observed. The mean flow rate on the North Equatorial Current was estimated to be 0.7 knot by the radioactivity. There was a weak discontinuity line along two atolls. A counter current toward the east reached about 1000 km from the atolls. Water temperature, chlorinity, and density on the northern side of the discontinuity line were 27.1 C, 19.23‰, and σ_t 22.51, respectively, at 50-m depth, while on the southern side the readings were 27.0 C, 19.04‰, and σ_t 22.29, respectively. The branching of radioactivity might be related to this discontinuity. Since counter current was located near

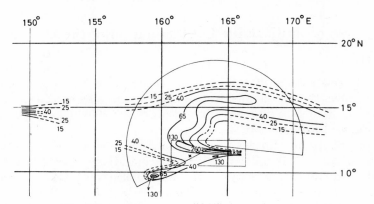

Figure 22.2 The horizontal distribution of radioactivity in the surface water near Bikini Atoll in 1954 (unit: 10^2 dpm/l).

the boundary between the North Equatorial Current and Equatorial Counter current, the flow rate might have been much slower.

The next year, from February to May 1955, the U.S. Atomic Energy Commission in cooperation with Scripps Institution of Oceanography of the University of California and the Applied Fisheries Laboratory of the University of Washington carried out a transoceanic expedition in the Pacific aboard the U.S. Coast Guard Cutter *Taney* for the purpose of tracing the widespread radioactivity which the *Shunkotsu-maru* had discovered.

Results of observation showed the activity up to 570 dpm/l in seawater and the highest value for plankton of 340 dpm/g. The contaminated area was located off the coast of Luzon Island of the Philippines, indicating the further westward drift of the activity with the North Equatorial Current, where weak activity was still detected.

In the summer of 1955 the North Pacific Expedition (NORPAC), which was carried out through the international cooperation of Canada, Japan, and the United States, detected seawater activity in a larger part of the western North Pacific (Miyake, 1957). The highest activity was located along the Kuroshio Current off the coast of Japan. We can regard these data, collected by radioactive tracers, as direct confirmation of the continuation of the Kuroshio and North Equatorial Currents. These results are summarized in Fig. 22.3.

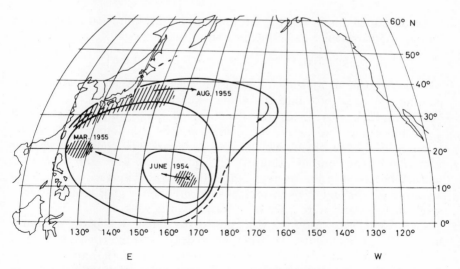

Figure 22.3 Radioactive contamination in the North Pacific 1954–1955 (shaded areas represent those of maximum contamination).

Assuming that the radioactive materials were transported by diffusion from a point source, the horizontal eddy-diffusion coefficient can be obtained from the change of the activity distribution with time. The relation between the eddy-diffusion coefficient, k, time, t, and lateral distance, Y, from the center of diffusion is as follows.

$$Y^2 = 2kt$$

From the observed value in Fig. 22.3 of Y and t, the value of k was calculated (Table 22.2).

Table 22.2. Horizontal Eddy-Diffusion Coefficient Calculated from the Time Change of Distribution of Radioactivity in the North Pacific Ocean

t (days)	Y (km)	k
60	890	0.8×10^9
270	2100	1.0×10^9
400	3200	1.4×10^9

According to Inoue (1950), the horizontal eddy-diffusion coefficient k for oceanic turbulence can be expressed as follows.

$$k = 0.01 L^{4/3} \qquad (10^{10} \text{ cm} > L > 10 \text{ cm})$$

where L is the magnitude of diffusion phenomena in the ocean. By putting the value of k into the above equation, $\log L = 9.1 - 8.4$ was obtained which corresponds with the observed value of $\log L$ of $7.9 - 8.5$.

By observing radioactivity in seawater, the dilution process which takes place in the oceans can be traced. Figure 22.4 shows the secular variation of the concentration of ^{90}Sr and ^{137}Cs in seawater from 1957 to 1967 observed in the western North Pacific. As described above, large amounts of ^{90}Sr and ^{137}Cs were deposited into water at Bikini-Eniwetok atolls in 1954, and were later transported to the west.

As a result, the western part of the North Pacific was badly contaminated with radioactivity. Dilution lowered the average concentration of ^{90}Sr from 2 pCi/l to 0.3 pCi/l from year to year from 1957 to 1961. The concentration of radioactive debris increased again after 1962 due to the fallout derived by nuclear testing by the United States and the U.S.S.R.

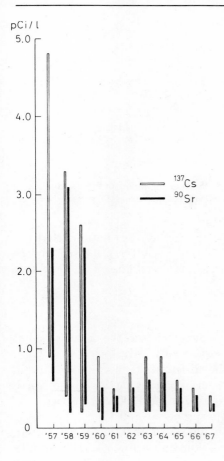

Figure 22.4 Secular variation of ^{90}Sr and ^{137}Cs in the western North Pacific. From Miyake et al., 1962.

in that year. Since the cessation of surface nuclear tests beginning in the summer of 1963, there was less radioactivity in 1964, and further dilution is in process.

It is interesting to compare the concentrations of ^{90}Sr and ^{137}Cs between the eastern and western parts of the North Pacific Ocean. Figure 22.5 shows the secular variation of ^{137}Cs concentration on the west coast of North America (Folsom and Sreekumaran, 1966). Radioactivity was very much lower in earlier years when higher activity was observed in the western North Pacific. But the activity along California coast increased gradually due to mixing with contaminated water and an increase in radioactive fallout. In 1964 radioactivity in the west and east became nearly equal, suggesting the homogeneous distribution of radioactive

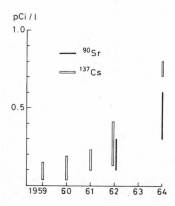

Figure 22.5 Secular variation of ^{90}Sr and ^{137}Cs in the eastern North Pacific. From Folsom and Sreekumaran, 1966.

materials in the midlatitude of the North Pacific. It is notable that the homogeneous, horizontal distribution of radioactivity at the sea surface between western and eastern waters was completed after a 10-yr mixing process.

Vertical

The vertical distribution of radioactive materials in the ocean was first observed aboard the *R. V. Shunkotsu-maru* by Miyake et al. (1955). Figures 22.6, 22.7, and 22.8 show the distribution of radioactivity in June 1954 in seawater on the vertical section perpendicular to the North Equatorial Current along the lines at 150, 570, and 1300 km, respectively, west of Bikini-Eniwetok atolls. In the profile at 150 km is shown the branching of radioactivity in the north and south directions. The dotted, vertical lines represent the latitudes of Bikini and Eniwetok. The activity

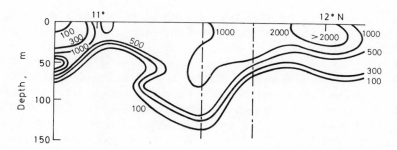

Figure 22.6 The distribution of radioactivity in seawater on the vertical section perpendicular to the North Equatorial Current along the lines 150 km west of Bikini Atoll (unit: cpm/l).

PATTERNS OF RADIOACTIVE CONTAMINATION OF THE OCEANS 575

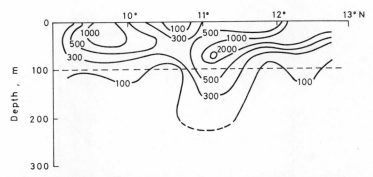

Figure 22.7 The distribution of radioactivity in seawater on the vertical section perpendicular to the north Equatorial Current along the lines 570 km west of Bikini Atoll (unit: cpm/l).

is separated on both sides of the dotted lines, and the distance between two maxima is about 100 km. There was radioactivity of 100 cpm/l at depths ranging from 80 to 120 m. Below the thermocline at 150 m only faint activity was observed.

Note that at a distance of 570 km west of the atolls, there is little change in activity compared with the section at 150 km, except that the activity shifted to the north. The depth of 100 cpm/l isolines was in accordance with that of the thermocline. At 1300 km from the atolls the activity was still observed within the mixed layer. Assuming that the average rate of flow was 0.7 knot, the amount of transport of radioactive materials through the 150-km section was estimated to be 1×15^5 Ci/hr. The same value of transport was obtained at the 570-km section, taking into account 2 weeks of radioactive decay.

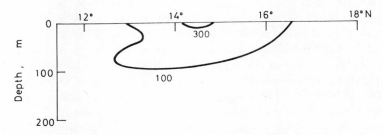

Figure 22.8 The distribution of radioactivity in seawater on the vertical section perpendicular to the North Equatorial Current along the lines 1300 km west of Bikini Atoll (unit: cpm/l).

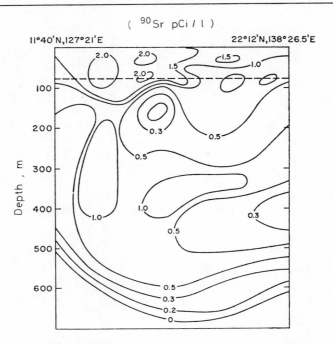

Figure 22.9 Vertical distribution of ^{90}Sr in the North Pacific, March 1955. The thermocline is shown as the broken line (unit: pCi/l).

It was found that during a few months' time after immediate fallout of radioactive debris, the large part of the radioactivity was confined within the mixed layer, flowing in a thin, narrow belt along the ocean surface.

As mentioned above, 8 months later the U.S. Atomic Energy Commission conducted the radioactive survey in the equatorial North Pacific. Using the data obtained by the United States expedition, the vertical distribution of radioactivity along the line from 11°40′N, 127°21′E, and 22°12′N, 138°26.5′E was drawn (Fig. 22.9). As shown in Fig. 22.9, the activity was higher in the surface layer but some of the activity already penetrated the thermocline (dotted line in Fig. 22.9) reaching a depth of 500 to 600 m. This suggests that a part of the activity which was present within the mixed layer 8 months before gradually descended across the thermocline and sank to a depth of 500 m due to vertical turbulent diffusion.

With respect to the vertical movement of the radioactive nuclides, in addition to the physical mixing, the biological effect must be considered. According to Steemann-Nielsen (1952), the organic productivity in the sea is 73 g-C/(m² yr) in the tropical Pacific. Assuming that the thickness of the euphotic zone is 100 m, the formation of organic matter as organic carbon is 0.7 mg-C/l, or 2.7 mg/l as planktonic organisms. When the concentration factor of a radionuclide in organisms is 10^3, the radioactivity in seawater in the euphotic zone which is taken up by organisms is 0.3 per cent of the total activity. This calculation suggests that decomposition of sinking marine life may not be the effective cause of transfer of radioactivity below the thermocline.

Ketchum and Bowen (1958) gave the following equation for the effect of migration of organisms on the transport of easily labile radioactive elements:

$$\frac{T_b}{T_p} = V_0 \frac{f_0}{A_z} \frac{d \times d_m}{t}$$

where T_b and T_p are the biological and physical transports, respectively; V_0 the volume of organisms per unit volume of water; f_0 the concentration factor of an element; A_z, the vertical eddy diffusion coefficient; d the depth of the thermocline; d_m the distance over which organisms migrate; and t the period of migration. When V_0 is 10^{-6} (1 cm³ of plankton in 1 m³ of water), d and d_m are, respectively, 100 and 500 m, and t is 1.7×10^7 sec (200 days), A_z is 10 cm²/sec and T_b/T_p will be about 0.0085 for the concentration factor of 10^3. This calculation also indicates that the biological transport is much less than the physical transport.

The penetration of ^{90}Sr and ^{137}Cs, which originated from radioactive fallout into the deep layer, was observed by Bowen and Sugihara (1960), Miyake et al. (1962), Higano et al. (1963), and Nagaya et al. (1964) in the Atlantic and the North Pacific Oceans. In the Atlantic, Bowen and Sugihara found that the concentration of ^{90}Sr at a depth of 300 to 700 m was about a half the surface value; ^{90}Sr was detected even at depths of 1000 to 1200 m, but no activity was found below 4000 km.

The vertical transport of ^{90}Sr and ^{137}Cs below the thermocline was demonstrated more clearly in the western North Pacific because of the higher concentration of these nuclides in the surface water than found in the Atlantic. Figure 22.10 shows the vertical distribution of ^{90}Sr and ^{137}Cs in the western North Pacific from 1958 to 1959. It is rather striking that these nuclides reached the layer as deep as 6000 m only 3 to 5 yr after the explosion tests of 1954.

It is shown in Fig. 22.10 that the vertical gradient is greater near the surface. This indicates the higher stability of the mixed layer and the difficulty of penetration of the nuclides through the thermocline. The

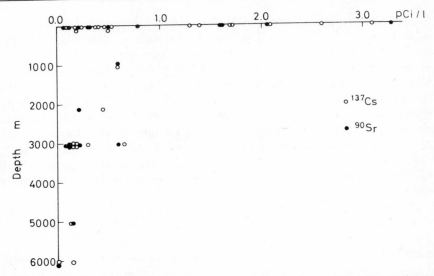

Figure 22.10 Vertical distribution of ^{137}Cs and ^{90}Sr in the western North Pacific (1958–1959).

gradient is much smaller below the thermocline, suggesting the higher rate of turbulent mixing in the deep layer than near the surface.

Note that the activity ratio of ^{137}Cs and ^{90}Sr averaged 1.4 in deep water, nearly that of surface water. The activity ratio of ^{137}Cs and ^{90}Sr in the fission product of ^{238}U is about 2.0. In the stratospheric fallout observed in Japan for a period of almost 10 yr, however, the mean ratio was 2.8. The higher ratio of ^{137}Cs/^{90}Sr in the stratospheric fallout and lower ratio in the Pacific may be due to the fractionation of ^{90}Sr and ^{137}Cs after detonation. The foregoers of ^{137}Cs and ^{90}Sr are rare gases, ^{137}Xe and ^{90}Kr with half-lives of 3.8 min and 33 sec, respectively. A larger portion of ^{137}Xe was injected into the stratosphere before it generated ^{137}Cs. On the other hand, a quantity of ^{90}Sr was trapped in the troposphere owing to the shorter half-life of ^{90}Kr.

Among sinking processes of ^{90}Sr and ^{137}Cs in the oceans, biological effect and precipitation, or settling effect, may be ignored. The concentration factors of ^{90}Sr and ^{137}Cs in organisms are small and these nuclides are present in seawater in the form of ions in solution. Thus penetration of these radionuclides into the deep layer is attributed mainly to the physical mixing process.

Assuming that the radioactive materials are homogeneously distributed in the wide area in the ocean and the depth of the ocean is 4000 m, the average vertical eddy diffusion coefficient below the thermocline to the bottom of the ocean can be calculated by using the following equation:

$$C = \frac{C_0}{\sqrt{\pi Dt}} e^{-Z^2/4Dt}$$

where D is the vertical eddy diffusion coefficient, Z the depth, t the time, C_0 the total amount of radionuclide which is added near the ocean surface at $t = 0$, and C the concentration of a nuclide at t and Z. By plotting log C against Z^2, a straight line is obtained, the inclination of which gives the vertical eddy diffusion coefficient D. The result of calculation showed that the average eddy diffusion coefficient between 200- and 4000-m depth is 200 cm²/sec for $t = 3$ yr.

In the same way, by using all the observational data on β activity obtained by Harley (1956) in the equatorial North Pacific aboard the cutter *Taney*, the vertical diffusion coefficient between 100- and 1000-m depth in the tropical area is estimated to be 40 cm²/sec.

Belayev et al. (1964) analyzed the ^{90}Sr and ^{137}Cs content in the Atlantic and calculated both the vertical eddy diffusion coefficient and the vertical rate of water flow at different depths. They found the diffusion coefficient at 500-m depth to be approximately 200 cm²/sec, decreasing to about 50 cm²/sec at 1000 m, increasing again to 150 cm²/sec at 2000 m and decreasing gradually toward the bottom to about 100 cm²/sec. The sinking of water occurs down to 500 to 1000 m (maximum rate 0.0007 cm/sec), and at lower depths there is a slow upwelling (0.0001 cm/sec). From these results they concluded that there is a considerably higher spreading rate of radioactive materials in the oceans than has been obtained by earlier estimates based on "the age concept" of the oceans.

By employing a tentative D value of 200 cm²/sec, Miyake et al. (1962) calculated the time change in the vertical distribution of nonradioactive materials placed at a depth of 200 m. Results of the calculation showed that the time which is required to attain practically homogeneous distribution at depths from 200 m to the bottom is about 100 yr (Table 22.3; Fig. 22.11).

In this connection, Broecker (1963) and Broecker et al. (1966) showed that their results of analyses of ^{90}Sr and ^{137}Cs in the deep layers of the oceans were not in accordance with the observations of Bowen and Sugihara (1958), Miyake et al (1962), and Russian workers.

According to their study, the significant amounts of these isotopes were, in general, not found below depths of a few hundred meters. They concluded that rapid deep penetration of ^{137}Cs and ^{90}Sr is not an ocean-wide phenomenon, and that their distributions are quite consistent with

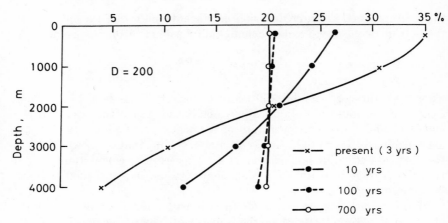

Figure 22.11 Time change of vertical distribution through diffusion.

vertical mixing rates required to explain the distribution of natural radiocarbon. However, as the studies by Broecker et al. (1966) were done in sea areas where radioactive contaminations are very small, some difficulties are apparent in interpreting the deep penetration in light of the very low activity in deeper layers of seawater.

Future Alteration of the Oceans by Artificial Radioactivity

It is obvious that man will have to depend largely on nuclear power in the future because of the increasing demand for energy and the decreasing stock of fossil fuels.

Table 22.3. Time Change of Vertical Distribution through Diffusion

	Depth (m)		
Time (yr)	200	2000	4000
3	1	0.58	0.11
10	1	0.83	0.47
100	1	0.98	0.93
500	1	0.996	0.985
700	1	0.997	0.989

Table 22.4. Energy Demand and ^{90}Sr to be Released at the Beginning of the Twenty-first Century

Present demand of energy	5×10^{10} MWH/yr
Expected demand at the beginning of the twenty-first century	10×10^{10} MWH/yr
Expected fraction of nuclear energy	20 per cent
^{235}U to be used	800 tons
Amount of ^{90}Sr released	5.6×10^3 MCi

It is known that the total demand of energy in the world has doubled in the last 30 yr. The present consumption of energy is estimated to be about 5×10^{10} MWH/yr. If we assume that the rate of increase of energy demand in future will remain constant, the energy consumption will be twice the present amount (1×10^{11} MWH/yr) at the beginning of the next century. By the twenty-first century nuclear power from fission processes is expected to comprise 20 per cent of the total energy supply. Calculations show that 800 tons/yr of fissionable materials will be necessary at the beginning of the next century, assuming that thermodynamic efficiency in the use of fission energy is 0.5 (Table 22.4).

Nuclear fission of 800 tons of fissile materials will produce 5600 MCi/yr of ^{90}Sr. This is 560 times the ^{90}Sr released by nuclear weapon tests through the end of 1962.

In the same way, the annual production of ^{90}Sr 100 yr from now is estimated to be 7×10^4 MCi.

The calculation of the potential steady-state situation in which radioactive wastes are uniformly introduced into the oceans was done by Craig (1959). He assumed an average residence time of the deep layer of 300 yr and a constant world fission rate for ^{235}U equal to 1000 tons/yr after 100 days of cooling, followed by the deposition of all fission products into the sea. His results show that the total activity in the surface layer will be almost equal to that of ^{40}K naturally present in seawater (330 pCi/l). The concentration of ^{90}Sr and ^{137}Cs will be 66 and 84 pCi/l, respectively.

Assuming a simple linear increase in the nuclear energy utilization, that is, an annual increase of 7×10^2 MCi ^{90}Sr and homogeneous disposal of radioactive waste into the oceans at average depths of 4000 m, Miyake and Saruhashi (1963) calculated the radioactivity level from the bottom of the thermocline to 1000-m depth, taking the decay constant of 28 yr for ^{90}Sr into account.

If we assume the leakage rate of the disposal package to be 5 per cent/yr, ^{90}Sr concentration near the ocean surface will be 540 pCi/l in another century, assuming that the vertical eddy diffusion coefficient is 200 cm^2/sec. In the case of nonpackaged disposal at the sea bottom, the activity level near the surface will be 840 pCi/l. Thus the package is not effective in preventing the diffusion processes from contaminating the oceans with radioactivity.

In the above calculation the higher value of the vertical eddy diffusion coefficient was used. If we employ the lower value, 10 cm^2/sec, for the vertical eddy diffusion coefficient, the activity decreases to 100 and 83 pCi/l for nonpackaged and packaged disposal, respectively.

From near the bottom of the oceans to 1000 m above the sea floor, the radio-strontium concentration will be 1600 and 3200 pCi/l, for vertical eddy diffusion coefficients of 200 and 10 cm^2/sec, respectively, when no package is used. In the case of packaged waste, corresponding values will be 770 and 1600 pCi/l, respectively.

The maximum permissible concentration of ^{90}Sr + ^{90}Y in drinking water is 800 pCi/l. Assuming that the concentration factor of ^{90}Sr in marine organisms is 10, the dose limit to radiation for man is $\frac{1}{10}$ the maximum permissible dose, and the ratio of marine-originated food to the total diet including water is $\frac{1}{10}$, the permissible concentration of ^{90}Sr + ^{90}Y in seawater will be 0.8 pCi/l.

It can be seen from these calculations that in order to keep the activity level of ^{90}Sr + ^{90}Y in all oceans below the permissible concentration of 0.8 pCi/l, packaged disposal to the sea floor of more than $\frac{1}{2000}$ of the total fission products produced must be avoided, even assuming homogeneous deposition of ^{90}Sr throughout the ocean. When we consider the localized disposal of radioactive waste, we have to consider a much smaller value than the above limit—for instance, $1/10^4$ to $1/10^5$.

When radioactive materials deposited from the seashore through pipelines or rivers into the surface layer of the oceans, radioactivity is transported by plume, coastal and tidal currents, and horizontal eddy diffusion. Figure 22.12 shows the distribution of ^{90}Sr and ^{137}Cs near the estuary of Columbia River (Park et al., 1965). The effect of radioactivity which was originated from the atomic plant at Hanford through the Columbia River extends as far as 300 to 400 km from the coast. The same results were obtained by Osterburg et al. (1966) with respect to ^{65}Zn and other radioactive nuclides along the Oregon coast.

If 0.1 per cent of the total radioactive waste is disposed evenly into the surface of the sea from the seashores of the world's continents, the concentration, for example, of ^{144}Ce within the distance of 1000 km from the

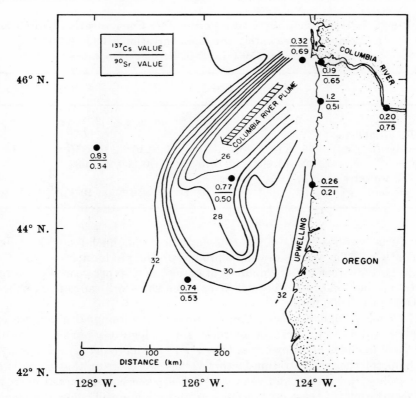

Figure 22.12 Salinity (‰), ^{90}Sr, and ^{137}Cs (pCi/l) in coastal waters off Oregon, U. S. A. From Park et al., 1965.

shore will reach 24 pCi/l at the end of this century, exceeding the maximum permissible concentration of ^{144}Ce in coastal waters of 10 pCi/l (National Academy of Science, 1962). The calculation was done under the assumption of homogeneous distribution in the mixed layer of radionuclides within the 1000-km belt along the coast of the world's oceans. The residence time of the mixed layer is assumed to be 5 yr. Needless to say, near disposing sites the concentration of radioactive nuclides will be much higher than obtained by the above calculation. Therefore, the upper limit of allowance of disposal of fission product will be very much smaller than 0.1 per cent, which is in good accordance with the previous calculation for ^{90}Sr.

Thirty years from now, the number of nuclear-powered ships in the world will reach about 500. In these ships, in addition to the fission

Table 22.5. Half-lives, Maximum Permissible Concentration (MPC) for Drinking and Seawater, and Concentration Factor (CF) of Radionuclides Contained in Primary Coolants

	^{60}Co	^{55}Fe	^{59}Fe	^{51}Cr	^{182}Ta
Half-lives	5.2 yr	2.9 yr	45 days	27 days	112 days
MPC, drinking water (μCi/ml)	5×10^{-5}	8×10^{-4}	6×10^{-5}	2×10^{-3}	4×10^{-5}
CF	200–10,000	10–10,000	10–10,000	1000	5
MPC, seawater (μCi/ml)	10^{-7}–10^{-9}	10^{-4}–10^{-7}	10^{-5}–10^{-8}	10^{-6}	10^{-5}

products, various kinds of neutron-activated radionuclides, ^{60}Co, ^{55}Fe, ^{59}Fe, ^{51}Cr, ^{182}Ta, and so on will be produced. The production rate of ^{60}Co may be a few tens to few hundreds of curies per ship per year. Therefore, total production of ^{60}Co in nuclear-powered ships will range from about 50,000 to 100,000 Ci/yr.

In Table 22.5, half-lives, the maximum concentration in drinking water and seawater, and concentration factors in biological organisms of some of these nuclides are shown (see Chapter 12 for more detail).

Some of the concentration-factors values are so great that there is a possibility that fishes and seaweed may become contaminated even if concentrations of these nuclides in seawater remain well below the maximum permissible concentration for drinking water. It is reported that a small amount of decontaminated primary coolant leaks out at the start of engines of nuclear-powered submarines. Still more dangerous is the disposal of spent exchange resins from nuclear-powered ships at sea. According to the observations of Hiyama (1960) some fishes, such as sardines, horse mackerel, and mackerel have a habit of swallowing exchange-resin particles.

Table 22.6 gives the amounts of radioactive nuclides (A, A') in exchange resin disposed at one time, the maximum permissible burden of these nuclides on one person (B), and the ratio of A/B and A'/B (Miyake and Hiyama, 1963). Note that A/B and A'/B are as high as 10^7 to 10^8 for ^{60}Co and ^{182}Ta.

In Table 22.7, a possible occurrence of active spots at which, for example, the concentration of ^{60}Co is higher than the maximum permissible level as a result of the disposal of spent resin is shown. Thirty years from now, as the number of nuclear ships increases to 500, thousands of active

Table 22.6. Amount of Radionuclides (RI) on Spent Resins and Maximum Permissible Body Burden (MPBB) on Man

	^{60}Co	^{55}Fe	^{59}Fe	^{51}Cr	^{182}Ta
RI on Resin (Ci)					
Savannah (A)	100	100	5	100	100
Nautilus (A')	10	—	0.5	0.3	—
MPBB on man					
(B, μCi)	1	300	2	80	2
A/B	10^8	3×10^5	3×10^6	10^6	5×10^7
A'/B	10^7	—	3×10^5	4×10^4	—

spots in a year will appear in some areas of the oceans, assuming disposal of spent resin at 6- to 12-month intervals and the radioactive materials spread within the surface layer of 10 m for a short time (U.S. National Academy of Science, 1959).

The present total catch of fishes in the world is 5×10^7 tons/yr. If one out of each million fishes eats the highly radioactive exchange resin, the radioactivity of about 50 tons/yr of fish will increase, supplying highly contaminated portions of about 200 g to 125,000 persons, assuming that the weight of edible parts comprises 50 per cent.

Because fishes can migrate very far, they can be caught in unexpected areas. The striking example of transport of radioactive materials by migratory fishes is shown in Fig. 22.13. In 1954 when the western part of the North Pacific was heavily contaminated by Castle Test at Bikini-Eniwetok atolls, large amounts of radioactive fishes were harvested by Japanese fishermen. At that time the Japanese Government regulated the maximum permissible radioactivity at 100 counts/min when indicated by a Geiger counter 10 cm above fish bodies. Fishes giving higher readings

Table 22.7. Contaminated Spots (^{60}Co) Induced by Spent Resin from 500 Nuclear Ships at the End of the Twentieth Century

Depth (m)	Area (km²)	Active Spots per Year
10	10–100	1000–3000

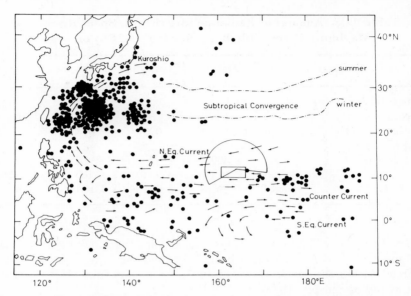

Figure 22.13 Distribution of fishing sites where considerably contaminated fishes were caught, 1954 (Japan Fisheries Agency).

had to be discarded. Figure 22.13 shows the fishing sites where the highly contaminated fishes were caught. From spring to summer in 1954, most spots were located near the equator, but from summer to autumn spots approached Japanese Islands up to 40°N.

This example clearly shows that radioactive fishes can migrate very far from contaminated sites. Therefore, even in the noncontaminated areas there is the potential danger of harvesting radioactive fishes as the use of the atomic power is progressively advanced in many countries.

References

Belayev, V. I., A. G. Kolesnikov, and B. A. Nelepo. 1964. Determination of intensity of radioactive contamination of the ocean based on new data relating to the exchange process (in Russian). *Proc. 3rd Int. Conf. Peaceful Uses of Atomic Energy, Geneva,* **14**:83–87.

Bowen, V. T., and T. T. Sugihara. 1958. Marine geochemical studies with fallout radioisotopes. *Proc. 2nd Intern. Conf. Peaceful Uses of Atomic Energy, Geneva,* **18**:434.

Bowen, V. T., and T. T. Sugihara. 1960. Sr-90 in the mixed layer in the Atlantic Ocean. *Nature,* **186**:71–72.

Broecker, W. 1963. Radioisotopes and large-scale oceanic mixing. In *The sea*, M. H. Hill (ed.), Interscience Publishers, New York.

Broecker, W. S., E. R. Bonebakker, and G. G. Rocco. 1966. The vertical distribution of cesium-137 and strontium-90 in the ocean. *J. Geophys. Res.*, **71**: 1999–2003.

Craig, H. 1959. Disposed of radioactive wastes in the ocean; the fission product spectrum in the sea as a function of time and mixing characteristics. *The biological effects of atomic radiation*, Natl. Acad. Sci. Publ. No. 551. Chap. 3, 34–42.

Folsom, T. R., and C. Sreekumaran. 1966. Survey of downward penetration of fallout in the ocean *in situ* absorption. In *2nd intern. oceanog. congr.*, Moscow, 1966, p. 120.

Harley, J. H. (ed.). 1956. *Operation Troll*, Health and Safety Laboratory, US-AEC, NYO 4656, 37 pp.

Higano, R., Y. Nagaya, M. Shiozaki and Y. Seto. 1963. On the artificial radioactivity in seawater. *J. Oceanog. Soc. Japan*, **18**:200–207.

Hiyama, Y. 1960. Radioactive contamination of marine products in Japan. Presented to U.N. Scientific Committee on the Effect of Atomic Radiation.

Inoue, E. 1950. The application of turbulence theory to oceanography. *J. Met. Soc. Japan*, **28**:420–424.

Ketchum, B. H., and V. T. Bowen. 1958. Biological factors determining the distribution of radioisotopes in the sea. *Proc. 2nd U.N. Intern. Conf. Peaceful Uses of Atomic Energy*, Geneva, No. (UN 402-OIC 724), 11 pp. (Contrib. No. 968 from Woods Hole Oceanogr. Inst.).

Miyake, Y. 1957. The distribution of artificial radioactivity in the equatorial region in the Pacific in the summer of 1956. *Proc. 9th Pacific Science Congress*, **16**:227.

Miyake, Y., and Y. Hiyama. 1963. Radioactive contamination of marine organisms due to the waste from nuclear-powered ships. *Kagaku (Science)*, **33**:492–495 (in Japanese).

Miyake, Y., and K. Saruhashi. 1963. Contamination of the ocean due to radioactive disposal. Read before the 1963 Annual Meeting of Oceanographic Society of Japan (unpublished).

Miyake, Y., K. Saruhashi, Y. Katsuragi, and T. Kanazawa. 1962. Penetration of Sr-90 and Cs-137 in deep layers of the Pacific and vertical diffusion rate of deep water. *J. Radiation Res.*, **3**:141–147.

Miyake, Y., Y. Sugiura, and K. Kameda. 1955. On the distribution of radioactivity in the sea around Bikini Atoll in June 1954. *Papers Meteor. Geophys.*, Tokyo, **5**:253–263.

Nagaya, Y., M. Shiozaki, and Y. Seto. 1964. Radiological survey of seawater of adjacent sea of Japan in 1963. *Hydrograph. Bull.*, **78**:63–67 (in Japanese).

National Academy of Science. 1959. Radioactive waste disposal from nuclear-powered ships. Publ. No. 658, 52 pp.

National Academy of Science. 1962. Disposal of low level radioactive waste into Pacific coastal waters. Publ. No. 958, 87 pp.

Osterberg, C. L., A. G. Carey, and W. G. Pearcy. 1966. Artificial radio-nuclides in marine organisms in the northeast Pacific Ocean off Oregon, USA. In *2nd int. oceanog. congr., Moscow,* 1966, p. 275.

Park, K., M. J. George, Y. Miyake, K. Saruhashi, Y. Katsuragi, and T. Kanazawa. 1965. Strontium-90 and Caesium-137 in Columbia River Plume, July 1964. *Nature,* **208**:1084–1085.

Steemann-Nielsen, E. 1952. The use of radio-active carbon for measuring organic production in the sea. *J. Conscil Perm. Intern. Exploration Mer.* **18**:117–140

23

A Model of Nutrient-Limited Phytoplankton Growth

RICHARD C. DUGDALE, *Professor of Oceanography, Department of Oceanography, University of Washington, Seattle, Washington*

JOHN J. GOERING, *Professor of Oceanography, Institute of Marine Science, University of Alaska, College, Alaska*

In designing marine sewage outfalls, efforts are customarily directed toward prediction of coliform bacterial counts at given locations, usually adjacent beaches (Levine et al., 1959). Although scientific evidence linking coliform counts to the incidence of infections acquired by bathers is apparently lacking for marine outfalls, state and local health authorities, using the only guidelines available to them, usually specify outfall performance in these terms. Aesthetic considerations also receive attention and the allowable concentrations of grease balls, rubber goods, and other visible evidence of disposed sewage at the sea surface are often specified. The possibility that the additions of primary nutrients such as nitrogen and phosphorus will result in enhanced growth of phytoplankton is occasionally considered, particularly with reference to the possible production of undesirable dinoflagellate blooms (red tides). In most instances, however, the added nutrients from sewage are believed to be quickly diluted, and therefore, do not greatly influence the growth of nuisance phytoplankton. The lack of concern over the production of unwanted phytoplankton blooms by sewage is partially attributable to the lack of a theoretical basis on which the nutrient-phytoplankton consequences could be incorporated into the design of outfalls.

The increasing use of the sea for municipal sewage disposal requires that we develop an understanding of the utilization of the nutrients released in a marine outfall. At the present time, effluent is often discharged as deep as practicable in hopes that the sewage field produced will remain buried. Generally, sewage burial is not completely achieved and when the diluted effluent reaches the surface, the intensification of nutrients can cause an increase in phytoplankton growth, the extent of which cannot presently be predicted. If the influence of nutrients on the growth of phytoplankton were predictable, it would permit man to control such blooms and to utilize sewage for productive purposes as suggested by Pomeroy (1967) and others, rather than to concentrate on unbeneficial disposal of our waste materials. Aquaculturists are currently studying the feasibility of increasing productivity in estuarine systems by adding extraneous nutrients from various natural sources such as deep ocean water. The feasibility of using municipal waste materials to increase estuarine productivity should be examined. For example, if the effluent were deliberately discharged into the euphotic zone it might be possible to confine the effects of the outfall to a small, predictable region and to control the growth and species composition of the resulting phytoplankton population. The resultant phytoplankton in these systems could be harvested and utilized to produce commercially valuable species of plants and animals.

In this chapter we discuss a model of nutrient-limited phytoplankton growth which may prove extremely useful in predicting the growth of phytoplankton resulting from nutrients entering the marine environment and thus be of assistance in designing marine sewage outfalls.

Nutrients in the Ocean

The successful control of phytoplankton growth in coastal systems fertilized with nutrients from sewage requires a thorough understanding of the natural physical, chemical and biological features in these systems.

The coastal systems of the ocean such as fjords, estuaries, and so on are, in general, areas of high fertility when compared with the open ocean. This high fertility is due in part to the increased supply of plant nutrients. Two major types of supply must be considered when examining any coastal system, and the significance of each will vary in each individual system. For example, circulation in estuaries is often a two-layered flow with the surface layers diluted by river water escaping seaward and the salt water entering near the bottom. In many instances, the river water supplies most of the nutrients. In other systems enrichment from the sea is the major nutrient source. Therefore, information on the circulation, nutrient levels in river water and seawater, and the rates of nutrient removal (primary production) and regeneration within an individual estuary must be considered when establishing a nutrient budget for the system.

On an ocean-wide basis, the nutrient budget (Table 23.1) has been summarized by Emery et al. (1955). It can be seen that nutrients supplied by continental runoff are of little significance when one considers the whole ocean. But as pointed out previously, on a local basis this supply can be of major importance. The estimated annual use of nutrients by phytoplankton is only about 1 per cent of the reserve in deep ocean water. It is also of interest to note that the amount added annually to the sediments is about equal to the supply from freshwater. Therefore, the ocean on a whole appears to be nearly in steady state with regards to the nutrients nitrogen, phosphorus, and silicon. Silicon is not often regarded as one of the primary elements required in biological systems. But in the pelagic oceans, diatoms are generally the most abundant phytoplankter and they do have an absolute requirement for the element if they are to reproduce.

Table 23.1. Nutrient Budget of World's Oceans

	Nitrogen	Phosphorus	Silicon
	(millions of metric tons)		
Reserve in ocean	920,000	120,000	4,000,000
Annual use by phytoplankton	9,600	1,300	—
Annual contribution by rivers	19	14	4,300
Dissolved	19	2	150
Suspended	0	12	4,150
Annual contribution by rain	59	0	0
Annual loss to sediments	9	13	3,800

From Emery et al. (1955).

General Scheme of Biodegradation of Organic Matter in Marine Environments

Biodegradable organic matter discharged into the sea is oxidized by microorganisms. The initial oxidation is accomplished by organisms entering with the effluent, and after dilution with seawater marine bacteria are probably the major oxidizers. The bactericidal properties of seawater are well documented (Ketchum et al., 1949). In the oxidation process, the dissolved oxygen in seawater is utilized as the electron acceptor, and when the rate of removal is greater than the rate of supply by diffusion and the photosynthetic activity of plants, the oxygen is depleted. When oxygen is depleted, anoxic microorganisms begin to stabilize the remaining organic matter using the nitrate ion first, and when it is depleted, the sulfate ion, as the electron acceptors. During the latter process, noxious hydrogen sulfide is produced. In all of the biodegradation reactions carbon dioxide, ammonia, and phosphate are released into the water, and become available for organic synthesis in algal growth.

The biodegradation of organic matter during which various nutrients are released in municipal sewage proceeds during collection, treatment, and transport to the marine outfall. The types of organic and inorganic materials released at the outfall will depend primarily upon the treatment process. The average composition of municipal waste and various purification processes are discussed by Føyn in Chapter 16.

The biochemical oxygen demand (BOD) has been generally accepted as a simple measure of the potential oxygen demand of biodegradable organic matter in an effluent. It represents the quantity of oxygen required to stabilize the contained oxidizable organic matter during 5 days of incubation at 20 C. The use of BOD measurements to assess the absolute oxygen demand of sewage discharged into the sea is questionable. Baalsrud (1967) showed that when a mixture of seawater and sewage, having a certain oxygen demand, was stored in the dark an oxidative breakdown occurred, thereby reducing the oxygen demand. However, when the mixture was inoculated with a few algae and placed in the light, algal growth gave rise to organic matter with an oxygen demand much greater than that originally found in the sewage. His experiments clearly demonstrates that the organic matter formed as a result of eutrophication potentially represents a much greater organic load than that added directly with sewage. Therefore, it appears necessary to clearly understand the secondary as well as the primary effects of sewage addition to seawater.

Generalized Models of Phytoplankton Growth

Eckenfelder (1966) reviews the mathematical models which attempt to explain the mechanism of BOD removal by biological oxidation processes during sewage treatment. Pretorius (1969) gives an excellent review of anoxic microbial growth kinetics. In these models the Michaelis-Menten kinetic parameters have been employed to define the microbial growth rate. However, as stated previously, production of organic matter by algae utilizing nutrients resulting from the oxidation of organic matter in sewage is of extreme importance when considering marine sewage disposal. In this chapter we concentrate on models describing such algal production.

Mathematical models describing nutrient-growth relationships in natural populations of algae are limited. Riley (1963) describes the most important of these and has been at the forefront of such modeling efforts. His earlier models were comprehensive, and solutions were difficult to obtain. Borrowing liberally from each other, Riley (1965) and Steele (1958) produced models simplified into a two-layered system described by Riley with more tractable mathematical expressions. However, the structure of these models is such that the effects of nutrients are largely lost from view. In a further variation, Dugdale (1967) proposed a model based upon the flow of a limiting nutrient through a compartmental system. This nutrient-oriented compartmental model may be useful in treating problems of disposal of effluents containing nutrients required in algal growth.

The paths of the primary nutrients are diagrammed in Fig. 23.1, where nitrogen is represented in various compartments, for example, phytoplankton, zooplankton, dissolved inorganic, and dissolved organic fractions. The size of each compartment at any given time is identical to the amount of the specific element contained, usually expressed on a unit volume or unit surface area basis. The arrows represent the transport

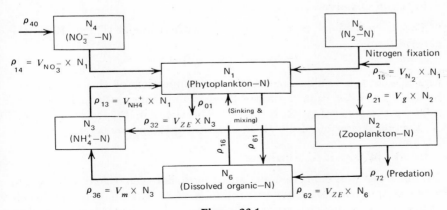

Figure 23.1

of material from one compartment to another. The rates are expressed by ρ's with two subscripts, the first representing the compartment receiving and the second the compartment contributing the material. $V_{NO_3^-}$, $V_{NH_4^+}$, and so on, fractional uptake rates, are essentially growth rates expressed in terms of nitrogen and have units of grams of nitrogen taken up [(grams N) $\times$ (unit time)], or simply unit time^{-1}. In the interests of clarity, not all possible pathways are indicated.

The model is intended for application under conditions of steady state or transient-nutrient limitation. Although some interactions may be expected, regulation at steady state will be upon one primary nutrient. For example, if the rate-limiting step is in the uptake of phosphorus, the uptake of nitrogen and silicon (if required) will be governed to some degree by the rate of phosphorus uptake. These considerations, based upon nutrient limitation, hold only where light is not relatively more scarce than nutrient. In the tropical Pacific Ocean, MacIsaac and Dugdale (1969) have shown the crossover between nutrient limitation and light limitation to occur approximately at the depth where the incoming radiation is reduced to 10 per cent of the surface value. Under light limitation the nutrient uptake rates are presumably regulated so that the nutrients enter the cell in approximately the ratios required. To obtain solutions for this model, valid mathematical expressions must be available for each of the important pathways. The processes fall into four basic categories: (a) influx of dissolved primary nutrients (ammonia, nitrate, phosphate, and silicate) and of unmineralized organic matter containing these elements, (b) uptake of primary and organic nutrients, (c) losses of phytoplankton from the euphotic zone, and (d) mineralization.

Influx of Dissolved Nutrients

These expressions take, in the two-layered model (Riley, 1965) the general form

$$P_{40} = m(N_0 - N_4) \tag{1}$$

where m is the mixing coefficient, the proportion of the upper layer replaced by water from the lower layer in unit time (units are therefore t^{-1}); N_0 the concentration of dissolved nutrient in the lower layer; and N_4 the concentration of dissolved nutrient in the upper layer.

Uptake of Dissolved Nutrients

This term was expressed in linear form between the range of zero nutrient and some arbitrary saturation level by Riley (1965) and Steele (1958) in the absence of experimental data. More recently, the expression used by Monod (1942) to describe microbial growth in his early work with chemostats has been shown by MacIsaac and Dugdale (1969) to apply to some heterogeneous populations of marine phytoplankton. Taking the form of the Michaelis-Menten expression for enzyme catalyzed reactions,

$$V_N = V_{N\max} \frac{N_4}{N_4 + K_{T(N)}} \qquad (2)$$

where V_N is the fractional uptake rate of nutrient species N by the population (units t^{-1}); $V_{N\max}$ the maximum fractional uptake rate of nutrient species N by the population under prevailing conditions; N_4 the ambient concentration of limiting nutrient; and $K_{T(N)}$ the transport constant, the concentration of the limiting nutrient at which $V_N = \frac{1}{2}V_{N\max}$. When V_N is plotted against N_4, a rectangular hyperbola is obtained. Expression 2 is commonly referred to as the Monod equation when it is applied to microbial culture theory, and some important consequences arise from its use in nutrient phytoplankton models. For example, in the elementary model described by Dugdale (1967) the relative abilities of different species or groups of algae to compete for the limiting nutrient are determined by the values of $K_{T(N)}$ and $V_{N\max}$. Very few values of $K_{T(N)}$ and $V_{N\max}$ for different species of algae are available. Recently, however, Eppley et al. (1969) tabulated values of K_T for ammonium and nitrate uptake obtained by themselves and others. The range is from 0.1 to 10 µg atom N/l, with more values falling at the lower end of the range than at the higher and in agreement with the K_T for both compounds measured by MacIsaac and Dugdale (1969) for natural oceanic populations. The specific uptake rates, or $V_{N\max}$, are largely lacking or unreliable. However, Eppley and Thomas (1969) found the specific nutrient uptake rates and specific growth rates to correspond closely. Their results were obtained with oceanic and neritic diatoms, and with flagellates at a temperature of 18 C. Thus it may be possible to combine the numerous available measurements of specific growth rates of algal species with $K_{T(N)}$ to produce characteristic nutrient uptake hyperbolae for various species of phytoplankton.

The values of K_T and $V_{N\max}$ are not true constants, even for a single species, and can be strongly influenced by changes in environmental conditions and by interactions between nutrient elements or between chemical species of the same nutrient element (see the section on nutrient interactions). For example, in one study the magnitude of K_T for ammonia oxidation by *Nitrosomonas* was found to increase by a factor of 10 with an increase of temperature from 8 to 30 C, and the effect of temperature was found to be even greater for nitrite oxidation by *Nitrobacter* (Great Britain Water Pollution Research, 1959–64). The effects of temperature on the growth rates of these nitrifying bacteria were also measured and the results are being applied to studies of nitrification by activated sludge.

Nutrient Interactions. As an example of the effect of one species of nutrient element upon another, the uptake of nitrate or nitrite by phytoplankton is known to be reduced in the presence of ammonia. At a relatively low concentration, approximately $2\mu g$ atom NH_4^+-N/l, virtually no nitrate or nitrite uptake occurs. Recent experiments using ^{15}N as a tracer indicate that the V_{max} for nitrate uptake is reduced in a linear fashion by increasing concentrations of ammonia, and that the K_T for nitrate uptake is unaffected. Such results suggest repression of the enzyme nitrate reductase. However, both the ammonia and nitrate uptake and nitrate reductase systems are poorly known in algae and the controlling mechanisms are likely to be complex. It should be mentioned that the rate-limiting step involved with a limiting nutrient may occur in the transport system at the cell membrane or in a subsequent, internal enzymatic step. In either case, it appears that some form of the Monod or Michaelis-Menten expressions is likely to hold. Fortunately, a considerable amount of work has been done on enzyme-control mechanisms, and the results are available to aid in interpreting the results of future experiments on uptake of nutrients by algae.

Many types of organic compounds are present in sewage and some can serve as sources of nutrients for marine phytoplankton. Hellebust (1965) reports that the marine diatom *Melosira nummuloides* takes up amino acids through active transport systems. Some amino acids are concentrated by a factor of more than two orders of magnitude when present in the medium at 10^{-4} M. Amino acids may contribute a substantial part of the cell's carbon and nitrogen at low rates of photosynthesis. However, amino acids do not support growth in the dark. It is suggested that this species is incapable of gluconeogenesis in the dark.

The existence of active uptake mechanisms for amino acids suggests the possibility of interaction between this organic source of nitrogen and the primary inorganic sources, nitrate, nitrite, and ammonia.

The uptake capacities and specificities of organic substrates vary from species to species. Hellebust (1970) conducted a general survey of the uptake capacities and specificities of some marine phytoplankton to find how widely distributed transport systems for organic substances are in marine photoautotrophs. Two members of the Chrysophyceae, *Cocco-*

lithus huxleyi and *Isochrysis galbana*, and two species of Chlorophyceae, *Dunaliella tertiolecta* and *Pyramimonas* sp., did not take up any of the seven organic substrates offered. Two dinoflagellate species, *Gymnodinium nelsoni* and *Peridinium trochoideum*, took up glucose and the amino acids lysine and alanine. Most of the diatoms studied utilized acetate at low rates. Two diatoms from a saltwater pond, *Cyclotella cryptica* and *Melosira nummuloides*, took up amino acids rapidly. The only open ocean diatom which took up amino acids to a measurable extent was *Cyclotella nana*.

The above information suggests that organic compounds are not often used by marine phytoplankton as a direct nutrient source. Thus the organic material entering in sewage must first be mineralized by bacteria before the primary nutrients are available for phytoplankton growth. However, it must be stressed that the capacity to use organic substrates varies widely in different species of phytoplankton. It is mandatory therefore that such information be available before attempting to control the growth of a species utilizing nutrients supplied by municipal or industrial sewage. It is also evident that the major chemical species contained in the sewage must be known before control of species composition and rate of phytoplankton growth can be successful.

Organic growth factors such as vitamins are required by marine algae. The exact requirements vary widely from species to species (Provasoli, 1963). Information on the exact amount of the various organic compounds found in sewage that are required for algal growth is lacking. However, it is generally assumed that sewage is abundant in the organic compounds (vitamins, etc.) and in the trace metals required by marine phytoplankton. We wish here to reiterate that *the gross chemical composition of sewage intended for use in controlling phytoplankton growth must be known before any success in regulating primary production will be realized.*

Losses of Phytoplankton from the Euphotic Zone

The loss of phytoplankton from the euphotic zone is controlled by zooplankton grazing, sinking rate of phytoplankton, and vertical mixing. As stated, the model of nutrient limitation is intended for application under conditions of steady-state or transient-nutrient limitation. At steady state, the growth rate of the phytoplankton is determined by the sum of the loss rates due to sinking, grazing, and mixing. The sinking rate of the phytoplankton appears to be a function of species, age, and physiological state. Different phytoplankton-zooplankton communities will probably stabilize in different relative proportions to give varying values for V_g (Fig. 23.1).

Mineralization

Mineralization is defined here as the release of soluble organic or inorganic plant nutrients by or from organisms or their decaying remains. The important pathways in the regeneration of nitrogen (ρ_{62}, ρ_{36}, ρ_{61}) are depicted in Fig. 23.1. The assumption that bacteria are the major agents of nutrient regeneration in the marine environment is currently being questioned. Recently a variety of other possible mechanisms for nutrient regeneration have been examined, and it has been demonstrated that in certain aquatic ecosystems a considerable quantity of nitrogen and phosphorus utilized by marine plants and animals is regenerated by processes other than direct bacterial action (Johannes, 1968). In some instances direct release of nutrients by zooplankton appears to be the major mechanism of nutrient regeneration while in other regions, such as the tropical ocean, the bacteria appear to be of prime importance (Dugdale and Goering, 1967). Therefore, it is necessary to clearly understand the controlling mechanisms involved in nutrient regeneration in the individual system under study before nutrient-regeneration terms caused by zooplankton grazing and putrifactive bacteria can be included in the nutrient-limitation model.

Real Models

Solutions for the model described in this chapter have not been included. Rather, we have described an approach which may be fruitful and for which some of the basic biochemical information currently is becoming available. Clearly a large amount of information must be supplied by physical oceanographers if a useful model is to be constructed.

Both the number of terms required to describe interactions between the forms of the limiting nutrient, and the requirement for stochastic terms to describe the changes in rate of supply of effluent, changes in relative concentration of nutrients, and changes in water circulation due to meteorological conditions, suggest that simulation using high-speed digital computers is the most practical route to meaningful solutions. An example of such a computer simulation model of phytoplankton growth near a marine sewage outfall is given by Dugdale and Whitledge (1970). Such models can be constructed at almost any level of complexity, and new terms may be added easily as more information is acquired. Simple simulation models such as described here are useful in planning research programs, in testing the results of such programs, and in teaching. For example, several versions are in use by graduate students at the

University of Washington using shore-based and shipboard IBM 1130 computers equipped with plotters to give rapid visual output of their results.

We have already discussed the advantages which may result from processing sewage in the euphotic zone of the marine environment in a manner analogous to that of freshwater oxidation ponds. Whether this goal can be achieved depends to a large extent upon having sufficient knowledge to avoid the accidental production of harmful dinoflagellate blooms. We do not have this information available as yet. However, there seems sufficient cause for optimism to expect that such information will be available within the next few years as a result of the kinetic studies of algal growth in progress in several laboratories. In the meantime, field experiments suggested by an examination of nutrient flow models must be approached very cautiously and in a way so that damage to shellfish populations and the human populations which consume them is precluded. An interim approach might be found in artificial ponds in which seawater and sewage effluent are mixed in appropriate proportions to obtain continuous algal growth. This method may also provide an accelerated means to an understanding of the causes of dinoflagellate blooms if laboratory studies prove to be too slow in forthcoming or unequal to the task. When this step has been accomplished, the way should be open for a more controlled processing of sewage discharged to the marine environment and perhaps the nutrients now wasted can be put to better use.

References

Baalsrud, K. 1967. Influence of nutrient concentration on primary production, In *Pollution and marine ecology*. T. A. Olsen and F. J. Burgess (eds.), Interscience Publishers, New York, pp. 159–169.

Dugdale, R. C. 1967. Nutrient limitation in the sea: dynamics, identification, and significance. *Limnol. Oceanog.*, **12**:685–695.

Dugdale, R. C., and J. J. Goering. 1967. Uptake of new and regenerated forms of nitrogen in primary productivity. *Limnol. Oceanog.*, **12**:196–206.

Dugdale, R. C., and T. Whitledge. 1970. Computer simulation of phytoplankton growth near a marine sewage outfall. *Rev. Int. Océanogr. Med. Tome*, **XVII**: 201–210.

Eckenfelder, W. W., Jr. 1966. *Industrial water pollution control*, McGraw-Hill, New York, 275 pp.

Emery, K. O., W. L. Orr, and S. C. Rittenberg. 1955. Nutrient budgets in the ocean. In *Essays in the natural science in honor of Captain Allen Hancock*, University of California Press, Los Angeles, Calif., pp. 299–309.

Eppley, R. W., J. N. Rogers, and J. J. McCarthy. 1969. Half-saturation "constants" for uptake of nitrate and ammonium by marine phytoplankton. *Limnol. Oceanog.*, **14**:912–920.

Eppley, R. W., and W. H. Thomas. 1969. Comparison of half-saturation constants for growth and nitrate uptake by marine phytoplankton. *J. Phycol.*, **5**:375–379.

Great Britain Water Pollution Research Board. 1959–64. *Water pollution research.* Her Majesty's Stationery, London.

Hellebust, J. A. 1965. Excretion of some organic compounds by marine phytoplankton. *Limnol. Oceanog.*, **10**:192–206.

Hellebust, J. A. 1970. The uptake and utilization of organic substances by marine phytoplankters. In *Symposium on organic matter in natural waters*, D. W. Hood (ed.), Institute of Marine Science, University of Alaska, Occasional Publication No. **1**:225–256.

Johannes, R. E. 1968. Nutrient regeneration in lakes and oceans. In *Advances in microbiology of the sea*, Vol. 1, M. R. Droop and E. J. F. Wood (eds.), Academic Press, New York, pp. 203–213.

Ketchum, B. H., C. L. Carey, and M. Briggs. 1949. Preliminary studies on the viability and dispersal of coliform bacteria in the sea. In *Limnological aspects of water supply and waste disposal*. Published by Amer. Assoc. Advan. Sci.

Levine, M., H. Minette, and R. H. Tanimoto. 1959. Characteristics and expeditious detection of bacterial indices of pollution of marine bathing beaches. In E. A. Pearson (ed.), *Proc. 1st int. conference on waste disposal in the marine environment*, Pergamon Press, New York.

Mac Isaac, J., and R. C. Dugdale. 1969. The kinetics of nitrate and ammonia uptake by natural populations of marine phytoplankton. *Deep-Sea Res.*, **16**:45–57.

Monod, J. 1942. *Recherches sur la croissance des cultures bacteriennes.* Hermann & Co., Paris, 2nd Ed., 210 pp.

Pomeroy, L. R. 1967. Sewage: a rose by another name. *Science*, **158**:579.

Pretorius, W. A. 1969. Anaerobic digestion III. kinetics of anaerobic fermentation. *Water Res.*, **3**:545–558.

Provasoli, L. 1963. Organic regulation of phytoplankton fertility. In *The sea*, Vol. 2, M. H. Hill (ed.), Interscience Publishers, New York, pp. 165–219.

Riley, G. A. 1963. Theory of food chain relations in the ocean. In *The sea*, Vol. 2, M. H. Hill (ed.), Interscience Publishers, New York, pp. 438–463.

Riley, G. A. 1965. A mathematical model of regional variations in plankton. *Limnol. Oceanog.*, **10** (Suppl.):R202–R215.

Steele, J. H. 1958. Plant production in the North Sea. *Scot. Home Dept., Marine Res.*, **7**:1–36.

Part VI

Implications of Man's Activities on Ocean Research Development

24

International and National Regulation of Pollution from Offshore Oil Production

ROBERT B. KRUEGER, *Partner, Law Firm of Nossaman, Waters, Scott, Krueger, and Riordan; Chairman, California Advisory Commission on Marine and Coastal Resources, Los Angeles, California*

Resources of the Continental Shelf

A very observable phenomenon of the evolution of both national and international law with respect to the sea is that its occurrence has had a direct, and perhaps necessary, correlation with the development of technology permitting a use and the need for that use. Before looking at regulatory aspects of offshore oil production, it would be helpful to look briefly at the need for offshore oil as evidenced by its development and the forseeable limits of this development. There are today approximately 30 nations which have established offshore oil and gas production with aggregate reserves of approximately 85 billion barrels, or 20 per cent of the world's total reserve figures. On a worldwide basis, current offshore production is about 6.5 million barrels per day, or 16 per cent of the world's total (Anonymous, 1970c). The Department of the Interior has estimated that by 1980 approximately 30 per cent of the oil requirements and 40 per cent of gas requirements from this country will come from our offshore (U.S. Department of Interior, 1968).

Looking at the United States alone, its outer continental shelf is at least 850,000 miles^2 (from established state limits to a depth of 200 meters) and may be as large as 1,329,000 miles^2 (between established state limits and a depth of 2500 m). Compared with the area of the uplands contained in the United States and its territories of 3,615,000 miles^2, its outer continental shelf is 23 per cent and 36 per cent as large, depending upon which measurement is used (U.S. Department of Interior, 1969; Nossaman et al., 1968). The presently proven reserves of oil and gas on the outer continental shelf are 4.3 billion barrels of oil and 34.2 trillion cubic feet of gas with prospective reserves of an additional 3 to 19 billion barrels of oil and 27 to 97 trillion cubic feet of gas. Sulphur reserves are believed to be approximately 37 million tons (U.S. Department of Interior, 1969; Nossaman et al., 1968). These figures do not include reserves for state offshore lands which have to date produced something in excess of 700 million barrels of oil and in areas, such as Prudhoe Bay, which have immense reserves (Anonymous, 1970a).

With due regard to the steadily growing demand for petroleum, it is likely that these sources will have to be further developed. If they are not, the resources on some other nation's offshore will have to be developed. The offshore is where a large part of today's and most of tomorrow's oil is located.

International Regulation

The requisite technology for offshore drilling has been developed only since the early 1940's, and to date the law which has consequently evolved has related primarily to resource exploration and exploitation. It is only in very recent years that sufficient offshore development on the continental shelf has taken place for the negative aspects of it, such as oil spillage and spatial interference with other uses, to become material and apparent. Thus in the Truman Proclamation (1945), in which the "natural resources of the subsoil and sea bed of the continental shelf . . . contiguous to the coasts of the United States" were claimed as being "subject to its jurisdiction and control," the principal purpose was to facilitate the development of petroleum resources. The Proclamation contained the premises that there was then a "long range world-wide need for new sources of petroleum and other minerals," that "such resources underlie many parts of the continental shelf off the coasts of the United States . . . and . . . with modern technological progress their utilization is already practicable or will become so at an early date," and that "these resources frequently form a seaward extension of a pool or deposit lying within the territory." It is noteworthy, however, that the Proclamation also stated that "self-protection compels the coastal nation to keep close watch over activities off its shores which are of the nature necessary for utilization of these resources." The Truman Proclamation is of continuing significance because the 1969 decision of the International Court of Justice (1969) in the North Sea Continental Shelf Cases held that the Proclamation continued to be the well-spring of the doctrine of the continental shelf which exists independently of the Convention on the Continental Shelf.* The doctrine enunciated in that decision was one of "sovereign rights for the purpose of exploring the sea bed and exploiting its natural resources."

The 1958 Convention on the Continental Shelf also had as its primary purpose the development of natural resources of the seabed and subsoil of the continental shelf. The basic rights of the coastal state in the shelf are stated as being "exclusive [and] sovereign . . . for the purpose of exploring it and exploiting its natural resources". The definition of the continental shelf also evidences this purpose in stating that it is to a depth of 200 m or beyond "to where the depth of the superjacent waters admits of the exploitation of the natural resources" (Convention on the Continental Shelf, 1958). The Convention does recognize that the waters above the shelf remain high seas, and states:

> The exploration of the continental shelf and the exploitation of its natural resources must not result in any unjustifiable interference with navigation, fishing or the conservation of the living resources of the sea, nor result in any interference with fundamental oceanographic or other scientific research carried out with the intention of open publication.

The Convention also requires that continental shelf installations and devices necessary for the development of natural resources and safety zones therefore shall not be established where the same would interfere with

---

* The Court based its decision on the Truman Proclamation, stating: "The Truman Proclamation [is] the starting point of the positive law on the subject, and the chief doctrine it enunciated, namely that of the coastal State as having an original, natural, and exclusive (in short a vested) right to the continental shelf off its shores, came to prevail over all others, being now reflected in Article 2 of the Geneva Convention on the Continental Shelf (1958)."

"recognized sea lanes essential to international navigation" (Convention on the Continental Shelf, 1958) and that within the safety zones the coastal state shall take "all appropriate measures for the protection of the living resources of the sea from harmful agents."

The companion 1958 Geneva Convention on the High Seas (1958) also provides that: "Every State shall draw up regulations to prevent pollution of the seas by the discharge of oil from ships or pipelines or resulting from the exploitation and exploration of the seabed and its subsoil."

Lastly, the 1958 Geneva Convention on Fishing and Conservation of the Living Resources of the High Seas (1958) a further companion measure, contains the following provision: "All States have the duty to adopt, or to co-operate with other States in adopting, such measures for their respective nationals as may be necessary for the conservation of the living resources of the high seas (Convention of the High Seas, 1958). The Geneva Convention on the Territorial Sea and the Contiguous Zone (1958), the remaining of the four 1958 conventions regarding the contiguous zone, does not add anything of significance to the foregoing.

Aside from these quite limited provisions there appears to be no international treaty or agreement providing for the regulation of water pollution resulting from offshore oil development. Further, except insofar as these provisions may constitute or evidence customary international law, there is no customary international law on the subject.* The International Convention for the Prevention of Pollution of the Sea by Oil (1954)† governs only the discharge of oil and oily substances by ships,

---

* The North Sea Continental Shelf Cases stated that Articles 1 to 3, which define the continental shelf and the rights of the coastal state therein, "were . . . regarded as reflecting, or as crystallizing, received or at least emergent rules of customary international law" (International Court of Justice, 1969).
† The United States became a party and implemented the same by the adoption of the Oil Pollution Act of 1961. In 1967 the General Assembly established an Ad Hoc Committee to Study Peaceful Uses of the Sea-Bed and Ocean Floor Beyond Limits of National Jurisdiction (1967), which was transformed into a permanent 42-member Committee in 1968. Such Committee was given a broad charter with respect to economic, technical, and legal aspects of the seabed beyond limits of national jurisdiction ("the continental shelf"), including the duty to "examine proposed measures of co-operation to be adopted by the international community in order to prevent the marine pollution which may result from the legal aspects of the seabed beyond limits of national jurisdiction (the continental shelf), including the duty to "examine proposed measures of co-operation to be adopted by the international community in order to prevent the marine pollution which may result from the exploration and exploitation of the resources of this area" (Committee on Peaceful Uses of Seabed, 1967). At the same time the General Assembly adopted a resolution encouraging the adoption by States of appropriate safeguards against the dangers of pollution and

and the scope of its injunctions were not broadened by the 1969 International Legal Conference on Marine Pollution Damage convened by the International Maritime Consultive Organizations (1969).

The growth of continental shelf claims being made under the Convention on the Continental Shelf and otherwise and of offshore oil and mining activity has led however, to a greatly increased international interest in the uses of the world's continental shelves and the areas beyond.

The United Nations Sea-Bed Committee, which met in New York in March of 1970 has been considering the issues of the extent of limits of national jurisdiction, the development of the seabed resources, and the preservation of environmental quality. In a 12-point program presented by the United States on March 6, 1970, it was proposed that the Committee adopt as two of its objectives the following:

8. To assure that exploration and exploitation of seabed mineral resources will be carried out in a manner that will protect human life, prevent conflicts between users of the seabed, safeguard other uses of the ocean environment against undue interference, avoid irreparable damage to the environment and its resources, and promote the use of sound conservation practices.

9. To provide terms and procedures governing liability for damage resulting from exploration and exploitation of seabed minerals so that damage will be adequately repaired or compensated.*

---

other hazardous and harmful effects that might arise from the exploration and exploitation of the resources of the seabed and the ocean floor, and the subsoil thereof, beyond the limits of national jurisdiction. Such resolution also directed that a study be made by the Secretary-General with a view to clarifying all aspects of protection of the living and other resources of the adjacent coasts against the consequences of pollution and other hazardous and harmful effects arising from various modalities of such exploration and exploitation.

* The remaining objectives were as follows:

1. To encourage exploration and exploitation of seabed resources.

2. To assure that all interested States will have access, without discrimination, to the seabed for the purpose of exploring and exploiting mineral resources.

3. To encourage scientific research and the dissemination of scientific and technologic information related to seabed resources.

4. To encourage the development of services, such as aids to navigation, maps and charts, weather information, and rescue capability.

5. To provide procedures for the assignment of rights to minerals or groups of minerals in specific areas under terms that protect the integrity of investments in seabed resource development, that encourage economic efficiency in the exploration and exploitation of seabed resources, that prevent a race for claims, and that discourage operators from seeking to hold large areas for purely speculative purposes.

6. To provide for a reasonable return on risk investment.

The same subject was also discussed in 1970 in the broader context of other forms of pollution of the human environment by the Preparatory Committee for the United Nations Conference on Human Environment scheduled for Stockholm in 1972 (U.S. Press Release, 1970).

There have been even more recent and significant developments. In April of 1970 legislation was introduced to the Canadian House of Commons that would establish an Arctic Waters Pollution Control Zone (1970) which would extend "seaward from the nearest Canadian land [above the 60th parallel] a distance of one hundred nautical miles," to the median line between Canada and Greenland (which is less than 100 miles) and so "all waters adjacent [to the 100-mile zone], the natural resources of whose subjacent submarine areas Her Majesty in right of Canada has the right to dispose of or exploit." Within such zone the discharge of any substance that would degrade the quality of the waters "to an extent that is detrimental to their use by man or any animal, fish or plant that is useful to man" is prohibited except as authorized by regulation. Further, within such zone Canada would assert the right to control all shipping, to prescribe standards of vessel construction and operation, and to prohibit free passage if deemed necessary. A rule of strict liability is imposed for any damage resulting from an unauthorized discharge whether by vessel or from natural resource development. A companion measure would establish a 12-mile territorial sea.

The Canadian proposal was challenged by the U.S. Department of State as constituting "unilateral extensions of jurisdictions on the high seas [which] the United States can [not] accept." Instead the United States proposed "international solutions . . . within the United Nations framework looking toward the conclusion of a new international treaty dealing with the limit of the territorial sea, freedom of transit through and over international straits, defining preferential fishing rights for coastal states on the high seas [and] for controlling pollution on the high seas" (U.S. Dept. of State Press Release, March 1970).

---

7. To provide revenue to benefit international community purposes, taking special account of the needs of the developing countries, and to meet the operating expenses of the international body established to administer its provisions.

10. To provide for the stability of rules, and yet for the flexibility to introduce modifications over time responsive to new knowledge and new developments.

11. To provide effective procedures for the settlement of disputes.

12. In the overall, to establish an international regime so plainly viable that States will in fact ratify the treaties establishing it. The Soviet Union presented a very similar position and set of objectives to the Seabeds Committee (U.N. Press Release, 1970).

President Nixon's Ocean Proposal

The reason for the strong, perhaps overly strong, reaction on the part of the United States became clear when on May 23, 1970, President Nixon announced a proposed new United States oceans policy which we here examine in some detail in view of its very significant potential effect on offshore resource development and coastal zone management, and consequently environmental protection. The Nixon proposal stated in part:

> The issue arises now—and with urgency—because nations have grown increasingly conscious of the wealth to be exploited from the seabeds and throughout the waters above, and because they are also becoming apprehensive about the ecological hazards of unregulated use of the oceans and seabeds. The stark fact is that the law of the sea is inadequate to meet the needs of modern technology and the concerns of the international community. If it is not modernized multilaterally, unilateral action and international conflict are inevitable.
>
> Therefore, I am today proposing that all nations adopt as soon as possible a treaty under which they would renounce all national claims over the natural resources of the seabed beyond the point where the high seas reach a depth of 200 meters (218.8 yards), and would agree to regard these resources as the common heritage of mankind.
>
> The treaty should establish an international regime for the exploitation of seabed resources beyond this limit. The regime should provide for the collection of substantial mineral royalties to be used for international community purposes, particularly economic assistance to developing countries. It should also establish general rules to prevent unreasonable interference with other uses of the ocean, to protect the ocean from pollution, to assure the integrity of the investment necessary for such exploitation and to provide for peaceful and compulsory settlement of disputes.
>
> I propose two types of machinery for authorizing exploitation of seabed resources beyond a depth of 200 meters.
>
> First, I propose that coastal nations act as trustees for the international community in an international trusteeship zone consisting of the continental margins beyond a depth of 200 meters off their coasts. In return each coastal state would receive a share of the international revenues from the zone in which it acts as trustee and could impose additional taxes if these were deemed desirable.
>
> As a second step, agreed international machinery would authorize and regulate exploration and use of seabed resources beyond the continental margins.
>
> The United States will introduce specific proposals at the next meeting of the United Nations Seabeds Committee to carry out these objectives (Weekly Comp. Presidential Doc., 1970).

President Nixon also stated that the proposed treaty would provide for a 12-mile limit for territorial seas and for free passage through international straits. The urgency with which the proposal is regarded by the

Nixon Administration is evidenced by the fact that on the following working day it was transmitted to the U.N. Seabeds Committee by U.S. Ambassador Phillips with the invitation to Committee members to discuss the same in preparation for the August 1970 meeting of the Committee (Press Release U.S.U.N., 1970).

The text proposal is essentially a liberalized version of that proposed by the Marine Sciences Commission in its 1969 report which would have confirmed unto coastal states exclusive jurisdiction to a depth of 200 m, or 50 miles from coastlines, whichever is further, and established an "intermediate zone" beyond to the 2500-meter isobath or 100 miles from coastlines, whichever is further (Commission on Marine Sciences, 1969). Within this zone coastal states would administer the resource, but proceeds from it would be paid to an international fund to be used for the benefit of the poor and undeveloped nations of the world.

In many respects the Nixon proposal is a clever one that could operate as the *modus vivendi* for achieving consensus on the troublesome jurisdictional issues over which there has been so much open disagreement the last few years between the developed and developing countries of the world. This factor was vividly illustrated by the fact that in December of 1969 the U.N. General Assembly, following extensive and heated debates in the United Nations Seabeds Committee and the U.N. First Committee, adopted a very important resolution over the active opposition of the United States and the U.S.S.R. and their usual supporting blocs. By a 65 to 12 vote with 30 abstentions, the General Assembly passed a resolution requesting the Secretary-General to determine "the desirability of convening at an early date a conference on the law of the sea to review the regimes of the high seas, the continental shelf, the territorial sea and contiguous zone, fishing and conservation of the living resources of the high seas, particularly in order to arrive at a clear, precise, and internationally accepted definition of the area of the seabed and ocean floor which lies beyond the national jurisdiction, in the light of the international regime to be established for that area" (Committee to Study Peaceful Uses, 1969).*

By a vote of 62 to 28 with 28 abstentions, again with active opposition of the U.S.S.R. the United States and their blocs, the General Assembly also passed a resolution providing that nations "are bound to refrain from all activities of exploitation of the resources of the area of the seabed and

---

* The original version of the resolution was introduced by Malta and called for the Secretary-General to determine the views of member states on the desirability of a conference "for the purpose of arriving at a clear, precise and internationally acceptable definition" of the area beyond limits of national jurisdiction (the "continental shelf") and the "prospective establishment of an equitable international regime" for such area (U.N. Doc., 1969).

ocean floor, and the subsoil thereof, beyond the limits of national jurisdiction" (Committee to Study Peaceful Uses, 1969a).*

The Nixon proposal expounds most of the principles of the 1967 proposal of Malta to the United Nations calling for the drafting of a treaty which would reserve the seabed and ocean floor "beyond limits of present national jurisdiction" as a "common heritage of mankind" and provide for their "economic exploitation . . . with the aim of safeguarding the interests of mankind [and using] the net financial benefits derived [therefrom] to promote the development of poor countries (U.N. Document, A6695, 1967). It accordingly can be expected to enjoy the high degree of popularity accorded the Maltese proposal among the smaller and lesser developed countries,† and by the same token possibly receive the whole-hearted criticism of the petroleum industry which in effect viewed such proposal as a "U.N. sellout."

It is, however, questionable whether the Nixon proposal warrants the criticism of extractive industries viewed from an operation standpoint. The proposal, if adopted by the world community, would bring about a stabilization of titles in presently uncertain situations, establish a means for acquiring concessions in deep sea areas that are presently nonexistent, maintain national control over the exploitation of resources in the "continental margins,"‡ and invest national concessions in areas beyond 200 meters with an international trusteeship character which could materially assist the present expropriation problem. Unless, therefore, international par-

---

* Resolutions of the U.N. General Assembly do not have a formal binding effect upon member states. Articles 10 through 17 of the United Nations' Charter which sets forth powers of the General Assembly provides merely that body may "discuss," "consider," and "recommend." On the other hand, resolutions of the Assembly can contribute substantially to the general body of customary international law (Higgins, 1963). With due regard to the interest that the General Assembly and its Committees have taken in this area, formal action by it could have considerable weight in establishing a rule of international law. As pointed out by U.S. Ambassador Christopher Phillips to the U.N. Seabeds Committee on March 6, 1970, however, Resolution 2467D did not evidence consensus but "sharp controversy and substantial division" (Press Release U.S.U.N., 1970).

† It also found a substantial amount of support in the United States, notably in a resolution proposed by Senator Pell that included its basic principles (S. Rep. 172, 176, 1967; Sen. Comm. on For. Rel. Rep., 1967). The proposal also found support in the Commission to Study the Organization of Peace, the 1967 World Peace Through Law Conference, the Center for the Study of Democratic Institutions and others (H.R. Rep. No. 999, 1967; Eichelberger, 1968; National Petroleum Council, 1969). There are indications however, that the petroleum industry is now prepared to take a less characteristic view of this type of proposal (Anonymous, 1970).

‡ The proposal does not define this term, but the Department of State has informally indicated that it would extend to the base of the continental rise. It would consequently extend slightly further seaward than the "intermediate zone."

ticipation in the proceeds of resource exploitation is *per se* undesirable, it would seem attractive to the extractive industries. Even here there is an interesting aspect to the proposal. While it would divert to developing countries very large revenues which the United States could expect to receive from offshore areas that are proven or highly potential for petroleum resources, this might well be compensated for by reducing direct contributions for economic assistance to these countries which have been quite large in the past.

Irrespective of the cleverness of the Nixon proposal from the standpoint of international politics and its efficiency with respect to mineral development, however, it could result in the mismanagement of or impingement upon nonmineral resources and uses. It should be emphasized that it is essentially another measure directed toward the exploitation of the natural resources of the seabed of the type which has predominated the world's thinking regarding the oceans in the past. Acknowledgment is dutifully paid to "ecological hazards" and the need to establish "general rules to prevent unreasonable interference with other uses of the ocean [and] to protect the ocean from pollution." Functionally, however, it could only serve to provide a further incentive, indeed a catalyst, for the exploitation of extractive resources of the seabed, particularly petroleum, both from the standpoint of the operator and the developing countries for whose benefit the exploitation will inure. In its present form the proposal does not contain any functional brakes necessary to permit the coastal state to slow or prohibit offshore development even on its "continental margins," if it thinks it desirable to do so in order to enhance other uses and resources of the offshore and protect its coastal zone.

The proposal would establish as international trust territory lands lying between mainland California and certain of the Channel Islands which lie more than 24 miles offshore (the sum of the two proposed 12-mile territorial seas combined) and are separated by waters more than 200 m deep. The same situation would also exist in similar areas off Alaska and perhaps in other areas of the United States. It is questionable whether the United States as trustee would be able to resist leasing the proven or highly potential oil properties which lie in these areas particularly where they requested so to do within the United Nations' structure. There recently has been a great deal of pressure to bring about the cessation of all offshore operations in the Santa Barbara Channel and severely curtail them elsewhere in offshore California.* Would not, how-

* On October 29, 1969 Senator Cranston introduced S. 3093 which would suspend all further federal leasing in offshore California provided that state law prohibits the issuance of oil and gas leases in offshore areas adjacent to the outer continental shelf.

ever, oil pollution in Southern California be quite acceptable in Madagascar, Tanzania, and the Maldive Islands Sultanate if it provided income for them? In fact, it is quite likely that pollution from oil development in their own offshore would be very acceptable to these countries, bringing to them as it would greater industrialization and economic growth. It is quite apparent in the international (and even national) discussions on the subject that the current concern over environmental quality is one relevant principally to affluent countries and people that can afford what is essentially a new luxury in an industrial society. Lesser developed countries and people would like to first enjoy the benefits of industrialization and technology before they begin to control its deleterious aspects. They also would like their first chance to pollute.

This coastal zone management problem could be corrected in large part by the United States asserting jurisdiction over such areas as inland waters under either an "historic bay" or a "straight baseline" concept where the international criteria for doing so are met.* Application of a

---

On February 26, 1970, Senator Muskie introduced S. 3516 which would require the Secretary of the Interior to assume operations with respect to all federal leases in the Santa Barbara Channel and terminate permanently all such operations in an orderly and safe fashion. S. 3516 would authorize actions against the United States to recover damages for the termination of such operations. In addition a number of complementary measures have been introduced in the California Legislature. Even if federal legislation on the subject is not adopted, it is highly unlikely however that the Secretary of the Interior will hold any lease sales in Southern California for some time to come.

* The Convention on the Territorial Sea and Contiguous Zone (1958) sets forth the criteria for the straight baseline measurement in Article 4:

1. In localities where the coastline is deeply indented and cut into, or if there is a fringe of islands along the coast in its immediate vicinity, the method of straight baselines joining appropriate points may be employed in drawing the baseline from which the breadth of the territorial sea is measured.

2. The drawing of such baselines must not depart to any appreciable extent from the general direction of the coast, and the sea areas lying within the lines must be sufficiently closely linked to the land domain to be subject to the regime of internal waters.

3. Where the method of straight baselines is applicable under the provisions of paragraph 1, account may be taken in determining particular baselines, of economic interests peculiar to the region concerned, the reality and the importance of which are clearly evidenced by a long usage. . . .

Article 7 of the Convention provides for a maximum 24-mile closing distance across bays treated as inland waters, but provides in paragraph 6 that its provisions "shall not apply to so-called 'historic' bays, or in any case where the straight baseline system provided for in article 4 is applied." As noted in United States vs Louisiana (1969),

straight baseline form of measurement would clearly seem appropriate in the case of Alaska with its deeply indented coastline and economic dependence on offshore fisheries and other resources. Other situations for favorable application exist in Maine, Massachusetts, Louisiana, and possibly Florida. In all of these areas and in California as well, strong cases exist for the claiming of portions of these areas as historic bays (United States vs. California, 1965).

The United States is currently in litigation with all of these states as to the location of their offshore boundaries, and this appears to account for the reluctance of the federal government to make international assertions of jurisdiction over these areas as inland waters.* Under the Submerged Lands Act of 1953 the coastal states were given ownership of all lands lying 3 miles seaward of the line of low tide or inland waters,

---

the provisions of the Convention embodied the principle of the United Kingdom vs Norway (1951) in which it was held that Norway properly could draw its baseline for measuring the territorial sea along the tips of thousands of rock ramparts which ring the mainland yet which do not qualify as bays. The Court held in United States vs California (1965), that the choice to use the straight baseline form of measurement is exclusively the federal government's. In the 1969 Louisiana decision, however, it was made clear that the disclaimer of the federal government with respect to this type of measurement would not be binding if inconsistent with its official international stance (United States vs Louisiana, 1969). A similar position was taken with respect to historic bays. Note that Canada has drawn straight baselines on both its east and west coasts in comparable situations: Maps 401 (Nova Scotia), 402 (Newfoundland), 403 (Labrador), 391 (Vancouver Island), and 392 (Queen Charlotte Sound) (Canadian Hydrographic Service, 1967 and 1969). In United States vs Louisiana (1969), the Court noted that "the straight baseline method was designed for precisely such coasts as the Mississippi River Delta area."

* In 1958 the United States filed suit against Alaska to enjoin state leasing in the Cook Inlet more than 3 miles from shore or from a 24-mile closing line drawn across the Inlet. The State seeks to establish that the entire Inlet is within its jurisdiction as a historic bay. There are many other potentially oil-rich areas, such as Bristol Bay, in which similar title disputes are foreseeable in Alaska. The highly convoluted and unstable Louisiana offshore has been referred to a special master to determine whether various water areas are inland waters on the basis of the application of the principles set forth in the Convention on the Territorial Sea and Contiguous Zone or on historic grounds. United States vs. Louisiana (1969). On April 26, 1968 the State of Maine accepted for filing an "application to record the staking out of a claim" accompanied by the required statutory fee. In response the United States filed complaint against the State of Maine and all other Atlantic Coast states which it stated, "in the exercise of the rights claimed by it, the State [of Maine] has purported to grant exclusive oil and gas exploration and exploitation rights in approximately 3.3 million acres of land submerged in the Atlantic Ocean in the area in controversy" and thereby put in issue all federal-state boundaries on that coast (Krueger, 1968). The issues presently remaining in dispute with respect to California are insignificant.

which were regarded as being a part of the state. If the federal government had made or were to make an assertion that an area were inland waters it would, therefore, be a concession of this state's ownership thereto and the area lying 3 miles beyond.

It may be that this essentially proprietary concern was sufficiently valid in the past to discourage the federal government from making international claims for which it had proper grounds. In light of the Nixon proposal it would no longer seem to be an intelligent deterrent, however. The Nixon proposal is sufficiently liberal in its concessions with respect to the continental shelf without a beneficence from the country's inland waters, particularly one that could come in part from the pocket of the United States coastal states.

The liberal posture toward settlement of international claims and interests contained in the Nixon proposal may very well be in the national interest, but it would seem incumbent upon the federal government to take equally progressive steps toward the settlement of the state-federal boundary disputes and in connection therewith make such international assertions as appear appropriate, even if the result would inure in part to the benefit of the coastal states.

The issue here extends not merely to ownership of offshore resources, but more importantly to jurisdiction over them. If coastal states do not have jurisdiction over areas that are of functional importance to them, they will be without the ability to coordinate and give priorities among all uses and resources that influence their coastal zone. There seems to be within the federal government a strong and growing recognition that coastal states are the proper repositories for coastal zone management responsibilities.* There would not, therefore, appear to be any valid

* The Marine Services Commission recommended the adoption of a coastal management act to establish policy objectives and authorize grants in aid to coastal states to plan and manage coastal waters and adjacent lands (Commission on Marine Science, 1969). Bills have been introduced in Congress (S. 2802, S. 3460, S. 3183, H.R. 14730, and H.R. 14731) to provide for grants to coastal states for designated state authorities to develop long-range plans for their coastal zones. After approval of the plans by a federal agency, the state authorities may also be given up to 50 per cent of the cost of implementing their plans. The coastal zone is described in the bills as being limited to the territorial sea or the seaward boundaries of the states, which would probably not cover areas, such as the Santa Barbara Channel, and which could prevent planning problems (House Comm. on Merch. Marine and Fisheries, 1969). There is strong congressional and executive support for the coastal zone management concept, and it is quite likely that federal legislation will ultimately be successful. The bills dealing with this concept call for the preparation of a comprehensive coastal zone plan by the coastal state on a matching fund basis. If the plan is then approved by the federal government as meeting federal policy objectives in the coastal zone, and the state is

long-range reason to minimize the jurisdictional position of the states in the offshore.

Similarly, the Nixon proposal in its present form does not give adequate recognition to the interests of the national coastal state over "other uses" of the seabed and ocean waters in the international trusteeship territory for such as traditional real property, recreational, and military uses. It would clearly seem that the nexus between the "continental margins" and the coastal nation which justifies coastal management in the case of natural resources of the seabed would also justify management and jurisdiction in the case of these other uses which could equally, perhaps more significantly, affect the condition of the coastal zone.* Already there have been attempts to build islands on the Cortes Bank which lies 120 miles off Southern California in an area over which the United States asserted jurisdiction as outer continental shelf of this country in 1967 under the Outer Continental Shelf Lands Act (Nossaman et al., 1968). It would be unfortunate if the authority of the federal government were left to "implied" or "inherent" powers arising under its sovereignty of the "freedom of the seas" as to which world opinion might change as quickly as it may have done as to the extent of the continental shelf (McDougal and Burke, 1962). In the future, particularly artificial islands and other fixed living habitats are likely to be of great significance and the coastal states' jurisdiction and control over them in the proposed international trusteeship zone should be well established.

It is recognized that the Nixon proposal is of necessity of preliminary and sparse nature at this point in time and that its present purpose is clearly to serve as a vehicle for future discussions which should develop detailed provisions on the many and complex subjects with which it deals. It will be unfortunate, however, if national and international discussions

---

determined to be institutionally organized to implement the plan, annual grants in aid to the coastal state for the cost of implementing the plan are to be authorized (Press Release, Office of the Vice President, 1969).

* Offshore islands and the other man-made coastal installations are proving to be very much in demand, if not necessary, in urbanized coastal areas. There have been a vast variety of such structures and installations proposed for continental shelf areas, including airports, floating cities, and hotels (Anonymous, 1969c; Nossaman et al., 1968). The fact that the Convention on the Continental Shelf does not expressly provide for such uses does not necessarily mean that a coastal state is without the power to make use of its continental shelf for such purposes, although a negative inference in this regard can be drawn from the language of the Convention. A valid case can be made that military installations are a permitted use by reason of the coastal nation's inherent right of self-defense (Franklin, 1961; McDougal and Burke, 1962). Note that nonmineral uses are not authorized under the Outer Continental Shelf Lands Act (1964).

on the proposal fail to develop provisions dealing with the foregoing problem areas. If they do not it is questionable whether it will meet the obvious interest of nations in offshore activities and pollution that affect their coastal zones that is evidenced by the recent Canadian legislation or the stated goals of saving "over two-thirds of the earth's surface from national conflict and rivalry [and protecting] it from pollution."

Quality of Offshore Oil Pollution

Since the Santa Barbara Channel oil spill and several of the other recent offshore oil accidents that have resulted in water pollution, there have been suggestions that offshore oil development itself is generically undesirable and should be severely curtailed and even prohibited. It is questionable whether offshore oil development should be judged except in the context of a particular area, under particular conditions, and at a particular time. Any assessment should determine the overall environmental and resource potential of the area and judge any proposed use in light of all relevant policy objectives. Decisions made on this basis could vary. An oil and gas lease that might be appropriate for offshore Louisiana might be inappropriate for the Santa Barbara Channel.

If offshore oil development is to be judged generically, however, it should be considered in the context of other forms of pollution and impingement upon the environment and in this light it is one of the least significant. Compared to other forms of pollution the oil spill resulting from offshore drilling is one of the most observable and traceable, yet one of the least permanently degrading to the environment. Despite the many scientific inquiries that have taken place to date, no evidence of any permanent adverse effect on living organisms in the Santa Barbara Channel has been found, (Neusal, 1969; Straughan and Abbott, 1969; Glude, 1969; Baldwin, 1969) and lawsuits are pending which would effect proper compensation to damaged property owners and others in the area if liability is established.* Compare in this regard the effect of the discharge of hard pesticides into the ocean which result in irreversible dam-

* On February 30, 1969 the State of California, the County of Santa Barbara, the City of Santa Barbara, and the City of Carpinteria, "on behalf of themselves and all other public entities and agencies of the State of California similarly situated," filed a complaint in the Superior Court of California for the County of Santa Barbara against the Union Oil Company, Mobil Oil Corporation, Gulf Oil Company, Texaco, Inc., and Peter Bawden Drilling, Inc. for injuries allegedly caused by oil drilling by the defendants off the coast of Santa Barbara (State vs. Union Oil Co., 1969). In addition a large number of damage suits have been filed by private parties. A rule of strict liability for oil well blowouts was established in California by Green vs. General Petroleum Corp (1928). It is quite likely that a rule of strict liability would apply to

age to some wildlife by means and through sources which are not readily observable or subject to being brought to account.*

With respect to liability for offshore oil spills, and particularly the Santa Barbara disaster, reference is made to an excellent article by Milton Katz of Harvard on "The Function of Tort Liability in Technology Assessment" (Katz, 1969). Mr. Katz concludes generally that in activities such as offshore oil development there has evolved a general doctrine of "enterprise liability" whether based upon a theory of "strict products liability, on responsibility for abnormally dangerous activities, on the doctrine of *Rylands v. Fletcher*, or on nuisance." He also points out specifically with respect to the Santa Barbara oil spill that there may be strict liability because of a California case holding this rule to be applicable to oil well drilling on the (grounds) that it is extra-hazardous (Green vs. General Petroleum Co., 1928). Lastly, Katz points out that liability may be sufficiently great in offshore oil spills so that offshore oil development may be discouraged unless some sort of governmentally assisted limitation is established.† The author concurs in all of these comments.

It has been suggested that petroleum development on the outer continental shelf of this nation is not necessary if import quotas were removed, particularly if other forms of price support to the domestic industry, such as the depletion allowance, were removed or lessened. From a standpoint of a strict economic analysis there is support for this proposition, although much time could be spent debating the merits and demerits from the standpoint of its effect on our domestic petroleum industry and the need for reserves in this country (Mead, 1970). Without becoming embroiled in these issues, however, it should be observed and emphasized

---

the Union Oil incident either on the basis of this case or for other reasons (Katz, 1969). In addition to the damage actions, there have been several proceedings for injunctive relief. County of Santa Barbara vs. Walter J. Hickel (1969): complaint for mandatory injunction seeking to enjoin the federal government and its oil lessees from engaging in further offshore drilling operations. County of Santa Barbara vs. Robert J. Malley (1969): complaint in the nature of mandamus, seeking to enjoin Malley, the District Engineer of the U.S. Army Corps of Engineers for Southern California, from granting further permits for offshore drilling structures and to require him to set aside certain permits previously granted (Alvin Weingand vs. Walter J. Hickel, 1969). Complaint for injunction (Outer Continental Shelf Act) seeking to enjoin further drilling on the Union Oil Co. A-21 lease as recommended by the Dubridge committee.

* Kennedy and Hessel, 1969. An Open Letter to Governor Ronald Reagan and the People of the State of California and the Summary on DDT. It concerns "hard" pesticides and makes recommendations for control and the ultimate ban of "hard" pesticides.

† The implications of this sort of liability are pointed up by the recent recognition by the Insurance Rating Board against any insurance with respect to offshore oil discharge (Anonymous, 1970).

that even if all of our petroleum needs were to be met from foreign imports and domestic onshore deposits (if there were unrestricted imports, there would be very few onshore reserves that would be competitive), the oil would need to be transported to this country by ships and other vessels which appear comparably as prone, and perhaps more so, to accident and oil spill as offshore production facilities and which in terms of quantity create tremendously greater pollution. The *Torrey Canyon* spillage illustrates this vividly. The wreckage of that ship resulted in 30 million gal of crude oil being released in the English Channel. By way of comparison the Santa Barbara oil spill, as estimated by the President's Panel (1969) involved only 1 to 3 million gal. With due regard to the present state of technology, therefore, offshore oil development would appear to offer substantially less of a pollution threat than foreign crude imports in large tankers.

It is foreseeable today that there will be seabed exploration and development techniques that will permit offshore fields to be developed without any visible or material impingement upon other uses and that pollution control techniques will be developed which will greatly reduce the chances of spillage. These new devices and techniques may be quite expensive, but if their use is made a condition to bidding for offshore leases, it would be properly reflected in the amount of the bids made and paid for by the federal government as a means of minimizing use conflicts. Would it not have been fairer to the federal government, oil lessees, and people of California if a portion of the $600 million bonuses received for Santa Barbara leases had been allocated instead to extraordinary safety measures and special installations to minimize the possibility of spillage and infringement with other uses?

National Regulation

This country has adopted the following broad policy objectives regarding the use of its outer continental shelf:*

1. Efficient resource management effecting the prudent use of resources through their intelligent management by the federal government.†

---

* Taken in the composite they would appear to comprise the "maximum benefit for the general public" objective of the Act creating the Public Land Law Review Commission (1964a). The objectives were identified during the course of the (Nossaman et al., 1968).

† The Truman Proclamation of 1945, the Marine Resources Engineering and Development Act of 1966, and the legislative history of the Outer Continental Shelf Lands Act of 1953 all evidence the objective of best effecting the prudent use of resources through their intelligent management by the federal government (Proclamation No. 2667, 1943–1948; Nossaman et al., 1969).

2. Encouragement of private participation—permitting qualified responsible representatives of the private sector to participate in the development of outer continental resources.*

3. The maximization of revenue to the federal government—effecting the greatest direct financial return to the resource owner.†

4. Encouragement of multiple use of resources—coordinating management of the various resources and uses of the continental shelf to minimize conflicts.‡

5. The advancement of knowledge and the development of technology—learning more about the offshore and its resources and achieving the technological capability to safely permit scientific exploration and resource development.§

* The Outer Continental Shelf Lands Act (1964) and the regulations promulgated pursuant thereto clearly contemplate that the development of minerals in the outer continental shelf be undertaken by qualified, responsible representatives of the private sector. The Marine Resources and Engineering Development Act (1966) also recognizes the desirability of "[t]he encouragement of private investment enterprise in exploration, technological development, marine commerce, and economic utilization of the resources of the marine environment."

† While there are indications in the legislative history of the Outer Continental Shelf Lands Act that the generation of revenue was a secondary consideration, its subsequent administration, particularly in recent years, clearly indicates that a basic policy objective has been to maximize revenue to the federal government from the sale of mineral leases. This was particularly manifest in the 1968 Santa Barbara lease sale, from which the federal government realized a profit of $600 million in bonus income at a time when it was sorely needed by the Johnson Administration (Nossaman et al., 1968.)

‡ This objective has been repeatedly acknowledged as necessary by many branches of the federal government. The Act for the classification of public lands which was passed contemporaneously with the law creating the Public Land Law Review Commission (1964b) defined "multiple use" as follows: "[T]he management of the various surface and subsurface resources so that they are utilized in the combination that will best meet the present and future needs of the American people; the most judicious use of the land for some or all of these resources or related services over areas large enough to provide sufficient latitude for periodic adjustments in use to conform to changing needs and conditions; the use of some land for less than all of the resources; and harmonious and coordinated management of the various resources, each with the other, without impairment of the productivity of the land, with consideration being given to the relative values of the various resources, and not necessarily the combination of uses that will give the greatest dollar return or the greatest unit output."

§ This policy is implicit in the Truman Proclamation and the administration of the Outer Continental Shelf Lands Act. It is made explicit by the Marine Resources and Engineering Development Act (1966) and is also evidenced in the Convention on the Continental Shelf. The Marine Resources and Engineering Development Act provides:

(b) The marine science activities of the United States should be conducted so as to contribute to the following objectives:

6. The protection of environmental quality—preserving and in some cases restoring the natural condition of the environment.*

Outer Continental Shelf Lands Act

The provisions of the Outer Continental Shelf Lands Act (1964) will be briefly examined in light of these objectives, more particularly in light of the last, because pollution, of course, involves consideration of environmental quality. The Outer Continental Shelf Lands Act is the federal vehicle for the leasing of all offshore lands designated "outer Continental Shelf" beyond the 3 mile territorial sea "of which the subsoil and seabed appertain to the United States and are subject to its jurisdiction and control."

Congress in passing the Act was aware that "continental shelf" as used in its geologic sense (S. Rep. No. 411, 1953; H. R. Rep. No. 413, 1953; Stone, 1968; Christopher, 1954) extended only to lands lying interior of the geologic slope normally lying above a depth of 200 m but the Act was not restricted to those lands. It is clear, therefore, that the Act applies to all lands properly claimed as continental shelf, continental slope or otherwise. For this reason the Act itself does not constitute an assertion of jurisdiction by the United States as to any particular offshore area and is best viewed as a legislative implementation of the 1945 Truman Proclamation (S. Rep. No. 133, 1953).

The extent of the lands over which the United States has or could assert jurisdiction to the offshore could be considerable and go far beyond the geologic continental shelf or the 200 set forth in the Nixon proposal. The doctrine of continental shelf which resulted from the Truman Proclamation arguably extends to the base of the continental slope, or approximately 2500 m.† Further, under the Convention on the Continental Shelf the coastal state may claim "to a depth of 200 meters or, beyond that limit, to where the depth of the superjacent waters admits

---

(1) . . . .
(2) The expansion of human knowledge of the marine environment.
(3) The encouragement of private investment enterprise in exploration, technological development. . . .

* As discussed infra, this policy objective is evidenced in a number of recent federal acts and in the administration of the Outer Continental Shelf Lands Act (Nossaman et al., 1968; §§4.79–484.).
† In the North Sea Continental Shelf Cases [1969] it is said that the "continental shelf is, by definition, an area physically extending the territory of most coastal states into a species of platform."

of the exploitation of the neutral resources." Commencing in 1961, the Department of the Interior has issued leases under the Outer Continental Shelf Lands Act covering areas 40 miles offshore and in waters as deep as 4000 ft and 26 miles offshore and in waters from 1200 to 1800 ft in depth.* In addition the Secretaries of the Interior and the Army asserted jurisdiction under the Act over an area approximately 120 miles off Southern California which is separated from the coastline by waters as deep as 6000 ft.† Lastly, the Secretary of the Interior has issued exploratory permits under the Act to conduct core drilling in the Gulf of Mexico in waters as deep as 3500 ft and on the Atlantic seaboard for waters as deep as 5000 ft and lying as far as 250 to 300 miles from the coast (Krueger, 1967). All of these acts are capable of being construed as an assertion of jurisdiction by the United States over the areas in question, notwithstanding the fact that exploration is not necessarily equated with exploitation for purposes of the Conventions's definition.‡ The Nixon proposal, then, clearly is a change of approach from the aggressive attitude with respect to offshore claims that the United States and many of the countries of the world have had in the past.§

---

* A 1961 lease of phosphate deposits lying approximately 40 miles seaward of Southern California was approved by the Solicitor as being authorized under the Outer Continental Shelf Lands Act on the grounds that it was applicable thereto "since the United States [by its ratification of the Convention] has now asserted rights to the seabed and subsoil as far seaward as exploitation is possible."
† The Secretary of the Army, who was given authority under section 4(f) of the Outer Continental Shelf Lands Act to "prevent obstruction to navigation [as to] artificial islands and fixed structures located on the outer Continental Shelf," formally advised the proposed island builders that their work could not be undertaken without the consent of the United States. Such action was taken pursuant to a letter from the Solicitor of the Department of the Interior to the Los Angeles, California, District Engineer, dated February 1, 1967, which stated in part: "It is our opinion that the Cortes Bank area is within [the Convention on the Continental Shelf] definition of 'continental shelf'."
‡ The permits involved appear to have been issued under section 11 of the Outer Continental Shelf Lands Act which applies to "geological and geophysical explorations in the outer Continental Shelf." There is no other statutory authority for the issuance of this type of permit. With due regard to the fact that section 2(a) of the Act defines the "outer Continental Shelf" as lands "subject to the [United States'] jurisdiction and control," there is, therefore, good reason to conclude that the permits constitute an assertion of jurisdiction over the areas of continental shelf by the branch of the federal government that has been entrusted with apposite administrative responsibility (Memorandum Opinion, 1961; Barry, 1968).
§ As of 1969, approximately 98 coastal nations had asserted general jurisdiction over offshore minerals; of that number, at least 37 appear to have done so in areas which appear to be in water deeper than 200 m (Ely, 1969).

One aspect of the Act that is quite noticeable today is that it contains no provisions indicating any real concern for or even awareness of values or uses other than mineral ones and that it provides no procedure for the weighing of and determination of priorities among such values and uses. This is perhaps understandable in light of the history of the Act. There is abundant evidence that the primary, perhaps controlling, purpose of the Act was to authorize and encourage the development of the vast reserves and highly potential prospects of oil and gas on the outer continental shelf. The Act itself states that the leasing power of the Secretary of Interior was authorized "[i]n order to meet the urgent need for further exploration and development of the oil and gas deposits of the submerged lands of the outer Continental Shelf."

The absence of a procedure to determine and resolve conflicts of multiple use on the outer continental shelf has, however, become less understandable with the passage of time and the obvious growth and importance of nonmineral uses and values. Today the public reaction to the Santa Barbara oil spill and the growing number of other incidents of offshore pollution would seem to render continued legislative inaction impracticable. Even if practicable, however, it is questionable whether it is excusable today in light of the clear identification of problems in this area. The State of California, in dealing with its offshore, has for many years recognized that special conditions may be present in particular areas which requires special leasing treatment.† It has further recognized that the public interest in offshore development is sufficiently great to warrant mandatory public hearings prior to the determination to lease and as a guide to the prescription of special terms.* The Outer

---

* The Cunningham-Shell Tidelands Act (1955), authorized the State Lands Commission, in offering tide and submerged lands for oil and gas leasing, to prohibit "a particular method of exploration, development or operation" if such method would result in "interfering with or impairing developed shoreline, recreational or residential areas" (Cal. Pub. Res. Code, 1956). The Commission was also instructed to require slant drilling from upland sites in designated parts of the state (Cal. Pub. Res. Code, 1956a).

† The Cunningham-Shell Tidelands Act (1955) required the Commission to publish notice of any proposed offering of tide and submerged lands for oil and gas lease, in which case any "affected city or county" could require a public hearing to be held with respect thereto (Cal. Pub. Res. Code, 1956a). That section further provides: "The Commission in determining whether the issuance of such lease or leases would result in such impairment or interference with the developed riverbank or shoreline, recreational or residential areas adjacent to the proposed leased acreage or in determining such rules and regulations as shall be necessary in connection therewith shall at said hearing receive evidence upon and consider whether such proposed lease or leases would (a) be detrimental to the health, safety, comfort, convenience, or welfare

Continental Shelf Lands Act permits the Secretary of the Interior to prescribe regulations determined "to be necessary and proper in order to provide for the prevention of waste and conservation of the natural resources of the outer Continental Shelf." The regulations imposed have always called for approval of the drilling plans for wells and for operations to be conducted in a way to minimize pollution. The ill-fated Union A-21 well which caused the Santa Barbara oil spill was, in fact, drilled under a "deviation" from these regulations which caused or substantially contributed to the problem encountered (Neushal, 1969). Following the oil spill the Secretary prescribed even more stringent conditions with respect to drilling and operating practices (Fed. Reg. 13544, 1969) and more significantly required the Bureau of Land Management to determine the effect of any proposed leasing upon the "total environment" and to develop such special leasing conditions as are "necessary to protect the environment and all other resources" (Fed. Regulations, 13549, 1969). In connection therewith the Director is permitted, but not required, to hold public hearings and to "consult with State agencies, organizations, industries, and individuals." When the new regulations were under consideration the petroleum industry was given the opportunity to comment upon them and did so. Its representatives uniformly opposed the concept of public hearings as being an invitation to delay. Many also, objected to the section dealing with the evaluation of the proposed leasing upon the environment (Anonymous, 1969a, 1969b).

This is a classic illustration of the gross insensitivity of the petroleum industry to the social temper and attitudes of our times—an insensitivity which has substantially contributed to, if not caused, today's precarious political situation regarding offshore drilling. The industry has traditionally had a close-mouthed, independent, and private attitude towards operations. A company makes its deal, drills its wells where it thinks it should, and sells what it gets. The public is not involved except as a shareholder. In fee-leasing situations and in early offshore leasing this philosophy presented no problems, certainly in dealing with the Depart-

---

of persons residing in, owning real property or working in the neighborhood of such areas; (b) interfere with the developed riverbank or shoreline, residential or recreational areas to an extent that would render such areas unfit for park purposes; (c) destroy, impair, or interfere with the esthetic and scenic value of such recreation, residential or park areas; (d) create any fire hazard or hazards, or smoke, smog or dust nuisance, or pollution of waters surrounding or adjoining said areas." Following the Santa Barbara oil spill the section was amended so as to require a public hearing on the matter in any case "within a city or county adjacent to such area [of proposed offering]" and required the Commission at such hearing to "propose . . . a plan for the control of subsidence and pollution which might occur as a result of the proposed oil and gas operation" (Krueger, 1958).

ment of the Interior, and many of the state leasing agencies whose employees had industry training or associations.

Today, however, even if the petroleum industry is still philosophically the same, few others are. As the public outcry in the Santa Barbara situation indicates, there are today many in this country, possibly even a majority, who feel that they have a legitimate interest in and "rights" with respect to the activities of industry, irrespective of how proper and conventional they may be from a contractual and legal standpoint. Further, and perhaps even more significantly, this philosophy, and the protests made as a part thereof, have found much support in the news media and a ready acceptance on the part of many influential people in government.

Even prior to the Santa Barbara catastrophe, it is doubtful that there were "safe" offshore areas in the country where oil development could be undertaken relatively without concern regarding the public's interest in other resources and values. Since this event, however, it is unlikely that the federal government or any of the coastal states would consider offshore leasing without considering what the effect of probable developments and possible accidents might have on the total environment. Moreover, each leasing agency must consider what steps can be taken to avoid confrontations with the public and resultant political ramifications. From this standpoint of neutralizing opposition and avoiding political criticism the evaluation study and public hearing procedure is extremely desirable, if not a necessity (Presidents Panel on Oil Spills, 1969).

Even when one keeps in mind the oilman's traditional mystique, it is difficult in the extreme to understand why the petroleum industry did not perceive the desirability of this procedure and use the opportunity to endorse it to gain goodwill. By the same token, it is difficult in the extreme to perceive why the petroleum industry opposed the stricter operating conditions and antipollution measures of the new federal regulations when many of the leading companies already followed the stricter standards as a matter of internal policy and it was obvious that there had to be some administrative recognition of the problem evidenced by the Santa Barbara oil spill.* The petroleum industry has shown only

---

* In *Fortune* (Davenport, 1970) oil company officials are quoted as saying: "The great California oil spill has tarnished the industry's reputation and is all the more exasperating to many oil executives because in their opinion it was foreseeable and preventable. They point out that thousands of wells have been sunk off the California and Gulf coasts without mishap. Union Oil, it is felt, stretched the rules of the game by not sinking a well casing deep enough in a notably unstable geologic area; and the fact that it had permission to do so from the federal government doesn't mend matters."

rudimentary recognition of the fact that the public needs to be prepared for and is perhaps more interested in developments off our coasts, than what brand of gas to buy.

Notwithstanding the massive influence which the Santa Barbara disaster has on state and federal offshore leasing and the repeated assertions made by opponents of offshore development that domestic production from our offshore is not needed, the industry has not yet presented its case to the public. It seems clear that the petroleum industry must convince the public and government that it needs to operate in this country's offshore and that it has the capability to avoid undesirable impingements upon other values and uses, if it is to continue to operate there and obtain necessary leases on a scale comparable to that of the past.

National Oil Pollution Legislation

There has been a substantial amount of recent federal legislation relevant to offshore oil pollution. The most recent and comprehensive national measure dealing with oil spillage, including pollution from offshore oil production, is the Water Quality Improvement Act of 1970 (Public Law No. 91-224). The Act repeals the prior basic offshore oil spill legislation, the Oil Pollution Act of 1924.* In its place, the new Act imposes certain obligations on owners and operators of fixed facilities, without question including oil wells (House of Representatives, 91-197, 1969), whether offshore or onshore, upon their discharging oil into the United States' navigable waters or contiguous zone, as defined in the 1958 Convention on the Territorial Sea and the Contiguous Zone. Such owners and operators must now notify federal authorities of the discharge (Stat. 92 §102), be subject to a civil penalty of up to $10,000 for each knowing discharge, and reimburse the United States for costs of removing the oil from the waters and shoreline, unless the discharge was caused solely by acts of God, war or third parties, or by federal government negligence (Stat. 94–95 §102). Similar obligations are imposed upon owners and operators of private vessels (Stat. 92, 94 §102).

The cost of removal for which they are liable are limited to $8 million in the case of oil discharge from a fixed facility, and $14 million or $100 per

---

* The Oil Pollution Act of 1924, before its repeal, prohibited the discharge of oil by any method from any ship within the United States' 3-mile territorial waters, except in cases of "emergency" or "unavoidable" accident. However, the 1924 Act did not apply to pollution from fixed offshore drilling operations. In addition, in 1966, Congress limited its application to grossly negligent or willful discharges.

gross registered ton, whichever is less, in the case of discharge from a vessel; however, liability is unlimited in both cases if willful misconduct or willful negligence is involved. A state or its political subdivisions may impose further liabilities with respect to oil discharges into state waters within the 3-mile territorial sea (Stat. 97 §102). Existing liabilities for property damage to public or private parties from discharges of oil or other hazardous pollutants remain unaffected by the Act (Stat. 97, 99 §102).

Under the new Act, an applicant for a federal permit to conduct any activity which may result in discharges into the United States' navigable waters, such as offshore oil and gas drilling, must provide certification from a state, interstate, or federal water pollution control agency, of reasonable assurance that applicable water quality standards will not be violated by the applicant's activity (Stat. 108 §103). Public notice of all applications for such certification is required and public hearings may be held thereon in the certifying agency's discretion. Existing permits for activities lawfully commenced prior to the Act's enactment requires no certification, but will terminate 3 yr after the Act's enactment if such certification is not provided (Stat. 109 §103).

The new Act also appropriates $35 million for a revolving fund to carry out removal by the United States of oil or other hazardous pollutants from waters and shorelines (Stat. 96, 99 §102). The Act specifies further that the President is to prepare a National Contingency Plan for removing discharged oil (Stat. 93 §102), and issue regulations defining harmful discharges (Stat. 96, 98 §102) and providing for removal of discharged pollutants and prevention of discharges. Federal contracts and grants are authorized for research into water quality control (Stat. 104–07 §102), and the Act calls for a federal study of the financial responsibility and liability limitations of owners and operators of facilities and vessels (Stat. 98 §102).

In addition the Nixon Administration has asked for ratification of the 1969 amendments to the 1954 International Convention for the Prevention of Pollution of the Sea by Oil* and for two new international conventions signed by the United States last year in Brussels.† The Amendment to the 1954 Convention deleted the provisions of that convention estab-

---

* Amendment adopted by the Assembly of the Inter-Governmental Maritime Consultative Organization in London (1969). Amendments will come into force 12 months after ratified by two-thirds of the governments which are parties to the Convention.
† International Convention Relating to the Intervention on the High Seas in Cases of Oil Pollution Casualties, and International Convention on Civil Liability for Oil Pollution Damage (1969).

lishing specified prohibited discharge zones. In its place, a provision was adopted prohibiting all discharge from tankers except where the tanker is more than 50 miles from the nearest land provided the rate of discharge of oil content does not exceed 60 l/mile and the total quantity of oil discharged on a voyage does not exceed 1/15,000 of the total cargo carrying capacity of the ship. The new Conventions would authorize the United States to take preventive action against vessels on the high seas which threatened iminent pollution danger to our coasts and would impose strict civil liability, with certain minor exception, upon the owner of vessels responsible for pollution damage regardless of the location of the vessel. There is, however, a limitation of such liability with respect to any one incident to approximately $125 per gross ton or $14,000,000, whichever is less, but such limitation would not apply when the pollution was due to the fault of the owner. The Administration has also instructed the Secretary of State to seek international agreement for the prescription of standards for the construction and operation of tankers. There is consequently evidence of the Administration's position that it will seek multilateral solutions to the problems dealt with unilaterally by Canada in establishing its Arctic Waters Pollution Control Zone.

Perhaps incipiently the greatest new development in federal regulation was the adoption of the National Environmental Policy Act of 1969 (Public Law No. 91-190). The Act provides in pertinent part:

The Congress authorizes and directs that, to the fullest extent possible: (1) the policies, regulations, and public laws of the United States shall be interpreted and administered in accordance with the policies* set forth in this Act, and (2) all agencies of the Federal Government shall—

(A) utilize a systematic, interdisciplinary approach which will insure the integrated use of the natural and social sciences and the environmental design arts in planning and in decision-making which may have an impact on man's environment;

(B) identify and develop methods and procedures, in consultation with the Council on Environmental Quality established by title II of this Act, which will insure that presently unquantified environmental amenities and values may be given appropriate consideration in decision-making along with economic and technical considerations;

---

* Section 2 defines the purpose of the Act as follows: "To declare a national policy which will encourage productive and enjoyable harmony between man and his environment; to promote efforts which will prevent or eliminate damage to the environment and biosphere and stimulate the health and welfare of man; to enrich the understanding of the ecological systems and natural resources important to the Nation; and to establish a Council on Environmental Quality."

(C) include in every recommendation or report on proposals for legislation and other major Federal actions significantly affecting the quality of the human environment, a detailed statement by the responsible official on—
  (i) the environmental impact of the proposed action,
  (ii) any adverse environmental effects which cannot be avoided should the proposal be implemented,
  (iii) alternatives to the proposed action,
  (iv) the relationship between local short-term uses of man's environment and the maintenance and enhancement of longterm productivity, and
  (v) any irreversible and irretrievable commitments of resources which would be involved in the proposed action should it be implemented.

It is difficult to predict what immediate impact the Council will have in its advisory capacity on the broader issues involved with due regard to its massive review responsibilities as to 102 statements. It was, however, this Council that recommended the cancellation of a large number of Santa Barbara leases and that motivated President Nixon's recent announcement to cancel a substantially lesser number of such leases (Anonymous, 1970). The Council has given its powers a liberal construction (Fed. Reg., 1970) and potentially this Act could be a major influence on many aspects of American life.

Conclusion

The feasibility and political acceptability of offshore oil operations in Southern California and elsewhere will depend almost entirely upon the petroleum industry itself. If it recognizes, and equally important gives evidence that it recognizes, the social impact of all its operations in the offshore upon other uses and values, adopts a plan of operation for offshore drilling that guarantees as far as is feasible the protection of such other values and uses, and establishes effective channels of communication with the rest of society, there is no reason that it cannot overcome, and overcome rather quickly, the present opposition. A very parallel case is presented in the town-lot drilling situation. For many years town-lot drilling was prohibited in most Southern California municipalities and many other parts of the country because operators showed only a capability of drilling and producing wells by conventional means. When industry showed that it had the technology and willingness to drill and produce by inoffensive and inconspicuous means, most of such prohibitions were lifted. If industry were to show that it was willing and able to phase out offshore oil platforms and other visually offensive coastal installations and to develop future leases from subsea stations with the strictest antipollution standards, offshore drilling could become socially acceptable.

The development of offshore oil in this country is presently necessary. It is also necessary that other uses and values be protected and it is doubtful that government of industry has put sufficient emphasis on this objective in the past. The present context is a viable one in which to effect sensible changes in our legal system. As unfortunate as it has been in many respects, the Union oil spill focused national and international attention on the problem of protecting the sometimes delicate ecological balances of the marine environment from the effects of mineral development. It very possibly served as a turning point for changing the thinking in many parts of the country from "how can we use it?" to "how can we protect it?" From the long range standpoint this may prove to be a gain for all of us. Hopefully the interest and proposals which have arisen out of this event will ultimately move us toward a more intelligent use of the world's coasts and oceans and their quite exhaustible resources.

References

Anonymous. 1969a. Oil industry opposes hearings on leases for offshore drilling, *N.Y. Times*, August 3, 1969.

Anonymous. 1969b. Watching Washington—offshore rules could be costly in many ways, *Oil Gas J.*, **67**:85.

Anonymous. 1969c. Urban expansion takes to the water. *Fortune*, September 1969, p. 131.

Anonymous. 1970a. Alaska oil to shake up the industry. *Oil Gas J.*, **68**:99.

Anonymous. 1970b. Nixon was urged to drop oil leases in much larger area. *L.A. Times*, June 13, 1970.

Anonymous. 1970c. Offshore oil enters new era in the 70's. *Offshore Mag.*, pp. 61–63.

Anonymous. 1970d. Nixons ocean trustee plan has some merit. *Oil Gas J.*, **68**:31.

Anonymous. 1970e. Oil firms denied spill insurance. *L.A. Times*, June 16, 1970.

Arctic Waters Pollution Prevention Act Bill C-202. April 1970. Sect. 3(1), (2), U.S. Dept. State telegram 091535z, Unclass. Ottawa 441.

Assembly of the Inter-Governmental Maritime Consultative Organization in London. 1969. I.M.C.O. Doc. AVI/Res. 175, Jan. 16, 1970.

Baldwin, R. 1969. A case history of the Santa Barbara oil spill, Public Land Law Review Commission, pp. 17–23.

Barry, F. J. 1968. The Administration of the Outer Continental Shelf Lands Act, 1967. *Natl. Res. Lawyer*, **38**:46.

California Pub. Res. Code. 1956. §6874. IBID. 1956a. §6872; 6873. amended 1969, 6871.2 and 6872.1.

Christopher, W. M. 1953. The Outer Continental Shelf Lands Act; key to a new frontier, *Stan. L. Rev.*, **6**:23–68.

Committee on Merchant Marine and Fisheries. 1969. Subcommittee on Oceanography, Hearings, Coastal Zone Management Conference, H.R. Rep. No. 91-14, 91st Congress., 1st Sess. p. 195 et seq.

Commission on Marine Science, Engineering and Resources. 1969. *Our nation and the sea*, pp. 145–147.

Committee to Study Peaceful Uses of the Sea-Bed and Ocean Floor. 1969.
A/RES/2340 (Sess. XXII, December 28, 1967)
A/RES/2467A (Sess. XXII, December 21, 1968)
A/RES/2574A (Sess. XXIV, December 15, 1969)
A/RES/2474D (Sess. XXIV, December 15, 1969a)
A/AC./138/SC 2/L. 52/Rev. 1, Annex/D, 16(Mar. 20, 1970).

Convention on Fishing on a Conservation of the Living Resources of the High Seas. 1958. [1966] 17(1) *U.S.T*: pp. 138–145, T.I.A.S. No. 5969.

Convention on the Continental Shelf. 1958. *U.S.T.*, **9**:471–8., T.I.A.S. No. 5578.

Convention on the High Seas. 1958. 13(2) *U.S.T.*:**2312**, T.I.A.S. No. 5200.

Convention on the Territorial Sea and the Contiguous Zone. 1958. [1964] 15(2) *U.S.T.*:1606–1614, T.I.A.S. No. 5639.

Cunningham–Shell Tidelands Act. 1955. California Statutes 1955. ca. 1724.

Davenport, J. 1970. Industry starts the big cleanup. *Fortune*, **91**:174.

Eichelberger, C. M. 1968. A case for the administration of marine resources underlying the high seas by the United Nations. *Natl. Res. Lawyer*, **45**.

Ely. 1969. Jurisdiction over Submarine Resources, Statement before Subcommittee on Oceanography of House Committee on Merchant Marine and Fisheries, Am. Bar. Ass'n Release. App. Bar. ASS'N RELEASE. APP.C. August 5.

Fed. Reg. 1970. Council on Environmental Quality, **35**:7390–93.

Fed. Reg. 1969. 13,544–48. Title 30—Mineral Resources, amending C.F.R. pt. 250 (1954), **34**:13,544–48.

Fed. Reg. 1969. 13,549. Title 43—Public Lands: Interior Subtitle—Leasing Acres. C.F.R. §3381.4, **34**:13,548–9.

Franklin, C. M. 1961. *The law of the sea: Some recent developments with particular reference to the United Nations Conference in 1958*. Naval War College Blue Book Series 65–67.

Glude, J. B. 1969. *Observations on the effects of the Santa Barbara oil spill on intertidal species*, Univ. So. Calif., Allan Hancock Foundation, Los Angeles, Calif.

Green vs. General Petroleum Corp. 1928. 205 Cal. 328, 270, p. 952.

Higgins, Rosalyn. 1963. *The development of international law through the political organs of the United Nations*, Oxford University Press, Oxford, England, 402 pp.

House of Commons Debates, April 22, 1970, pp. 6170–6172.

House Rep. No. 999. 1967. House Comm. on For. Aff. Subcom. on Int'l. Organizations and Movements. The United Nations and the issue of Deep Sea Resources, 90th Cong., 1st Sess., 80–113.

House Rep. (Public. Wrks. Comm.) No. 91-127. 1969. Water Quality Improvement Act of 1969. 91st Cong., 1st Sess. 60 pp.

House Rep. Report No. 413. 1953. 83rd Cong., 1st Sess. 2, pp. 6–7.

International Convention for the Prevention of Pollution of the Sea by Oil. 1954, [1961]. 12(3) *U.S.T.*:2989. T.I.A.S. No. 4900.

International Convention on Civil Liability for Oil Pollution Damage, 1970. Brussels, 9. Int. Legal Materials 24, 45.

International Court of Justice. 1969. North Sea Continental Shelf Cases. I.C.J. 3.

International Maritime Consultative Organization. Jan. 10, 1969. Rept. to the Legal Comm. by Legal Comm. Working Group, II-3d Sess.

Juridical Regime of Historic Waters, Including Historic Bays. 1962. 2 Y.B. Int'l. Comm'n. 1/13, U.N. Doc. A/CN. 4/143.

Katz, Milton. 1969. The function of tort in liability in technology assessments. *U. Cin. L. Rev.*, **38**:587, 602, 645 et seq.

Kennedy, Donald, and J. Hessel. 1969. The Biology of Pesticides. Cry California, 4: 2–10.

Krueger, R. B. 1958. State tidelands leasing in California. U.C.L.A., Law Rev., **427**:448.

Krueger, R. B. 1968. The development and administration of the outer continental shelf lands of the United States. *Proc. 14th Ann. Rocky Mt. Min. L. Inst.*, **643**:687–88.

Krueger, R. B. 1967. Mineral development on the continental shelf and beyond. Dep't. Int. U.S.G.S. Release No. 94229–67.

Marine Resources and Engineering Development Act. 1966. 33 U.S.C. §1101(b)(3).

McDougal, M. and W. Burke. 1962. *The public order of the oceans*, Yale Univ. Press, New Haven, pp. 718–720.

Mead, W. J. 1970. The system of government subsidies to the oil industry. *Natl. Res. J.*, **112**:122.

Memorandum Opinion from Assoc. Solicitor. 1961. Dep't. Interior to Director BLM, M36615/94127-61.

National Petroleum Council. 1969. Petroleum resources under the ocean floor.

Neushal, M. 1969. Final Report Dealing with the Early Stages of the Santa Barbara Oil Spill, FWPCA Con. No. 14-12-516, Santa Barbara, Calif.

Nossaman, Waters, Scott, Krueger, and Riordan. 1968. *Study of the Outer Continental Shelf Lands of the United States*, **II**:4-A-9.

Nossaman, Waters, Scott, Krueger, and Riordan. 1969. *Study of the Outer Continental Shelf Lands of the United States*, **I**:342–344.

Oil Pollution Act of 1961. Pub. L. No. 87-167; 75 Stat. 402.

Outer Continental Shelf Lands Act. 1964. 43 U.S.C. **1331–43**:8432–8438.

Outer Continental Shelf Lands Act. 1953. 43 U.S.C. §1337a.

Outer Continental Shelf Lands Act. 1964. 43 C.F.R. §3386.1.

President's Panel on Oil Spills. 1969. Offshore mineral resources—a challenge and an opportunity, 2nd Rept.

President's Panel on Oil Spills. 1970. The oil spill problem.

Press Release Office of Vice President, 1969. October 19.

Press Release U.S.U.N. May 25, 1970. No. 70.

Proclamation No. 2667, 1943–1948. 3 C.F.R., 67.

Public Land Law Review Commission. 1964a. 43 U.S.C. §1391.

Public Land Law Review Commission. 1964b. 43 U.S.C. §1415(b).

Public Law No. 91-190, Title I, Declaration of National Environmental Policy. 83 Stat., p. 852.

Public Law No. 91-224, Title I, Water Quality Improvement. 84 Stat., p. 91.

Santa Barbara vs. Walter J. Hickel. 1969. No. 69-636-AAH. U.S. Dist. Ct. April 4.

Santa Barbara vs. Robert J. Malley, 1969. No. 69-1986-ALS, U.S. Dist. Ct. October 3.

Senate Comm. On Foreign Rel. Rep. 1967. Governing the use of ocean space. 90th Cong., 1st Sess., 1–7.

Senate Rep. No. 133. 1953. 83rd Cong., 1st Sess., 2.

Senate Rep. No. 411. 1953. 83rd Cong., 1st Sess., 2, 4–7, pp. 211–224.

Senate Rep. Nos. 172, 176. 1967. 90th Cong., 1st Sess.

State vs. Union Oil Co. 1969. No. 84594 Santa Barbara County Superior Court. February 20.

Statutes:
    84 Stat. 92 §102, adding FWPCA, 11(b)(5).
    84 Stat. 92 §102, 1964. Amendment to Federal Water Pollution Control Act.
    84 Stat. 92, 94, §102, adding FWPCA, 11(b)(4)–(5), 11(f)(1).
    84 Stat. 93 §102, adding FWPCA, 11(c)(2).
    84 Stat. 94–95, §102, adding FWPCA, 11(f)(2)–(3).
    84 Stat. 96, 98, §102, adding FWPCA, 11(j)(1), 12(a).
    84 Stat. 96, 99, §102, adding FWPCA, 11(k), 12(h).
    84 Stat. 97, §102, adding FWPCA, 11(a)(2).
    84 Stat. 97, 99, §102, adding FWPCA, 11(a)(1), 12(e).

84 Stat. 98, §102, adding FWPCA, 11(p) and 12(c).
84 Stat. 104–07, §102, adding FWPCA, 16–19.
84 Stat. 108, §103, amending FWPCA, 31(b)(1).
84 Stat. 109, §103, amending FWPCA, 21(b)(7).
Stone, O. L. 1968. United States legislation relating to the continental shelf. *Int. Comp. L.* **Q.**, **17**:102.
Straughan, Daliz, and B. C. Abbott. 1969. *The Santa Barbara oil spill: ecological changes and natural oil leaks*, 6, Allan Hancock Found., Los Angeles, Calif.
Truman, H. S. 1945. Title Proclamation No. 2667. (Policy of the United States with respect to the natural resources of the subsoil and seabed of the continental shelf.) 3C.F.R. (1943–1948 Comp.) September 28, pp. 67–68.
United Kingdom vs. Norway. 1969. Fisheries Case. I.C.J. 11, pp. 73 and 77.
U.N. Doc. A/C.L. 473. October 1969. Rev. 2, December 2, 1969.
U.N. Doc. A/6695. August 18, 1967.
U.N. Press Release. March 6, 1970. HE/1/Rev.I SG/SM/1220/HE/2, March 9, 1970.
United States vs. California, 1965, 381 U.S., p. 139.
United States vs. Louisiana, 1969, 394 U.S., pp. 11, 23, 72.
U.S. Dept. of the Interior. 1968. United States petroleum production through 1980, pp. 14–18.
U.S. Dept. of the Interior. 1969. Petroleum and sulphur on the U.S. continental shelf.
U.S. Dept. State Press Release. March 1970.
U.S. Press Release. 1970. No. 27, Rev. 1, p. 6.
Weekly Comp. Presidential Doc. 1970. May 25. pp. 677–678.
Weingard, Alvin, vs. Walter J. Hickel. 1969. No. 68-1317-EC. U.S. Dist. Court, July 10.

25

Toward a Public Policy on the Ocean

E. W. SEABROOK HULL, *Editor, Ocean Science News, 1056 National Press Bldg., Washington, D.C.*

The evolution of a national public policy on the ocean has been difficult and painfully slow in coming—if, indeed, it has been achieved even now. For over a decade the logical need for such a policy has not been questioned. What has been lacking is the emotional impetus. Also political compulsion and feasibility have been elusive. And while the development of a public policy on the ocean has thus been deferred and repeatedly pushed back, the task of doing so has grown more difficult, the circumstances more complex. Such a policy should have been established and pursued with determination in the early 1960s. A nationally accepted policy and major on-going programs are more readily adapted to normally changing times and circumstances than they are initiated in the midst of fundamental change. It is in such a period of fundamental change that the nation and, indeed, the whole human community finds itself today. If the author of this chapter may quote himself:

Until recently there was virtually nothing man could do which would irreparably damage his environment. He was not a maker of cataclysms; only God through the forces of nature could bring about such things.

Until recently there was no limit to the extent to which our terrestrial environment could provide man's needs; no limit to the extent to which it could absorb his wastes. Indeed, as natural life and nutritional cycles go, within broad limits one creature's waste matter is another's food.

Witness the complete reciprocity between animals, on the one hand, and plants, on the other. It is a beautiful balance of energies and processes with built-in compensators and a wide tolerance, therefore, to variance and disruption. Only excesses in the extreme can exceed the capacity of the system to absorb, supply, adjust and offset . . . .

Now, through the sheer (and rising) quantity of his numbers, through the level of his energy requirements, and through the sheer volume of his activities, man is quite capable of destroying his planet's ability to support human life . . . . Similarly, not only is man capable of imposing disaster on his environment through premeditated action, he is also capable of destroying its capacity to support him and his industry solely through inaction—through a lack of positive concern for the impact on his environment of his industry, his communal living, his recreation, his transport and his waste disposal.

This basic circumstance, the rising pace of man's activities and the rising quantity of his numbers, not only threatens the viability of his planet, but presents man, in his communities, with serious problems of considerable economic and political import. The obsolescence of his cities, problems of rapid transit, intercity transport, antipoverty programs, educational costs, SST, and so on, confront him on both national and local levels with increasing demands on his resources.

Simultaneously, he feels a compulsion—sometimes defined as a logical need, but often merely an indefinable, almost instinctive feeling—to attend to matters extending beyond his own national bounds, even beyond his planet, as with the space program. Urban renewal, modern transport, the poverty war—all of these things are increasingly costly in terms of both budgeted dollars and available human and material resources. So are global programs of ocean research and exploration, space research and exploration, a truly *World* Weather Watch, planetwide environmental research and management, and programs to assure a nutritional sufficiency for everyone.

What are emerging are (*a*) two distinct levels of human endeavor and (*b*) a rate of demand on man's wealth and resources which threatens to exceed his capacity to provide. Simply put, there are (*a*) urgent matters of national (local) interest and matters of national (political) survival; and (*b*) there are urgent matters of *mankind-wide* interest, matters of earth-ecology (and, therefore, human) survival. Together they threaten to exceed the capacity of even the largest, most wealthy nations to cope individually. Neither the United States nor the U.S.S.R., for example, alone can adequately tend simultaneously to its domestic needs and to such *mankind-wide* matters as spaceflight, ocean exploration, environmental management, and human nutrition.

It is this evolving requirement for a fundamental change in national outlook that is the latest factor complicating efforts to develop a public policy on the ocean. In terms of logic and supporting documentation, the need and benefits of such a policy and its attendant programs and Federal structures have been enumerated repeatedly. There has been little controversy over the fact that the nation does indeed need such a policy. The problem has been in translating this awareness and "agreement in principle" into definable, discrete action. Among the factors preventing this translation has been the competition for national resources—more specifically the competition for funds. It becomes increasingly apparent that if we as a *nation* and we as *mankind* are going to meet the requirements both of survival and of the continuing improvement of our lot, then we must change the way we think and the way we have been doing things.

Matters of purely national concern will continue to be handled on a national basis, but as these draw more and more heavily on national resources there will be less and less left for application to matters of concern for all mankind. The limits of economic capability, though sometimes difficult for us to define and enumerate, are nevertheless there and unforgivingly rigid in their imposition on us of limits as to how much we can and cannot do. There does not seem to be any choice but that matters of *universal* concern will have to be funded and handled on a *universal* basis. This means that matters of broad, global concern, such as ocean exploration and spaceflight, are going to require a joint cooperative effort among many nations—if that effort is to be of any meaningful magnitude and if it is to enjoy any useful degree of continuity and growth.

Now, with that preface, let us look at how a public policy on the ocean for the United States has evolved to date, what some of the incentives and some of the constraints have been, how they have interacted, and what they portend for the future. For, fundamental change or not, a public policy on the ocean must evolve. Energetic programs of research and exploration must be developed and pursued. An understanding of the ocean and the development of an acceptable and realistic regime for man's activities in the ocean are basic to his future.

Space vs. Ocean: an Invidious Comparison

For a public policy to be developed, adopted, and energetically implemented, there must be a reason—a national need, for example. The technological capability for its achievement must be in hand, the money must be available, and it must be politically feasible. This last means that the policy must have widespread public support and that that support must be translated through the public's political representatives into national action. This means that the particular policy must win out over competitive policies and programs. And if the policy and its objectives are to have continuity and continuing public support, they must be ably administered. This implies, at least, that an effective administrative structure must be established. The ease with which this can be done, then, is a factor in determining with what ease a public policy will evolve.

Unlike the onset of the space age on October 4, 1957, and its culmination in the Apollo lunar landings in the late 1960s and early 1970s, oceanography, marine science, or whatever you care to call it has had none of the political prerequisites for strong Federal policy and financial support, for an organized priority effort, or even for a steady, continuous policy. Let's examine some of these factors.

Impetus

There have been a number of incidents which have underscored the limits of our oceanic knowledge and capabilities, some of them tragic, all of them costly. Within a space of a few years the nuclear attack submarine *Thresher* was lost with all hands in 8400 ft of water, and *Scorpion* was lost in 10,000 ft of water. There was the grounding of the *Torrey Canyon* off the coast of Cornwall and the discharge into the local marine environment of its 119,000 tons of Kuwait crude oil. There was the fracturing of the ocean floor in the Santa Barbara Channel and the concomitant release of crude oil there. An even greater marine disaster, perhaps, though in no way so dramatic, has been the continuing near-catastrophic decline of the United States as a world fishing power—from first to fifth place in a few years. Both dramatic and sinister was the explosive population increase of the stinging, coral-eating starfish crown of thorns, *Acanthaster planci*. Its relentless and devastating expansion throughout the Indian and Pacific Oceans is an example of how little we understand the ocean environment, how readily its delicate balance can be upset, and how ill-prepared man is to cope. These and other matters demonstrate beyond question the nation's growing involvement with the ocean and the pressing need for us to improve our position of knowledge with respect thereto.

But none of these events has produced the sense of urgency and consternation that *Sputnik* generated in the middle 1950s—neither singly, nor in the aggregate. Indeed, *Sputnik* was but one factor facilitating the development of a public policy on space. *Sputnik* was the kicker. Man had sent a machine from his planet. He had created an artificial moon.

Certainly this was something of profound, historic significance. Not only that, but from the United States point of view it had been done first by a nation which we considered unfriendly, the chief threat to our national security, and a nation whose technology we publicly disdained. So, in Washington, where it counted, *Sputnik* had a shock effect. It gave impetus to the development of a space policy.

Ocean No Vacuum

Additionally, space was free ground. Not only was it a vacuum physically—there was no one and nothing out there—but it was a vacuum administratively and in terms of industry as well. There was no Washington bureaucracy that had to be displaced before an organization to administer the nation's space effort could be set up. And there was no space industry. Indeed, an aircraft industry with more capacity than business welcomed the space effort as the answer to a board of directors' prayers.

For decades into the future it was difficult for anyone to imagine much that would be done in space that would not be done by government. In areas other than communications satellites, it was difficult to attach a profit motive to space activities. The military staked its claim in space with Inter-Continental Ballistic Missiles, submarine-launched Polaris-Poseidon missiles, spy-in-the-sky satellites, and eventually, perhaps, orbiting laboratories and weapons pods.

Therefore, the shock of the Soviet "first" made it easy for the United States to decide to do "something" in space. Relatively speaking, it was also easy to establish the National Aeronautics and Space Administration. There were brief squabbles initially between the Army and the Air Force over missile roles and missions, and later between Air Force and NASA over primacy of role and funding—in *manned* space flights mainly. Though noisy at the time these differences were quickly and conclusively resolved.

But the ocean was not then, and is not now, "free ground" devoid of vested interest, either public or private. One way or another, and to a greater or lesser extent men have made their way by the sea for thousands of years. The sea was never an abstract "thing out there," like space, which most of us can see but never touch or enter into. The sea is a part of man's natural environment. He takes food from it. He recreates on and in it. He explores it. He seeks to build his homes and his cities by it. He fights on it and within it. He dumps his wastes into it. More recently he has begun to farm it and to occupy it.

Some of the nation's earliest government agencies concerned themselves with the sea; and, though perhaps with newer names, many survive to this date. The Revenue Service established in 1790 is what we now know as the U.S. Coast Guard. The Navy was authorized into being in 1798, as was the Marine Hospital Service (now the U.S. Public Health Service). The year 1807 saw the establishment of the Coast Survey; we now call it the U.S. Coast and Geodetic Survey. In 1824 it was the U.S. Army Corps of Engineers. In 1830 the Depot of Charts and Instruments was founded; it is now U.S. Navy Oceanographic Office. And, in 1871 there was established the U.S. Fish Commission, later to become the Fish and Wildlife Service encompassing the U.S. Bureau of Commercial Fisheries and the U.S. Bureau of Sports Fisheries and Wildlife.

Ever since World War II there has been a large, established bureaucracy in Washington which considers the ocean its own special realm. While some of these agencies (see Table 25.1), such as the U.S. Atomic Energy Commission, are comparatively new, many have been around for a long time. All are manned by people, to many of whom change is anathema and security resides in the status quo. All have historic interests in specific areas of endeavor, and are often parochial, therefore, in their outlook. And, concomitantly, each agency had its system of suppliers, its industry, as well as those segments of the economy it was created to serve.

As a result, when a decade or two ago man decided that he should know a lot more about the ocean than he did, when he felt that he should have a public policy and a central administration for implementing it, he did not enjoy the luxury of an administrative vacuum in which to insert a new and energetic ocean administrative structure. Instead, he came face to face with an *establishment* of long-standing vested interests and powerful, suspicious political alliances. It was not as easy as space.

Behind in Space: Ahead in the Sea

Additionally, the launching of *Sputnik* (and scarcely a month later *Mutnik* with a dog in orbit) put the United States second in the space race from the very beginning. This made it essential to the restoration of both our international image and our national self-confidence to publicly establish a goal in space which (*a*) would be far enough in the future to enable us to make up for lost time, (*b*) would be of such significance that its accomplishment would have a profound and dramatic impact on both world and United States opinion, and (*c*) was a project of such magnitude and complexity that the U.S.S.R. could not readily, at least, match it.

Table 25.1 Federal Agencies with Major Marine Responsibilities

```
                              President
                                 |
                                 | (dashed)
                                 |
              National Council on Marine Resources and
                     Engineering Development
```

- President
  - Defense
    - Navy
    - Army Corps of Engineers
  - Commerce
    - Environmental Science Services Administration
    - Maritime Administration
  - Interior
    - Geological Survey
    - Federal Water Quality Administration
    - Bureau of Commercial Fisheries
    - Bureau of Sport Fisheries & Wildlife
    - Bureau of Mines
    - Bureau of Land Management
    - Bureau of Outdoor Recreation
    - National Park Service
    - Office of Saline Water
  - National Science Foundation
  - Health Education and Welfare
    - Public Health Service
    - Office of Education
  - Atomic Energy Commission
  - Transportation
    - U.S. Coast Guard
  - Smithsonian Institute
  - State
    - Agency for International Development
  - National Aeronautics and Space Administration

It was easy, therefore, for Congress and the White House to establish the goal of landing American men on the moon before the end of the decade—in 1958. NASA was organized and funded. The U.S.S.R. beat the United States with a man into orbit. Rather than discouraging us it gave new urgency to our efforts. There was never any real doubt that then-present and succeeding Administrations and Congresses would see America's space effort through to that first big goal. And so it was.

With the ocean, however, things were different. There was no question but that in the 1950s and 1960s America's position was a good solid first in the ocean—in terms of both its oceanographic research and its military posture. The U.S.S.R. and some other nations were putting out a greater effort in the area of high-seas fisheries, but otherwise the United States enjoyed primacy. A series of lesser events does not substitute in shock value for a single event like *Sputnik*. And, it is difficult to instill a sense of urgency to "catch up" when, for the most part, one is already ahead.

Continuity

Additionally, the space program, once underway, had built-in momentum. This gave it continuity. The program itself possessed major milestones of dramatic impact in themselves—man in orbit, Mars probes, Venus probes, unmanned lunar landers, and so on. And almost continuously there were above, for all to see, one or more satellites—many of which reflected quite brightly in the early evening sky. If you wished, you could listen to their beeping chatter over shortwave ratio. These, along with a heavily funded and elaborately manned public relations program, plus (increasing with time) pressure of heavy prior capital investment, were sufficient to assure a continuity and degree of public interest, without which the necessary support is not forthcoming.

Few of these ingredients of continuity and widespread public support have been available to the nation's marine sciences effort. There was, for a time, one possibility, but the application of the spoils system of political favor and poor management combined to destroy the project before it ever really got off the ground. This was Project Mohole, a program which would culminate in drilling a hole in the bottom of the ocean through the first and second crustal layers of the earth finally to sample the mantle. It probably never would have had the public impact of *Sputnik* and other space firsts, but scientifically it was both an exciting and a big program. And properly handled, it had a high potential for public appeal.

Phase I of Project Mohole had been well and inexpensively executed, and drilling just over 600 ft into the ocean floor in 12,000 ft of water, it had produced much new geological knowledge, including the first sample of basalt from the earth's second crustal layer. Phase II, however, became more a matter of political patronage and less a matter of good science. Management at both the Government and the contractor level left much to be desired. Costs were afflicted with galloping escalation, and Congress finally lost patience—but not before some tens of millions of dollars had been spent and the name of big ocean science badly sullied.

History of Ocean Policy Development

When man first examined the ocean and made some logical deduction about it is not known. In terms of who first applied the scientific method to the ocean and wrote about it, it can be argued reasonably that Aristotle is the patron founder of a branch of oceanography—more specifically, biological oceanography. His observations with respect to *cetacea*, eels, octopi, and certain other marine animals were remarkably astute and, as far as they went, largely correct.

The modern-day effort, however, to establish a public policy for the United States in the ocean probably really began with the appointment of the third National Academy of Sciences Committee on Oceanography (NASCO) in the late middle 1950s and the issuance in 1959 of its 12-volume report "Oceanography—1960-to-1970" (Anonymous, 1959). It defined oceanographic research, virtually for the first time on such an exhaustive scale, in terms of its meaning to man and his activities—including defense, raw materials resources, energy, fisheries and mariculture, weather and climate forecasting, health and recreation, transport, and—simply—new scientific knowledge. It urged a Federal expenditure rate of $1 billion for the decade ahead.

In that same year the U.S. Senate passed a resolution unanimously approving the NASCO report. The whole Congress passed legislation removing statutory limitations on the oceanographic research activities of the U.S. Coast Guard, U.S. Coast and Geodetic Survey, and U.S. Geological Survey. The President's Scientific Advisory Committee established a Subcommittee on Oceanography. In July 1960 the science advisor to Congress, Edward Wenk, Jr., produced "Ocean Sciences and National Security" which was released by the House Science & Astronautics Committee, Rep. Overton Brooks (D., La.), chairman (Wenk, 1960). This was Congress' analysis and "answer" to the NASCO report. Among other things, Dr. Wenk concluded that NASCO estimates of funding needs were low by a factor of 4.

In 1960, too, the Interagency Committee on Oceanography (ICO) was established as a permanent coordinating group for Federal activities in the ocean. Assistant Secretary of the Navy (R & D), James H. Wakelin, Jr., served as the first ICO chairman, and an energetic scientist from the Office of Naval Research, Robert B. Abel, was named as its very effective Executive Secretary. The membership of the ICO included all departments and agencies with oceanographic interests. ICO's purpose was to optimize the national effort by coordinating planning among the several agencies so as to minimize duplication and waste. It is important to remember that while the ICO was not directly a policy-making body, it contributed indirectly to the policy-making process by providing a formalized and continuing forum for discussion and the exchange of ideas. Also, the ICO had no budget-requesting authority. Individual agencies submitted their respective budget requests separately to Congress. In terms of retaining overall program integrity through the legislative process, this was a less-than-optimum arrangement. Table 25.2 shows ICO membership.

Table 25.2. Membership of the Interagency Committee on Oceanography—ICO; Individual Membership, ex officio, by Title

Navy Department—Chairman—Assistant Secretary of the Navy for Research and Development

Commerce Department—Director, U.S. Coast and Geodetic Survey

Interior Department—Director, Bureau of Commercial Fisheries

Treasury Department—Commandant, U.S. Coast Guard

State Department—Office of International Scientific Affairs, appointed by agency

National Science Foundation—Director's appointee

Health, Education and Welfare Department—Public Health Service appointed by agency

Atomic Energy Commission—appointed by agency

Smithsonian Institution—appointed by agency

Office of Science and Technology (observer)

Bureau of the Budget (observer)

National Academy of Sciences (observer)

Executive Secretary (throughout ICO's life): Robert B. Abel.

In 1961 the U.S. Navy brought out the first of its series of TENOC reports, "Ten-Year Program in Oceanography—1961-to-1970" (Anonymous, 1960), which proposed a decade spending level on oceanographic research of $800 million for the Navy alone. These reports, which were usually classified, detailed Navy ocean research expenditures and related various efforts to military payoff areas.

In 1963, 2 yr later, the Marine Technology Society was chartered. ICO brought out ICO Pamphlet No. 10, "Oceanography—The Ten Years Ahead, A Long Range National Oceanographic Plan—1963-to-1972 (Anonymous, 1963)." It proposed decade expenditures on Federal oceanographic research of $2.8 billion, of which Navy would account for $1.3 billion. Notice the steadily rising estimates of what adequate decade ocean research expenditures should be. Note also that while each successive plan updated, refined, and modified previous proposals, there were still no formal suggestions for major changes in the Federal organizational structure for ocean research management. These were all action plans, not Federal reorganization plans.

In 1964 NASCO—now a permanent organization—issued the booklet *Economic Benefits from Oceanographic Research* (Anonymous, 1964). This was a first effort to apply cost/benefit analysis techniques to oceanographic research—as a possible means of providing economic justification for the effort. Many economists roared in protest at much of the methodology employed in the study. Nevertheless, its message was clear qualitatively, if not precise quantitatively: an investment in ocean research and exploration would pay handsome dividends. Specifically, the NASCO figures were as follows: an investment in oceanographic research of $1389 million over a 10-to-15 yr period would yield benefits worth $6064 million, or a "cost/benefit ratio"($1389 million/$6064 million) of 4.4.

In 1966 the President's Science Advisory Committee Panel on Oceanography (PSACPOO) issued its report "Effective Use of the Sea (Anonymous, 1966a)," once again detailing the things the nation should do in the ocean and why it was all worth while. In the same year NASCO issued "Oceanography 1966: Achievements & Opportunities," which was more of the same (Anonymous, 1966b). It was in this year, too, that Congress passed the Marine Resources and Engineering Development Act, the Sea Grant College and Program Act, and an act authorizing the Bureau of Commercial Fisheries to build and operate pilot plants for the production of Fish Protein Concentrate (FPC).

The passage of the Marine Resources and Engineering Development Act was one end result of several years' efforts by the Congress to get the nation's ocean research program off dead center. It was passed against the wishes of the White House, largely through the efforts of the House Merchant Marine & Fisheries Committee's Subcommittee on Oceanography and the Senate Commerce Committee. It was the first concrete step toward possible reorganization of the Federal ocean structure.

It established the National Council on Marine Resources & Engineering Development, a kind of upgraded, demilitarized ICO with Hubert H. Humphrey as Chairman and Ed Wenk, Jr., as Executive Secretary. Since issuing his report on oceanography as Congress' Science Advisor, Dr. Wenk had established the Science Policy Branch of the Library of Congress. Earlier, as an engineer with Southwest Research Institute, he had worked out the first design parameters for the research/work submersible *Aluminaut*. Thus he brought both policy and technology capabilities to the Council.

This Act also established simultaneously the Commission on Marine Science, Engineering and Resources which would examine the nation's present and future role in the ocean, recommend national policies, propose specific programs of research and exploration, relate these to funding levels, and recommend an optimum Federal administrative structure. If ever there was a blue-ribbon ocean commission, this was it. Chosen to head it up was Ford Foundation Director Julius A. Stratton, and serving as Executive Director was U.S. Bureau of the Budget's Samuel A. Lawrence, and as Deputy Director, University of Rhode Island geographer and Head of the Law of the Sea Institute, Lewis M. Alexander.

The Council was the interim manager and coordinator of the nation's Federal ocean activities. Until the Commission completed its study and reported and until Congress and the White House acted on its recommendations the Council would be *it*. Everything looked orderly and proper, but it was going to be a long, sometimes turbulent, interim period. The established ocean agencies, almost without exception, saw both the Council and the Commission as direct threats to their traditional ocean roles and missions. Their attitude could scarcely be described as cooperative; the fighting was not long in starting. Tables 25.3 and 25.4 show the Commission and Council memberships. Keep in mind that the Commission was a blue-ribbon *ad hoc* group, while the Council was a full-time coordinating group, whose members were *ex officio*.

The Sea Grant College and Program Act introduced a new dimension onto the Federal ocean scene. Credit for the concept goes back some years to Athelstan Spilhaus, who likened his Sea Grant College idea in some ways to the Land Grant College concept of just about a century earlier. More specifically, the Sea Grant Act was aimed at formalizing and

Table 25.3. Membership of the Commission on Marine Science, Engineering and Resources

Chairman	Julius A. Stratton, Chairman, Ford Foundation
Vice Chairman	Richard A. Geyer, Head Department of Oceanography Texas A & M University

Members

David A. Adams
Commissioner of Fisheries
North Carolina Department of
Conservation and Development

Carl A. Auerbach, Professor of Law
University of Minnesota

Charles F. Baird
Undersecretary of the Navy

Jacob Blaustein, Director
Standard Oil Co. of Indiana

James A. Crutchfield, Prof. of Economics
University of Washington

Frank C. DiLuzio, Assistant Secretary
Water Pollution Control, Department of Interior

Leon Jaworski, Attorney
Fulbright, Crooker, Freeman, Bates and Jaworski

John A. Knauss, Dean
Graduate School of Oceanography
University of Rhode Island

John H. Perry, Jr., President
Perry Publications

Taylor A. Pryor, President
The Oceanic Foundation

George E. Reedy, President
Struthers Research and Development Corp.

George H. Sullivan, MD
Consulting Scientist
General Electric Reentry Systems

Robert M. White, Administrator
Environmental Science Services Administration
Department of Commerce

Congressional Advisors: Senators Norris Cotton and Warren G. Magnuson
Representatives Alton A. Lennon and Charles A. Mosher

Executive Director: Samuel A. Lawrence

Deputy Director: Lewis M. Alexander

Assistant Director: Clifford L. Berg

Table 25.4. Organization of the National Council on Marine Resources and Engineering Development. All Members ex officio

Chairman:
  Vice President of the United States

Members:
  Secretary of State
  Secretary of the Navy
  Secretary of Interior
  Secretary of Commerce
  Secretary of Health, Education & Welfare
  Secretary of Transportation
  Chairman, Atomic Energy Commission
  Director, National Science Foundation

Observers:
  Administrator, National Aeronautics & Space Administration
  Secretary, Smithsonian Institution
  Administrator, Agency for International Development
  Director, Bureau of the Budget
  Chairman, Council of Economic Advisors
  Director, Office of Science and Technology

Executive Secretary:
  Edward Wenk, Jr., from start until January 31, 1970
  Enoch L. Dillon, since January 31, 1970

giving program continuity to nonmilitary federal funding of university ocean research and technology activities. With the formation of the Marine Resources Council, the old ICO was disbanded and its very able Executive Secretary Bob Abel moved over to take charge of the Sea Grant Program, which was placed administratively under the National Science Foundation.

The Sea Grant Act in all probability marks the first real establishment of any kind of national public policy on the oceans. It was designed to give universities growing financial support on a long-term continuously funded basis. Specifically, it called for the following moves:

1. Initiating and supporting programs at sea grant colleges and other suitable institutes, laboratories, and public or private agencies for the education of participants in the various fields relating to the development of marine resources;

2. Initiating and supporting necessary research programs in the various fields relating to the development of marine resources, with preference given to research

aimed at practices, techniques, and design of equipment applicable to the development of marine resources; and

3. Encouraging and developing programs consisting of instruction, practical demonstrations, publications, and otherwise, by sea grant colleges and other suitable institutes, laboratories, and public or private agencies through marine advisory programs with the object of imparting useful information to persons currently employed or interested in the various fields related to the development of marine resources, the scientic community, and the general public.

And it defined a "sea grant college" as follows:

The term "sea grant college" means any suitable public or private institution of higher education supported pursuant to the purposes of this title which has major programs devoted to increasing our Nation's utilization of the world's marine resources.

The concept had popular public appeal: it addressed itself to the problems and aspirations of the academic community; it provided incentive for establishment of industry-university partnerships; it materially extended the scope of Federal support for science traditionally exercised by the National Science Foundation. It only remained for the Administration and its Bureau of the Budget and the Congress to allow it adequate funding—enough to permit initiation of sea grant college programs and, in turn, to provide for their continuing growth. The budget aspect of national policy implementation has never been easy, but the Sea Grant Program has usually fared rather better than other ocean efforts in this regard. It is not only a case of success breeding success, but also of the political appeal of the program.

The FPC legislation should have been an important step in the establishment of a public policy with regard to food from the sea, but unfortunately this was not soon forthcoming. Quite evidently the protein health of the nation's poor is important politically only to the extent that it does not jeopardize the financial interests of such politically powerful groups as the dairy lobby, the grain lobby, and so on. Even after the FPC Act was passed it took interminable months of procrastination by the U.S. Food and Drug Administration and finally a direct White House order by Lyndon B. Johnson before the FDA reluctantly, and with many strings attached, approved FPC for human consumption (Anonymous, 1967). Among the "unfair competition" restrictions FDA placed on FPC for human consumption were (*a*) a *much* lower bacteria plate count than that required for milk for human consumption; (*b*) an economically unrealistic, low permissible level of fluorine content; (*c*) an absolute restriction on its inclusion as a protein additive in manufactured foods, such as breakfast cereals; (d) a restriction that it could only be manufactured from fresh, whole hake; and (e) the ludicrous limitation that it could only be sold in packages of no more than one-pound capacity and only to the ultimate consumer, that is, the housewife. Legally, FPC could be manufactured and sold; practically it was still under wraps.

In January 1969, after having labored mightily and well, the Stratton Commission issued its report "Our Nation and the Sea—A Plan for National Action (Anonymous, 1969a). Backed up by three supporting volumes, the reports of its technical panels, this work remains today the most complete and most authoritative data base and analysis of the nation's stake in the ocean available anywhere. Its only limitation is that it chose to ignore the issue of the nation's defense stake in the ocean. This is unfortunate, because the Navy's role in the establishment and implementation of national policy in the oceans is considerable. Politically, however, it was probably an astute move—for the Navy traditionally has fought bitterly both above board and behind the scenes *any* move which it thought would erode its role in national ocean policy. Whether that move was in the best long-term national interests did not matter: if top Navy brass considered it not in the Navy's best short-term interest, every procedural step, every administrative trick, and every lobbying effort would be brought to bear to block it.

In leaving the Navy's roles and missions and the various Navy offices and agencies out of its considerations, the Commission undoubtedly increased the chances of constructive action being taken on its recommendations. The Navy could be dealt with later.

Recommendations were numerous and specific, covering virtually every aspect (except military) of the national interest in the ocean, and can be found by examining the Commission documents themselves or the numerous reports and critiques written thereon. Key to the Commission recommendations were those concerned with reorganization of the Federal structure in Washington, namely for establishment of a National Oceanic and Atmospheric Agency (NOAA) which would incorporate the roles, missions, buildings, personnel, equipment, and budgets of the Coast Guard, Bureau of Commercial Fisheries, Environmental Sciences Services Administration, National Sea Grant Program, U.S. Army Corps of Engineers Lake Survey, National Oceanographic Data Center, and marine and anadromous fisheries activities of the Bureau of Sports Fisheries and Wildlife.

This would create a single ocean agency of 55,000 employees, 11,000 of whom would be scientists or engineers and 22,000 of whom would be specially trained technicians. It would bring together under single administrative control 15 marine biological, 6 technological, and 12 physical environmental science laboratories. The super ocean agency would have over 300 oceangoing ships and smaller craft, numerous aircraft, and satellite sensing and communications capabilities. The Commission went on to outline a 10-yr budget for the NOAA of some $8 billion, suggesting that the Navy's oceanography and ocean technology budget—quite separate and apart from NOAA's—would probably equal at least as much.

Simultaneously, the Commission also recommended formation of a National Advisory Committee for the Oceans—fashioned in makeup and purpose after the old National Advisory Committee for Aeronautics, which preceded the National Aeronautics and Space Administration and paced man's first 50 yr of powered flight (Emme, 1961). The NACO would advise the director of NOAA in carrying out his functions and report annually to the President and the Congress on the "progress of government and private programs in achieving the objectives of the national ocean program." The NACO would number 15 individuals in all, drawn from outside government and appointed by the President with the advice and consent of the Senate.

Without treading on its prerogatives by suggesting a solution to the problem, the Commission also pointed out that Congress' organization for treating with ocean matters was no less chaotic than that at the administrative level. Indeed, no fewer than 60 committees and subcommittees of the House and Senate get a crack at the nation's socalled "National Oceanographic Program" during the authorizing and appropriating procedures—during which time they can and do reduce, increase, and otherwise alter the program (see Table 25.5). This is disconcerting to say the least—after a major effort has been made to coordinate the many parts into a reasonably efficient and coherent whole. Often a single element of one agency's budget may be serving the purposes of several agencies and programs—one of the major benefits of coordination (otherwise money would be wasted doing the same thing over and over again). If some congressional committee cuts a project because a Congressman does not think *his* agency has any business doing that particular thing or because it hurts *his* pet industry, not one but many programs are hurt.

The Commission report remains an excellent work; it is still a major reference as to the why's, wherefore's, and statistics of the nation's stake in the ocean. When it came out—largely because it was such a fine job and so well received—virtually everyone breathed a long sigh of relief. After a decade of frustrating study and procrastination, here at last was an intelligent, well-thought-out plan of action—one which would assure

Table 25.5 Congressional Committee Jurisdiction over Executive Agencies in Marine Science

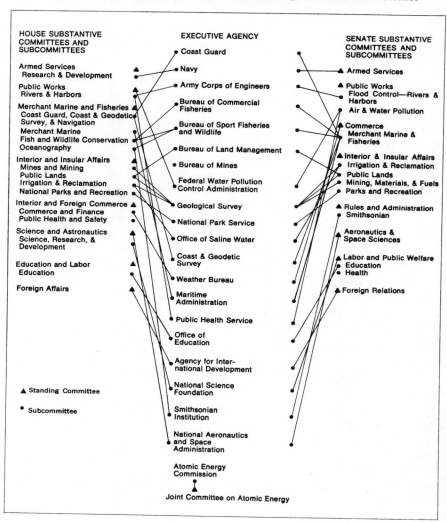

the nation's primacy in the ocean for the foreseeable future and one which was almost universally acceptable. Only the Navy, predictably, viewed the prospect of a single monolithic ocean agency as a threat to its senior ocean role, and started laying the groundwork to defeat the proposal. The same Navy which in 1915 sponsored establishment of the National Advisory Committee for Aeronautics, which was to pace the progress of world aviation for virtually half a century, in the late 1960s and early 1970s bitterly opposed comparable action designed to pace man's peaceful advance into the sea. Somewhere along the line the Navy had become petty and short-sighted, placing vested self-interest above national interest.

The biggest danger to the Commission proposals, however, came not from the Navy but from the vicissitudes of national politics. Virtually the entire buildup to the Commission recommendations had taken place under Democratic administrations—first John F. Kennedy and then Lyndon B. Johnson. It was a Democratic Congress that initially passed the Marine Resources and Engineering Development Act, and it was a Democratic President (Johnson) who appointed the members of the Commission. The official issuance of the Commission report in January 1969 coincided with the inauguration of Republican President Richard M. Nixon. The big question in ocean circles, of course, was, what would the Nixon Administration do about the Commission recommendations? The answer wasn't long in coming.

The Nixon Administration—initially at least—considered that the oceans were a Democratic trick and that all the studies, reports, analyses, and so on, of the preceding decade were some kind of Trojan seahorse. For a while it looked as though there would be a new decade of ocean-organization studies, this time sponsored by the Republicans. Another factor entered the picture: as chairman of the National Council on Marine Resources and Engineering Development, Spiro T. Agnew demonstrated neither an intelligent appreciation of the needs of the environment (ocean or otherwise) nor the sense of history and statesmanship necessary to guide man to a position of compatability with his planet. His predecessor, Hubert H. Humphrey, came into the ocean picture cold but soon developed a deep appreciation of man's oceanic dependence and destiny. His election to the White House would have boded well for the nation's oceanic interests. Thus far the same cannot be as readily said of either Agnew or Nixon.

Indeed, the Nixon Administration seems to go to considerable extremes merely to come up with something *different* from what was recommended under the Democrats. It is both sad and frightening when the nation's, and indeed man's destiny, is founded upon such shallow bases.

The Nixon Administration recommendations, put forth in the summer of 1970 and approved by Congress, proposed establishment of (a) a National Oceanographic and Atmospheric Administration and (b) an Environmental Protection Administration. NOAA is a separate agency within the Commerce Department and would include Environmental Science Services Administration, Bureau of Commercial Fisheries, Lakes Survey of U.S. Army Corps of Engineers, the anadromous fisheries activities of the Bureau of Sports Fisheries and Wildlife, the ocean data buoy activities of the U.S. Coast Guard, the National Oceanographic Data Center, the Naval Oceanographic Instrumentation Center, and the Office of Sea Grant Programs.

The environmental agency is an independent agency and assumes the responsibilities, budgets, personnel, and facilities of the Interior Department's various water pollution programs, the Department of Health, Education and Welfare's air pollution and solid waste management activities, the Department of Agriculture's and the Food and Drug Administration's pesticides standards and controls activities, and the Atomic Energy Commission's radiation regulatory functions.

One result of this bi-agency approach is, in effect, to put two potentially competing agencies in (a) the environmental research and management business and (b) the ocean resources research and management business. It amounts to a kind of large-scale musical chairs, using Federal agencies instead of chairs, and roles and missions instead of individual players.

The net result of NOAA and EPA may amount to a step in the right direction, but insofar as the nation's nonfederal oceanographic laboratories are concerned, it leaves much to be desired. Among the problems that the laboratories, WHOI, SIO, Miami, URI, Washington, and Alaska, for example, face are the matters of continuity and unity of funding and program support from the Federal agencies. Before NOAA, in the main the laboratories had to deal with a quasi-independent Office of Sea Grant Programs, the National Science Foundation, and the Navy, primarily the Office of Naval Research.

Additionally, of course, they obtained miscellaneous funding from countless other agencies—BCF, HEW, AEC, and ESSA, for example—but the bulk of support for *scientific* programs and the facilities and ships necessary in support thereof came from NSF, ONR, and Sea Grant. Under first the ICO and later the Marine Resources Council in an overall kind of way, support was coordinated—assuming Congressional second-guessing did not obviate that coordination.

Now, however, while it may look like a neater administrative package to Washington bureaucrats, it may be just more of a nightmare to the oceanographic laboratories. For, instead of three main agencies with which to deal, there are in effect *five:* still a quasi-independent Office of Sea Grant Programs (transferred over to NOAA), NOAA, the National Science Foundation, EPA, and the Office of Naval Research. It looks as though science, once again, falls in the category of Washington oversight.

Finally, it is doubtful if the invention of NOAA and EPA did anything really constructive toward rationalizing the hodge-podge of Congressional ocean responsibilities.

In the summer of 1970 President Nixon began to get oceanography out from under the Democratic "cloud" and to apply the Republican label instead. It was *his* NOAA which was approved. It was *his* Deep Seabeds Regime which was presented before the U.N. Seabeds Committee in Geneva. Earlier, through the person of the Vice President, *he* had put the Republican stamp on a series of Coastal Zone initiatives and on the IDOE (International Decade for Ocean Exploration). Unfortunately, it often is on such considerations that programs (and nations) rise or fall.

Further confounding efforts to arrive at a national ocean policy is the difference in viewpoint, approach, and enthusiasm found between Congress on the one hand and the White House on the other. And it really does not seem to matter whether the same party is in control in each case—or whether, say, the Democrats control Congress and the Republicans control the White House. When Johnson was President and the Democrats held sway at both ends of Pennsylvania Avenue, there was very little agreement between the President and Congress on a national policy for the oceans—little more, indeed, than when Nixon became President and the Democrats still controlled Congress.

Support or opposition with respect to a strong national ocean program does not seem to be a matter of partisan politics. Both support and opposition are bipartisan. Rather, the difference seems to depend more on the level of statesmanship and vision, the sense of destiny and ecological awareness found among individual administrators and politicians. Those with vision appreciate the significance of the ocean to man's future economic progress and well-being and, more fundamentally, to his very survival. Those who take the narrow, short-term view generally are those who line up in opposition and include those with a specific financial stake in maintaining the *status quo*—such as the real-estate lobby with respect to the preservation of coastal marshlands, the dairy interests when it comes to FPC, and thermal polluters versus estuarine fisheries.

Thus the history of the development of a national public policy on the oceans has been largely one of procrastination and lack of development.

There was and remains no Sputnik equivalent in the oceans to spur the nation to a priority national ocean program. This does not mean that there will not be one or that it will not be Russian. The Kremlin places greater primacy on the ocean than does the United States. In the political power struggle that now besets the world, this fact alone could assure first position for the U.S.S.R. within the decade.

The Key Points of Public Ocean Policy

Upon the development and execution of a sound global policy with respect to the ocean depends the future of the planet as a viable life-support system: without the complete functioning, living ocean, the earth ecological system simply malfunctions. And it would not be a matter of gradual, scarcely noticeable decay through centuries. Rather it would be a matter of days, months, or a very short number of years at the longest.

Everyone is familiar with the role of the ocean in modifying the weather and controlling climate. What may be less familiar are some of the other ways in which the ocean plays a fundamental role in the basic oxygen-carbon life cycle of the world's flora and fauna. The oceans constitute far and away the largest proportion of the surface area of earth. Water absorbs gases, and, indeed, thereby the ocean (*a*) removes more of the carbon dioxide respired by man and his machines and his fellow creatures than any other single mechanism; (*b*) concomitantly it converts more carbon dioxide back into oxygen than all the plants of the land masses combined; and (*c*) absorbs other gases from the atmosphere, thereby purging it of deadly pollutants.

The primary carbon dioxide-to-oxygen converters are the microscopically tiny phytoplankton of the sea. These are accustomed to living within very narrow environmental limits of salinity, temperature, and pH ("partial hydrogen"). The salinity of the ocean, or—put more generally—its chemistry can be and is being altered by the carbon dioxide, lead oxide, sulfur oxides, and other materials produced by the combustion of fossil fuels. It is being altered, too, by various types of municipal, industrial, agricultural, and military (nuclear) wastes.

We do not know the role of trace elements in planktonic life cycles, except minimally. We do not really know the level of impact at which irreversible reactions will begin to take place. The nature of the chain reaction which might take place also for the moment eludes us. Chemical (or other) annihilation of the phytoplankton would greatly modify or terminate life on this planet within a very short period of time.

The ocean is a great absorber of carbon dioxide. But the amount of carbon dioxide in the water determines its pH—the more carbon dioxide, the more acid the water becomes. Within the ocean there is also a natural buffering system, based upon the alkaline components bicarbonate and carbonate ions which are in equilibrium with calcium carbonate going in or out of solution, which serves to bring pH back to within the narrow limits required by the phytoplankton. This question is unanswered: what happens when the jet planes, cars, factories, furnaces, and so on, spew forth carbon dioxide at a rate greater than the buffering system can readily accommodate? (See Chap. 11.) The answer may be all too simple: a short circuit in the earth's life support system with effects that are not now fully predictable, but would doubtless be disastrous.

The ocean holds more heat than any other single major surface feature of the planet. Again, it is the sink that moderates excesses. How much *new* heat can we release into the system by burning carbon fuels and releasing nuclear energy and how much can we block reradiation of terrestrial heat before we short the system simply by over-energizing it? What about environmental alterations not so great as to cause major malfunctions but great enough so as to destroy traditional ocean population balances?

The ocean is key to the continuation of life as we know it. It is also key to survival. The overriding objective of a public ocean policy, therefore, whether national or global must be maintenance of the viability of the ecosystem—for without this there are no other considerations. It should be the primary objective of public policy to study and to monitor the ocean and its interaction with the land and atmosphere. Therefore, the first order of public policy for the ocean is environmental management.

The other two major areas are (*a*) resource management and (*b*) defense. In this discussion resources are considered as properties and characteristics of the ocean which, in relation to man's needs and aspirations, are productive. Such resources include, for example, living resources (food), mineral resources (hydrocarbon energy sources, raw materials, building materials, and so on), recreation, transportation, real estate (area), energy (tidal, for example), and a sink (dump) for liquid, gaseous, and solid wastes of all kinds. Defense, by its very nature, is not productive. It may prevent something from being destroyed, but it does not build. It may be a necessary use of monetary and human resources, but it is neither an efficient nor a preferred one. Further, it has the potential for being highly destructive of the environment—sometimes in ways which "for reasons of military security" must be hidden from the public eye. Defense, therefore, is not included among "resources," but is treated separately.

The most natural reason for the development of a resource, oceanic or otherwise, is that conditions of supply, demand and technology make it economically attractive. This assumes a condition of what the economists call "perfect competition"—an ideal condition which is seldom, if ever, achieved. More often other factors mar this perfection. In the case of ocean resources, to mere considerations of supply and demand may be added matters of defense, domestic and international politics, and balance of payments. The oceans could contain reserves of strategic materials, which, though otherwise uneconomic to produce, are needed for one's own defense—or merely in order to prevent their recovery by an enemy. Matters of national prestige may be involved in ocean resource policy decisions, and usually are. And, while it may be uneconomic to exploit a given ocean resource under normal conditions (if it means spending domestic currency and paying a higher price, rather than hard-to-get foreign exchange and paying a lower price), prestige by itself could be sufficient reason to proceed with ocean resource development. Quite obviously when such factors of imperfect competition enter the picture, some form of Government subsidy, direct or indirect, usually is necessary.

Whereas, however, individual resources and/or classes of endeavor may require specific consideration to determine the exact nature and extent of Government involvement, such decisions are not necessary in determining basic Federal policy with regard to the fundamental questions of environmental management, resource management, and defense. Let us consider these prime areas one by one.

Ocean Environmental Management

Involved in environmental management are (*a*) study of the environment, (*b*) monitoring the environment, (*c*) establishment of both national and international environmental regulations, and (*d*) enforcement of same. Study of the ocean environment requires scientific research. This is the basic acquisition of knowledge of the single most fundamental feature of our planet. Its funding is primarily a government function. Profit is too far separated from basic research of this nature for industry to begin to play more than a very minor role in such activities, and if profit is ultimately involved the data understandably are held as proprietary, whereas the need with which we are concerned here is a *public* need. The suggestion has been offered, however, that special tax incentives be permitted industry for providing financial aid to basic research by the nation's university laboratories. This might take the form of removing, say, twice the grant from corporate revenues subject to income taxes and would be even more attractive were 1.1 or more times the grant deductable directly from the corporate tax bill.

In the long run, however, the burden of financial support for basic research must continue to be a direct taxpayers' expense. And for the taxpayer to get the best return on his investment, such research should be funded on a continuing multiyear basis—so that institutions and individual researchers alike can plan intelligently and economically. A major feature of the Sea Grant College program is such multiyear planning. This aspect of the Sea Grant program *must* not be compromised, and the philosophy should be extended to other areas of federally supported research.

The fundamental public policy requirement here, however, is a bipartisan recognition *both* on Capital Hill and in the White House that the earth is in danger and that the single most important area of research and management in regard thereto is the ocean. Such a formal declaration is necessary as a first step towards assuring both the level and the predictable continuity of support necessary to a successful and useful program of oceanographic research.

Another area of ocean science public policy involves the balance of effort between the academic community on the one hand and the federal ocean laboratory structure on the other. It is this author's strong view that government laboratory research should be limited to applied research concerned with the provision of government services, such as reporting and predicting oceanic conditions, mapping and charting, and assurance of safety of life and property at sea. And even that should be parceled out quite liberally to industry under private contracts. *Basic* research, however, should be concentrated mainly in the academic community. Not only does it ensure the continuation of a strong university system, but it is the condition most favorable to freedom of research and carries with it the added advantage of serving the interests of postgraduate education which, generally, such research in Federal laboratories does not.

Additionally, while professors may remain in one university post for some years, the flow of graduate students through the system is constant, putting forth new and enthusiastic scientists to the outside world and bringing new people and new ideas into the academic community. In contrast, the Federal laboratory, with its civil service tenure tends to become stagnant and both less receptive to and less productive of new ideas.

Public policy on ocean science, therefore, should be founded on a firm national conviction of urgent necessity and a high level of continuing multiyear funding. The primary emphasis of this funding should be within the university system and should strive as far as possible to keep funding of basic research separate from direct ultimate payoff considerations, that is, *applied* research and development.

National research of the world ocean should be a coordinated part of a total global effort. Adequately done, it is too costly, in talent as well as money, for any one nation to undertake alone. The need is universal; the effort should be likewise. And it should not be couched in the limiting concepts of "a year" or "a decade" of international ocean exploration. The mechanisms and institutions for continuing coordinated oceanographic research should be established on a permanent basis.

Monitoring the environment requires the systematic measurement of the ocean, the ocean-atmosphere interface, and the ocean-land interface on a continuing basis for purposes of (*a*) basic research, (*b*) reports and predictions of natural oceanographic phenomena, and (*c*) reports and predictions of abnormal phenomena, such as serious pollution threats. Individual nations would continue to monitor their own coastal zones and contiguous parts of the high seas conditions of which bear directly on national well-being—though these data, too, would be fed readily and promptly into the international pool of knowledge.

The total global ocean-monitoring job must be international. Single nations cannot afford to do the global job alone and unilaterally. As in the case of the scientific study of the ocean, the monitoring job must be *internationally* organized, *internationally* funded and *internationally* executed. New institutions are going to be necessary, or present ones must be materially altered and expanded. Again, as in the case of scientific research, there must be a national, indeed a global conviction that the effort is necessary, even urgent.

The other aspects of environmental management—establishment of regulations and their enforcement—must be forthcoming through international conventions. The alternative is a proliferation of unilateral, uncoordinated national pollution control actions, such as Canada's Arctic declarations of the summer of 1970 (Koers, 1970a). These regulations and their enforcement must transcend traditional concepts of international law enforcement which rely mainly on voluntary acceptance by the affected nation. The international authority must be supranational and possess commensurate enforcement powers—similar in some respects to the supranational nature of the European Economic Council (Koers, 1970b). Pollution of the ocean can lead to destruction of the ocean's life-support capabilities. Its protection, therefore, transcends considerations of national convenience. It will be difficult for many to accept this subjegation of national sovereignty, but it may be the only alternative to the extinction of the human race, along with all other of the higher orders of animal and plant life.

It is evident that in this area of scientific research, environmental monitoring, and pollution regulation and enforcement, a rather extraordinary degree of national leadership is going to be required—initially and forthrightly from the White House. This means not only initial statements couched in high-sounding phrases, but also a strong follow-up and follow-through of the words with action. It should be noted here, too, that studies are not a substitute for policies, either their determination or their implementation. Policies may require studies as a sound underpinning of fact. National policies are ennunciated by legally constituted national leaders. The recommendations of study groups do not constitute policies. In the past there has been an unfortunate, misleading tendency to equate the one with the other.

Resource Management

With exceptions where matters such as vital national interest and humanitarian considerations prevail, resource development—ocean or otherwise—follows the profit motive. This means that demand for a given product or service must be strong enough to support a price (willingness to pay) higher than (to allow for profit) the cost of production based on current technology and the accessibility and quality of raw materials. Basically, then, while considerations of long-term national interest may favor development of a viable offshore or deep-sea technology, industry should bear the main costs of exploitation. Special depletion allowances or other tax allowances may be provided in cases where the Government wishes to encourage development—as was done in the early days of petroleum exploration.

Government should not, however, directly engage in activities which are traditionally those of industry. The Federal Government might, for example, research basic deep ocean engineering principles—even as the NACA researched basic principles of flight and propulsion. It should not, however, endeavor to develop offshore oil drilling equipment, deep ocean mining gear, and so on. This is the role of competitive free enterprise, and to the company that performs the best should go the chance to profit therefrom. The Federal Government should conduct bathymetric surveys and produce bathymetric charts. Detailed geological exploration leading to future bidding on concessions and the ultimate production of minerals from the sea, again, is the traditional role of industry. The lines of demarcation between the traditional roles of industry and those of Government are clearly and well drawn on land; it is not difficult to draw oceanic analogies.

This author is of the opinion that it is in the national interest to encourage within private industry the development of a sound and growing oceanic technology. It not only enhances the nation's competitive position in the world marketplace, but also it provides a sound industrial base for possible future defense purposes. And should signs suddenly develop that the ocean ecology is materially malfunctioning, the chances of remedying that malfunction are materially improved by the existence of the technology, rather than having to await (and gamble on) its crash development.

There are several ways in which such technological development can be encouraged. One is to maximize government work contracted out to private industry while minimizing in-house projects. The trend in the decades of the 1960s and 1970s was in exactly the opposite direction. Another, is to provide tax incentives for oceanic enterprise across the board—rather than just for certain categories of enterprise. If there is a big user market for ocean goods and services, a sound and competitive industrial and technological base will evolve for their supply.

In consideration of mineral resources, there are those which are clearly on the continental shelves and which fall clearly within the sovereign jurisdiction of the contiguous coastal state, and there are those which clearly lie beneath the high seas and remain the "common heritage of all mankind." The United States has accepted this principle and has proposed a regime which draws the line between the two types of oceanic resource. This is President Nixon's proposed regime for the exploitation of the seabed resources of the deep ocean and is covered elsewhere in this book. Recognition of the two types of resource and acceptance of an international convention for formalizing the distinction is an essential part of the development of a public policy for the oceans.

Mineral resources, generally speaking, are nonmigratory and are matters of consumptive concern mainly to man's modern technology. Living resources, in many instances, do not stay put and are matters of consumptive concern mainly to man's individual health and survival. The one would fill his factories; the other would fill his stomach and nourish both his body and his mind.

In any event, the profit motive is less involved, perhaps, in the management of the ocean's living resources than it is in the management of nonliving resources. They say that an army marches on its stomach. Similarly, much of the power of world politics lies in filling the stomachs of the poor. But the food resources of the ocean are more than just stomach fillers. They are a vast, constantly regenerating supply of some of the highest quality protein available from any source. A small amount of Fish Protein Concentrate, say 5 per cent by weight, added to wheat flour gives the wheat flour the food and nutritional value of "steak and potatoes." Much of the world suffers from protein malnutrition which, in the young, stunts both physical and mental growth, resulting in permanent damage or retardation. In adults, it limits physical energy, mental acumen, and generally robs individuals of any drive or ambition.

Nations are poor because their people cannot work and produce enough to better themselves—including improvement of their diet. Their people cannot work and produce more because they suffer from protein malnutrition. It is a vicious circle from which there is no escape unless a higher quality diet is provided continuously and over a long period of time. The lowest cost source of high quality proteins is the ocean—fish in the form of FPC. Preeminence in high-seas fisheries is a matter both of profit and of national prestige. More importantly, perhaps, high-seas fisheries in the aggregate constitute a global resource of tremendous human significance.

Again we have a case where global considerations transcend those of individual nations. A viable world fisheries operating at a maximum sustainable yield is necessary if catastrophic food riots on a global scale are to be avoided a decade or so hence. This requires effective resource management, and effective management requires not only a sound basis of knowledge and an appreciation of the need, but also (again) the subordination of supreme national sovereignty to a supranational authority endowed with a comprehensive and realistic set of regulations and the sanctions with which to make their administration effective.

Carrying the animal protein supply-demand picture to its ultimate conclusion, it is obvious that fisheries alone, even effectively administered, cannot fill the protein gap. The potential for maraculture exceeds that of traditional concepts of fisheries. National policy and international policy should stress (a) the development of a national maraculture comparable in size, technology, and productivity to our national agriculture and (b) the evolution of international fish farming of migratory high seas species. The technology of maraculture should be freely exported and imported. If anything, the United States is backward in this area and should welcome knowhow from other lands.

The concept of "open range" fish farming is not new, but with the exception of some salmon-farming techniques, it has not really been tried.

Considerable research would be required before, perhaps, it could become effective for other than anadromous species. However, the potential of farming a common resource for the common good, with the whole world ocean as the farm, boggles the imagination.

Whatever the techniques and programs, protein from the sea may be more important to future human survival than any other single major food source. Public policy must not only protect the resource but must materially enhance it. On the high seas it must entail a willingness to accept decisions in favor of long-term human interest—sometimes at the expense of short-term national interest. Domestically it must entail adequate laws and enforcement to protect the productivity of the coastal (estuarine) environment. And it should entail a national effort to develop mariculture to a level comparable to the administrative, technological, and county agent structure now serving the dry-land farmer.

Defense

In the sense that the ocean is largely opaque to all forms of energy other than sound and that modern nuclear submarines operating within its depths are difficult, if not impossible to find, modern ballistic missile-carrying submarine forces are important constituents in the overkill standoff which so far has prevented any of the major world contenders for supreme power from pushing the button.

Other than the provision of a retaliatory force which is virtually immune to detection, the role of sea-power in world affairs has changed little in modern times—except that it has gone from a two- to a three-dimensional phenomenon. It is not only how far-ranging and how impressive a naval fleet may be in public display, it is increasingly now a matter also of how deep that fleet can extend its reach and how effectively it can operate at depth.

The U.S.S.R. in the early 1960s made the fundamental policy decision —spurred on by America's Polaris threat—that it would stress seapower in *all* of its aspects, ranging from oceanography and fisheries, to missile-firing submarines and an impressive array of modern surface warships. By the early 1970s the U.S.S.R. was threatening America's primacy in the seas; and while U.S.S.R. production and deployment of seapower grew, in the United States it was being cut back. The high cost of the Vietnam War and the competition for funds to solve domestic problems were given as reasons.

So long as war remains the ultimate means of settling international disputes, seapower will play a major role in world politics. If the United

States is to maintain its leadership role in world affairs, it cannot afford to let the U.S.S.R. or any other major power gain a material advantage in naval strength. This is a matter simply of national survival policy.

Ultimately, however, with continued scientific study of the ocean and with man's continued advance into the sea for peaceful resource exploitation, the ocean will become increasingly less opaque. It is not hard to visualize the time in, say, the 1980s or beyond when *all* underwater activities are routinely tracked and monitored—not for military purposes but for matters of commercial search, rescue, and traffic control. In such a time the ocean will have lost much of the singular advantage for military purposes it enjoys today. The nation should give some thought now to such a time. Perhaps it will signal the onset of an era when all that remains for man to do is to revert to political machinery (as an alternative to oblivion) as the ultimate means for the settlement of disputes among nations. Indeed, perhaps, the sovereign ego of nations itself will have become little more than a remnant vestige of the past—itself fast disappearing.

Establishment of a constructive public policy on the oceans requires no more basic studies. Rather it requires more than anything else concerted leadership at the top. Republican or Democrat, it does not matter. The ocean, like man's survival, is nonpartisan. Without such leadership there will be no effective public policy for ocean research and management. Without such a policy, man will be deprived of the resources of the ocean, and in his growing numbers it is doubtful if, as a species, he can long survive this deprivation.

More important even than these fundamental considerations, the ocean is the heart mechanism of the earth's life support system. Modern technological man is quite capable of upsetting its delicate mechanisms and of causing a catastrophic and irreversible malfunction with the resulting elimination of life as we know it from the planet.

Conversely, the institution (finally) of a strong public policy, ably administered and effectively executed, can assure a profitable, healthy, exciting, and challenging world future for all.

References

Hull, E. W. 1969. *Critique and study of the government of the state of Florida and its capability for meeting the ocean challenge,* Florida Council on Marine Sciences and Technology, Miami, Fla.

Aristotle. *History of animals; On the parts of animals; On the generation of animals;* other works.

Anonymous. 1959. *Oceanography 1960 to 1970*, NAS—NRC, Washington, D.C.

Wenk, Edward, Jr. 1960. *Ocean science and national security*, Report of the Committee on Science and Astronautics, U.S. House of Representatives, 86th Congress, 2nd Session; Government Printing Office, Washington, D.C.

Anonymous. 1960. *TENOC—1961–1970: ten year program in oceanography*, U.S. Department of the Navy, Washington, D.C.

Anonymous. 1963. *Oceanography—the ten years ahead, a long range national oceanographic plan*, Pamphlet No. 10 of the Interagency Committee on Oceanography, Washington, D.C.

Anonymous. 1964. *Economic benefits from oceanographic research*, Committee on Oceanography of the National Academy of Sciences (NASCO), Washington, D.C.

Anonymous. 1966a. *Effective use of the sea*, Report of the Panel on Oceanography, President's Science Advisory Committee, U.S. Government Printing Office, Washington, D.C.

Anonymous. 1966b. *Oceanography 1966—achievements and opportunities*, Report of the Committee on Oceanography, National Academy of Sciences, Washington, D.C.

Anonymous. 1967. *Rules and regulations; Food and food additives; Fish protein concentrate*, Federal Register, Washington, D.C., February 2, 1967.

Anonymous. 1969a. *Our Nation and the sea*, Report of the Commission on Marine Science, Engineering and Resources, U.S. Government Printing Office, Washington, D.C.

Anonymous. 1969b. *Marine resources and legal-political arrangements for their development; Industry and technology; Science and environment*, Panel Reports of the Commission on Marine Science, Engineering and Resources (3 Volumes), U.S. Government Printing Office, Washington, D.C.

Emme, Eugene M. 1961. *Aeronautics and astronautics—an american chronology of science and technology in the exploration of space 1915–1960*, National Aeronautics and Space Administration, U.S. Government Printing Office, Washington, D.C.

Koers, Albert. 1970a. *Canadian arctic and the northwest passage—a legal position*, The Nautilus Papers series, Nautilus Press, Inc., Washington, D.C.

Koers, Albert. 1970b. *The european economic community and the sea*, Proc. June 1970 Symp., The law of the sea: The U.N. and the Sea, Law of the Sea Institute, Kingston, R.I.

26

Uses of the Ocean

DONALD W. HOOD, *Professor of Marine Science, Institute of Marine Science, University of Alaska, College, Alaska*

C. PETER McROY, *Assistant Professor of Marine Science, Institute of Marine Science, University of Alaska, College, Alaska*

The great variety of present and future uses of the sea impose stresses on the oceans of such kind and magnitude that the most dedicated and intelligent efforts of reasonable people of the world must now attempt to resolve the many conflicts of interest that have and will occur. Historically, the oceans have been used mainly for fisheries, limited mineral extraction, transportation, warfare, recreation, and waste disposal. These are not vastly different uses than those of the present or expected future; however, the intensity has increased and consequently their use for waste disposal, whether direct (dumping) or indirect (atmospheric lead or insecticides), has now reached an alarming level. It is now time that we stop trifling with the world ocean contamination problem and learn in detail the processes involved to assure ourselves that the many uses of the oceans' resources for the benefit of man will be possible for our and future generations.

Many publications on the uses of the sea have recently appeared. It would be redundant, even if physically possible, to treat this topic exhaustively in this chapter. Rather we have chosen to summarize the literature on this subject with special emphasis being given to point out the importance of the multiple use concept in dealing with ocean resources whether these be on the coasts of sovereign states or in the international waters of the open sea.

The multiple-use concept simply stated is the use of coastal resources for maximum public interest. With which public are we concerned?—the local public, people who are only employed there, live in houses nearby, or commute into the area for jobs? Is it the public who uses the products of the local region who may live as far off as transportation lines may reach? Is it the user public who may come to the area to fish, hike, hunt, swim, or use a marina? Perhaps it is the future public who may have far different ideas about maximum public interest of the resource. Or it might be the public who have foregone their maximum benefit by having been earlier preempted through other uses. There is little question that all of these interests must be weighed in deciding the use of each specific resource. Most important of all is to have available a firm basis-scientific, sociological, and economic knowledge-for use in making decisions and setting priorities.

Limitations to the Multiple-Use Concept

Multiple use of marine resources for maximum public interest is a concept of great value. There are areas of use in which severe limitation or complete prohibition would be the most prudent, however. The possible long-range irreversible effect is an important issue. This may occur from certain aspects of waste-disposal, over exploitation of fishery resources, mineral extraction, or real estate development. No uses should escape review and discussion on the greatest overall benefit by an appropriate (knowledgeable) public body.

It is now clear that many kinds of waste materials should be kept out of the oceans or, at least, their concentrations minimized. Many chapters in this book have dealt directly on specific problems of concern: heavy-metal wastes such as chromium, copper, mercury, lead; insecticides and other organic molecules of the nonbiodegradable type such as DDT and polychloraromatic hydrocarbons; high levels of radioactive materials particularly the long half-life members such as ^{90}Sr and ^{137}Cs resulting from fallout (which can only be avoided by elimination of fission energized atomic explosions) or effluents from reactors or fuel processing plants; indiscriminate dumping of unwanted dangerous materials such as unexploded war materials, dangerous gases, and chemicals with unknown destiny or potential reactions, particularly on the continental shelves; overuse of near-shore areas for disposal of debris that may, though unseen to surface dwellers, litter the entire ocean floor with unuseable wastes and prohibit future development of a useful, but perhaps yet unknown, resource; and the unwise disposal of refuse from individual, municipal, government, and industrial functions into any regions of the seas that would prohibit the growth and development of desirable biological species. A worldwide irreversible effect, whether or not harmful, should be scrutinized severely. Irreversible ocean reactions may lead to irreversible biological effects with worldwide consequences.

If it were possible to prevent polychloraromatic hydrocarbons, fallout radioactivity, lead from industrial use, or other potentially harmful materials from reaching the sea, the fear of future ocean pollution could be greatly alleviated. The simple act of not putting it in would suffice. Those materials, however, largely reach the ocean fortuitously through transport by winds, scavenging by rainfall, and runoff from rivers. The appearance of these materials in the ocean is not desirable, yet the only successful way of preventing them passage is at their source—at the user or at the manufacturer. Because of these difficulties, the oceans' maximum, essential uses may be impaired. In fact, continued enlargement of our highly technological society either in direct ocean use or inadvertantly by its interaction with other environmental factors, will certainly decrease or eliminate some presently accepted important uses of the ocean. The excessive quantities of DDT in silver salmon of Lake Michigan which rendered them unfit for sale is a case in point. This has not yet happened in the ocean, but if man were using petrels or cormorants or brown pelicans as a food resource, the contamination levels might already have been reached preventing their consumption. Species presently used by man for food are also in danger; and misuses of the ocean, even though sometimes fortuitous, must be corrected to preserve the ocean, our last resource, from unwanted, unneeded, and unwise destructive forces.

If properly conserved and managed, the ocean can supply many needs of man for the forseeable future. We describe many of these uses briefly in the following sections of this chapter. One use is not intended to dominate the others, although some may stand out because of more extensive discussion or greater available information.

Historical Uses of the Seas

Man has a long history of association with the sea. References to associations with the ocean exist in the most ancient of recorded cultures. Today the oceans are receiving more and more attention and are being utilized in more and more ways. The record of history reveals that the basic ideas or principles for nearly all current uses of the seas are not new or unique to twentieth century man, but extend back hundreds or even thousands of years (Table 26.1). The current wave of marine ex-

Table 26.1. Some Historical Uses of the Seas

Use or Harvest	Approximate Beginning	Location
Seaweeds	800–600 B.C.	China
Pearls	Thirteenth Century	China
Precious corals	Third Century B.C.	Greece
Shellfish	1000 B.C.	China
Sponges	Third Century B.C.	Greece
Fish	Stone Age	Egypt
Turtles	Sixteenth Century	West Indies
Whales	1200	European Coast
Seals	8000 yr ago	Bering Sea
Fur seals	1784	South Atlantic
Salts	2200 B.C.	China
Military ships	Fourth Century B.C.	Mediterranean
Military divers	Third Century B.C.	Greece
Diving suit	Thirteenth Century	England
Diving bell	1707	England
Submarine cable	1848	North Atlantic
Submarine photography	1856	North Atlantic
Modern submarine vessel	1890–1900	North Atlantic
Submarine oil and gas	1959	North Sea
Scientific study	Fourth Century B.C.	Greece

From Tressler and Lemon (1953); Hull (1964); Loftas (1969); and Mero (1964).

ploitation is certainly a good deal more complex and intense than those of previous times. The history of the many uses of the sea is included in accounts by Tressler and Lemon (1953), Loftas (1969), and Hull (1964).

One of the oldest products of the sea is derived from seaweeds. Seaweeds were used by ancient man for food, medicine, and fertilizer. Anthropological studies of the north coast of the Mediterranean Sea have shown that seaweeds were used to make mattresses by stone age man. In a sense this use has persisted until recent times in the form of eelgrass (*Zostera* sp.) mattresses and furniture stuffing in parts of Europe. An old Chinese record from 600–800 B.C. indicates the use of seaweeds in cooking. The foodstuff Nori, the red seaweed *Porphyra tenera* Kjell., was cultivated in Hawaii and Japan in 1670.

Surely fishing is one of the early attractions of the sea to man. The fish spear is an ancient object in archaeologists' collections; hook and line date back more than 10,000 yr. We also know that nets and fish traps had been in use for thousands of years before the time of Homer, and that reels and netting were used in ancient Egypt. Porpoise were first used to drive fish into nets thousands of years ago in India. In addition, fish cultivation near Venice may date to the time of the Roman Empire.

Perhaps the most ancient of all fisheries is for shellfish. Oyster culture was practiced in China several thousand years ago and the Romans also practiced oyster culture. Aquaculture is hardly a new concept although it is currently a popular one.

Crabs, shellfish, and lobster have been harvested in Europe for centuries. The early colonists to America even imported lobsters before they developed their own fisheries. Abalone were prized by the ancient Chinese, and certain tribes of American Indians used the shells as money and ornaments. Even turtles have been harvested for centuries; by 1620 restrictions and conservation measures were placed on turtle harvesting areas around Bermuda.

It is difficult to find a taxa of marine organisms that have not been harvested for at least centuries. Even such restricted fisheries as sponges, precious coral, and pearls are very old.

The exploitation of marine mammal populations is an old practice. The Basque whalers have hunted since 1200 A.D. Fur seal harvests began in the Southern Hemisphere rather recently—about 1784. The Russians began exploitation of the Alaska sea otters in 1745. But long before that time a well-developed culture of Eskimos and Aleuts harvested seals and whales on the Bering Sea Coast of Alaska (Laughlin, 1967). The tools and weapons for this trade have been recovered from middens dating back 4 to 8 thousand years. This culture persisted and flourished unchanged until very recently (within 50 yr), when acculturation became more intense.

The Phoenicians are considered to be one of the earliest peoples to explore by sea. Others are the Polynesians and Vikings. Coastal charts of parts of the Mediterranean date back to 5000 B.C. The era of exploration by sea is over, but the exploration of the sea itself has only begun. The origins of travel and transportation of materials by boats are traced to the civilizations on the shores of the Mediterranean. Surely boats were developed independently many times by civilizations around the coast of the world ocean. The skin boat of the Arctic Eskimo dates back at least 4 to 5 thousand years.

Using the sea in wars was a well-established practice in the Mediterranean Sea long before the time of Christ. In addition to using ships for transporting armies and for actual battles between vessels, submerged obstacles were used to stop enemy ships and divers were used to run a blockaded port. The practice of diving was used thousands of years ago in the Mediterranean where pearls and sponges were recovered from shallow water. A kind of diving suit was first developed in England in the thirteenth century, and by the seventeenth and eighteenth centuries experiments with devices for taking man into the seas, the forerunners of the submarines, became more and more frequent. The first submarine, developed in about 1620, is credited to the Dutchman Cornelius van Drekel.

The science of the sea, oceanography, is generally considered a recent addition to the scientific disciplines, and well it is. Nonetheless, the first scientific study of marine life is credited to Aristotle in 4000 B.C. He described and classified many types of marine animals and outlined the life history of several species. The Dark Ages did little for science, but from about the fifteenth century man's knowledge of the sea has progressed continuously.

Since the last half of the nineteenth century the uses of the sea have developed rapidly as a result of the industrial and technological revolution. For the most part these uses are the result of historical ideas originating hundreds and thousands of years ago. The exceptions are essentially products of technology. Perhaps the only truly new use of the ocean is actually one of effect rather than use. Only in modern times have we been able to affect large portions of the ocean by indirect and inadvertent contamination.

Current Uses of the Seas

The oceans are now the source of hundreds of products derived from living and nonliving resources. Tressler and Lemon (1953) provide an excellent, though dated, account of these products. A more current treatment of ocean resources is compiled by Firth (1969).

Fisheries

Food ranks high as one of the major uses of the sea. The world catch of fisheries species now exceeds 50×10^6 metric tons and is continuing to increase at the rate of 8 per cent/yr. The maximum sustainable yield of world fisheries is thought to be about 100 to 150×10^6 metric tons. The present world catch is largely restricted to the coastal zones and continental shelves. Large areas of the ocean yield only a low fish catch (Table 26.2). Three areas of the ocean stand out—the northeast Atlantic (the British Islands, the North Sea), and west central Pacific (the waters

Table 26.2. World Catch of Fish by Ocean Area for 1967

Area	Live Catch (metric tons $\times 10^3$)
NW Atlantic	4.0
NE Atlantic	10.2
Mediterranean and Black Seas	1.1
N Pacific	6.4
West Central Atlantic	1.3
East Central Atlantic	1.6
W Indian Ocean	1.3
E Indian Ocean	0.7
SW Atlantic Ocean	1.3
SE Atlantic Ocean	2.5
SW Pacific Ocean	0.4
SE Pacific Ocean	11.2
World Ocean	53.3

From FAO Yearbook of Fisheries Statistics (1967).

of the East Indies and Japan), and the well-known upwelling region of the west coast of South America. These three regions comprise some 60 per cent of world fisheries (excluding whales).

The world fish catch includes some 500 species of marine organisms (Table 26.3). This is a small percentage of the total number of species in

Table 26.3. Proportion of Major Groups of Species Harvested from the Sea

Type of Group	Number of Species Harvested	Total Number of Marine Species
Fishes	275–300	~1900
Invertebrates	130	~160,000
Seaweeds	50	~4500
Mammals	25	124
Amphibians	3–5	5

From Tressler and Lemon (1953); Dawson (1966); Morris (1965); and Russel-Hunter (1968).

the sea, but probably a large proportion of the potentially harvestable fishes and invertebrates. The biomass contribution of each group of species to the world catch shows that the herring, sardine, and anchovy fisheries are now the major food source from the sea (Table 26.4). These fishes and the others used in fish meal are far more important than the contribution of the quality protein fish such as salmon.

The whale for some time ranked highest in the world fish catch, but since 1963 the whale fishery has experienced a drastic decline as a result of overexploitation (Ehrlich and Ehrlich, 1970). Of the four countries engaged in antarctic whaling at that time, only two remain: Japan and the U.S.S.R.; and these nations are shifting their efforts to the North Pacific and Bering Sea.

Table 26.4. World Live Catch of Marine Species by Groups of Species (metric tons $\times 10^3$)

	1965	1966	1967
Salmon	475	498	436
Smelts	28	30	34
Capelin	281	521	513
Flounders, halibut, sole	960	1,090	1,200
Cod, hake, haddock	6,750	7,260	8,150
Redfishes, bases, conjers	3,190	3,220	3,140
Jacks, millets	2,110	2,090	2,030
Herring, sardines, anchovy	16,980	18,740	19,680
Tunas, bonitos, skipjacks	1,200	1,320	1,330
Mackerels, billfishes, etc.	1,670	2,000	2,680
Sharks, rays, chinaeras	400	420	440
Other fishes	7,620	7,960	8,290
Crustaceans	1,190	1,280	1,350
Molluscs	2,880	2,950	3,080
Other invertebrates	44	44	50
Blue, fin, sperm, sea whale	6,468	57,891	51,593
Winke and pilot whales	10,432	9,039	7,951
Porpoises	2	6	7
Seals and walruses	—	4	3
Turtles	4	5	12
Peals, shells, corals, sponges	7	9	11
Aquatic plants	720	750	800

From FAO Yearbook of Fisheries Statistics, 1967, Vol. 25.

The North Pacific and Bering Sea are regions of intense annual exploitation. Seals, fur seals, and walrus are hunted in addition to the whales. The only commercial American operation is the harvesting of fur seals on the Pribilof Islands in the Bering Sea. By treaty, a portion of this harvest is given to the Soviets and the Japanese. Other seals are harvested by the Soviets in the ice edge of the Bering Sea. An estimated 15,000 ribbon seals are annually harvested by the U.S.S.R. (Burns and Kenyon, 1968). Natives of Alaska hunt seals and walrus from shore camps for meat, hides, and ivory.

Seaweeds are currently the basis of an industry that uses their chemical extractives, and continues to be an important food in oriental countries (Booth, 1969). The important chemical extract agar is derived from red algae. The processing of the red algae for this product was for a long time limited to Japan; it is now produced in Chile, Portugal, the U.S.S.R., New Zealand, South Africa, Spain, Canada, and the United States. The red algae also include the main edible species, and many of these are cultured and harvested from natural areas. The brown algae, principally kelp, comprise the other major group of plants harvested from the sea. These seaweeds are used as sources of alginic acid, seaweed meal, and fertilizer. Harvesting is done by the United States, Canada, Great Britain, and Norway.

The annual harvest of seaweeds in the world is increasing (Table 26.5). The demand continues to exceed the supply and this will stimulate the search for new seaweed areas along the coasts of the world. Japan is currently the major producer, largely as a result of its harvests of edible seaweeds which exceed the sum of all world industrial uses.

Aquaculture

With the development of increasingly more effective methods for fish harvesting and expanding scientific knowledge come increasingly more experiments to farm certain species in controlled areas of the sea. In view of over-exploited species and an increasing demand for food, the potential of controlled farming seems great.

The development of aquaculture in coastal waters will effect a change in attitudes in using parts of the sea as a common property resource. Bardach (1968) states: "Aquaculture resembles agriculture rather than fisheries in that it does not rely on a common property resource but presumes ownership or leased rights to such bases of production as ponds or portions of, or sites in, bays or other large bodies of water."

In a detailed study of world aquaculture practices in marine and fresh-

Table 26.5. World Seaweed Harvest, Fresh Weight (metric tons $\times 10^3$)

	1963	1964	1965
Argentina	8.3	10.2	19.9
Canada	22.5	19.8	25.0
France	—	12.9	14.8
Japan	425.7	361	395
Korea	42.9	54.5	N.A.[a]
Mexico	19.5	23.3	17.0
Morocco	2.1	2.2	2.6
Norway	56.9	63.6	74.2
Philippines	0.1	0.1	0.3
Portugal	4.2	4.8	2.7
Ryukyu Islands	0.1	0.1	0.1
Scotland	14.5	18.4	21.2
Spain	4.3	5.3	N.A.[a]
Taiwan	0.8	0.8	1.1
Tanzania	0.3	0.4	0.2
United States	3.3	2.2	2.4

From Booth (1969).
[a] Data not available.

water, Ryther and Bardach (1968) concluded that aquaculture could make a significant contribution to the world food supply. They found that numerous species of marine organisms are currently the subject of aquaculture projects (Table 26.6). The yields from certain organisms reach levels higher than those known for any marine organisms in nature. Well-managed oyster culture in Japan has yielded 58 tons/ha as opposed to the 0.01 to 5 tons/ha in the United States. The potential is indeed great but the practice currently is limited to high quality fish and invertebrates which are the only possible economic competitors under current techniques. A recent review of coastal zone farming practices is presented by Iversen (1968).

Insect Control

On the Atlantic coast of the United States insect control has long been a problem (Spinner, 1969). Before 1920 insects in the central zone were controlled by ditching marshes, by spraying oils or pyrethrums, and by filling mosquito areas. During the 1930s many miles of salt marshes were

Table 26.6. Marine Organisms Used in Aquaculture

Species	Yield [kg/(ha)(yr)]	Method	Location
Plaice, sole		Stocking from hatchery	Great Britain
Shrimp, crab, abolone, sea bream, puffer fish, salmon		Stocking from hatchery	Japan
Lobster, salmon		Stocking from hatchery	United States
Oysters	9–5,000	Cultivation of natural populations	United States
Oysters	400–935	Cultivation of natural populations	France
Oysters	150–540	Cultivation of natural populations	Australia
Oysters	58,000	Cultivation of natural populations	Japan
Cockles	12,500	Cultivation of natural populations	Malaya
Mussels	2,500	Cultivation of natural populations	France
Mussels	125,000	Cultivation of natural populations	Philippines
Mussels	300,000	Cultivation of natural populations	Spain
Porphyra (red algae)	7,500	Cultivation of natural populations	Japan
Undaria (red algae)	47,500	Cultivation of natural populations	Japan
Shrimp	1,250	Cultivation of natural populations	Singapore
Shrimp	6,000	Stocking of cultivation, intensive feeding	Japan

From Ryther and Bardach (1968).

drained. After 1947 the use of insecticides became the accepted method of control. Recently, the knowledge of the role of estuaries in the ecology of the coastal zone and the effects of DDT on organisms have led to a reconsideration of this practice and the outlawing of organic insecticides in many states. Control now involves filling swamps, ditching salt marshes, and using transient chemicals. The flooding of mosquito nursery areas is becoming the most accepted control technique. Delaware, for example, has impounded 3130 acres. It would seem that this technique is compatible with certain other uses of coastal areas, especially aquaculture, and would provide high return from an otherwise expensive control measure.

Wildlife Wetlands

In the United States recognition has been growing of the importance of natural coastal wetland areas to fisheries, wildlife, and public recreation. Recently a zoning plan was prepared for the entire Atlantic coast of the United States (Spinner, 1969). This plan outlined all current land uses along the coast and suggested guidelines to local, state, and federal planners that would minimize the destruction of excessive amounts of marine resource habitat.

New ecological information has emphasized the value of wetlands but how much area and which areas are essential for nursery areas, wildlife, recreation, and scientific research cannot yet be decided. The balance between conservation and development in specific estuaries has not been reached.

In one sense a national program to preserve wildlife wetlands of the coastal zone has been in operation for years. The National Wildlife Refuge and National Park programs have reserved several areas along the coast of the United States.

In the past 5 yr the first underwater area was reserved as a National Park. This is a part of the coastal islands and reefs in the Florida Keys. In addition to the national program, many state and local parks and reserves have been established. Undoubtedly, there is an inadequate number of these areas, and a plan resembling that for the Atlantic coast is urgently needed for the Pacific coast. In addition, the United Nations should address itself to this problem on a global basis and should assist nations in planning for development of future marine resources.

Recreation

In the United States there has been a phenomenal increase in outdoor recreation since 1947. The numbers of visitors to parks has more than doubled, as has the number of boats for use in water sports. The numbers are expected to continue to increase over the next decade. More public parks, new beaches, new marinas, and a variety of other ocean-oriented services will be needed. The dollar value of ocean-oriented recreation was about $14 billion in 1968 and involved an estimated 112 million people (Bigler and Winslow, 1969). Monetarily, this use of the ocean resources ranks first among all others in the United States and is easily the most important use socially as well. The most popular forms of marine recreation include swimming, fishing, boating, water skiing, surfing, and skin diving. The value of recreation is expected to reach $29 billion by 1980. Bigler and Winslow state: "The main areas where new technology and approaches can be brought to bear to enhance the recreational use of the marine environment can be categorized as (a) access, (b) preservation, (c) management (including both physical and living resources), (d) development of new recreation activities, (e) design of new equipment and facilities, (f) marketing, and (g) institutional arrangements." These authors also call for a recreation scientist to complement the other oceanographic scientists.

In addition to the above uses some note should be given to the most popular marine recreation activity—ocean watching. For many people, especially in large seacoast cities, the adjacent sea is a wilderness area of the highest magnitude, having all the redeeming qualities of a pristine forest. The city of Miami Beach epitomizes what an ocean view can do for a locality.

Energy

The dynamic oceans are an immense source of mechanical energy. The endless Pacific swell approaches the shore at about 30 knots and crashes on the beach releasing an average of 40 kW/m (Loftas, 1969). Dramatic examples of the action of waves exist upon every coast. Tales of destruction by storms or the devastation of a Tsunami reveal the tremendous potential energy of sea waves (Anonymous, 1970a). So far the energy of the sea has not been utilized; it has, however, been directed for sediment transport.

Currently there is more promise of harnessing the energy of the sea through the development of hydroelectric plants using the power of the tides. The first station to derive energy from the tides was built at the Tance Estuary in France in 1966. This station is designed to produce 240 mW of electricity from a tidal difference of 13 m at spring tides.

The other other operational station was opened in 1968 at Kislogubsky near Murmansk, U.S.S.R. This arctic station is an experimental project producing 800 kW. Other tidal power stations have been planned in the United States, Canada, and the U.S.S.R.

An essential for tidal power is the range of the tides which determines the hydrostatic head for power generation. Only about 20 to 25 locations in the world have the required range of tides for power generation and of these only about half have been seriously considered.

Minerals and Mining

The sea and its floor have long been known to contain immense quantities of minerals. Presently the cost of mining most of these minerals is far greater for sea than for land sources. Nonetheless, active programs of mineral exploration currently extend throughout the world ocean and adjacent seas (Marine Science Affairs, 1970; Chapter 21, this book).

The sea as a source of minerals offers some distinct advantages over the land (Mero, 1964). The distribution of a marine mineral is generally widespread and consequently available to most coastal nations without the political-economic constraints that operate to localize and nationalize land minerals. This aspect of the sea will become increasingly important as the technology for ocean mining develops, making resources more available.

Currently most mining of the sea is limited to beaches and continental shelves. Only one deep sea mining operation now exists (Table 26.7). These operations involve several minerals, including gold and diamonds, that are extracted from the unconsolidated marine sediments. The sub-bottom deposits being mined include sulphur, magnetite, and petroleum and natural gas. Petroleum accounts for more than 90 per cent of the value of marine mining and constitutes about 16 per cent of the world petroleum supply (Marine Science Affairs, 1970).

A limited amount of recovery of minerals and freshwater from seawater is now practiced. Common salt is the most important extractive and accounts for about 29 per cent of all salt mined. Desalination is fast reaching the practical level as a source of freshwater in areas of high freshwater demand and limited sources of supply (Othmer, 1969).

Depletion of land sources coupled to developing technology can only lead to enhancing the value of marine minerals. It is now time to initiate studies of effects of certain types of mining on other ocean resource uses.

Drugs and Pharmaceuticals

The application of medical sciences to the seas has barely begun. Isolated programs of research on toxins, drugs, and pharmaceuticals from marine organisms exist but a long-term coordinated effort is lacking (Pell

Table 26.7. Marine Minerals Currently Being Mined

Deposit and Mineral Type	Mining Location
Beach Deposits	
Diamonds	SW Africa
Gold	Alaska
Magnetite	Japan
Cement rock sands	California, Gulf of Mexico, Iceland
Ilmenite, zircon, magnetite	Ceylon, India, Australia, Brazil
Chronite, gold, platinum	Oregon
Silica, sand, gravel	California
Titanium	Florida, New Zealand
Continental Shelves	
Calcareous shells	California, Gulf of Mexico
Phosphorite	California
Tin	Thailand, Indonesia
Platinum	Alaska
Sulphur	Gulf of Mexico
Magnetite	Gulf of Finland
Iron ore	Newfoundland
Coal	Japan, England, Nova Scotia
Petroleum and natural gas	North Sea, Gulf of Mexico, California, Alaska, Persian Gulf
Deep Sea Floor	
Manganese nodules	Blake Plateau, Atlantic Ocean
Sea water	
Sodium	California, Sweden, U.S.S.R.
Chloride	China, India, Japan, Turkey, Philippines
Magnesium	Texas, California
Bromine	North Carolina, Texas, California
Freshwater	Kuwait, Hong Kong, California, Cuba

From Mero (1964); Pell and Goodwin (1966); Maine Science Affairs (1970).

and Goodwin, 1966; Marderosian, 1968). The possible contribution of this field to medical science seems immense and worthy of effort by national and international agencies (Freudenthal, 1968).

Ocean Technology

The progress and expansion of ocean-oriented activities that has occurred in the last 20 yr has been dependent on and has instigated an associated development of ocean technology and engineering. This field is complex and diverse and relates to all uses of the oceans—including transportation, communications, navigation, ships, subsea vehicles and habitats, photography and television, underwater equipment and tools, satellites, bouy systems, ocean structures, marine materials, seafood processing equipment, fishing gear, recreational equipment, and military technology. The literature on all of this is voluminous. An adequate survey of this entire field demands consultation of original source material (Brahtz, 1968; Bretschneider, 1969; Meyers et al., 1969; Wooster, 1969). In this chapter only more recent trends and developments can be covered, and then only in a cursory way to indicate the complexity of the problem of ocean use. An excellent article on the new ocean technology recently appeared (Bascom, 1969).

Shipping and transportation by sea have continued to increase since 1956. The increase for 1967 to 1968 resulted in a gain of 672 ocean-going merchant ships; the world total merchant fleet in 1968 was 18,208 vessels of which 64 per cent are registered in ten nations (Clark et al., 1970). One of the significant recent developments in this area of ocean technology is the container system of cargo handling. This simple innovation promises to greatly accelerate the movement of goods by sea as well as reduce the turn-around time for a vessel. Widespread use of these units will require changes in port cargo handling facilities and in vessel design.

Another major development in transportation is the construction of supertankers such as the *Universe Ireland*, a 312,000-ton vessel that carries oil from the Persian Gulf to Ireland. Tankers of this size have proven to be economical but demand special port facilities. The ability of a ship to navigate in the open sea has been a problem for a long time. Currently, through the use of electronic devices, a guidance systems, and orbiting satellites, a ship that is 1000 miles from shore can locate its position to within 0.1 mile (Loftas, 1969; Badgely et al., 1969).

In recent years a variety of unconventional surface vessels, as well as a number of subsea vessels have been developed (Loftas, 1969; Hunley, 1968). The development of hovercraft and hydrofoils have added a greater dimension to surface transportation; both are much faster than conventional ships. A large hydrofoil has not yet been developed for commercial application, but the U.S. Navy has an operational 71-ft hydrofoil gunboat. The British have constructed a number of successful hovercraft for commercial use. The largest is a 128-ft long ferry that carries 250 passengers and 30 autos. The ship crosses the English Channel three times faster than conventional ferries. Another unconventional ship is the oil tanker *Manhattan* that has been refitted for experimental work in Arctic ice (Anonymous, 1969).

The building and design of submarine vehicles for commercial exploitation and ocean research has proliferated since the days of the *Trieste* (Sweeney, 1970). Marine Science Affairs (1970) lists some 60 vessels built in the United States for undersea work. In addition to these craft there have been several experimental undersea habitats constructed that have permitted men (and women) to live and work for periods of several weeks at depths as great as 600 ft (Clifton et al., 1970; Stachiw, 1968).

The undersea activities of man, spurred largely by military and mining goals, have developed a variety of accompanying tools and equipment (Hromadik, 1968; Spiess, 1968). Underwater photography and television are some of the more important instruments, and refinements have made them highly reliable tools. The variety of underwater tools continues to grow as the oil and mining companies extend their operations into the sea. More and more devices designed for work on land are being adapted to the sea. As an example, a company has developed and used a submarine tractor-dredging machine for obtaining offshore sand deposits (Bascom, 1970).

Working and living in the sea have resulted also in the development of new diving equipment, permitting greater comfort and agility at depth (Bond, 1969). Furthermore, interest in ocean equipment has spurred considerable research and development of materials suitable for saltwater. Glass, plastics, and fiberglass have been the miracle materials of the oceans (LaQue, 1968). In addition, new high-strength steel and aluminum are being developed and used; new experiments with ferro-cement material for boat hulls appear promising.

The exploitation of continental shelf oil reserves rivals the military in its effect on the progress of ocean technology. Hundreds of offshore drilling platforms have been built and have demonstrated success despite formidable environmental conditions. An outcome of the continental shelf drilling is the *Glomar Challenger*, a deep-sea drill ship, that has successfully drilled in depths greater than 10,000 ft. It is even possible to reenter a drill hole at this depth (Anonymous, 1970b).

The military effort for many years sponsored nearly all marine technology, development, and research. Unfortunately, the most important military developments of the day are likely to remain secret for some time. One of the more publicized programs is the development of submarines and detection equipment. Today the submarine has become a highly sophisticated, nuclear-powered, missle-carrying weapon. These efficient machines have directed a considerable amount of Navy effort to the development of submarine locating devices. One result of the airborne detection system is the use of hydrophone bouys that broadcast to aircraft. This and numerous other military programs have feedback to industry and research. The U.S. Navy continues to spend more than half of the federal ocean research and development budget (Marine Science Affairs, 1970).

Waste Disposal

Waste disposal in the oceans has long been practiced. In the past pollution resulting from it was of little concern because of the apparent limitless capacity to absorb and assimilate these wastes. It has now been found that, as with rivers and lakes, every body of water has a limited capacity to absorb and neutralize inflowing materials. In the case of the oceans, as indicated in the introductory chapter, the capacity for waste assimilation is great but the time lag required for exhaustive mixing is excessive. No better example is now apparent than that discussed by Brocker et al. in Chapter 11 concerning fossil-fuel-produced carbon dioxide distribution between the ocean and the atmosphere. At equilibrium each molecule of CO_2 gas which remains in the atmosphere would be matched by 60 molecules mixed in the oceans. Thus a production of an equal amount of CO_2 from fossil fuels to that now present in the atmosphere would increase the atmospheric content by less than 2 per cent. Yet because of the sluggish mixing of the oceans these authors predict a doubling of the atmospheric CO_2 content sometime at the beginning of the next century. These and other data already indicate that to use the potential capacity of the ocean for waste disposal will require that much more attention be given to creating favorable mixing situations either by taking advantage of natural processes that might be found in basins with high tidal velocities, such as Cook Inlet, the Bay of Fundy, or to enhance mixing by mechanical means, such as dispersal from moving barges (Ketchum and Ford, 1952; Hood et al., 1958) or through dispersers properly engineered to give the desired dilution before release into the ocean.

Wastes Reaching the Sea by Direct Addition

Wastes enter the oceans from many sources.* With increased coastal urbanization and general population increases the pressure on the oceans for waste disposal is becoming enormous. The highly industralized society of the United States makes this country responsible for about 20 per cent of the worlds' total coastal additions. Estimates of various kinds of waste disposed at sea in various regions of the United States is shown in Table 26.8.

Effluents for sewage disposal plants, industrial outfalls, runoff from land, and that arising from spillage in waterways all reach the oceans unless degraded, precipitated, or adsorbed while in transit. A quantitative estimate of the total amounts reaching the oceans has not been made. The exercise of measuring these amounts would be difficult and probably of limited value at the present time since most of the effects of the waste occur near the point of discharge. If local waste disposal problems are properly handled a relatively low level of waste would reach the open ocean reservoir.

The common fertilizer-type nutrient wastes resulting from decomposition of human sewage which reach the sea through rivers is probably not a serious contaminant because of the small contribution it makes. Some influence on productivity might occur near river mouths or in estuaries which may, depending upon the circumstances, be detrimental. More serious problems arise from industrial wastes composed of types of materials listed in Table 26.9. Here lies an area of considerable concern. New chemical substances are produced commercially at the annual rate of 400 to 500. Some of these are toxic and nondegradable, and will find their way into the ocean. We are just now recognizing the serious problems with DDT and its closely allied compounds, the polychloraromatics. These are also airborne contaminants. Is it possible that air pollutants or contaminants are as serious to the world ocean contamination as are water pollutants or contaminants carried by the surface water drainage systems? There is increasing evidence that this may be so.

Inert plastics, which the synthetic organic chemist struggled so long to perfect, are excellent in food and commodity preservation, but offer serious problems in their ultimate disposal. On this problem technologists

---

* A most unwise use of the ocean may be exemplified by the use of a region on the continental shelf off California as a munition dump. This is a region of known phosphorite deposits. Collier Carbon and Chemical Corp., operating on a lease with the U.S. Government, had to abandon efforts to mine the phosphorite because of the hazards from unexploded shells. This is truly unwarranted exclusion of the multiple-use concept and cannot be allowed to continue if proper resource utilization is realized.

Table 26.8. Estimated Amounts and Costs of Wastes Barged to Sea in 1968[a]

Wastes	Pacific Coast Disposal		Atlantic Coast Disposal		Gulf Coast Disposal	
	Tons	Cost (dollars)	Tons	Cost (dollars)	Tons	Cost (dollars)
Dredging spoils	7,320,000	3,175,000	15,808,000[b]	8,608,000	15,300,000	3,800,000
Industrial wastes (chemicals, acids, caustics, cleaners, sludges, waste liquors, oily wastes, etc.):						
Bulk	981,000	991,000	3,011,000	5,406,000	690,000	1,592,000
Containerized	300	16,000	2,200	17,000	6,000	171,000
Garbage and trash[c]	26,000	392,000				
Miscellaneous (airplane parts, spoiled food, confiscated materia etc.)	200	3,000				
Sewage sludge			4,477,000[d]	4,433,000		
Construction and demolition debris			574,000	430,000		
Totals	8,327,500	4,577,000	23,872,200	18,894,000	15,996,000	5,563,000

From Marine Disposal of Solid Wastes, an Interim Summary, October 1969. Dillingham Corp., Applied Oceanography Division Under Contract to Bureau of Solid Wastes Management, Department of Health, Education, and Welfare, October 1968.

[a] Does not include outdated munitions.
[b] Includes 200,000 tons of fly ash.
[c] At San Diego dumping of 4700 tons of vessel garbage was discontinued in Nov. 1968.
[d] Tonnage on wet basis. Assuming average 4.5 per cent dry solids, this amounts to approximately 200,000 tons/yr dry solids being barged to sea.

Table 26.9. Types of Industrial Wastes

Origin	Type of Waste
Slaughter houses	Organic compounds
Fish and fish-meal factories	
Breweries	
Fat and margarine industry	
Sugar and starch factories	
Fruit and beet sugar	
Fairy products	
Fuel and petrochemicals	
Dye-stuffs	
Chemical pulp mills, paper mills	Organic and inorganic compounds
Fiber and plastics industry	
Textile industry	
Leather industry	
Soap industry	
Soap works	
Photographic chemistry	
Paint and explosives	
Iron and other metals	Inorganic compounds
Mining	
Electroplating industry	
Cement and ceramics	
Fertilizer	
Acid manufacture	

are faced with a dilemma. Burning the plastics creates air pollution; burying them causes serious solid waste problems. The answer probably lies in the use of environmental degradable plastics that would hydrolyze, react with ambient light, or be consumed by bacteria or other natural degrading agents.

The long-term chronic effects of certain fractions of crude petroleum are still unknown and the quantities of this material reaching the ocean has now reached over a million tons annually (see Chapter 13). We know less of the synthetically modified hydrocarbons, the petrochemicals. The amounts which reach the oceans is difficult to determine. Some headway has been made in assessing the extent of the petrochemical additions to the environment by Parker, as discussed in Chapter 15. However, the diversity of products and by-products from this industry and the unfortunate lack of control of effluent discharge into coastal water near major centers of petrochemical activity makes it virtually impossible at the present time to give valid data on the kinds and amounts of compounds discharged or their ecological impact. Fortunately most of the compounds discharged are water-soluble and of low molecular weight. They probably most severely affect local waters, with deep oceans escaping harmful effects.

In the industrial wastes we find the compounds which may lead to world ocean pollution, but it is also these wastes that can be effectively removed from the effluent waters if the potential hazard is recognized and appropriate action taken. The problem of long-term accumulation of nonbiodegradable or physiologically highly sensitive compounds has not been adequately addressed and is in serious need of careful scientific attention.

Wastes Reaching the Sea Through the Atmospheric Exchange

Each year 350,000 tons of dry cleaning solvents evaporate into the air, accompanied by one million tons of gasoline lost through evaporation. Much of this eventually reaches the oceans. The lead problem has been capably discussed by Patterson in Chapter 9. About 10,000 tons reach the ocean on an annual basis. There is a vast difference in the problem of contamination by these three substances. Perchloroethylene, the most common ingredient of dry cleaning fluids, is relatively nontoxic and decomposes through bacterial action. Gasoline will behave similarly. The transport of these and lead to the sea surface is by similar routes—atmospheric and aerosol form—to eventually be washed out by rain and falling dust particles, or by direct interaction with the sea surface. Chlorinated aliphatic hydrocarbons (not to be confused with chlorinated aromatic hydrocarbons) and gasoline will decompose. Lead is an accumulative toxic metal and will be in the ocean hundreds of years before being removed through reactions with the sediments.

Carbon monoxide, produced chiefly by internal combustion engines on land, appears to originate also from the ocean (Swinnerton et al., 1969). Possible origins of this gas are photochemical oxidation at the surface, or production from algae, green plants, and siphonophores. Some escapes through leakage or as a result of decomposition of the organisms. The amounts so produced provide a flux to the atmosphere. There is no evidence of a high accumulation of this gas either in the atmosphere (0.075–0.44 ppm) or the oceans (10 to 450 ml/l $\times$ 10^6).

A proper ocean use is that of waste disposal. This has occurred through geological time. Man has now begun to exceed the oceans' capacity to assimilate these wastes. It is this use of the oceans that unquestionably provides the most hazardous threat to worldwide ocean pollution. It is this threat that needs the most careful scrutiny in avoiding serious impingement of man on the ocean environment.

Decision-making Processes in Ocean Use

In beginning the discussion of this subject it is important to point out the great disparity between coastal, continental shelf, and deep-sea uses of the ocean. These three regions require entirely different approaches. The coastal zone is that narrow strip of land and water along the shore of the oceans, tidal estuaries, and very large lakes whose use is greatly influenced by its proximity to land and sea. This is where people work, play, and live, and the importance of the local region to local citizenry makes the local population the first segment of society to be involved in management of uses in the coastal region, commenserate of course with the concept that if a larger segment of the population benefits then the local interests must yield to the larger area pressures. In these cases, as must always be the case in a democracy, the individual benefit must yield to the greater benefit of a group, which in turn must yield to the larger population segment. Finally national needs must have dominance over all others, but only after due process in establishing these needs. Problem-solving in these cases often becomes one of definition and assessment of benefits. For example, is a pulp mill, which must occupy beach property for its mill site, which requires transportation lanes and storage facilities for raw materials and finished materials, and which, by necessity, may change the area from one of recreational to industrial use, of greater public benefit than the benefit to what may be relatively few local sports enthusiasts? Again the problem is one of defining the public interest and must of course relate to how much of the total areas resources are effected by such a user group.

The continental shelf or that region that reaches from the shore to the edge of the abyssal deep has a broader, but less intimate, interest of a large segment of the population even to the extent of state or sovereign boundaries. It is most important in an industrial and military sense, and its uses for public benefit can best be determined in terms of economic and security considerations. The conflict with other countries as to jurisdictional rights is most vigorous in this zone. The present legal problems now being considered and fought in the world courts are discussed by Krueger in Chapter 24.

The deep sea and the sea bottom are of ever-increasing importance.

Technological advances show that the once unreachable sea depths are now within reasonable access as shown by the U.S. National Science Foundation deep drilling program of *Glomar Challenger*. Oil provinces not previously dreamed of in the deep sea are now highly probable and techniques for deep extraction of this resource are rapidly advancing. This matter of using the ocean floor has been discussed by the Commission on Marine Science (1969), commented on by Miron (1969), and evaluated from a legal point of view by Kruegar (Chapter 24). Conflicts of interest in both uses and in sovereign rights will continue to occur. But it is imperative that the high seas will never be used in a way that would jeopardize the ocean's quality as a resource for the maximum public interest of all world people.

The decision-making processes in each of the above regions will vary greatly. The coastal zone area is used as an example in this discussion. Obviously the use of the coastal zone should be initiated locally at the city, county, or borough level, but the authority should rest with the state. Where conflicts occur it should rest with a region of states or the Federal government. The magnitude of the project often determines the bodies involved in decision-making. A sewer outfall or a marina for a small coastal town may be located by the Corps of Engineers and the State Department of Health, and may well have limited consideration for other ocean uses. Far too often, however, these decisions have been made with little knowledge of the oceanographic influences involved. Many past mistakes could have been avoided with rudimentary knowledge of local currents, rates of dispersion, or littoral drift of sediment. We can no longer afford to make off-the-cuff decisions of this level of importance. Minimum criteria for such uses must be set and adhered to if maximum benefit from ocean uses is to be a reality. It is evident therefore, that in every state or group of states criteria for zoning and ocean uses must be established and a coastal zone plan initiated. Many coastal states are currently forming such groups, and it now appears that (*a*) zoning must be initiated on presently available information, (*b*) a detailed survey to obtain an inventory of coastal resources, to include oceanographic parameters, must be obtained to give a basis for a specific-use designation, and (*c*) rezoning must occur continuously to be current with new information and developed ocean uses. These functions must be carried out and administered by a coastal zone authority through a developed program of physical, biological, social, and economic research coupled with a system of public hearing. Far too often decisions are made far short of the above due processes and often to the detriment of the public interest.

An example of the types of problems that have arisen without coastal zoning or coastal management authority can be illustrated by the pipeline terminal development of Port Valdez, Alaska. In an effort to get the huge petroleum reserves out of the Arctic slope to the energy-hungry contiguous 48 states, Port Valdez was chosen by the oil companies with very little consideration of the factors discussed above. This is not to put blame on any segment of the population. Rather, it points out the need for multiple-user consideration in establishing facilities of this magnitude. Few states, Alaska included, now have a functioning body with authority to assign resource uses. This lack will continue to cause much unneeded consternation that could be avoided both by user and public had coastal management existed.

I (D. W. H.) am writing this chapter while at the Wyatt House in Valdez, Alaska, where I can look from my window across 3 miles of Port Valdez to the site of the terminal of the Trans Alaska Pipe Line. Within 3 yr the super tankers will probably come to the harbor facilities across the bay at the rate of about 2 each day to handle the 2 million barrels/day flow of Prudhoe Bay crude oil that has crossed the 800 miles of arctic tundra, frozen ground, and tiaga forests to reach this terminal. Limited storage will be provided in a tank farm nestled in a saddle of a sharply rising 4000-ft mountain.

Port Valdez is one of the pristine wonders of Alaska. It is a tidal fjord estuary about 20 miles long and 3 to 5 miles wide, surrounded by precipitious mountains, waterfalls, and visible glaciers. The town of Valdez, inundated by the 1964 Alaska earthquake, has now been rebuilt on the north shore of the inlet. Its main occupation has been tourism for which it is properly famous. The Port enjoys a good run of silver, pink, and chum salmon during later summer months. This, coupled with the exotic beauty of the surroundings, guarantees its continued success at attracting tourists who come by excursion boats (the first one of the season came in today, May 8th), ferries, airlines, and various road vehicles down the Richardson Highway from the interior of Alaska.

Port Valdez opens in the southwest through the Valdez Narrows into Prince William Sound. Prince William Sound has a prodigenous quantity of only partially exploited renewable resources ranging from the invertebrates—tanner, king, and dungeness crab; razor, butter, and little neck clams; and scallops—to the teleosts of commercial importance—including salmon, halibut, herring, flatfish, ocean perch, cod, and hake—to mammals—seals, sea lions, whales, and even bear and deer, which frequent the grassy beaches and freshwater streams.

This truly is a confrontation of powerful and important public interests.

Great concern has been expressed on worldwide news coverage about the hazards and woes of the pipeline itself which passes through the arctic down to the ice-free port of Valdez. The task of building such a line is indeed formidable; and the effort, the largest ever undertaken by private enterprise, must involve some innovative techniques to cope with frozen ground, natural hazards such as earthquakes, maintenance of spawning grounds for anadromous fish, and preservation of terrain for the migrating caribou and other big game animals. Difficult as this problem may be, the maintainance of multiple use of the heavily stressed coastal zone may prove to be even greater.

In cases of this type, which use or uses should have precedence? What are the values? How are they equated? Are they necessarily incompatible? The scenic view across the bay from my window will undoubtedly change. The coming and going of large tankers can itself be interesting, but it will be different. The docks and storage tanks will be miniscule compared with the background mountains, but they will appear as some of man's greatest structures. They need not deter appreciably from the overall scenic value of the total area; and from the tourist's point of view they may even supply new reasons for coming to Valdez. These two uses do not seem to be inherently incompatible.

What of the renewable resources of this region? There are many questions to be answered. The tankers will bring with them from the "lower 48" journey large quantities of seawater ballast necessary to make them seaworthy in crossing the North Pacific. This ballast water will be processed to meet water quality standards for disposal into the waters of Port Valdez. Most of these hydrocarbons will be quickly degraded, for hydrocarbon-utilizing organisms exist in all natural waters awaiting the addition of petroleum hydrocarbons for their source of energy. Were this not so the ocean would be covered by a continuous slick as a result of over 10^6 metric tons (Moss, Chapter 13) of petroleum being added to the sea each year. There is some question of the effects of low levels of petroleum on the indigenous phytoplankton's photosynthetic capacity and on the secondary producers' (the zooplankton) rate of growth and reproduction. This has not been adequately examined at this writing and should be the subject of an early investigation under simulated natural conditions. Should the present allowable levels in ballast water be too high, it is relatively easy technologically (through construction of suitable diffusers) to reduce the concentration level entering the bay or to disperse the ballast water in such a way that only waters below the euphotic zone would receive the waste. Additional oceanographic data are needed to determine that the circulation and flushing rates of the system are such that the hydrocarbon levels affecting any known biological process will not be reached under normal operational procedures.

An inherent difficulty with petroleum tanker operations is the occasional accidental spill. This, in a well-planned operation routinely operating as indicated above, is the nemisis of any oil-loading facility. Spillage, coupled with the ever-present hazard of catastrophic events in which large volumes of oil may be lost, comprises the real concern in resource-use conflicts.

The first group of animals to suffer from a petroleum spill are the coastal birds, followed by mammals, beach organisms (especially the egg or larval stages of spawning fish or invertebrates), and finally the organisms, plant or animal, that occupy the intertidal regions of beaches.

Between me and the proposed tanker facility at a distance of about 200 m is a spit of land used by large numbers of arctic terns as a nesting colony. These birds feed on the biological plenty of the fiord waters and would be most affected by a surface slick. The birds' feet and feathers would become oily. Oil would be transmitted to the eggs or young, bringing havoc to the reproduction and rearing rate, particularly if such spills occurred in the early part of the season. Similar sensitive periods occur for other forms of life in the area. Some effects, such as that on the eggs and fry of the intertidal spawning salmon, would be very subtle and could occur anytime in the 6 to 8 winter months during which the eggs lie buried in the stream-bed gravels. It is not yet known what the effects of petroleum are on salmon eggs, primary productivity, larval or adult invertebrates, or seaweed—but all this must be learned. It is unthinkable that these resources must turn toes up in the interest of the nonrenewable resources.

It is well known that there are serious effects to organisms and property from surface oil contamination of the ocean waters, particularly in the coastal regions. Adequate controls and clean-up procedures to prevent these effects from occurring are now available or can be developed. Technology can conceivably curb, control, and clean up small oil spills on the sea surface. Surely these techniques will be operational in Valdez, and problems arising from accidental spillage will be minimized. Some damage to the biota will undoubtedly occur from time to time, but it must in each case be corrected or restored to prevent more major damage to a significant portion of important organisms.

There yet remains the most serious problem of all—that of a major tanker or oil spillage accident. To help ensure against this, advanced technology in harbor control and surveillance of the most sophisticated type are needed. All available assistance for the safe operation of tankers in these waters is essential. This should involve weather, wave, and current prediction, and positioning reconnaissance of all area vessels by a harbor authority. In addition, an oil spill alert force that can respond to catastrophies with effective means of control is essential.

Where does the liability for damage lie when all the above fail is a question of much discussion. There is little question but that the liability lies with the user that causes the damage. In the case of oil spills, does blame lie with the oil companies who own the cargo, or the carriers who transport it? Still more difficult is the question of what are the liabilities. How are damages repayed? Can the colony of terns on the near spit be replaced, and if so, for how much? How can a run of salmon damaged by oil spillage be reconstituted? Can a reduction in biological productivity of the area be adequately assessed, and if so, can it be reestablished to its former levels and at what cost? These questions provide material for research and study for industry (both oil and fisheries), university, local, state, and federal government scientists and technologists in the immediate years ahead. Data on the levels and exigencies of the present resources must be gathered and analyzed. Means must be discussed for replenishment of natural resources should damage occur. The cost of damage, it must be realized, is more than monetary. Disruption of the balance of nature by man can no longer go ignored. The living organism earth, and with it man's survival, is at stake.

The difficult question to answer in the case of the Port Valdez is to what extent can multiple use of this resource be consummated. The same question could be asked of many locations all over the world. Many people have been convinced that one must give up biological resources, recreation, beautiful scenery, and other values in the interest of a bludgeting economic pressure. Far too often this has been true. Experience without the vision of improved attitudes, despair without hope of technological advances, pressure of economic development without adherence to environmental warning—all lead one to pessimism. Yet the future of the world depends upon change.

Recently a committee of the National Academy of Science and National Academy of Engineering drew up a priority list and an estimated minimum effort required to initiate an effective coastal wastes management program. Their recommendations estimate 20 man-years needed for research and investigation for improving waste discharge and receiving water monitoring programs, 720 man-years for physical processes and interactions, 450 man-years for chemical factors, and 1280 man-years for studies of biological effects. If one uses the rule of thumb of $50,000/

man-year, a total of about $140 million will be needed for research in this area in the next 5 yr. One need not be a national or state budget expert to realize that these kinds of funds are not presently forthcoming (Nas-Nae, 1970).

There is no question but that we are at the crossroads; we must choose our destiny. Either we plunge recklessly into a sea of known and potential hazards, with almost certain irreversible impact as the result of continued loss of air and water and ever-increasing discharge of waste loads, or we go into a carefully plotted course toward waste reuse, closed cycling, and minimal discharge to the environment except for intended, nondamaging, or even beneficial materials.

The latter choice is the only one tenable. The costs will be high: the advancement in standard of living (creation of wastes) must cease and the philosophy of economic gain through increased per capita consumption will have to be eliminated. *Not to live compatibly with the living organism earth is not to live at all;* the oceans are *our last frontier.* There is yet time to eliminate the smog, clean the rivers and estuaries, reclaim the lakes, and keep the ocean clean. But it is now only a few minutes to midnight. We on the earth must respond now—in this decade—or we will lose our destiny.

References

Anonymous. 1969. Retrofitting the *Manhattan* for the Arctic. *Ocean Ind.,* **4**:71–72.

Anonymous. 1970a. New concept for harnessing ocean waves. *Ocean Ind.,* **5**:62–63.

Anonymous. 1970b. Re-entering a hole in 10,000 feet of water. *Ocean Ind.,* **5**:28–29.

Badgely, P. C., L. Miloy, and L. Childs (eds.). 1969. *Oceans from space,* Gulf Publishing, Houston, 234 pp.

Bardach, J. E. 1968. Aquaculture. *Science,* **161**:1098–1106.

Bascom, W. 1969. Technology and the ocean. *Sci. Amer.,* **221**:199–217.

Bascom, W. 1970. Underwater dredge. *Ocean Ind.,* **5**:16–18.

Bigler, A. B., and D. F. Winslow. 1969. *Marine recreation: problems, technologies, and prospects to 1980,* Marine Technology Society, pp. 43–60.

Booth, E. 1969. Seaweeds of commerce. In *Encyclopedia of marine resources,* Frank E. Firth (ed.), Van Nostrand Reinhold, pp. 626–630.

Bond, G. F. 1968. Undersea ambient environmental habitation and manned operations. In *Ocean engineering,* J. F. Brahtz (ed.), John Wiley, New York. pp. 487–492.

Brahtz, J. F. (ed.). 1968. *Ocean engineering*, John Wiley, New York, 720 pp.

Bretschneider, C. I. (ed.). 1969. *Topics in ocean engineering*, Gulf Publishing, Houston, 420 pp.

Burns, J. J., and K. W. Kenyon. 1968. Proposal to negotiate an international meeting on the conservation of ice seals in Bering and Chukchi Seas. *Amer. Sci. Mamal.*, Marine Mammal Comm., 7 pp. In press.

Clark, E. W., H. W. Haddock, and S. J. Jones. 1970. *The U.S. merchant marine today*, The Labor-Management Maritime Committee, Washington, D.C., 243 pp.

Clifton, H. E., C. V. W. Mahnken, J. C. Van Derwalker, and R. A. Waller. 1970. Tektite I, man-in-the-sea project; marine science program. *Science*, **168**:659–663.

Commission on Marine Science. 1969. *Our nation and the sea*, U.S. Government Printing Office, Washington, D.C., 305 pp.

Dawson, E. Y. 1966. *Marine botany*, Holt, Reinholt, & Winston, New York, 371 pp.

Ehrlich, P., and A. Ehrlich. 1970. The food-from-the-sea myth. *Saturday Rev.*, SR/APR **4**:53.

Firth, F. E. (ed.). 1969. *The encyclopedia of marine resources*, Van Nostrand Reinhold, 740 pp.

Food and Agriculture Organization. 1967. *Yearbook of fisheries statistics*, Vol. 25, United Nations.

Freudenthal, H. D. (ed.). 1968. *Drugs from the sea*, Marine Technology Society, 297 pp.

Gribbin, A. 1968. *Sea horizons*, The National Observer, Princeton, N.J., 205 pp.

Hood, D. W., B. Stevenson, and L. M. Jeffrey. 1958. Deep sea disposal of industrial wastes. *Ind. Eng. Chem.*, **50**:885–888.

Hromadik, J. J. 1968. Deep-ocean installations and fixed structures. In *Ocean engineering*, J. F. Brahtz (ed.), John Wiley, New York. pp. 310–349.

Hull, S. 1964. *Bountiful sea*, Prentice-Hall, Englewood Cliffs, N.J., 340 pp.

Hunley, W. H. 1968. Deep-ocean work systems, pp. 493–552. In *Ocean engineering*, J. F. Brahtz (ed.), John Wiley, New York, 720 pp.

Iversen, E. S. 1968. *Farming the edge of the sea*, Fishing News (Books) Ltd., London, 301 pp.

Ketchum, B. H., and W. L. Ford. 1952. Rate of dispersion in the wake of a barge at sea. *Trans. Amer. Geophys. Union*, **33**:680–685.

LaQue, F. L. 1968. Materials selection for ocean engineering. In *Ocean engineering*, J. F. Brahtz (ed.), John Wiley, New York. pp. 288–632.

Laughlin, W. S. 1967. Human migration and permanent occupation in the Bering Sea area. In *The Bering land bridge*, D. M. Hopkins (ed.), Stanford Univ. Press, Stanford, Calif., pp. 409–450.

Lawrence, L. G. 1967. *Electronics in oceanography*, H. W. Sams, Indianapolis, Ind., 288 pp.

Loftas, T. 1969. *The last resources*, Hamilton, London, 256 pp.

Lowman, F. G. 1969. Radionuclides of interest in the specific activity approach. *Biosci.*, **19**:993–999.

Marderosian, A. D. 1968. Current status of drug components from marine sources. In H. D. Freudenthal (ed.), *Drugs from the sea*, Marine Technology Society. pp. 19–66.

Marine Science Affairs. 1970. Annual report of the President to the Congress on marine resources and engineering development. U.S. Government Printing Office, Washington, D.C., 284 pp.

Mero, J. L. 1964. *Mineral resources of the sea*, Elsevier, 312 pp. Houston.

Meyers, J. J., C. H. Holm, and R. F. McAllister. 1969. *Handbook of ocean and underwater engineering*, McGraw-Hill, New York.

Miron, G. 1969. The management of the mineral resources of the ocean floor. *Stanford Journal of International Studies*, **4**:32–45.

Morris, D. 1965. *The mammals*, Harper & Row, New York, 448 pp.

National Academy of Sciences, National Academy of Engineering. 1970. *Waste management concepts for the coastal zone*. Nas-Nae. Washington, D.C., 126 pp.

Othmer, D. F. 1969. Desalination of seawater. In *The encyclopedia of marine resources*, F. E. Firth (ed.), Van Nostrand Reinhold, New York. pp. 162–169.

Pell, C., and H. L. Goodwin. 1966. *Challenge of the seven seas*, W. Morrow, New York, 306 pp.

Russell-Hunter, W. D. 1968. *A biology of lower invertebrates*, Macmillan, New York, 181 pp.

Ryther, J. H., and J. E. Bardach. 1968. The status and potential of aquaculture, particularly invertebrates and algae culture prepared for National Council on Marine Resources and Engineering Development, P.B. 177767 (Clearinghouse Fed. Sci. Tech. Info., Springfield, Va.).

Seymour, A. H. (ed.). Radioactivity in the marine environment, NAS–NRC, Washington, D.C. In press.

Spiess, F. N. 1968. Oceanographic and experimental platforms. In *Ocean engineering*, J. F. Brahtz (ed.), John Wiley, New York. pp. 553–587.

Spinner, G. P. 1969. Wildlife and wetlands of the Atlantic coastal zone. Amer. Geographical Soc., 80 pp.

Stachiw, J. D. 1968. Hydrospace—environment simulation. In *Ocean engineering*, J. F. Brahtz (ed.), John Wiley, New York. pp. 633–711.

Sweeney, J. B. 1970. *A pictorial history of oceanographic submersibles*, Crown Publishers, Long Island City, N.Y.

Swinnerton, J. W., V. J. Linnebom, and R. A. Lamontagne. 1970. The ocean: a natural source of carbon monoxide. *Science*, **167**:984–986.

Tressler, D. K., and J. M. Lemon. 1951. *Marine products of commerce*, Reinhold, New York. 782 pp.

University of California. 1965. California and use of the ocean. Institute of Marine Resources, Univ. of Calif., La Jolla, Calif.

Vogt, J. R. 1969. Radionuclide production for the nuclear excavation of an Isthmian Canal. *Biosci.*, **19**:138–139.

Wooster, W. S. 1969. The ocean and man. *Sci. Amer.*, **221**:218–234.

Author Index

Aase, Jim, 334
Abbott, B. C., 617
Acree, F., 269, 270
Adams, A. S., 219
Adams, J. A. S., 16
Albenesius, E. L., 336
Alekin, O. A., 17, 18
Alexander, L. T., 330
Alexander, P., 360
Allen, J. S., 330
Allen, S. W., 496
Allison, F. E., 513
Allsup, J. R., 436
Anatasia, R., 266
Anderson, D. W., 263, 275, 277, 278, 280
Anderson, E. C., 154
Anderson, E. K., 422
Anderson, E. R., 120
Anderson, H. W., 515
Anderson, J. J., 211, 214
Anderson, J. P., 270
Anderson, J. S., 450
Andrews, H. L., 341
Angelovic, J. W., 356-358
Aristotle, 643
Armstrong, F. A. J., 472
Arnold, J. R., 154
Arons, A. B., 102, 103
Arrhenius, G. O. S., 195
Ashworth, De B., 462
Aten, A. H. W., Jr., 362
Atkins, W. R. G., 472
Atkinson, L. P., 207, 209, 212, 213
Austin, Carl F., 558

Azuma, T., 343

Baalsrud, K., 592
Bäckström, J., 476
Badgely, P. C., 682
Bagley, G. E., 266
Bailey, T. E., 269
Baker, M., 468
Balassa, J. J., 251
Balchelor, G. K., 101, 125, 126, 128
Balcius, J. F., 479
Baldi, E., 463
Baldwin, R., 617
Balgood, W. A., 17
Ballou, N. E., 343, 344
Bandt, H. J., 475
Banks, R. C., 280
Baptist, J. P., 343
Bardach, J. E., 547, 675-677
Barnes, C. A., 214, 345
Barrows, H. L., 513, 515
Barry, D., 271
Barry, F. J., 622
Barsom, G. M., 136
Barth, T. F. W., 194, 195
Barthel, W. F., 267, 268
Barthelemy, M. H., 423
Bascecu, M., 185
Bascom, W., 682, 683
Bates, C. G., 505
Baumgartner, A., 506
Beasley, T. M., 333, 334
Beck, R. A., 171, 172
Becker, G. F., 193

Beland, F. A., 266
Belayeu, V. I., 579
Bell, F. H., 541
Bellanca, S. C., 463
Beninson, D., 350
Bennett, C. F., 253
Benson, B. B., 213
Berg, W., 462, 476
Bernhard, M., 349
Beroza, M., 270
Bertine, K. K., 21-24, 38, 40, 47, 49, 62, 67, 248
Beverton, R. J. H., 545
Beyer, F., 451
Biannuci, G., 462
Biedermann, G., 227
Bien, G. S., 32, 50
Bigeleisen, J., 432, 434
Bigler, A. B., 679
Bitman, J., 280
Black, W. A. P., 473
Blackburn, M., 350, 353
Blakely, J. P., 335
Blanchard, D., 102, 103
Blumer, M., 421, 424
Boato, G., 440
Boëtius, J., 476
Bohm, G., 23
Bolin, B., 154-157, 299
Bollen, W. B., 520
Bolton, G. C., 267, 268
Bond, G. E., 2
Bond, G. F., 683
Bonebakker, E. R., 345, 365, 579, 580
Bonham, K., 355-358
Booth, E., 675, 676
Bormann, F. H., 16, 509, 513
Bowden, K. F., 89, 103, 107, 130, 138, 143
Bowen, H. J. M., 466
Bowen, V. T., 245, 329, 332, 345, 349, 470, 472, 473, 577, 579
Bowles, P., 91, 130, 137
Bowman, M. C., 269, 270
Braarud, T., 454
Bradford, A., 271
Bradley, W. F., 171
Brady, D. N., 352
Brahtz, J. F., 682
Brar, S. S., 13, 466
Brazhnikova, L. V., 17, 18

Bretschneider, C. L., 494, 682
Brey, M. E., 171
Briggs, M., 592
Brinkhurst, R. O., 463
Broecker, W. S., 154, 157, 291, 292, 296, 297, 301, 305, 307, 311, 329, 330, 345, 365, 436, 579, 580
Broenkow, W. W., 203, 209, 210, 212-214
Broker, F., 402
Brongersma-Sanders, M., 215
Brook, N. H., 144
Brooks, J. H., 347
Brooks, R. R., 354, 470, 471, 473
Broquet, D., 470
Brown, E., 268
Brown, J. M., 341
Brown, M. D., 478
Brown, R. M., 331, 358
Brown, V. M., 350
Brownell, G. L., 479
Bryan, G. W., 350
Bryan, K., 148
Budyko, M. I., 315
Buelow, R. W., 451
Buglio, B., 468
Burdick, G. E., 273
Burke, W., 616
Burnham, C. D., 466
Burns, J. J., 675
Burns, R. H., 91, 130, 137
Burrell, D. C., 172, 175, 178
Burton, J. D., 343, 345
Burton, W. M., 253, 329
Butler, P. A., 275, 517
Button, D. K., 424, 427
Buyanov, N. I., 357
Byrne, John V., 538

Cadle, R. D., 12
Cahill, W. P., 270
Calder, J. A., 434, 436-438
Calder, K. L., 120
Calvert, S. E., 51
Cancio, D., 350
Carey, A. G., 350, 582
Carey, C. L., 592
Carlisle, D. B., 470, 472
Carpenter, J. H., 134, 345
Carroll, D., 174
Carslaw, H. S., 140

Carsola, A. J., 493
Carter, H. H., 103, 130, 138, 141, 147, 159
Caspers, H., 210-212
Cavalloro, R., 472
Cazianis, C. T., 470
Cecil, H. C., 280
Cerrai, E., 347
Champagnat, Alfred, 401
Chan, L. H., 22, 23
Chang, P. S., 465
Chapman, D. W., 516
Chen, J. H., 263
Cheng, H., 347
Cheng, T. C., 457
Childers, R. C., 479
Childs, L., 682
Chipman, W. A., 343, 354
Cholak, J., 249
Chow, T. J., 245, 247, 248, 250, 251, 253-255
Christ, C. L., 238
Christensen, R. E., 277
Christopher, W. M., 621
Clark, E. W., 682
Clarke, F. W., 193
Clayton, H. H., 319
Clayton, R. N., 440
Clendenning, K. A., 463, 477
Clifton, H. E., 683
Cline, J. D., 209, 211-214
Colby, D., 273
Cole, D. W., 507, 509-513
Cole, H., 271
Collias, E. E., 214
Conney, A. H., 277
Contois, D. E., 32, 50
Conway, E. J., 193, 248
Cook, F. D., 441
Cooke, R. F., 421
Cooper, A. C., 516
Cooper, J. A., 479
Cooper, R. D., 479
Cooper, W. W., 76
Coote, A. R., 212
Cope, O. B., 517
Copeland, B. J., 185, 508
Copper, J. W., 113
Corbella, C., 463
Corbett, M. K., 269
Corcoran, E. F., 348

Corliss, W. R., 341
Corrsin, S., 95, 100, 101, 119, 125-127, 139
Corwin, J. F., 79
Cory, L., 271
Costin, M., 137
Cottam, Clarence, 462, 517
Coulson, J. C., 265
Countryman, C. M., 508
Cox, M. D., 148
Crabtree, A. N., 265
Craig, H., 154-156, 291, 434, 440, 581
Crandall, C. A., 475
Crank, J., 302
Cromarite, E., 266
Cronin, J., 346
Crosby, D. G., 279
Crosby, R. M., 115
Cross, F. A., 351
Cruickshank, M. J., 558
Crutchfield, J., 542, 544
Csanady, G. T., 143, 147
Culshall, N., 48
Curl, H., Jr., 346, 350, 470
Curtis, T. G., 159
Cutshall, N., 343, 346

Dansgaard, W., 321
Darwin, C., 271
Davenport, J., 625
Davidson, B., 148
Davis, E. M., 358
Davis, J. B., 422
Davis, J. D., 263
Davis, J. D., 266, 278, 280
Davis, J. J., 331, 339
Davitaia, F. F., 81
Dawson, E. Y., 673
Day, C. G., 115
Dean, H. J., 273
Dean, J. M., 351
DeBary, E., 253
Debyser, J., 172
Defant, A., 108, 109, 121
Degens, E. T., 183, 433
Degurse, P. E., 263
Delaney, A. C., 78, 79, 84, 271
deLoture, R., 538
De Padovani, I. Oliver, 468-470
DeStefani, G., 462
Deuser, W. G., 433

AUTHOR INDEX

DeVega, V. Roman, 348, 468-470
deVos, R. H., 265, 266
Dianova, E. V., 425
Dice, S. F., 509-511
Diecidue, A. T., 470, 471
Dietz, R. S., 171
Dils, R. E., 516
Dole, M., 440
Dolph, R. E., 519
Donaldson, L. R., 355-358
Dorrestein, R., 156
Doudoroff, P., 474, 475
Douglas, I., 19
Douglas-Wilson, I., 256
Dudley, N. O., 466
Dugdale, R. C., 211, 593-595, 598
Duke, T. W., 266, 346, 347
Dunlop, H. A., 541
Dunster, H. J., 363
Duprey, R. L., 82, 84
Durum, W. H., 21, 24, 248
Duursma, E. K., 468
Dyrness, C. T., 514-516

Earl, J., 251, 253
Eckart, C., 92
Eckelmann, W. R., 436
Eckenfelder, W. W., Jr., 592
Eckhardt, D. L., 12
Edmondson, W. T., 454, 514, 517
Edvarson, L., 343
Edward, C. A., 270
Ehrlich, A., 674
Ehrlich, P., 674
Eichelberger, C. M., 611
Eisenbud, M., 340
Ellison, T. H., 110
Elson, P. F., 475
Ely, N., 546, 622
Emery, K. O., 591
Emme, Eugene M., 651
Enderson, J. H., 266
Engelstad, O. P., 521
Engle, J. B., 539, 540
Engler, A., 505
Eppley, R. W., 594
Epstein, S., 440
Eriksson, E., 10-12, 154, 299, 466
Escher, A. R., 516
Estesen, B. J., 270

Ettinger, M. B., 249
Evans, A. G., 340
Ewing, W. M., 157

Fabricand, B. P., 171
Fairbridge, R. W., 422
Farb, P., 531
Farrell, R. P., 450
Fedorov, A. F., 357
Fedorov, K. N., 112, 113
Feely, H. W., 329
Ferguson, G. J., 310, 312
Feth, J. H., 16
Fina, L. R., 515
Firth, F. E., 672
Fischer, H. B., 138
Fisher, D. W., 509, 513
Fjeld, P., 271
Fleming, R. H., 121, 122, 153, 269, 272, 296
Fofonoff, N. P., 115
Folsom, T. R., 121, 129, 345, 573, 574
Fonselius, S. H., 211
Ford, H. T., 441
Ford, J. H., 267, 268
Ford, W. L., 684
Forrester, W. D., 159
Foster, R. F., 352, 356-358, 360
Fowler, T. K., 336, 351
Fox, D. L., 468
Foxworthy, J. E., 136
Føyn, E., 450
Franklin, C. M., 616
Frear, D. E. H., 271
Frederick, L., 346
Fredriksen, R., 516
Freeman, N. L., 540
Freiling, E. C., 344
Freke, A. M., 362, 363
French, M. C., 280
Frenkiel, F. N., 144
Freuchen, P., 536
Freudenthal, H. D., 682
Frey, P. J., 519
Friedlander, G., 432
Friedman, G. M., 171
Friedman, L., 438
Friend, J. P., 148
Fries, G. F., 280
Fruton, J. S., 463

AUTHOR INDEX

Fuglister, F. C., 106
Fuhremann, T. W., 270
Fuhrmann, G. F., 427
Fujiki, M., 466
Fukai, R., 347, 468, 470
Funk, J. W., 515

Gade, H. G., 214
Gambell, A. W., 509
Game, P. M., 78
Garcia, R. J., 468-470
Garrels, R. M., 16, 54, 171, 172, 177, 238
Garrett, R. L., 275
Gaucher, T. A., 549
George, M. J., 582, 583
Gerard, R., 137, 157
German, W. H., 419
Gessel, S. P., 509-511, 513, 519
Gibbs, R. J., 17, 18, 20
Gibson, D. T., 423
Gifford, F., 146
Gilfillan, S. C., 246
Gill, C., 402
Glude, J. B., 617
Goering, J. J., 211, 598
Goldacre, R. J., 426
Goldberg, E. D., 75-81, 84, 86, 129, 195, 198, 220, 245, 248, 259-262, 265, 267, 269, 271, 272, 276, 277, 344, 467-470, 473
Goldschmidt, V. M., 193, 195, 219, 291
Goloskova, E. M., 357
Goodnight, C. J., 475
Goodwin, H. L., 681, 682
Gorbman, A., 355
Gordon, L. I., 440
Graber, F. M., 479
Grant, H. L., 111, 115, 116, 127
Grant, V. E., 345
Gray, J., Jr., 329
Greendale, A. E., 343, 344
Gress, F., 263, 278, 279, 280
Griffin, J. J., 75, 76, 78-81, 84, 86, 259-262, 265, 267, 269, 271, 272, 276, 277, 467
Grim, R. E., 171
Grimanis, A. P., 462, 470
Gripenberg, S., 209
Grissinger, E. H., 267, 268
Groen, P., 133, 155

Grosch, D. S., 359
Gross, Grant M., 184
Gross, M. G., 48, 345
Grzenda, A. R., 519
Guinn, V. P., 463
Gunnerson, C. G., 456
Gunther, F. A., 517, 520, 521
Gustafson, P. F., 13

Haagen-Smit, A. J., 80
Haddock, H. W., 682
Haffty, J., 21, 24
Hagemann, F., 329
Hall, L. C., 159
Halley, E., 192
Hamaguchi, H., 347
Hamaker, H. C., 214
Hamon, B. V., 113
Hannum, J. R., 269
Hanson, W. C., 339
Hanya, T., 468
Harding, S. T., 320
Hariss, R. C., 16, 25
Harleman, D. R., 159
Harley, J. H., 579
Harris, E. J., 273
Harrison, W. W., 478
Harvey, H. W., 468
Hasselrot, T. B., 475
Hattori, D. M., 13, 466
Haushild, W. L., 339, 346
Hawthorne, J. C., 267, 268
Hayashi, K., 352
Heath, R. G., 276
Hecht, A., 145
Hedgpeth, J. W., 503
Heezen, B. L., 157
Heide, F., 22, 23
Heidel, S. G., 248
Heilman, P. E., 513
Hellebust, J. A., 596
Helm, J. M., 268
Hemley, J. J., 237
Henderson, C., 263, 426
Henle, R. C., 339
Hennessen, J. A., 479
Henry, A. J., 505
Herlinveaux, R. H., 212
Herman, S. G., 265, 275
Hessel, J., 618

AUTHOR INDEX

Hibbert, A. R., 505-508
Hickey, J. J., 273, 275, 277, 280
Higano, R., 577
Higgins, E., 517
Higgins, Rosalyn, 611
Hildebrandt, P. W., 352
Hinze, J. O., 108, 124
Hiyama, Y., 568, 569, 584
Hobbie, J. E., 424
Hoering, T. C., 433, 441, 442
Holberg, A. J., 423
Holeman, J. N., 19
Holland, H. D., 238
Holm, C. H., 682
Holme, N. A., 425
Holmes, A., 193
Holmes, D. L., 265
Holt, J. J., 545
Honstead, J. F., 352, 367
Hood, D. W., 3, 468, 684
Hoover, M. D., 516
Horn, M. K., 219, 220
Hosohara, K., 468
Howard, P., 171
Howard, T. E., 554, 558
Howe, M. R., 113
Howells, H., 352
Hromadik, J. J., 683
Hudswell, F., 91, 130, 137
Huggett, R. J., 76, 80, 259-262, 265, 267, 269, 271, 272, 276, 277, 467
Hughes, B. A., 127
Hughes, D. F., 277
Hull, S., 670
Hume, J. D., 490, 493
Hummel, K., 170
Hunley, W. H., 682
Hunt, E. G., 261, 263, 270
Huynh-Ngoc, L., 468

Ichi Kuni, H., 468
Ichiye, T., 128, 152
Ikuta, K., 465
Imbimbo, E. S., 171
Inglis, A., 263
Inoue, E., 572
Irsa, A. P., 438
Irukayama, K., 466
Isaacs, J. D., 129, 365, 366
Ishiwatari, R., 468

Ivanov, V. N., 358
Ivanov, Y. A., 152
Iversen, E. S., 676

Jaeger, J. C., 140
James, M. S., 355
Jaworowski, Z., 251
Jefferies, D. F., 347
Jefferies, D. J., 279, 280
Jeffrey, L. M., 3, 468, 684
Jehl, J., 263, 278
Jenkins, C. E., 334
Jenkins, R. L., 558
Jennings, D., 346
Jensen, S., 260, 263-266
Jernelöv, A., 520
Johannes, R. E., 598
Johannesson, J. K., 463
Johnels, A. G., 260, 263-266, 462, 476
Johnsen, S. J., 321
Johnson, D. G., 25, 245
Johnson, D. W., 490, 492, 498
Johnson, M. J., 423, 424, 426
Johnson, M. W., 121, 122, 153, 269, 272, 296
Johnson, N. M., 16
Johnson, R. R., 519
Johnson, V., 48, 346
Johnson, W. L., 263
Joly, J., 193
Jones, L. G., 422
Jones, R. W., 237
Jones, S. J., 682
Joseph, J., 128, 130, 131, 133, 134
Joyner, Timothy, 559, 560
Judd, J. M., 350
Judson, S., 18-20
Jukes, T. H., 260, 274
Junge, C. E., 10, 11, 78, 253
Junkins, R. L., 339

Kai, F., 466
Kallio, R. E., 423
Kamata, E., 12-14
Kameda, K., 569, 574
Kamenkovich, V. M., 120
Kanabrocki, E. L., 13, 466
Kanamori, S., 84
Kanazawa, T., 332, 345, 573, 577, 579, 582, 583

Kan-No, H., 185
Kanwisher, J., 171, 172
Karweit, M. J., 145
Katsuragi, Y., 332, 345, 573, 577, 579, 582, 583
Katz, B., 137
Katz, M., 474
Katz, Milton, 618
Kawasaki, K., 468
Kaye, S., 249
Keeling, C. D., 154, 155, 157, 291, 308, 309, 312, 435
Keith, J. O., 261, 263, 270
Keller, W. D., 17, 174
Kennedy, Donald, 618
Kennedy, J. W., 432
Kennedy, V. C., 36, 42, 43
Kent, R. E., 120
Kenyon, K. W., 675
Ketchum, B. H., 201, 203, 213, 480, 577, 592, 684
Kettredge, J. S., 468
Kharkar, D. P., 21, 22, 24, 38, 40, 47, 49, 53, 62, 67, 248
Kilezhenko, V. P., 357
Kilmer, V. J., 513, 515
Kimball, J. F., Jr., 348
Kinsman, B., 98
Kirven, M. N., 264, 265, 267, 268
Kirwan, A. D., 122
Kleinkopf, M. D., 21, 23
Kleinman, M. T., 333
Klement, A. W., 330, 331
Klienert, S. J., 263
Kline, J. R., 13
Knauss, J. A., 116
Knowles, L. I., 174, 175
Knull, J. R., 208, 209
Koch, J. R., 423
Koch, P., 462
Koczy, F. F., 184, 245
Ködderitzch, H., 22
Koeman, J. H., 265, 266
Koers, Albert, 660
Koide, M., 344
Kolbe, R. W., 86
Kolesnikov, A. G., 579
Kolmogorov, A. N., 115
Kondo, T., 466
Konovalov, G. S., 21, 22

Korringa, P., 454, 462, 465, 472
Kosourov, G., 330
Kosyreff, V., 277
Koyama, T., 12-14, 468
Kozuma, H., 468
Krajewski, B. K., 333
Krammes, J. S., 516, 519
Kreitzer, J. F., 276
Krey, P. W., 333
Krishnaswami, S. K., 427
Krouse, H. R., 441
Krueger, R. B., 604, 614, 616, 619-622, 624
Kuenzler, E. J., 350
Kullenberg, G., 121, 135, 136, 172
Kupchanko, E. E., 427
Kuroda, P. K., 76

Laevastu, T., 468
Lal, D., 79
Lamontagne, R. A., 688
Landström, O., 22, 23
Langford, J. C., 334
Langway, C. C., Jr., 321
LaQue, F. L., 683
Laughlin, W. S., 671
Laventure, R. S., 177, 180
Lazrus, A. L., 251
Ledford, R. A., 263
Lee, R. E., 466
Lemon, J. M., 670, 672, 673
Lense, A. H., 558
Le Petit, J., 423
Lerman, J. C., 310
Lerz, H., 23
Letey, J., 516
Levin, W., 277
Levine, M., 589
Lewis, F., 342
Lewis, J., 343, 353, 365
Lewis, J. M., 347
Li, Y. H., 291, 296, 297, 305, 307
Libby, W. F., 331
Lichtenstein, E. P., 270
Lieberman, J. A., 515
Likens, G. E., 16, 509, 513
Linekin, D. M., 479
Linnebom, V. J., 688
Lipper, R. I., 515
List, R. J., 330
Little, J. B., 359

Liu, O. C., 457
Livingstone, D., 17, 19, 21
Lloyd, R., 474, 475
Lodge, J. P., 251
Loftus, T., 670, 679, 682
Longley, H., 340
Lorange, E., 251
Low, K., 343
Lowe, J. I., 266
Lowman, F. G., 334, 344, 346, 350, 356, 468-470
Ludwig, F. L., 253
Lumley, J. L., 110
Lundholm, B., 466
Lyerly, R. L., 335

Macek, K. J., 273, 276
MacEwan, G. F., 133
Machta, L., 329, 330
MacIsaac, J., 594, 595
Mackensie, F. T., 16, 54
Mahnken, C. V. W., 683
Mair, B. J., 423
Mamayev, O. I., 120
Mamuro, T., 343
Manabe, S., 313, 314, 316, 317
Manheim, F. T., 2
Manigold, D. B., 268, 270
Marderosian, A. D., 682
Marshall, J. S., 359
Marter, W. L., 340
Martin, D. J., 263, 276, 278
Martin, F. T., 519
Martin, P. S., 530
Martin, W. E., 334
Mason, J. O., 457
Masuzawa, J., 152
Mathews, W. H., 206, 210
Matson, W. R., 198
Matsuda, H., 352
Matsunaga, Y., 212
Matsunami, T., 343
Mauchline, J., 339, 340, 343, 348, 350, 358
Mayer, M., 432
McAllister, R. F., 682
McCarthy, J. J., 595
McClin, R., 468-470
McColl, J. G., 512
McCrone, A. W., 174
McDougal, M., 616

McDowell, L. L., 267, 268
McKee, J. E., 461
McKinney, T. F., 171
McLean, W. R., 457
McMullen, C. C., 435, 439
McNeil, W. J., 516
Mead, W. J., 618
Meade, R. H., 2, 19
Melcalf, R. L., 81
Menzel, D. B., 263, 276, 278
Menzel, R. G., 331
Menzie, C. M., 267
Merlini, Margaret, 472
Mero, J. L., 670, 680, 681
Merrill, C. W., 249
Meyer, C., 237
Meyer, M. W., 330
Meyers, J. J., 682
Mikkelsen, V., 172
Miller, J. M., 432
Miloy, L., 682
Miner, J. R., 515
Miner, N. H., 506
Minette, H., 589
Miron, G., 690
Mitchell, C. T., 422
Mitchell, J. M., 317, 318, 320, 321
Mitchell, R. L., 473
Mitchell, W., III, 335
Miyake, Y., 332, 343, 345, 358, 441, 442, 569, 571, 573, 574, 577, 579, 581-584
Moberly, Ralph, Jr., 492
Moghissi, A. A., 340
Moilliet, A., 111, 115, 127
Möller, F., 313, 314, 316, 317
Møller, J., 321
Monin, A. S., 100
Monod, J., 595
Montgomery, R. B., 103, 104, 121, 153, 154, 157
Monti, R., 463
Mook, M. G., 310
Moore, C. E., 13
Moore, D. G., 520
Moore, E. E., 466
Moore, W. S., 23, 435
Moretti, G. P., 463
Morgan, J. J., 239
Morris, D., 673
Moss, James E., 382, 383, 385, 387, 388,

394, 397, 400
Moubry, R. J., 268
Müller, G., 347
Munk, W. H., 114, 116, 120, 149, 150
Munro, I. A. H., 256
Murdoch, M. B., 353
Muroga, T., 212
Murozumi, M., 247, 250, 251, 253
Murphy, B. L., 457
Murphy, G. I., 543
Muse, L., 468
Myrdal, G. R., 268

Nagaya, Y., 577
Naito, H., 21
Nakai, N., 12, 13, 14
Nan'niti, T., 116
Neev, D., 320
Nelepo, B. A., 579
Nelson, D. M., 13, 466
Nelson, I. C., 339
Nelson, J. L., 48, 339, 346
Netsky, M. G., 478
Neumann, G., 103
Neushal, M., 617, 624
Newton, M., 519
Nicholls, G. D., 470
Nicholson, H. P., 519
Nielson, J. M., 346
Nishihara, S., 462
Nishikawa, Y., 212
Nishioka, Y. A., 268
Noddack, I., 469
Noddack, W., 469
Norris, L. A., 519
Norris, R. M., 492
North, W. J., 422, 463, 477
Noshkin, V. E., 245, 332, 345, 466
Nossaman, 604, 619, 620, 621

O'Brien, N. R., 178
Ohle, W., 503, 514
Okubo, A., 102, 103, 130, 133-135, 138, 141, 143-145, 147
Okubo, K., 475
Okubo, T., 475
Olcott, H. S., 263, 264, 276-278
Olson, E. A., 292
Olson, F. C. W., 128
Olsson, M., 260, 263-266

Ondrejcin, R. S., 336
O'Neil, J. R., 440
Ophel, I. L., 350
Oppenheimer, C. H., 468
Oppenheimer, J. H., 277
Orr, W. L., 591
Osborn, F. J., 516
Osterberg, C. L., 48, 339, 343, 346, 350, 351, 582
Othmer, D. F., 680
Otterlind, G., 260, 263-266
Ottoboni, A., 279
Overall, M. P., 558
Overstreet, R., 150
Ozmidov, R. V., 107, 110, 115, 130, 133-135

Padan, J. W., 554, 558
Pales, J. C., 291, 308, 309, 312
Palmer, H. E., 333, 334
Palumbo, R. F., 355, 357
Panofsky, H. A., 110
Papadopoulou, C. P., 470
Park, K., 582, 583
Parker, F. L., 337, 341, 367
Parker, P. L., 434, 436, 437
Parkin, D. W., 78, 79, 84, 271
Parr, A. E., 122
Parsons, D. A., 267, 268
Parsons, T. R., 468
Patric, J. H., 519
Patterson, C., 245-251, 253-256
Patterson, C. C., 427
Patterson, R. K., 466
Pauszek, F. H., 30, 31, 54
Peakall, D. B., 265, 277, 279, 280
Pearcy, W. G., 350, 582
Pelishek, R. E., 516
Pell, C., 680, 681
Pentelow, F. T. K., 455
Perdriau, J., 186
Perkins, H. T., 127
Perkins, R. F., 519
Perkins, R. W., 13, 339, 346, 470, 479
Persson, P. I., 462, 476
Peterle, T. J., 267
Peters, B., 79
Petterson, O., 536
Phelps, D. K., 468-470
Phillips, N., 116

Phillips, O. M., 107, 109, 112, 116
Pickard, G. L., 206, 210
Pickering, Q. H., 426
Pierce, R. S., 509, 513
Pierson, W. J., 102
Pingree, R. D., 113
Podymakhin, V. N., 357
Poldervaart, A., 248
Polikarpov, G. G., 341, 343, 345, 348-350, 358, 359, 363
Polzer, W. L., 16
Pomeroy, L. R., 590
Pories, W. J., 479
Porter, C. R., 340
Porter, R. D., 276
Post, R. F., 336
Potts, G. R., 265
Powers, M. C., 171
Poznaniyu, K., 171
Preston, A., 348, 360, 367
Pretorius, W. A., 592
Price, T. J., 346, 354, 355
Prichard, D. W., 103, 115, 120, 121, 134, 141, 159
Pringle, B. H., 451
Proudman, J., 151
Provosoli, L., 597
Pytkowicz, R. M., 205

Rae, K. M., 174, 175
Rama, M. Koide, 248
Rancitelli, L. A., 13, 470, 479
Ratcliffe, D. A., 274
Rattray, M. Jr., 150
Ray, P. H., 493
Redfield, A. C., 201, 202, 204, 205, 208, 213
Reesman, A. L., 17
Reiche, P., 264, 265, 277
Reichel, W. L., 266, 267
Reid, J. L., 153
Reid, R. O., 103
Reigner, I. C., 505, 519
Reimann, B. E. F., 78, 79
Reimers, R. S., 434, 436, 438
Reinert, R. E., 273
Reinhart, K. G., 516, 519
Renig, W. C., 340
Revelle, R., 129, 154, 288, 344
Rex, R. W., 76, 77, 493

Reynolds, H. G., 516
Reznikoff, P., 249
Rheinheimer, G., 456
Rice, T. R., 348, 350, 354
Ricebrough, R. W., 76, 80, 259-269, 271, 272, 275-280, 467
Rich, L. R., 516
Richards, F. A., 50, 201, 203, 206-214, 350
Richardson, A., 265
Richardson, L. F., 127
Richardson, W. S., 115
Richet, C., 474
Richter, D. H., 237
Riekerk, H., 519
Riel, G. K., 345
Riesen, W. F., 231
Riley, G. A., 51, 68, 593, 594
Riley, R. A., 148
Riordan, 604, 619-621
Rison, L. J., 248
Rittenberg, S. C., 591
Ritter, D. F., 18-20
Rob, C. G., 479
Robbins, R. C., 12
Roberson, C. E., 16
Robertson, A. R., 150
Robertson, C., 419
Robertson, D. E., 470
Robinson, E., 253
Robinson, J., 265, 275
Robinson, R. A., 227
Rocco, G. G., 329, 330, 345, 365, 579, 580
Rogers, J. N., 595
Rohlich, G. A., 450
Romanowitz, C. M., 558
Rona, E., 23, 468
Roos, B., 457
Rose, D. J., 337
Roskam, R. Th., 465
Ross, E. E., 466
Rothacher, J., 516
Rothstein, A., 427
Rowe, P. B., 508, 516
Royce, 541
Rudd, R. L., 275
Rumsby, M. G., 354, 470, 471, 473
Russell, E. S., 542
Russell-Hunter, W. D., 673
Ruzicka, J. H. A., 265, 267
Ryther, J. H., 261, 266, 272, 547, 676, 677

Sackett, W. M., 435
Saffman, P. G., 120, 138, 140
Sakuri, Y., 185
Salo, E. O., 516
Sanders, J. E., 171
Saruhashi, K., 332, 345, 358, 573, 577, 579, 581, 582, 583
Sasaki, M., 185
Saunders, R. L., 475
Sax, N. I., 426
Schafer, L. J., 249
Schaffel, S., 53
Schaffer, M. B., 344
Schalk, M., 490, 493
Schelske, C. L., 343, 353, 354, 365
Schindler, P. W., 231, 239
Schmitt, R. A., 344
Schönfeld, J. C., 133, 155
Schreiber, B., 347
Schreiber, R. W., 265, 268, 278
Schroeder, H. A., 251, 478
Schubert, J., 473
Schulz, K. R., 270
Schulze, J. A., 268, 270
Schutz, D. F., 16, 62, 66, 345
Schwatz, H. L., 277
Schweiger, G., 474
Schweitzer, G. K., 343
Scott, 604, 619-621
Scott, M. R., 22, 44, 69, 248
Scotten, J. W., 79
Seitz, H., 148
Sendner, H., 128, 130, 131, 133, 134
Seraichkas, H. R., 457
Serat, W., 271
Seto, Y., 577
Seymour, A. H., 343, 356, 358
Sharma, G. D., 172, 176, 178
Sheridan, W. L., 516
Shigematsu, T., 212
Shimizu, M., 568
Shiozaki, M., 577
Shishkina, O. V., 171
Short, N. M., 16
Shuert, E. A., 135
Sibley, F. C., 264, 267, 268, 279
Sierp, F., 458
Siever, R., 171, 172
Silker, W. B., 21, 22
Sillén, L. G., 219, 227, 236, 237, 344

Silverman, S. R., 434
Simmonds, S., 463
Simmons, J. H., 265
Simpson, A. C., 427
Simpson, L. B., 533
Sisefsky, J., 343
Sjöstrand, B., 462, 476
Skea, J., 273
Skopintsev, B. A., 209
Sladen, W. J., 267
Slowey, J. F., 468
Small, L. F., 351
Smith, D. C., 514, 515
Smith, J. L. B., 535
Smith, R. H., 344
Smith, V. K., 269
Smith, W. D. C., 343, 353, 365
Sobleman, M., 270, 281
Soldat, J. K., 352, 360
Somerville, B. T., 105
Soper, T., 402
Spann, J. W., 276
Spar, J., 329
Spence, R., 558
Spencer, D. A., 275
Spiess, F. N., 683
Spinner, G. P., 676, 678
Sprague, J. B., 475
Springer, P. F., 517
Sreekumaran, C., 345, 573
Stachiw, J. D., 683
Starkey, J. C., 174
Steele, J. H., 593, 594
Steemann-Nielsen, E., 577
Steen, M. E., 122
Stefánsson, U., 50
Sterling, T. O., 249
Stevenson, B., 684
Stevenson, Bernadette, 3
Stevenson, R. A., 470, 471
Stewart, N. E., 253
Stewart, N. G., 329
Stewart, R. W., 110, 114-116
Stickel, L. F., 76
Stimson, P. B., 115
Stokes, R. H., 227
Stommel, H., 102, 103, 106, 110, 112, 113, 128, 150, 154, 156, 157
Stone, E. C., 495, 516
Stopper, W. E., 519

Storey, H. C., 505, 508
Strain, W. H., 479
Straughan, Daliz, 617
Strickland, J. D. H., 468
Strøm, K., 206, 210-212
Strutt, R. J., 325
Stumm-Zollinger, E., 238
Suess, H. E., 154, 288, 309, 312
Sugawara, K., 14, 15, 21, 22, 84
Sugihara, T. T., 245, 329, 332, 345, 577, 579
Sugiura, Y., 204, 569, 574
Sumino, K., 466, 470, 476, 477
Surks, M. I., 277
Sutton, D., 472, 473
Suzuki, K., 185
Sverdrup, H. U., 121, 122, 153, 154, 269, 272, 296
Swallow, J. C., 106
Swank, W. T., 506
Sweeney, J. B., 683
Swinnerton, J. W., 688
Sylvester, R. O., 503, 514
Szule, T., 342

Tabushi, M., 212
Tait, R. I., 113
Takahashi, T., 291, 296, 297, 305, 307, 316
Takenouti, Y., 116
Tamm, C. O., 513
Tanimoto, R. H., 589
Tanner, James, C., 401
Tarrant, K. R., 267, 268, 272, 467
Tarrant, R. F., 516-518, 520
Tarzwell, C. M., 517
Tatsumoto, M., 245, 248, 251, 255, 427
Tatton, J. O'G., 265, 267, 268, 272, 467
Taylor, D. M., 422
Taylor, G. I., 110, 122
Teasley, J. I., 519
Templeton, W. L., 339, 340, 343, 348, 350, 358
Ten Noever de Brauw, M. C., 265, 266
Thode, H. G., 435, 439
Thomas, A., 474
Thomas, E. A., 503
Thomas, L. K., 353
Thomas, W. H., 32, 50, 595
Thompson, J. F., 341
Thompson, M. F., Jr., 479
Thompson, T. G., 468

Thompson, W. C., 494
Thompson, W. F., 540, 541
Thorade, H., 152
Tibby, R. B., 136, 185
Ting, R. Y., 348
Tipton, I. H., 478
Tonolli, L., 463
Tonolli, V., 463
Tressler, D. K., 670, 672, 673
Triulzi, C., 347
Tsuruta, T., 468
Turekian, K. K., 16, 21-25, 38, 40, 44, 47, 49, 62, 66, 67, 245, 248, 345
Turkevich, A., 329
Turner, J. S., 102, 103, 121

Ufret, L. S., 470
Ulfvarson, U., 466
Urey, H. C., 432
Urry, W. D., 23

Vaccaru, R. F., 211, 213
Valissa, V. T., 330
Vander Elst, E., 350
Van Derwalker, J. C., 683
Van Eyck, L., 214
Vasey, B., 516
Verber, J. L., 451
Vine, A. C., 129
Vinogradov, A. P., 469
Vogel, J. C., 310, 433
Vogel, W. M., 111, 127
Vogt, J. R., 334
Volchok, H. L., 329, 330, 332
Volchok, N. L., 245
Vollenveider, R. A., 463, 464
Voorhis, A. D., 107, 127
Vorshilova, A. A., 425

Wada, E., 441, 442
Wagman, J., 466
Walford, Lionel A., 532
Walker, K. C., 462
Walker, T. M., 273
Waller, R. A., 683
Wallis, J. R., 515
Ward, E., 350
Ware, G. W., 270
Warkentin, B. P., 177, 180
Watanabe, T., 185

AUTHOR INDEX

Waters, 604, 619-621
Watson, D. G., 339
Wayson, E. C., 339
Watson, K. S., 450
Watts, G. W., 492, 497
Wayne, L. G., 80
Webb, D. C., 106, 107
Webster, F., 115, 116
Weidemann, H., 131
Weigel, R. L., 159
Welander, A. D., 356, 358
Welander, P., 150
Welch, R. M., 277
Wellman, R. P., 441
Wendt, I., 433
Wenk, Edward, Jr., 643
Wenner, C. G., 22, 23
Werby, R. T., 11
West, J. A., 516
Westermark, T., 462, 472
Westlake, W. E., 517, 520, 521
Westley, R. E., 539, 540
Westoo, G., 462
Wetherald, R. T., 313, 314, 316, 317
Whipple, R. T. P., 91, 130, 137
White, J. C. Jr., 356-358
Whitledge, T., 598
Whitlock, D. W., 436
Widmark, G., 266
Wiebe, A. H., 426
Wiemeyer, S. N., 276
Wilkins, C. H., 115
Willen, D. W., 515
Willets, D. B., 519
Williams, P. M., 206, 210, 437
Williams, R. B., 353

Willis, J. N., 346, 347
Willis, V. M., 348
Williston, S. H., 84
Wilson, A. J., 266
Windom, H., 75, 79, 80, 81
Wingate, D. B., 262, 517
Winslow, D. F., 679
Wirth, T. L., 263
Wischmeier, W. H., 514, 515
Wiseman, W. J., 115
Wolfe, D. A., 347, 353, 354
Wolfsberg, M., 434
Woodhouse, A. F. B., 105
Woods, J. D., 111-113, 116
Woodwell, G. M., 517, 519, 520
Wooster, W. S., 207, 682
Worthington, L. V., 106, 107
Wright, R. T., 424
Wuhrmann, K., 450
Wurster, C. F., 262, 275, 277, 517
Wüst, G., 152

Yamagata, N., 331, 362
Yasui, M., 116
Yoshikawa, K., 343
Youngberg, C. T., 516
Yudelson, J. M., 144

Zattera, A., 349
Zavitowski, J., 519
Zaystev, Y. P., 358, 359
Zellner, A., 542, 544
Zhelezhova, A. A., 172
ZoBell, Claude E., 382, 398, 400, 456
Zon, R., 505

Subject Index

Abalone, ancient harvesting, 671
Academic community, role in oceanographic research, 659
Acetaldehyde in surface waters, 79
Acetate, uptake by diatoms, 597
Acetone in surface seawater, 79
Activity, scales of chemical, 226-228
   versus concentration, 226-228
Adsorption of, metals on particles, 35
   trace elements from seawater, 44, 48
   trace metals in estuaries, 40, 44, 48
Advection, 89
   horizontal, steady state produced by, 151-153
   horizontal-vertical diffusion model, 153-154
Aerosols, 10
   chloride, 10
   composition, 10-11
   continental component, 14
   heavy metals in, 466-467
   lead in, 251-252
   marine, 10-14
   tropospheric, 78
Agriculture, management practices and future implications, 520-521
   practices, 503-523
Aitken particles, 10
Alaska, North Slope, impact of oil development, 490
Alaskan earthquake, 498
Aldrin, toxicity to fish, 268
Aleuts, harvesting seals, 671

Algae, agar from, 675
   brown, 675
   industrial utilization, 675
   mathematical models for growth, 592-593
   nori (*Porphyra tenera* Kjell), 670
   nutrient limited growth, 589-599
   red, 675
Alkalinity, components of seawater contributing to, 296
   excess removal in seawater, 54
   in inorganic reactions, 50
   in Long Island Sound, 53, 55
   specific, 53-55
   specific versus chlorosity, 53-54
Alkyls, lead, 247, 251
Aluminum, in fresh water, 21
   interlayered clay, 36
Amazon, estuary, organic material in surface waters, 79
   River, 20
American Bureau of Shipping, 385
Amino acids, uptake by, *Cyclotella cryptica* and *Melosira nummuloides*, 597
   Melosira nummuloides, 596
Ammonium, in forest soils, 512
   in lake Lago d'Orta, 464
   in nitrogen-carbon-phosphorus utilization model, 203
   oxidation by *Nitrosomonas* sp, 595
   stable isotope relations in, 441
Amoebae, toxicity by oil, 426
Anaerobic muds, 40
Andes, 38

SUBJECT INDEX

Anisotrophy, 110-119
Anoxic, conditions, 205
  denitrification under, 206
  properties of, 207
  sulfate reduction under, 206-208
  sulfide-bearing waters, 207
  sulfide formation, 208
  systems, 210-215
    Black Sea, 210
    Cariaco Trench, 213-214
    Lake Nitinat, 213
    locations in world, 211-214
    polluted, 214-215
    Saanich Inlet, 213
Ansonia, Connecticut, 27
Antarctic Ocean, silicious diatom deposits, 51
Antimony in fresh water, 22, 24
AOU (Apparent Oxygen Utilization), 204
Aquaculture, 547-549
  marine organisms used in, 676-677
  as use of the ocean, 675-676
Aragonite, degree of saturation in ocean, 306
Arctic oil transport, 426
Aromatic hydrocarbons, 423
Arrenius equation, 425
Arsenic in, fresh water, 22
  insecticides, 462
  rain water, 84
Artemia salina (brine shrimp), 359
  radiation effect on population, 359
Ascidians, vanadium in, 470
Aswan Dam, Egypt, 161
Atmosphere, lead in, 253
  precipitation, 11
  transport by, 11, 75
    of DDT and PCB, 271
    time parameters, 80
Atmospheric, composition, carbon dioxide, 309
  carbon monoxide, 688
  dust, in Bay of Bengal, 78
  hydrogen sulfide, 13
  in Japan, 14
  oxides of nitrogen, 13
  sulfur and oxides, 12, 13
  contamination, input sources, 82-83
    rate from U. S. and world sources, 82-83
    from volatile organic compounds, 82-83

dust, lead in, 253
  size distribution, 78
Azodrin, 260

Bacteria, in air-sea interfaces, 456
  concentration by filter feeders, 456-457
  Escherichia coli in sewage, 455-456
  in sewage, 455-456
Bald Eagle (*Haliaietus leucocephalus*), effect of chlorinated hydrocarbons on population, 273
Baltic Sea, development of communities, 535-536
Barbados Islands, dust, 84-87
Barbados in Carribean Sea, 76, 78
Barium, 35
  barium-140-lanthinum-140, 329
  concentration in ocean surface water, 254
  in Connecticut rivers, 28, 29
  in Connecticut streams, 30
  in Neuse river, 39
Barrow, Alaska, beach erosion, 490
Batelle Northwestern, oil spillage study, 399, 401, 404
Beach, borrow, at Barrow, Alaska, 493
  equilibrium, effect of structures on, 495
  erosion, 489-498
    Barrow, Alaska, 490
    continental shelf, 494
    Corps of Engineers, responsibility for, 493-494
    ecological effects, 498-499
    effect of land management practices on, 514-517
    hurricanes, 494
    protection against, 494-495
    as related to ecology, 498
  sands, 491-492
Bengal, Bay of, atmospheric dust, 78
  clay minerology of dust, 86
  dust burden, 86
  winds, 86
Benzene, toxicity, 426
Berkshire Mountains, 26
Beryllium in fresh water, 21
Bicarbonate, 16, 18
  clay mineral reactions, 54
  concentration in Connecticut Rivers, 16, 25
  in forest soils, 512

SUBJECT INDEX 715

ion, concentration in ocean, 296
Bikini Atoll, radioactive contamination of
    land crabs, 352
    stronium-90 in, 352
Bikini-Eniwetok, Castle tests, 569-570
    fish kill from nuclear explosion, 354
    U. S. Atomic Energy Commission nuclear
        test series, 569-570, 574-575
Biochemical oxygen demand, 449, 458
    increase by algae growth, 592
Biocides, annual application in U. S., 518
    DDT, 259-260
    effect on ion transport, 514
    pollution control by government, 275
    retention by soils, 517-520
    U. S. annual production, 259
Biodegradation, 4
    kinetics of oil, 423-426
    of oil, 421-428
    of petroleum, 397-401
        factors controlling, 399-400
    of sewage, 453-454
Biological cycling of, carbon, 435-436
    CO_2 in soils, 512
    heavy metals, 471-473
    lead in oceans, 248, 255
    radionuclides, 347-354
Biosphere, concentration factors, 470-471
    concentration of heavy metals by, 469-473
    cycles of heavy metals, 471-473
Bismuth in, human tissues, 478
    man, 478
Bison, 531
Black Sea, 184
    anoxic systems in, 210
Boron, concentration in ocean, 296
    in fresh water, 21
Brazos River, 37
    sediment, 37
Brine shrimp (*Artemia salina*), resistance to
    radiation, 357-358
    toxicity of heavy metal to, 475
British sparrow hawk (*Accipiter nisus*),
    effect of chlorinated hydrocarbons
    on population, 274
Bromine, in fresh water, 22
    from seawater, 558
Brown algae, 675
Bubbles, 10

Bureau of Mines, offshore developments,
    554, 558
Bureau Veritas, feasibility of super tankers,
    385
Buthraldehyde in surface seawater, 79

Cactus (*Agave americana*), cultivation in
    Mexico, 534
Cadmium, 461
    in air, 466
    in human tissue, 478
    tolerance of organisms to, 474
    toxicity to rainbow trout, 474
Calcite, degree of saturation in ocean, 306
Calcium, calcium-144 in clams, 353
    carbonate, degree of saturation in oceans,
        306-307
        effect of solution on CO_2 exchange,
        306
        in sediments, 304-305
    in Connecticut streams, 28, 29
    distribution and transfer in ecosystem,
        510
    effect on heavy metal toxicity, 474
    in Neuse river, 37
    from seawater, 558
Canary Islands, 78
Cape Thompson, Alaska, harbor studies, 491
Carbohydrates in sewage, 447
Carbon, cycle, in soils, 512
    stable isotopes ratio in, 435-436
    dioxide, 287-322
        alkaline components of CO_2 system,
        296
        alkalinity, 296
        atmospheric changes with time, 308-309
        atmospheric concentration and global
        temperature, 313-317
        chemical reactions with seawater, 294-296
        climatic effect of geographic distribution, 317
        correlation between CO_2 production
        and, mean annual temperature, 317-318
        rainfall with time, 319-320
        cycling in soils, 512
        effect of, atmospheric concentration on
        climate, 312-317
        calcium carbonate on exchange

SUBJECT INDEX

rate, 306
exchange, between ocean and atmosphere, 291
between surface and deep water of ocean, 299-304
coefficient, 296
with ocean surface layer, 296-304
from fossil fuels, 289-290
fraction reaching ocean with time, 303
"greenhouse effect", 312-315
infra-red absorption by, 313-314
Mauna Loa data, 309
from organic matter decomposition, 209
partial pressure considerations in equilibrium models, 237
predicted distribution among reservoirs, 304, 306-308
production from, coal by decade, 289
fossil fuel, 289, 292
lignite by decade, 289
liquid hydrocarbons, 285
natural gas by decade, 289
rate to atmosphere from man, 82
rate of increase of production, 290
reservoirs on earth's surface, 291
residence time, in box model, 156
system, 228
uptake by, biosphere, 292-293
ocean, 293-306
ocean surface, 294
in fresh water, 21
gases, 75
monoxide, 12
in atmosphere and ocean, 688
rate to atmosphere, 82
reservoir on earth's surface, 291
stable isotope abundance in nature, 432
Carbon-12/carbon-13 ratios, 433-436
in effluents of sewage plants, 436
in seawater, 433-434
Carbon-14, atmospheric–ocean exchange, 300-304
in the oceans, 326-327
Carbon-14/carbon, isotopic equilibrium, 311
in tree rings with time, 309-310
Carbonate ion, concentration in ocean, 296
Cariaco Trench, 209-210
Carroll glacier, 173
"Castle Test", 569-570

Catechol, 422
Cation, exchange, calcium, 174
capacity, 36
of clay in streams, 43
in glacio-marine sediments, 172-184
with illite and chlorite, 178
of magnesium, 174
of potassium, 174
in sediments, 347
with silt in streams, 43
of sodium, 174
with stream sand, 43
transport in soils, 513
Cattle, feed lots, source of pollution, 515
Cerium-141, 329
Cerium-144 in oceans, 329, 582-583
Cerium-144-promethium-144 in sediments of Ligurian sea, 347
Cesium, adsorption on marine clays, 347
in fresh water, 22, 24
Cesium-137, amount on earth's surface with time, 333
biological activity of, 329
in clams, 353
distribution in, Columbia River plume, 582
North Pacific, 572-574
leeched from land mass, 331
in ocean surface waters, 332
vertical distribution in North Pacific, 576-578
vertical transport, 344-345
Chaparral, Southern California, effect of land erosion, 495-496
S/S Charles Pratt, 396-398
Chemical, equilibrium, 219, 229-240; *see also* Equilibrium, models
gradient, 39
oxygen demand, 449
reactivity, as related to residence time, 197-198
Chemicals, industrial, new chemical additions to wastes, 685
Chesapeake Bay Institute, 115
Chicago, Illinois, heavy metals in air, 466
Chloride, atmospheric, 10, 11
in Connecticut river, 29
in fresh water, 21
ion ratios in rain, 12
Chlorinated, aliphatic hydrocarbons, 3

SUBJECT INDEX

hydrocarbons, 259-281
  in atmospheric dust, 271
  in Barbados dust, 271
  DDT, 259-260
  effect on, bird population, 261-263, 267, 273-274
    reproduction in birds, 273
  in fresh water fish, 263
  in marine fish, 263-266
  pesticides, 259
  polychlorinated biphenys, 260, 263
  in rain, 272
  residues in petrels and shearwaters, 261-263, 267
  retention by soil, 269-270
  in river waters, 268-269
  transport rates, 80
  volatility, 270-271
Chlorosity, Connecticut River, 31
  in Connecticut streams, 26
  in Long Island Sound, 30, 31, 58
  relation to specific alkalinity, 53, 54
Chromium, adsorption on minerals, 41
  chemical behavior in ocean, 468
  chromium-51 on sediments, 346
  concentration factors in mollusks, 471
  concentration in rivers, 44
  desorption, 48
  in English atmosphere, 466-467
  in fresh water, 21, 24
  in marine organisms, 470
  in plants, 478
  toxicity to kelp, 477
Cincinnati, Ohio, heavy metals in air, 466
Clams (*Mercenaria mercenaria*), radiation tolerance, 355-356
Clay minerals, 41, 45-47
  alteration by ionic species, 171
  in Bay of Bengal dust, 86
  with bicarbonate, 54
  cation exchange in streams, 43
  illite, 36
  incorporation of organic molecules, 183
  in inorganic reactions, 50
  kaolinite, 36
  montrorillonite, 36
Climate, effect of atmospheric CO_2 concentration on, 312-317
  geographic variations with CO_2 concentration, 317

Coagulation of sewage, 452
Coal, CO_2 from, 289
  from the ocean floor, 558
Coastal, engineering, 489-499
  beach erosion, 489-496
  research needs in coastal management, 694-695
  zone, authority, need for, 690-691
  management problems, 690
Coastline, importance of wetlands to, 678
  modification of, 538-539
  radioactive contamination of, 567-568
Cobalt, adsorption of, 40, 45
  chemical behavior in oceans, 468-469
  cobalt-60 transport on sediments, 48
  concentration in rivers, 44
  in Connecticut river, 29
  contamination in Nawgatuck River, 27
  desorption, labeled, 48, 49
  in fresh water, 21, 24
  in Long Island Sound, 56, 59, 62, 63, 66
  in marine organisms, 470
  in Neuse River, 40
  in petroleum, 463
  in plants, 478
  in streams, 38
  supply to oceans, 49
  toxicity to rainbow trout, 474
  vertical transport in ocean by organisms, 574-575
Codfish, North Atlantic, development of fishery, 537-538
Columbia River, 48, 582
  distribution of strontium and cesium in plume, 582
Colville River, Alaska, 422
Commerce, maritime, 537
Commissions, intergovernmental, related to fisheries, 532-533
Communities, city problems, 636
  development of seafaring, 535-537
Composition of the oceans, calculated from law of mass action, 229-232
  mass balances, 219, 220
Concentration factors, 347-348, 351, 584, 568
  of elements in marine organisms, 568
  for elements in primary coolants, 584
  of heavy metals, 470-471
  of significant radionuclides, 351

in shell fish, 470
Congressional committees, jurisdiction over marine science, 652
Connecticut, River, 25
  silicon content of, 51
  trace metals in, 28, 33
  state of, 25
  streams, 51
Conservation, 20th century attitudes, 532-533
Contaminant additions, continuous source, 144-147
  horizontal mixing of, 129
  instantaneous source of, 128-144
  plume diffusion models for, 144-147
  vertical mixing of, 129
Contamination, definition, 1
  distribution of, 122-127
  human, 19
  major areas of concern, 669
  maximum limits of ocean, 668-669
  radioactive, horizontal distribution, 569-574
    in North Pacific, 571
    vertical distribution in North Pacific, 574-580
  world ocean problems, 667
Contiguous seas, 548
Continental, dust, 10, 76
  margins, 50
  precipitation, 11
  shelf, convention on (1958), 605-606
    council on environmental quality, 628-629
    expanse of U. S. offshore, 604
    National Environmental Policy Act (1969), 628-629
    national resource regulation, 619-626
    national technology, 619
    North Sea Continental Shelf Cases, 605-608
    offshore oil resources of, 603-617
    Oil Pollution Act of 1924, 626
    Outer Continental Shelf Lands Act (1964), 620-624
    relation to beach erosion, 494
    Truman Proclamation, 604-605
    Water Quality Improvement Act (1970), 626-627
Cook Inlet, effects of Alaskan earthquake on, 498
  oil spills, 421, 423
Copper, in air, 466
  effect on diatoms, 464
  in fish kill, 464-465
  in fresh water, 21
  "Green oyster" phenomenon, 465
  as naphthenate, 462
  organic complexes of, 468
  in plants, 478
  in pollution of coastal area, 464-465
  as sulfate, 462
  tolerance of organisms to, 474
  toxicity to, kelp, 477
    rainbow trout, 474
    yeast, 427
  toxic synergism with zinc, 475
  vertical distribution in Gulf of Mexico, 468
Coriolis effect, 106, 114
Cormorants, shell thinning phenomenon in, 278
Corp of Engineers, responsibility for beach erosion, 493-494
Coweeta watershed, N. C., 506-507
Crocodile, 531
Currents, Antarctic Circumpolar, 104
  characteristics, 103-107
  Coriolis effect on, 106
  Equatorial Countercurrent, 104
  Gulf Stream, 104, 106
  Kuroshio, 104, 106, 152
  laminar flow, 95
  measurement by, Doppler-shift, 115
    hot wire, 115
  North Equatorial, 104
  Oyashio, 152
  South Equatorial, 104
  turbulent flow, 95-96
Cyclic salts, 11, 196

Daigo Fukuryu-Maru (Lucky Dragon V), 569-570
Daphnia, radiation effect on population, 359
Darwin Bay, 214
p,p'-DDE, 259-281
  biodegradation, 263, 265
  retention by soils, 270
  in the sea, 259

SUBJECT INDEX

shell thinning phenomenon in, American egret, 279
  American kestrel, 276-277
  ashy petrels, 278-279
  cormorants, 278
  mallard ducks, 276-277
  murres, 278-279
  pelicans, 277-278
DDT, 259-281
  from agricultural crops in Africa, 76
  in atmospheric dust, 271
  atmospheric transport, 271
  in Barbados dust, 271
  p,p'-DDE from, 259-281
  distribution in environment, 267
  in fish reproduction, 273
  in freshwater fish, 272
  in marine fish, 263-266, 272
  pollution control by government, 275-276
  position of industry on, 281
  in rain, 272
  residues in petrels and shearwaters, 261-263
  retention by soil, 269-270, 519
  retention and transport, 517-520
  in river waters, 268-269
  shell thinning phenomenon in, American egret, 279
    American kestrel (*Falso sparverius*), 276-277
    ashy petrels, 278-279
    cormorants, 278
    mallard ducks, 276-277
    murres, 278-279
    pelicans, 277-278
  solubility in fresh water, 269
  volatility of, 270-271
  in Wisconsin, 276
  world production of, 266
Defense, historical use of oceans, 671-672
  as involving the oceans, 664-665
Denitrification, 202
Density, laminar layers, 112-113
  vertical gradient, 104
Denticula elegans in sediments, 86
Denudation, 18
Derby, Connecticut, 27
Desorption, of metals from sediments by sea water, 41, 44
  trace metals in estuaries, 40, 44, 48

Diagenesis, hydro, 172
Diatoms, *Cyclotella cryptica* in amino acid uptake, 597
  Cyclotella nana in amino acid uptake, 597
  deposits in Gulf of California, 51
  Melosira nummuloides in amino acid uptake, 597
  ooze in Antarctic, 51
Dieldrin, 265, 271
  in Antarctica wild life, 265
  in atmospheric dust, 271
  in Barbados dust, 271
  in rain, 272
Diffusion, 89
  eddy, 118-119, 141
  as effected by wave numbers, 125
  experiments, deep water, 135-136
    upper mixed layer, 135
  Fickian, 91
  horizontal, 120, 133
    advection solutions, 136-144
    Cape Kennedy experiments, 141
    dye patch characteristics, 137-140
    radially symmetric solutions, 130-136
    shear-diffusion solutions, 141-143
  molecular, 95
  Richardson's equation for, 127-128
  turbulent, 120
  vertical, 120
    steady state produced by, 151-153
Dinoflagellates, *Gymnodinium nelsoni* in organic uptake, 597
  Peridinium trochoideum in organic uptake, 597
Discharge, water, effect of forest fire on, 508, 516
  effect of vegetative composition on, 505-506
  as related to forest cover, 507-508
Disease from sewage to man by shell fish, 456-457
Dispersion, 90; *see also* Diffusion
Disposal of wastes, *see* Wastes, disposal
Dissolved organic matter, acetaldehyde in surface waters, 79
  acetone in surface waters, 79
  butyraldehyde in surface water, 79
  ethyl alcohol in surface waters, 79
  formaldehyde in surface waters, 79
  model for decomposition, 201-202

uptake by organisms, 596-597
  see also Amino acids; Organic, matter
Dissolved solids, concentration in, Connecticut streams, 25
  runoff, 17
  streams, 24
  flux, 17
  in Neuse River, 34
Domestic wastes, enhancement of production in the sea by, 455
Drainage basin, see Watershed
Drilling, offshore oil, social impact of, 629-630
Drugs from the ocean, 680-682
Dry cleaning solvents, manufactured per year, 84
Ducks, mallard, shell thinning phenomenon in, 276-277
Dust, atmospheric, 84
  in Bay of Bengal, 78, 86
  continental, 10
  fall in high Carcausus, 81
  flux, in atmosphere, 80-81
    versus Russion economy, 81
  glacial snow fields as annual recorder, 80-81
Dysprosium in fresh water, 23

Ecology, effect of beach erosion on, 498
  major ocean problems, 656-658
Eddy diffusion, near Cape Kennedy, 141
  coefficient of, 118-119
  in estuaries, 121
  Fourier modes of, 109-110
  in the ocean, 572
  in open ocean, 121
  from radioactivity distribution in North Pacific, 572
  in thermocline, 121
  vertical coefficient, 579
Eelgrass (*Zostera* sp.), 670
Egret, American (*Casmerodius albus*), shell thinning phenomenon in, 279
Electricity from tides, 679
Endrin, 265, 267-268
  in brown pelicans, 265, 268
  toxicity to fish, 267
  transport in streams, 267-268
Energy, Gibbs considerations, 221-225
  from the ocean, 679-680

projected requirements, 581-582
Environment, damage to, 635-636
Environmental management of the oceans, 658-661
Environmental quality, impact at Port Valdez, 691-694
  major problems in the oceans, 656-658
  need for international considerations, 657
EPA (Environmental Protection Administration), formation of, 654
Equilibrium, chemical, 219, 229-240; see also Equilibrium, models
  constants, 225-226
    formulation of, 228
    stable isotope effect on, 433
  models, 229-240
    climatic effects, 240
    complete, 237
    constants for system $H_2O-CO_2-MgO-CaO$, 231
    dynamic, 240
    effect of residence time, 239
    improved, 234, 236
    partial pressure CO_2 considerations in, 237
    phase, diagrams for system $H_2O-CO_2-CaCO-MgO$, 232
      diagram for system $H_2O-CO_2-CaO-MgO-Mg(OH)_2$ and magnesian calcites, 233
    rule relationships, 229-230
    simple, 233-234
    steady state, 238
    states, condition for, 223
Erbium in fresh water, 23
Erosion, chaparral, Southern California, 495-496
  effect, on coastal environment, 514-517
    of forest fires on, 516
    of logging on, 515-516
    of vegetation, 495, 496
  land management practices on, 514-517
Escherichia coli, survival in sea water, 456
Eskimos, harvesting of seals by, 671
Estrogen in shell thinning phenomenon, 277
Estuaries, Amazon, 79
  interstitial water in sediments of, 40
  Long Island Sound, 26, 50-51
  mixing in, 92, 102
  Muir Inlet, Alaska, 175

SUBJECT INDEX 721

phytoplankton in, 353
productivity from municipal wastes, 590
Queen Inlet, Alaska, 173
radioactivity in, 352-354
Taku Inlet, Alaska, 175-176
Tance, France, 679
Thames, England, 32
zooplankton in, 353
Estuarine environments, 39
Ethyl alcohol in surface waters, 79
Eulerian diffusion, 100, 102, 117; see also Diffusion
Euler's theorem, 222
Euphotic zone, phytoplankton loss from, 597
Europium in fresh water, 23
Eutrophication from sewage, 592
Evaporation, effect of vegetation on, 505-506
Exchange reactions, 42-43
   isotope effect on, 432-433; see also Cation, exchange

Fallout, radioactive, 566-567
   wastes from, 566
   stratospheric over Japan, 578
   types of, 566
Farming, ecological effect of, 533-534
Fats in sewage, 447
Fatty acids in sewage, 452
Fayetteville, Arkansas, 76
FDA (U.S. Food and Drug Administration), reluctance to approve fish protein concentrate, 649-650
Feed lots, cattle, as source of pollution, 515
Ferric oxide, adsorption of metals to, 41, 45-47
Fertilizers in drainage waters, 521
Fish, contamination by radionuclides, 567
   marine, contamination by DDT and p,p'-DDE, 263-266
     species in world catch, 674
     toxicity by aldrin, 268
   world catch, 673
Fisheries, eumetric fishing methods, 545-546
   historic, 535-537
   jurisdictional zones, 546-547
   need for regulation on, 544-546
   optimum catch, 546

   radioactivity in products, 352
   sustained yield, 529, 533
   U. S. decline in, 638
   world catch by species, 674
   world-wide activities, 672-675
Fjords, Muir Inlet, Alaska, 175
   Oslo, 203
   Queen Inlet, Alaska, 173
   Taku Inlet, Alaska, 175-176
Flocculation of sewage, 452
Florida Straits, 79
Fluorine in fresh water, 21
Forests, fertilization, 513
   fires, effect on erosion of, 516
   management, 503-523
     practices and future implications, 520-521
   National, establishment of, 532
   precipitation runoff from, 505
   relation of stream flow to cutting of, 506-507
   salmon spawning as related to logging of, 516
Formaldehyde in surface waters, 79
Fossil fuels production, effect on composition of the ocean, 656
   production by decade, 288
Fourier formulation, components in turbulent motion, 90
   transform of a random function, 98, 124
FPC (fish protein concentrate), congressional act, 649
Fungicides, 461-462
   arsenic in, 462
   mercury in, 461-462
Fusion reactors, 336-337

Gadolinium in fresh water, 23
Galena, Oregon, 245-246
Gallium in fresh water, 22
Gases, atmospheric, exchange with ocean, 79
   dissolved, effect on heavy metal toxicity of, 474
Gas exchange, CO_2 in surface ocean water, 298
Genetics, radiation effects on, 359
Geneva, 1958 Conventions on High Seas, 606
Geologic age, model for the ocean, Barth

model, 194-198
Goldschmidt model, 193-194
Halley model, 192-193
Jolley model, 193
Germacides, heavy metals in, 461-462
Germanium in fresh water, 22
Gibbs free energy considerations, 221-225
  equilibrium constant, 225
Glacial, debris, 16, 25
  firn, stratigraphy of, 81
  snow fields, dust analysis, 80-81
Glaciers, Carroll, 173
  lead in, 251
Glomar Challenger, 683, 690
Gold in, fresh water, 23
  human tissues, 478
Golden Eagle (*Aquila chrysaetos*), effect of chlorinated hydrocarbons on population, 274
Goldfish, toxicity of mercury compounds to, 476-477
Government laboratories, role in oceanographic research, 659
Gravity waves, effect on turbulence, 111-112
Great Britain, maritime nation development, 536-537
  Roman influence on development, 536-537
Greenland, lead in snow, 250-253
  $^{18}O/^{16}O$ ratios, 321
  temperature variations with time as measured by $^{18}O/^{16}O$, 321
Ground water, change in downstream composition, 34
Growth, limitation, kinetics of, 424, 425, 427
  by nutrients and light, 594
  specific growth rates, 595
Gulf of California, silicious diatom deposits, 51

Haddock, effect of fishing pressure on, 542
  history of fishery, 541-542
Halibut, catch per unit effort, 541
  history of fishery, 540
Halmyrolysis, 170
Hanford atomic energy plant, 352
  reactor, 48
    isotopes released from, 339

Heat, pollution by, 657
Heavy metals, in blood, 473
  cycles in organisms, 471-473
  cycling in the hydrosphere, 467
  in fungicides, 461-462
  in germicides, 461-462
  in industrial wastes, 461
  Interior Department program in, 558
  in man, 477-478
  in marine environment, 464-477
  in the ocean, the problem, 467
  organic complexes, stability of, 473
  in plants, 478-479
  pollution from the atmosphere, 466-467
  in terrestrial waters, 463-464
  toxicity to biota, 473-477
  uptake and accumulation mechanism, 472-473
  see also specific element
Heptachlor epoxide in Antarctica wildlife, 265
Herbicides, effect on water quality, 519
Herring, fluctuation in Baltic Sea, 536
High seas, jurisdictional zone, 546
History of ocean uses, 669-672
Holmium in fresh water, 23
Homer, Alaska, effect of Alaskan earthquake, 498
Housatonic River, 25, 27, 51, 65
  silicon content of, 51
Houston Ship Channel, 436-437
  Del-^{13}C of organic matter, 437
Hovercraft as ferry in English Channel, 683
Hubbard Brook, N. H., watershed, 509
Hudson Laboratories, 57
Hunting, ecological effect, 529-533
Hurricanes, effect on beach erosion, 494
Hydrocarbons, addition to sea from photosynthesis, 422
  aromatic component oxidation, 423
  biological source of, 422
  chlorinated, 3, 259-286
    in atmospheric dust, 271
    in Barbados dust, 271
    effect on, bird population, 261-263, 267, 273-274
      reproduction in birds, 273
    in fresh water fish, 263
    in marine fish, 263-265
    PCB, 260

SUBJECT INDEX

in pesticides, 259
in rain, 272
residues in petrels and shearwaters, 261-263
retention by soil, 269-270
in river waters, 268-269
volatility, 270-271
see also p,p'-DDE; DDT
CO_2 from, 289
normal hydrocarbon oxidation, 424
oxidation mechanism, 423, 424
not from petroleum, 422
in plants, 422
rate to atmosphere, 82
steady state in sea, 424, 425
toxicity to marine organisms, 426-427
see also Petroleum
Hydrogen sulfide in atmosphere, 13
Hydrosphere, heavy metals in, 467-469

ICO (Interagency Committee on Oceanography), membership on, 644
proposed 1963-1972 expenditures, 645
IDOE (International Decade for Ocean Exploration), 655
Illite, 36
adsorption of metals, 41, 45-47
IMCO (Intergovernmental Maritime Consultative Organization of U.N.), resolution on waste oil separators, 394-395
Industrial wastes, heavy metals in, 462
person equivalents of types, 458
types of, 687
and world ocean pollution, 688
Infra-red light, absorption by, atmosphere, 314
CO_2, 313-314
water vapor, 314-315
Inland waters, jurisdictional zone, 547
Insecticides, 517-520
annual application in the U. S., 518
arsenic in, 462
Interfaces, bacteria in air-water, 456
reactions at oxygen-sulfide, 207-208
Interior Department, heavy metals program, 558
International Biological Program (IBP), 504
Interstitial waters, 175-185
in estuarine sediments, 40
major ions in, 177

major and minor element distribution in, 179
manganese in, 177
migration of, ions in, 183-184
nitrate in, 185
phosphate in, 184
silicate in, 185
sulfate in, 184
sodium in, 180-181
Iodine, in fresh water, 22
iodine-131 concentration by carnivorous fish, 355
Ion exchange, 35-36, 39, 42-43, 48
capacity, 35, 36, 42-43
radioactive tracer studies, 48
with sediments, 182
Ionic strength of seawater, 228
Ions, land release to aquatic systems, 509
Irmiger Sea, water characteristics, 107
Iron, 461
in blood, 473
chemical behavior in ocean, 468, 469
concentration factors in mollusks, 471
in coprecipitation, 38-39
tolerance of organisms to, 474
wastes, 3
Iron-55, biological activity of, 329
distribution in earths surface, 333-334
Isentrophic processes, 153-154
Isotope ratios, 431-442
Isotopes, stable, abundance in nature, 432
chemical effects, 432-435
Del-^{13}C in biological samples, 434
exchange in carbonic acid equilibrium, 432, 433
in geochemical processes, 431-435
as indicators of contamination, 435
kinetic effect, 432-435
of oxygen, 440
of sulphur, 439
Isotrophy, 130-131

Kaolinite, 36
adsorption of metals by, 41, 45-47
Kelp (*Macrocystis pyrifera*), effect of mercury, copper, nickel, chromium and zinc on, 477
hydrocarbons in, 422
Kerosene, 424
Kestrel, American (*Falso sparverius*),

shell thinning phenomenon in, 276-277
Kinetics of oil biodegradation, 423-426
Krypton-85, 336-337

Labrador Sea, water characteristics of, 107
Lago d'Orta, 463-464
　ammonium in, 464
　copper pollution in, 464
Lagrangian diffusion, 100-102, 117
Land use, insecticide application related to, 518
Lateritic soils, 16
Lead, 245-258
　aerosol production, 252
　in aerosols, 251-252
　　from coal, 251
　alkyl, 247, 251
　annual loss to environment, 249
　in Antarctica, 251
　biological removal constant, 255
　in Byzantine Empire, 246
　concentration in, Atlantic, Mediterranean and Pacific Oceans, 254
　　soils, 249
　cycling in oceans, 247-248
　distribution in oceans, 245, 253-254
　in fresh water, 23
　in gasoline, 250
　in glaciers, 251
　in Greenland snow, 251
　histopathological effects to *Lebistes reticulatus*, 475
　historical development of use, 245-247
　in human tissues, 478
　industrial use, 245-247
　interaction with suspended solids, 248-249
　lead-210 dating technique, 81
　lead-silicon ratios, 248
　mechanism of removal from oceans, 248, 255
　in metallurgy, 245
　in municipal sewage, 248
　production in world, 246-247
　in rain, 251, 255
　residence time in atmosphere, 253
　in Roman Empire, 246
　seasonal variation, 253
　in snow, 250-253

　in storm waters, 248-249
　supply to ocean, 248
　toxicity, 255-256　427
　transport to oceans, 246-247
　in urban atmospheres, 253
Legislation, oil pollution, 626-629
Liability for damage of ocean resources, 694-695
Lignite, CO_2 from, 289
Lithium in fresh water, 21
Living resource exploitation, 529-550
　multiple use concept, 546-547
　over fishing, 539-540
Living resources, management, 652
Lloyd's, super tanker feasibility, 385
Logging practices, effect on, salmon spawning, 516
　sedimentation, 515
Long Island Sound, 26, 50
　alkalinity in, 53
　cobalt in, 56, 59, 62, 63, 66
　exchange with Atlantic Ocean, 51
　fresh water inflow, 51
　nickel in, 56, 60, 62, 64, 65
　sampling stations, 56, 57
　silicon, budget, 51-53
　　supply, 51
　silver in, 56, 61, 63, 64, 67
　trace metal additions, 27
Lop Nor, China, 76
Lutetium in fresh water, 23

Magnesium, in fresh water, 21
　from sea water, 558
Mammals, extinction in North America, 530-531
Man, ability to meet world problems, 636
　bismuth in, 478
　cadmium in, 478
　gold in, 478
　heavy metals in, 477-478
　lead in, 478
　mercury in, 478
　requirements for survival, 637
　silver in, 478
　the super predator, 529-531
　tin in, 478
　titanium in, 478
Manganese, chemical behavior in ocean, 468, 469

concentration factors in mollusks, 471
concentration in rivers, 44
in fresh water, 21
mining from sea floor, 558
organic complexes, 468
in petroleum, 463
in plants, 478
in streams, 38
tolerance of organisms to, 474
toxicity to rainbow trout, 474
vertical distribution in Gulf of Mexico, 468
Manganese-54, 329
adsorption on sediments, 346
in clams, 353
in estuarine scallop (*Asquipecten iradians*), 343
ion exchange on stream sediments, 48
Manganese oxide, 38, 39
adsorption of metals to, 41, 45-47
SS *Manhattan*, 416-419, 683
Mariculture, 663; see also Aquaculture
Marine, aerosols, 10, 32
mining, 553-554
placer deposits, 554
research needs, 560
technological advances, 553-554
worldwide, 556
organisms, concentration factors of elements, 568
effect of radiation on, 354-360
radiation, effects on populations, 358-359
sensitivity of eggs to, 358-359
role in translocation of radionuclides, 350-354
somatic effects of radiation, 355-358
used in aquaculture, 676-677
Material transport, 16, 75
latitudinal dependence, 76-77
Matsushima Bay, effect of waste disposal, 185
Mauna Loa volcano, carbon dioxide data, 309
Maximum permissable concentration (MPC) of radionuclides, 584
Mediterranean Sea, water characteristics of, 107
Melosira granulata in sediments, 86
Mercury, in air, 466

as alkoxyo-compounds, 462
as arly-compounds, 462
atmospheric, 84
effect of compounds on fishes, 476
in fresh water, 23
in fungicides and germicides, 461-462
in human tissue, 478
in marine organisms, 470
in organic complexes, 468
in sea water, 466
tolerance of organisms to, 474
toxicity to, goldfish, 476-477
kelp, 477
rainbow trout, 474
salmon, 476
Meriden, Connecticut, 26
Metal pollution, 29
Metals, heavy, see Heavy metals
Methyl-ethyl ketone in surface water, 79
Mexico, Gulf of, 36
disposal of chlorinated aliphatic hydrocarbons in, 3
interaction with Mississippi River, 50
Michaelias-Menten kinetics, 593, 595-596
effect of temperature on rate constant, 595-596
Michaelis constant, 424
Michigan, Lake, water level problems, 490
Microbiology of, petroleum, 397-401
petroleum oxidation, 423
Middletown, Connecticut, 30
Military sea transportation service, 393
Minamata disease, in Agano River, Japan, 466
from mercury in fish, 465-466
in Minamata Bay, Japan, 466
Mindanao deep, temperature and salinity distribution, 149-150
Mineral, production from oceans of U. S., 555-556
resources, management, 652
Mineralization of nutrients, 598
Mineralogy, stream clay, 42
Minerals, placer deposits, 554
recovered from the oceans of the U. S., 555
research needs, 560
weathered, 16
Mine residues, as source of heavy metals, 463-464

Mining, marine, 553-561
 ecological effects, 559-560
 of the ocean, 680-681
 placer deposits, 554
 research needs, 560
 worldwide, 556
Mississippi River, 36, 38
 interaction with the Gulf of Mexico, 50
Mixing, basic equations, 117-128
 box model solutions, 154-158
  circulation of deep ocean water, 157-158
  cyclic, 156-157
  exchange rates, 155-156
 in a cup of coffee, 92-94
 dynamically active, 91
 by eddy-diffusion, 572
 effect of heated water discharge, 159
 horizontal, 91
  distribution of properties, 147-149
  from specified sources, 128-147
 mechanism of, 92-103
 oceanic, 89
  by vertical eddy-diffusion, 579
 "Salt Finger" phenomenon, 102-103
 velocity gradients, 94-97
 vertical, 91
  distribution of properties, 149-151
  from specified sources, 128-147
  model for ocean, 299
  steady state produced by, 149-151
 see also Diffusion
Models of ocean processes, anoxic system, 210-215
 box, 154-158
 carbon dioxide partial pressure in equilibrium, 237
 for decomposition of organic matter, 201-202
 diffusion, 153-154
 equilibrium, 229-240
 of nitrogen-carbon-phosphorus utilization, 203
 nutrient flow, 593
 nutrient limited phytoplankton, 589-599
 for ocean age, 192-198
 for photosynthesis, 203
 practical use of, 598-599
 by radioactivity distribution, 565-586
 real biological, 598-599
 for residence time of elements, 192-198
 for respiration, 203
 two-layered growth, 593-594
 vertical mixing, 299
Mohole drilling project, 642-643
Molecular diffusion, Eulerian framework, 100, 102
 of heat, 103
 Lagrangian framework, 100-102
 of salt, 103
 see also Diffusion
Molluscicides, copper sulfate in, 462
Mollusks, concentration factors for heavy metals in, 471
 mercury in *Hormomya mutabilis,* 465-466
Molybdenum, adsorption, 40-41
 desorption, 48
 in fresh water, 22, 24
Monsanto Chemical Company, 266
Montmorillonite, 36
 adsorption of metals by, 41, 45-47
Montrose Chemical Corporation, 268
Moscow River, oil degradation, 425
MPC (maximum permissible concentration), as derived from conventional and specific activity, 366
 in drinking water, 362
 in occupational exposure, 362
 in seawater, 363
 specific activity approach, 364
Muir Inlet, Alaska, 175
Multiple use concept, 546-547
 a case study, 691-694
 limitations of, 668-669
 of ocean resources, 667-669
Municipal wastes, 445-458
 bacteria in, 455-456
 effect of disposal on the ocean, 452, 455
 enhancement of production in the sea by, 455
 parasites in, 457
 phosphorus in, 448-450
 physical chemical properties of, 446-450
 productivity as related to estuaries, 590
 purification, 450-452
 sludge disposal, 450-451
 viruses, 455-456
 water quality, 449
Murres (*Uria aalge*), shell thinning phenomenon in, 278-279
Muscles, distribution and potential, 547-548

SUBJECT INDEX

killed by copper, 465

NASCO (National Academy of Sciences Committee on Oceanography), pulications, 643
National Bureau of Standards, isotope reference standards, 433
National Council on Marine Resources and Engineering Development, creation of, 646, 648
Natural gas, CO_2 from, 289
Naugatuck River, 26, 27, 65
Naviface, 104
Navy Department, "Rock Site" project, 558
Nematodes, Anisakis in herring, 457
   in sewage, 457
Neodymium in fresh water, 22
Neuse River, 25, 33-34
   composition, 33-41
   dissolved solids, 34
New Hampshire, 16, 28
New Haven, Connecticut, 26
New London, Connecticut, 26
Nickel, chemical behavior in oceans, 469
   concentration factors in mollusks, 471
   concentration in rivers, 44
   in fresh water, 21
   in Long Island Sound, 56, 60, 62, 64-65
   in petroleum, 463
   in plants, 478
   toxicity, 427
   to kelp, 477
   to rainbow trout, 474
Nile River, 169
Nissho Marn, 385
Nitinat, Lake, specific alkalinity, 209
Nitrate, distribution in the oceans, 148
   in forest soils, 512
   reductase, 596
   stable isotope relations, 441
Nitrogen, cycling in douglas fir, 511
   Del-^{14}N in reservoirs, 441
   dioxide, 13
   distribution and transfer in ecosystem, 510
   fixation effect on isotope ratios, 441
   in fresh water, 21
   leeching from soils, 513
   in the ocean, 591
   stable isotope, abundance in nature, 432

   effects, 441
Nitrosomonas in ammonia oxidation, 595
Nixon, Richard M., ocean proposal, 609-617
NOAA (National Oceanic and Atmospheric Agency), 650-651, 654-655
   agencies included, 654
   attitude toward marine mining, 561
Noble gases, 79
Nome, Alaska, placer mining, 558
Nori (*Porphyra tenera* Kjell), radio isotopes in, 352
   usage of, 670
NORPAC (North Pacific Expedition), 571
North Carolina, 25
Northeast trade winds, dust transport by, 78
North Equatorial current, 570, 574
   in transport of radioactivity, 570
North Haven, Connecticut, 26
North Pacific, cesium-137 distribution, 572-574
   radioactive contamination, 571
   strontium-90 distribution, 572-574
   vertical distribution of, cesium-137, 576-578
   strontium-90, 576-578
North Sea, development of municipalities, 535-536
North Sea Continental Shelf Cases, 605-608
Northwest passage, voyage of the *SS Manhattan*, 416-419
NSF (National Science Foundation) in oceanographic funding, 654-655
Nuclear, excavation, radio isotopes resulting from, 334
   fuel, wastes from reprocessing, 340
   powered ships, 338
   power plants, schematic design, 335
   reactors, Hanford and Windscale production models, 338-340
   radionuclides released annually, 337
Nutrients, budget in oceans, 591
   interaction of, 596
   limited phytoplankton growth, 589-599
   metabolic pathways, 593
   in the ocean, 590-591
   phytoplankton growth limitation by, 594
   "preformed," 204
   regeneration by, bacteria, 598
   zooplankton, 598

SUBJECT INDEX

uptake by phytoplankton, 594-595
    see also specific nutrient

Ocean, exploration, political difficulties in, 639-640
    uses, philosophy for, 691-694
    waste disposal, 685-688
Oceanic mixing, *see* Mixing
Oceanographic research, academic community function, 659
    government laboratory, 659
    need for continuity, 659
Oceanography, historical development, 672
Oceans, alteration by artificial radioactivity, 580-586
    budget of nutrients, 591
    changes, 2
    defense policy, 664
    dispersal, 3
    dissolved load supply, 18
    environmental management, 658-661
    exploration, 637-643
    federal agencies with responsibilities, 640-641
    heavy metals, the problem, 467
    jurisdictional zones, 546-547
    major ecological problems, 656-658
    multiple use concept, 667-669
    nutrients in, 590-591
    public policy concerning, 635-665
    research continuity in, 642
    surface area per human, 3
    uses of, 667-695
    volume, 3
        per human, 3
    waste, assimilation by, 2
        disposal in, 3
Offshore oil development, 603-630
Oil, crude, *see* Petroleum
    lake, Alaska, 422
    offshore production, 603-630
        resources of the continental shelf, 603-617
        pollution from offshore production, 603-630
        Canadian legislation governing, 608
    seeps, 382, 422
    Santa Barbara Channel, 414
    slicks, mechanism of degradation, 425, 426
    spills, Batelle Northwestern study of, 399, 401, 404
        Cook Inlet, 421, 423
        dispersion of, 159
        effect of burning, 410
        emulsification, 410-411, 425
        liability for, 694-695
        Santa Barbara Channel, 617-619, 625-626
        in Los Angeles and Long Beach areas, 396-397
    Oil Pollution Convention (1954), 405
    Santa Barbara Channel accident, 413-415, 422, 423, 638
        status of our knowledge concerning, 415-416
    terrestrial, 421-422
    Torrey Canyon, 381
        accident, 401-403
        English Channel, 619
    treatment of, 408-413
        use of ship's engines for dispersal, 410
        from U. S. vessels (1966), 396-397
ONR (Office of Naval Research), in oceanographic funding, 654
Organic, carbon, in eastern rivers, 36
    stable isotopes in ocean, 436
    chemicals, 84
        U. S. annual production, 83
        volatile compounds manufactured in U. S. annually, 83
    complexes, of copper, 468
        of manganese, 468
        stability of heavy metal, 473
        of zinc, 468
    detritus, 50
    growth factors for algae, 597
    material in sewage wastes, 453, 454
    matter, 38, 39
        based on elemental composition of plankton, 203
        biodegradation of, 201, 592
        carbon dioxide production from decomposition, 209
        decomposition as related to other chemical parameters, 201-202
        dissolved, *see* Dissolved organic matter
        model for decomposition, 201-202
        release of ammonia, amino acids, urea and uric acid from decom-

position, 208
uptake by organisms, 596-597
Organisms, vertical transport of elements by, 50; *see also* specific types
Oslo, Norway, 447
Fjord, 203
Osprey (*Pandion haliaetus*), effect of chlorinated hydrocarbons on population, 273
Outer Continental Shelf Lands Act (1964), 620-624
Oxidation, by β and ω pathway, 423
Oxygen, biochemical oxygen demand (BOD), 592
Del-^{18}O in reservoirs, 440
depletion of dissolved, 201-202
distribution in, Oslo Fjord, 453
ocean, 148
effect on heavy metal toxicity, 474
in near-shore areas, 202
rates of oxygen-consuming processes, 214
stable isotope, abundance in nature, 432
effects, 440
utilization by, organisms, 39
biodegradation of sewage, 453-454
Oysters, *Crassostrea virginica*, radiation tolerance, 355-356
culture in ancient China, 671
effect of, copper on color, 465
sewage on growth, 454
in Nobeoka Bay, Japan, 465
Pacific, 539
production in U. S., 540

Paleosalinities, 172
Pamlico Sound, North Carolina, 34
Paper Mill effluents, Del-^{13}C values of, 438
Paraffin hydrocarbons, *see* Hydrocarbons
Parasites in sewage, 457
Parks, National, establishment of, 532
Particles, stream borne, 48
Particulate matter, transfer from atmosphere to oceans, 78
PCB (polychlorinated biphenyls), 260-263
atmospheric transport, 271
biodegradation, 263, 265
industrial use, 265
residues in petrels and shearwaters, 261-263
in shell thinning phenomenon, 280

Peat, adsorption of metals by, 41, 45-47
Pelican, brown, pesticide residues in, 261-263, 265, 267
shell thinning phenomenon in, 277-278
white (*Pelecanus erythrorynchos*), pesticide residues in, 261-263, 265, 267
shell thinning phenomenon in, 277
Peregrine falcons (*Falco peregrinus*), effect of chlorinated hydrocarbons on population, 273
Permafrost, 490
Pesticides, aerial transport of, 76
chlorinated hydrocarbons in, 259, 462
DDT in, 259-260
effect, on aquatic biota, 520
of U. S. treatment, 462
pollution control by government, 275
transport rates, 80
Petrels, ashy (*Oceanodroma homochroa*), shell thinning phenomenon in, 278-279
pesticide residues in, 261-263, 267
Petroleum, 381-441
ballast, water containing, 388
biodegradation of, 421-428
biodegradation in the oceans, 397-401
cleaning of ships, 387-388
dispersion of slicks, 159
domestic and world consumption, 417
effect of burning, 410
emulsification of, 410-411
general observations on pollution by, 405-406, 421-422
heavy metals in, 463
international regulations pertaining to offshore oil production, 604
leakage from sunken vessels, 412-413
legislation relevant to offshore oil production, 626-630
"load-on-top" principle, 388-389
metal porphyrins in, 463
microbiology of, 397-401
movement by tank ships, 383-385
offshore drilling, 414-415
oxidation kinetics, 423
pollution research needs, 406-407
quantities moving by sea, 383
sources to the ocean, 382-383
stable isotope ratios as indicator, 431-442
stable isotopes in crude, 436

tanker, development, 383-386
  operations, 386-389
  treatment of spills at sea, 408-413
  use of ship's engines for dispersal, 410
  wastes external to tanker, 389-390
  world offshore production volume, 603-604
Pharmaceuticals from the ocean, 680-682
Phase rule relationships in equilibrium models, 229-230
Phenol, 423
  toxicity, 426
Phosphate, distribution in the ocean, 148
  growth limitation kinetics, 426, 427
Phosphine, 207
Phosphorite,
  mining from, sea floor, 558
  ocean, 685
Phosphorus, distribution and transfer in ecosystem, 510
  in fresh water, 21
  mining from sea floor, 558
  in the ocean, 591
  in sediments, 50
  in sewage, 448-450
Photosynthesis, effect on atmospheric CO_2, 293
  model for, 203
  oxygen isotope effects in, 440
Phytoplankton, in estuaries, 353
  loss from euphotic zone, 597
  mathematical models for growth, 592-593
  nutrient limited growth, 589-599
Placer deposits, 554
Plaice, effect of fishing pressure, 542
  history of fishery, 541-543
Plastics, ultimate disposal of, 685-688
Polar Seas, water characteristics, 107
Policy, public (international), environmental management, 658-661
  environmental problems concern, 656-658
  toward oceanographic research, 660
Policy, public (national), 635-665
  political concerns, 655-656
Pollution, atmospheric modification by, 12
  definition, 1
  effect on coastal marine biota, 539
  industrial, 29
  international conference on (1926), 394
  offshore oil, national legislation, 626-629

  from production, 603-630
  oil, liability for, 694-695
    present state of control, 391-397
    by ships other than tankers, 397
  petroleum, general observations, 405-406
    research needs, 406-407
  see also specific item
Porphyrins, metal, availability to biota, 463
  in petroleum, 463
Porpoise, use in fish herding, 671
Port Valdez, Alaska, impact of oil terminus, 691-694
Potassium-40, distribution and transfer in ecosystem, 510
  in the oceans, 326
Prandtal number, 102
Praseodymium in fresh water, 22
Precipitation, 9, 11
  composition in Japan, 14, 15
  continental, 11, 12
  effect of vegetation on, 505-506
  of sewage, 452
President's Science Advisory Committee, panel on oceanographic publications, 645
Primary productivity, 261
  enhancement by sewage in the sea, 455
  see also Photosynthesis
Prince William Sound, 691
  effect of Alaskan earthquake, 498
Protein, from muscles, 547-548
  in sewage, 447
Public policy (international), see Policy, public (international)
Public policy (national), 635-665; [see also Policy, public (national)]
Puget Sound, agricultural use of basin, 521

Quartz, 50
  in plankton, 78
  in sediments, 77, 85
  transported by winds, 76, 77
Queen Inlet, Alaska, 173
  sediments of, 173-174
Quinnipiac River, 24

Radiation, damage, to brine shrimp, 357-358
  to eggs of marine organisms, 358-359
  genetic effects, 359-360

SUBJECT INDEX

to marine, invertebrates, 357
  organisms, 354-360
maximum permissible concentrations, 361-367
hazards, modes of exposure to, 360-361
Radioactive, contamination, horizontal distribution, 569-574
  in North Pacific, 571
  fallout, classification of, 329
  by nuclear excavation, 334
  pattern of, 329-330
  wastes, from cooling water, 565
    from exchange resins, 566
    from fallout, 566
    from fuel reprocessing plant, 566
    handling of, 565-567
    from nuclear, plant, 566
      submarines, 566
      weapons tests, 566
    packages for ocean disposal, 565-566
Radioactivity, adsorption of isotopes on sediments, 346
  artificial distribution and cycling in the oceans, 342-354
  artificial sources in the oceans, 327-342
  barium-140-lanthanum-140 on biological surfaces, 329
  behavior in the ocean, 567
  cerium-141, 329
  cerium-144, 329
    in the oceans, 582-583
  cesium-137, 329
    distribution in North Pacific, 572-574
  in Columbia river, 346
  concentration of isotopes by organisms, 348, 351
  cycling of isotopes by organisms, 347-354
  discovery of, 325
  discrimination of ^{65}Zn by organisms, 349
  disposal in the ocean, 325
  effect on organisms in the ocean, 354-360
  in estuaries, 352
  exposure to, 360-361
  fall out, from nuclear explosions, 327-334
    leeching from land mass, 331
  in fish, 352
  future alteration of the oceans, 580-586
  genetic effects, 359-360
  horizontal transport, 345
  induced in seawater by underwater nuclear explosion, 331
  international viewpoints on waste disposal, 367
  isotope translocation by organisms, 350-354
  low level wastes, 334-335
  maximum permissible concentration (MPC), 361-367
    for marine environment, 362-364
    in seafoods, 362
    in seawater, 363
    specific activity approach, 364
  mechanism of accumulation of isotopes by organisms, 354
  from nuclear reactors, 337
  ocean models based on, 565-586
  physical-chemical state of radionuclides in seawater, 343-344
  projected amount in oceans, 582-586
  radiation damage to marine organisms, 355
  radionuclides, from fallout in the sea, 327-334
    induced from underwater nuclear explosion, 331
    occurring naturally, 326
  rate of accumulation of radioisotopes by organisms, 349
  in seaweed, 352
  translocation by organisms, 354
  transport by winds, 75
  vertical transport, in ocean, 345
    of isotopes, 344-345
  see also specific isotope
Radioisotopes, application in medicine, 342; *see also* specific isotope
Radionuclides, *see* Radioactivity; specific isotope
Radium, concentration in ocean surface water, 254
Rain, composition, 12
  correlation between CO_2 concentration and rainfall with time, 319-320
  in Japan, 14, 15
  lead in, 251, 255
  trace metals, 13, 14
Rainbow trout (*Salmo gairdneri*), radiation tolerance, 356
  toxicity of heavy metals to, 474
Rare-earth isotopes, vertical transport, 345

Recreation in marine environment, 678-679
Red algae, industrial use, 675
Redox potential, in anoxic system, 208-209
Red Sea, water characteristics of, 107
"Red Tide," effect of sewage wastes on blooms, 454-455
Residence time, 191-199, 219
  of Al, Fe, and Ti, 199
  carbon in soil organic material, 293
  of elements in the oceans, 195-199
  of lead in, atmosphere, 253
    oceans, 255
  model for elements, 192-199
  relation to equilibrium state, 239
Resource management, dichotomy of, 496-497
  government versus industry, 661-664
  of living and non-living resources, 652
Respiration, model for, 203
Reynolds, number, 110
  rules, 99
Rhone River, composition of, 38
Richardson number, 110, 114
Richardson's 4/3 law, 128
Rio Maipo, 38
Rivers, Amazon, 20
  runoff, 17
  United States, 20
Riverwater, composition, 19-23
  in Japan, 14
  volume, 18
Romans, influence on development of Great Britian, 536-537
Rubidium in fresh water, 22, 24
Russia, dust fall versus economy, 81
Ruthenium-103-rhenium-103^m, 329
Ruthenium-106-rhenium-106, 329
  adsorption to sediments, 347
  in clams, 353

Saanich Inlet, 210
St. Joseph's Bay, Florida, 185
Salmon, effect of heavy metal pollution on fryling in rivers, 475-476
  effect of logging on spawning, 516
  effect of sedimentation on spawning, 516
  Salmo solar, heavy metal, toxicity to, 475
  toxicity of mercury compounds to, 476
Salt, cyclic, 11, 196
  nuclei, 10

  obtained from the sea, 670
  sea composition, 11, 12
"Salt fingering," 113, 121-122
Samarium in fresh water, 22
Sand, 36
  cation exchange in streams, 43
  transport, 492-493
San Francisco Bay, smog, 84
Santa Barbara, natural oil seeps, 422
  oil spill, 413-415, 422, 423, 638
Sardine (*Sardinops caerula*), history of fishing, 543-544
  seasonal catch on Pacific coast, 543
Savannah, USS, radioactive wastes from, 338
Scandinavia, precipitation and runoff, 11
Scandium, chemical behavior in oceans, 469
  in fresh water, 21
Scavengers, trace metal, 39
Scheveningen and Ijmuiden, Holland, fish kill by copper, 464-465
Schmidt number, 102
Scorpion, USS, 342
  catastrophe, 638
Sea disposal by barges in 1968, 686
Sea Grant College and Program Act, 647-649
Sea otter, harvest by Russians, 671
Sea salt, composition, 11, 12
  obtained from the sea, 670
Seabed, jurisdiction of, 604-617
  territorial jurisdiction of, Canadian proposal, 608, 617
  U. N. Committee on Peaceful Uses of, 606-607, 609-611
Seafood, species harvested from the sea, 673
Seas, present uses of, 672-689
Seawater, bactericital properties, 592
  mining for sodium, magnesium, calcium and bromine, 558
  radionuclide interaction, 567-568
Seaweed, brown, 675
  industrial utilization, 675
  radioisotopes in *Porphyea umbilicalis,* 352
  red, 675
  world harvest, 676
Sediment, load, 18, 20
  streams, 35
  transport, 18
    seasonal effects, 492-493

Sedimentation, effect of logging on, 515
  effect on salmon spawning, 516-517
  effect of seafloor mining on, 559
  rate of at 30°N, 80
Sediments, adsorption of radionuclides, 346
  under anoxic waters, 210
  atmospheric dust content, 78-80
  cycle in the oceans, 170
  glacio-marine, 172-184
  interrelation with biota, 184-186
  interstitial water ionic ratios, 171-172
  ion exchange with, 182
  magnesium and iron-rich chlorites in, 182
  mid-Atlantic ridge, 84-85
  montmorillonite in, 84
  nutrient-rich, 169
  as related to petroleum, 397-398
  seawater interaction with, 171-184
    after deposition, 177-183
    during transport, 174-177
    post burial, 183-184
  and solid waste disposal, 184
  transport by rivers, 170
Selenium, adsorption, 40, 47
  in air, 466
  desorption, 48, 49
  in fresh water, 22, 24
  supply to oceans, 49
Sewage, 445-458
  bacteria, 455-456
  communicable diseases from, 456-457
  from cattle feed lots, 515
  effect, of disposal on the ocean, 452-455
    on light transmission, 452-453
  enhancement of production in the sea by, 455
  favorable effects on the sea, 455
  ocean outfalls, 590
  parasites, 457
  per capita effluent, 446-447
  phosphorus in, 448-450
  physical chemical properties of, 446-450
  productivity as related to estuaries, 590
  purification of, 450-452
    chemical process, 450
    electrolytic process, 450
  sludge disposal, 450-451
  viruses in, 455-456
  water quality of, 449
Shearwaters, pesticide residues in, 261-263, 267
Shell Chemical Company, 260, 268, 270
Shell fish, concentration factors for heavy metals, 470
  quahaugs (northern) and polio virus, 457
  transfer of disease to man, 456-457
Shell thinning phenomenon, 276-280
  biochemical mechanism of, 277-280
  by DDT in, American egret, 279
    American kestrel, 276-277
    ashy petrels, 278-279
    cormorants, 278
    mallard ducks, 276-277
    murres, 278-279
    pelicans, 277-278
Shipping, as use of the ocean, 682-684
Ships, ballasts of, 388
  S/S Charles Pratt, 396-398
  classification societies for, 385
  cleaning of, 387-388
  conditions of tanker fleet, 404-405
  Glomar Challanger, 683
  history of oil tanker development, 383-386
  International Load-Line Convention (1933), 396
  "load-on-top" principle, 388-389
  SS Manhattan, 416-419, 683
  Nissho Maru, 385
  nuclear powered, 338
  oil pollution by ships other than tankers, 397
  oil tankers, 383-397
  petroleum leakage from sunken vessels, 412-413
  projected nuclear powered, 583-584
  responsibility of U. S. Coast Guard for, 392-393
  suggestions for handling casualties at sea, 408-413
  supertankers, 385
  tanker, operations, 386-389
    regulations, 392
    T-2, 393
  Torrey Canyon accident, 401-403
  Universe Island, 385, 682
  wastes, external to tanker, 389-390
    from nuclear powered, 566
Shrimp, brine, toxicity of heavy metals to, 475

Sierra Nevada, 16
Silicious diatom deposits, 51
Silicon-silicate, adsorption on clay minerals, 50
  biological removal, 51
  in Columbia river estuary, 51
  in Connecticut streams, 27
  cycle in, fresh water, 32
    Long Island Sound, 50
    oceans, 50
  in fresh water, 21, 32
  in Neuse River, 36
  in the ocean, 591
  in sediments, 50
  soluble, 17
Silt, 36
  cation exchange in streams, 43
Silver, adsorption, 40, 46
  concentration in rivers, 44
  in Connecticut River, 29
  contamination in Naugatuck river, 27
  desorption, 48, 49
  in fresh water, 22, 24
  in human tissue, 478
  in Long Island Sound, 56, 61, 63, 64, 67
  in streams, 38
  supply to oceans, 49
  in western U. S. rivers, 38
Smoke particles, rate to atmosphere, 82
Snow, lead in, 250-253
Sodium, in fresh water, 21
  "mined" from sea water, 558
Soils, 38
  bauxitic, 16
  lateritic, 16
Solid waste disposal, 184
  effect on, biota, 185
    natural sediments, 185
  in the oceans, 668-669
Soot, to atmosphere, 12
South America, 18
Space exploration, 637-643
  U. S. impetus for, 640-642
Spanish Sahara, dust from, 78
Sputnik, 638-639
Stable isotopes, *see* Isotopes, stable
Standard equilibrium states, 226-228
Starfish, coral-eating *Acanthaster plancia,* 638
Stratton Commission (Commission on Marine Science, Engineering and Resources), organization of, 646, 647
  "Our Nation and the Sea" publication, 650
Stream clay, mineralogy, 42
Streams, bedload, 18, 19
  composition, all elements, 19-23
    trace metals, 24
  dissolved load, 17-18
  flow as related to forest cover, 507-508
  sediment load, 18
  sediments, 35
  surface runoff, 16
Stronium in, Connecticut streams, 28-29
  fresh water, 22
  Neuse River, 38
Stronium-89 in rain on Barbados, 76
Stronium-90, amount on earth's surface with time, 333
  biological activity of, 329
  in clam shells, 569
  distribution in, Columbia River plume, 582
    North Pacific, 572-574
    the ocean, 329-330
  latitudinal distribution in the soil, 330
  leeched from land mass, 331
  in ocean surface waters, 332
  projected release to meet energy demand, 581
  as radiation source in the sea, 358
  in rain on Barbados, 76
  vertical distribution in North Pacific, 576-578
  vertical transport, 344-345
  worldwide deposition, 332
Submarine, *Trieste,* 683
Submarine, nuclear, radioactive wastes from primary coolants, 566
Submersibles, *Trieste,* 683
Sulfate, atmospheric, 12-14
  in Connecticut River, 30
Sulfate-reducing bacteria in sediments, 40
Sulfide formation, 202
Sulfides, precipitation of metals, 209
Sulfur, atmospheric, 14
  bacteria, 209
  Del-^{34}S in reservoirs, 439
  dioxide, atmospheric removal, 80
  in fresh water, 21

oxides, rate to atmosphere, 82
stable isotope, abundance in nature, 432
  effects, 439
Supertankers, 385
Survival, requirements for man's, 637
Suspended, load, in streams, 18-19
  trace elements in, 35, 38
  in U. S. rivers, 36-42
  sediments, cation exchange with, 174-176
  trace elements in, 37
Sustained biological yield in fisheries, 529, 533

Taku Inlet, Alaska, 174
  exchangeable cations in suspended sediments, 175-176
  mineralogy of sand and silt fractions, 176-177
Talc minerals in atmospheric dust, 79
Tance estuary, France, energy from, 679
Temperature, effect of CO_2 on global temperature, 313, 317-318
  effect on heavy metal toxicity, 474
  variation with time as measured by $^{18}O/^{16}O$, 321
  world changes with time, 318
Tennessee Valley Authority, watershed studies, 507-508
TENOC (Ten Year Program In Oceanography), reports of, 645
Terbium in fresh water, 23
Tertiary deposits, 34
Tetra-ethyl lead, rate to atmosphere of combustion products, 82
Thames, estuary, England, 32
  River, 26
Thermocline, 104, 112, 116
  eddy diffusion in, 121
  radioactive penetration through, 576-577
Thermodynamic principles, activity scales, 226
  adiabatic systems, 220
  chemical equilibrium, 229-240
  closed systems, 220
  first law of thermodynamics, 221
  Gibb's considerations, 221-225
  open systems, 220
  spontaneous reactions, 223
  standard states, 226
Thorium, in fresh water, 23
  in the ocean, 326
Thresher, USS, 342
  catastrophe, 638
Thulium in fresh water, 23
Tin in human tissues, 478
Titanium, in fresh water, 21
  in human tissue, 478
Toilet tissue in ocean, 2
Torrey Canyon, 401-403
  accident, 381
  oil spill, 638
Toxicity, of aldrin to fish, 268
  of benzene, 426
  of Ca to rainbow trout, 475
  of chromium to kelp, 477
  of cobalt to rainbow trout, 474
  of copper to, kelp, 477
    rainbow trout, 474
    yeast, 427
  effect of, calcium on heavy metal, 474
    temperature on heavy metal, 474
  of endrin to fish, 267
  of heavy metals to, biota, 473, 477
    brine shrimp, 475
    sea urchins, 475
  of hydrocarbons, 426-427
  of lead, 255-256, 427
  of mercury to goldfish, 476-477
  of nickel to, kelp, 477
    rainbow trout, 474
  of oil to amoebae, 426
  of zinc to, kelp, 477
    rainbow trout, 474
  synergism with copper, 475
Trace elements, concentration control, 66-70
  in Connecticut streams, 25-32
  distribution patterns in Long Island Sound, 66
  in suspended load, 35, 38
  transport, 37
Trace metals, Andes Rivers, 38
  contamination of Naugatuck River, 27
  in rain, 13, 15
  in streams, 24
  variations in freshwater, 24
Tracers, of physical ocean processes by radioactivity, 568-580
  stable isotope ratios in sewage, 436
  stable isotopes in petrochemicals, 436

Trans-Alaska pipeline, 691
Transfer processes, atmosphere to oceans, 78-80
Transmissivity, effect of sewage on, 452
Transport, of biocides, 260-261
 of material, 16, 75
 of radioactivity by north equatorical current, 570
 river, 9
Transportation, by oceans, historical, 671
 as ocean use, 482, 682-684
 oil in the arctic, 426
 of petroleum, 421
Triassic Valley, Connecticut, 25
Trieste, 683
Tritium, concentration in ground water, 331
 from fusion reactors, 337
 in the oceans, 326-327
 from reprocessing plants, 566
 in Savannah river, 340
 from ternary fission, 336
Truman Proclamation (1945), 604-605
Tuna, radioactivity in, 352
Tungsten in fresh water, 23
Turbulence, 89, 96, 108-117
 in deep waters, 116-117
 processes responsible for, 114-115
 as related to vorticity, 108-110
Turbulent diffusion, 89, 90, 120
 coefficient of, 100

United States, Coast Guard, responsibility for ships and oil pollution, 392-393
 House of Representatives, Science and Astronautics Committee, publication, 643
Universe Island, 385, 682
Uranium, in fresh water, 23
 in the oceans, 326
Urchins, sea, heavy metals in, 470
 toxicity of heavy metals to, 475
Urea, effect on soil colloids, 520
Uses of the ocean, aquaculture, 675-676
 currents, 672-689
 decision making processes regarding, 689-695
 drugs and pharmaceuticals, 680-682
 energy from, 679-680
 fisheries, 672-675
 historical, 669-672

 mining, 680-681
 multiple use concept, 667-669
 recreation, 678-679
 shipping, 682-684
 waste disposal, 685-688

Vanadium in, ascidians, 470
 fresh water, 21
 petroleum, 463
 plants, 478
Viruses, concentration by filter feeders, 456-457
 polio virus thru shell fish, 457
 in sewage, 455-456
Visicol Chemical Corporation, 267
Vitamins, requirements by marine algae, 597
Volatile organic compounds, manufactured in U. S. annually, 83
Volatiles, excess in seawater, 195

Wagon Wheel Gap, Colorado, controlled watershed, 505
Wallingford, Connecticut, 26
Waste disposal, of chlorinated aliphatic hydrocarbons at sea, 3
 cost of barging to sea, 686
 as related to mixing energy, 684
 use of the ocean, 685-688
Wastes, added directly to the sea, 685-688
 barged to sea in 1968, 686
 biodegradation of, 4
  petroleum, 397-401
 coastal management research needs, 694-695
 cycling in the environment, 261
 degradability, 4
 disposal by sea, 686
 effect of discharge on coastal environment, 539
 industrial, types of, 687
  and world ocean pollution, 688
 municipal, 4, 445-450
  bacteria, 455-456
  effect of disposal on the ocean, 452-455
  enhancement of production in the sea by, 455
  parasites in, 457
  phosphorus in, 448-450
  physical chemical properties of, 446-450
  productivity as related to estuaries, 590

purification of, 450-452
sludge disposal of, 450-451
viruses in, 455-456
water quality and, 449
to the ocean by atmospheric exchange, 688-689
oil, 394-395
person equivalents of types, 458
radioactive, aerospace applications, 341-342
  from cooling water, 565
  disposal of high level wastes, 340-341
  from exchange resins, 566
  exposure to, 360-361
  from fallout, 566
  from fuel reprocessing plants, 566
  from fusion reactors, 337
  handling of, 565-567
  international viewpoints on disposal, 367
  isotopes released from Hanford reactors, 339
  from major accidents, 342
  from nuclear, power plants, 336, 566
    ships, 338
    submarine, 566
    weapons tests, 566
  packages for ocean disposal, 565-566
  from radioisotope applications in medicine and industry, 342
  from reprocessing of nuclear fuel, 340
  from *USS Savannah*, 338
solid, ocean disposal, 668-669
sulfuric acid in sea, 3
Water, discharge, effect of forest fire on, 508, 516
  effect of vegetative composition on, 505-506
  as related to forest cover, 507-508
masses, Antarctic, bottom water, 150
  deep water, 106
  common water, 157
  "layering" structure, 113
  North Atlantic deep water, 106
  tracing with artificial radionuclides, 345, 569-580
quality, effect of agricultural and forest crops, 504-520
  effect of herbicides on, 519
  insecticide relation to, 519

and municipal wastes, 449
vapor, effect on light penetration, 314-315
Watersheds, Cedar River, Washington, 509, 512
  contribution of forest lands to flow, 505-506
  Coweeta, N. C., 506-507
  Hubbard Brook, N. H., 509
  management, Wagon Wheel Gap approach, 505
  nutrient release from, 509
  Tennessee Valley Authority, 507-508
Waves, gravity, turbulence by, 111-112
Weapon tests, nuclear, radioactive wastes from, 566
Weathering, 9, 15
  aerosol supply, 14
  by carbon dioxide, 170
  supply of beach sand from 491-492
Wetlands and coastal ecology, 678
Whales, harvesting historically, 671
Wildlife refuges, establishment of, 532
Will-o'-the-wisp (*Ignus fattus*), 207
Winds, global systems, 76-78
  northeast trade, 84-85
Windscale plant, adsorption of radionuclides on sediments, 346-347
Wind transport, of materials, 75
  processes, 84-87
World, human survival in, 636

Xenon-133, 337

Yeast, concentration by filter feeders, 456-457
  toxicity of copper to, 427
Ytterbium in fresh water, 23
Yttrium in fresh water, 22

Zinc, in air, 466
  chemical behavior in ocean, 468-469
  in fresh water, 21
  histopathological effects in *Lebistes reticulatus*, 475
  in Nobeoka Bay, Japan, 465
  organic complexes of, 468
  in plants, 478
  as sulfates and oxides in pesticides, 462
  tolerance of organisms to, 474
  toxicity to, kelp, 477

rainbow trout, 474
toxic synergism with copper, 475
vertical distribution in Gulf of Mexico, 468
Zinc-65, 329
   adsorption on sediments, 346
   in clams, 353
   discrimination over stable isotopes by organisms, 349
   in fecal pellets, 350
   ion exchange on stream sediments, 48
   labeled sediment transport, 48
   vertical transport by organisms, 351
Zirconium-95-neobium-95, 329
   adsorption to sediments, 347
   in fecal pellets, 350
Zooplankton in estuaries, 353